S0-BWX-442

# The Historical Development of Quantum Theory

Erwin Schrödinger as an Artillery Officer in World War I, *circa* 1916.
(Courtesy of *Zentralbibliothek für Physik*, University of Vienna.)

Jagdish Mehra
Helmut Rechenberg

# The Historical Development of Quantum Theory

VOLUME 5

## Erwin Schrödinger and the Rise of Wave Mechanics

Part 1

Schrödinger in Vienna and Zurich
1887–1925

Springer-Verlag
New York Berlin Heidelberg
London Paris Tokyo

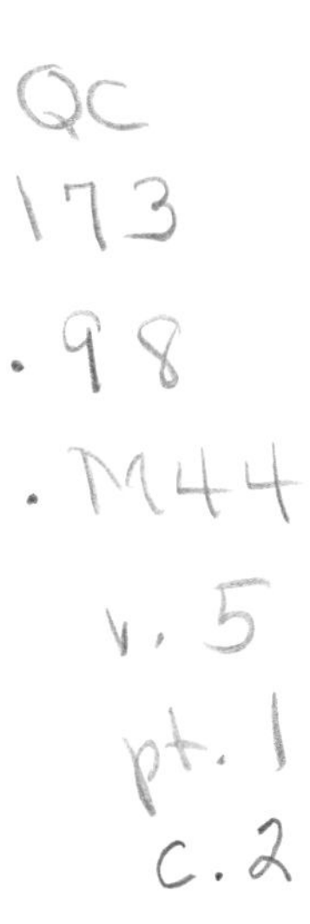


Library of Congress Cataloging in Publication Data
Mehra, Jagdish.
Erwin Schrödinger and the rise of wave mechanics
(Part 1. Schrödinger in Vienna and Zurich, 1887–1925)
(The historical development of quantum theory/
Jagdish Mehra and Helmut Rechenberg; v. 5)
Bibliography and Index are included in Part 2.
1. Quantum theory—History. 2. Schrödinger, Erwin,
1887–1961. 3. Physicist—Austria—Biography.
History. I. Rechenberg, Helmut, joint author.
II. Title. III. Series: Mehra, Jagdish.
The historical development of quantum theory; v. 5.
QC173.98.M44 vol. 5 530.1'2'09s 86-5445
[530.1'2'09] AACR2



Typeset by J. W. Arrowsmith, Ltd., Bristol, England.
Printed and bound by Arcata Graphics/Halliday, West Hanover, Massachusetts.
Printed in the United States of America

9 8 7 6 5 4 3 2 1

ISBN 0-387-96284-0 Springer-Verlag New York Berlin Heidelberg
ISBN 3-540-96284-0 Springer-Verlag Berlin Heidelberg New York

**Dedicated to the Memory of**
**Erwin Schrödinger (12 August 1887–4 January 1961)**
**on the Centenary of His Birth**

# Contents—Part 1

# Abbreviated Contents—Part 2

## The Creation of Wave Mechanics; Early Response and Applications 1925–1926

# Foreword

The date of 12 August 1987 marks the hundredth anniversary of Erwin Schrödinger's birth. Schrödinger's name is linked with an equation which is, by far, the most-often quoted in the scientific literature of the twentieth century. The Schrödinger equation occurs regularly in papers not only by physicists working in a variety of fields—from atomic and molecular physics to solid state, nuclear and elementary particle physics—but nearly as often in papers on topics of theoretical chemistry—both organic and inorganic—and occasionally even in biological papers. Even more justly than Einstein's popular mass-energy relation, $E = mc^2$, the Schrödinger equation should be taken as the most characteristic equation of twentieth-century science.

The Schrödinger equation is more than just a mathematical relation. It denotes a complete mathematical and physical method of describing atoms, molecules and submicroscopic particles, appropriately called undulatory mechanics or wave mechanics. This theory explicitly involves previously unknown aspects of matter, especially in the domain of atomic and subatomic phenomena. Historically, the establishment of wave mechanics—the initial steps towards it, as well as its final formulation—can be linked almost exclusively with the name of Erwin Schrödinger, though a few important contributions to it were also made by a small number of other physicists. Being a very complex, highly cultivated and erudite person, Schrödinger took a rather sophisticated path to his atomic theory, which emerged rather late in his professional career. Therefore, without a review of the scientific work and ideas of the young Schrödinger, his publications on wave mechanics in early 1926 must appear an unbelievable wonder. The reconstruction of the origins of wave mechanics demands a detailed and careful analysis of Schrödinger's development as a person and as a scientist, one whose interests extended far beyond the borders of physics and science in general. Consideration of everything that Schrödinger did—experimental and theoretical physics, statistical physics and meteorology, colour theory and physiological optics, biology and philosophy, as well as poetry—becomes inevitable, because all these things were somehow intimately interwoven in his mind.

In order to present the extensive material pertaining to the makeup of Schrödinger's personality and his scientific development, we have divided the present volume into two parts, each containing two chapters. Part 1, entitled 'Schrödinger in Vienna and Zurich 1887–1925', deals with the prehistory of wave mechanics. In Chapter I, on 'Schrödinger in Vienna 1887–1920', we discuss Schrödinger's Viennese background, his youth and studies at the University of Vienna, his early scientific work (on kinetic theory, atmospheric electricity, the atomic structure of solids and other topics), his duties as an artillery officer during World War I, and scientific work during the war years (on statistical physics, general relativity, atomic and quantum theory), and his concern with philosophy. In Chapter II, on 'Waves and Quanta: Preludes to Wave Mechanics', we concentrate, after summarizing the recognition of the dual nature of light and reviewing the side-stages of the development of quantum theory, mainly on the scientific development of Erwin Schrödinger in the 1920s. We discuss his arrival in Zurich, Switzerland, and the main ideas that he brought to his new environment. At first, during the early years in Zurich, quantum theory and the analysis of atomic structure did not lie at the centre of Schrödinger's scientific interests—though he did consider selected problems of atomic structure, besides his work on the radiation problem and physiological optics at that time. Statistical mechanics was a far more important occupation and a continuous field of research for Schrödinger. He had been interested in this subject from his student days in Vienna, and now in Zurich he published, from 1922 to 1926, a series of papers on it. These investigations on statistical mechanics must be viewed as contributing a crucial step, apart from Louis de Broglie's conception of matter waves, to the genesis of wave mechanics.

Part 2, entitled 'The Creation of Wave Mechanics; Early Response and Applications 1925–1926', is again divided into two chapters (III and IV). In Chapter III, on 'The Creation of Wave Mechanics', we make an attempt to recreate the steps that, in our view, led to the actual rise of wave mechanics in the hands of Schrödinger: the scientific exchange with Planck and Einstein on the fundamental questions of statistical mechanics, his initial steps towards a relativistic hydrogen equation, his correspondence with Wilhelm Wien and the development of the nonrelativistic hydrogen equation, the fundamental conceptual sources of undulatory mechanics, the foundation of undulatory mechanics, and the initial reception of the new theory by Wilhelm Wien and Arnold Sommerfeld. In Chapter IV, on 'Early Response and Applications', we discuss Schrödinger's early extension of the wave equation to atomic problems, the continuation of correspondence with Wien and Sommerfeld, the response to wave mechanics from Planck, Einstein and Lorentz, the establishment of the formal equivalence of Schrödinger's wave mechanics and the quantum mechanics of Born, Heisenberg, Jordan and Dirac, by Schrödinger himself and by Pauli and Carl Eckart. We also discuss the early applications and generalization of the wave mechanical scheme by Erwin Fues (diatomic molecules), Ivar Waller, Gregor Wentzel, Carl Eckart and Paul Epstein (Stark effect and the hydrogen spectrum), Max Born (collision processes in wave mechanics), Werner Heisenberg (the helium problem) and Paul Dirac (symmetry properties of wave functions and quantum statistics). Further, we discuss the initial steps towards the interpretation of the new atomic mechanics taken by Schrödinger (and the reactions of certain physicists to it), including his visits to Berlin and Munich to present his theory, his discussions there with Planck, Einstein, Wien and Sommerfeld, and to Copenhagen for his discussions with Niels Bohr. We close the last section with further fundamental applications and developments of wave mechanics in the second half of 1926 (such as the contributions of Schrödinger himself, Charles Manneback, Fritz Reiche, Walter Gordon, Gregor Wentzel, Oskar Klein, Vladimir Fock, Janos Kudar, Théophile De Donder, Frans van den Dungen, J. Robert Oppenheimer, Waldemar Alexandrow, Øyvind Burrau, Ralph Kronig, Isidor Rabi and others), which served to increase the fame of its creator who would soon succeed Max Planck in the chair of theoretical physics in Berlin.

For the purpose of our account of the prehistory, the birth and rise of wave mechanics, we have first made use of the primary sources: Schrödinger's published papers, his unpublished notebooks and memoranda, as well as his scientific correspondence (with Pauli, Sommerfeld, Wien, Planck, Einstein, Lorentz, and others); we have also made use of interviews and discussions with, and the recollections of, many physicists, including Born, Dirac, Heisenberg, Eckart, Debye, Fues, Uhlenbeck, and others, and we have benefited from other secondary sources including historical articles, dissertations and accounts of the reminiscences concerning Schrödinger's life and scientific work.

We wish to thank Professors Ilya Prigogine, Abdus Salam, Willis E. Lamb, Jr. and Eugene P. Wigner for their continuous encouragement during the writing of this work.

We take great pleasure in presenting this result of our efforts to readers in the scientific community.

25 December 1986

Jagdish Mehra
Helmut Rechenberg

# Acknowledgments

Erwin Schrödinger's scientific papers have been reprinted recently in *Erwin Schrödinger: Gesammelte Abhandlungen*, Volumes 1–4 (Vienna, 1984). Many of his essays and selections from other writings have been collected together and published as books. Selected letters exchanged by Schrödinger with Max Planck, Albert Einstein and Hendrik Antoon Lorentz were edited in *Briefe zur Wellenmechanik* (1963) and in an English translation, as *Letters on Wave Mechanics* (1967). Copies of large parts of his unpublished diaries, manuscripts, notebooks and notes on physical, philosophical and other topics, as well as his scientific correspondence and an interview with his widow, Frau Annemarie Schrödinger, have been filed on microfilms deposited in the *Archives for the History of Quantum Physics* at the American Philosophical Society, Philadelphia, Pa. We are deeply indebted to Schrödinger's daughter, Frau Ruth Braunizer, Alpbach (Tyrol), Austria, for her generous permission to quote from Schrödinger's correspondence and his published and unpublished writings.

Copies of Erwin Schrödinger's scientific correspondence with Albert Einstein are contained in the duplicate of the *Einstein Archives*, Sealy G. Mudd Library, Princeton, N.J. We are grateful to Charles E. Bloom and Ehud Benamy of the American Friends of the Hebrew University, Inc., New York, N.Y., for handling the arrangements to obtain permission to quote from Einstein's writings and correspondence; these materials are published by permission of the Hebrew University of Jerusalem, Israel.

Schrödinger's correspondence with Arnold Sommerfeld and Wilhelm Wien is contained in the Sommerfeld and Wien Collections at the *Deutches Museum*, Munich, Federal Republic of Germany. We are grateful to the *Deutches Museum* for permission to quote from this correspondence.

We are indebted to the *Stiftung Preußischer Kulturbesitz*, Berlin, Federal Republic of Germany, for making available the Schrödinger–Landé correspondence contained in the *Alfred Landé Collection*.

The scientific correspondence of Wolfgang Pauli up to 1939 has been published in *Wolfgang Pauli: Wissenschaftlicher Briefwechsel/Scientific Correspondence*, Volume I (1979) and Volume II (1985). For permission to quote from Pauli's correspondence, we are indebted to Frau Franca Pauli, Victor F. Weisskopf and Springer-Verlag.

For permission to make use of the resources of the *Niels Bohr Archives* in Copenhagen, Denmark, and to quote from them, we are grateful to Aage Bohr.

We have drawn upon the reminiscences and the conversations and discussions of one of us (J.M.) with several quantum physicists (including Max Born, P. A. M. Dirac and Werner Heisenberg). We have also made use of the interviews of the Sources for the History of Quantum Physics with Max Born, Peter Debye, Paul Dirac, Carl Eckart, Erwin Fues, Werner Heisenberg, George E. Uhlenbeck, and others. For help in making use of the materials contained in the *Archives for the History of Quantum Physics*, we are grateful to the Manuscript Librarian, American Philosophical Society, Philadelphia, Pa.

Ample literature dealing with Erwin Schrödinger and wave mechanics exists, especially when compared with that on the matrix branch of quantum mechanics. In addition to articles in various scientific books and journals, several doctoral dissertations have also been devoted to the subject. Of the latter, we mention two: Linda A. Wessels' *Schrödinger's Interpretations of Wave Mechanics* (Indiana University, 1975), and Paul A. Hanle's *Erwin Schrödinger's Statistical Mechanics* (Yale University, 1975). William T. Scott's book, *Erwin Schrödinger: An Introduction to His Writings* (Amherst, Mass., 1967), contains Schrödinger's bibliography and an introductory essay on his life and on his scientific and popular writings. We have benefited from these accounts.

We are grateful to the *Instituts Internationaux de Physique et de Chimie* (*Solvay*), *Université Libre de Bruxelles*, Brussels, Belgium, and *Alfried Krupp von Bohlen und Halbach-Stiftung*, Essen-Bredeney, Federal Republic of Germany, for grants in aid of the publication of this volume.

JAGDISH MEHRA
HELMUT RECHENBERG

# Part 1
# Schrödinger in Vienna and Zurich
# 1887–1925

# General Introduction

In issue No. 4 of *Annalen der Physik*, whose editing was completed on 13 March 1926, there appeared a paper written by Erwin Schrödinger in Zurich bearing the title '*Quantisierung als Eigenwertproblem.* (*Erste Mitteilung*)' ('Quantization as a Problem of Eigenvalues. Part I', Schrödinger, 1926c). Four weeks later, Wolfgang Pauli from Hamburg wrote in a letter to Pascual Jordan in Göttingen:

> Today I want to write neither about my Handbook-Article [on quantum theory, which occupied Pauli at that time: Pauli, 1926b] nor about multiple quanta [in radiation theory, which occupied Jordan then]; I will rather tell you the results of some considerations of mine connected with Schrödinger's paper "*Quantisierung als Eigenwertproblem*" which just appeared in the *Annalen der Physik.* I feel that this paper is to be counted among the most important recent publications. Please read it carefully and with devotion. (Pauli to Jordan, 12 April 1926; English translation of the German text in Van der Waerden, 1973, p. 278)

In the following months Schrödinger published five further papers extending his method to what he called wave mechanics and applied it to a variety of problems in atomic theory (Schrödinger, 1926d, e, f, g, h). Many years later, Max Born said in an obituary of Schrödinger: 'What is more magnificent in theoretical physics than his first six papers on wave mechanics?' ('*Was gibt es Großartigeres in der theoretischen Physik als seine ersten sechs Arbeiten zur Wellenmechanik?*': Born, 1961a, p. 85).

Born, as well as Jordan and Pauli belonged among the pioneers who had developed, since the middle of 1925, a new version of atomic theory which they called 'quantum mechanics'. This development had emerged in a more or less direct way from the quantum theory of atomic structure, introduced by Niels Bohr in 1913 and extended afterwards by him and Arnold Sommerfeld to describe a large number of atomic systems.

## From the Bohr–Sommerfeld Theory to Quantum Mechanics

The original scheme of Bohr and Sommerfeld represented a mixture of classical dynamics and quantum requirements or conditions, the latter determining the stationary states of atoms and molecules. However, the spectra emitted by those objects did not obey the laws of classical electrodynamics, where the electron is supposed to move around the atomic nuclei due to classical dynamics. The frequencies emitted and absorbed were given rather by the (classical) energy differences between the stationary states, or more accurately by those energy differences divided by Planck's quantum-theoretical constant $h$. Thus the ingenious mixture of classical theory and quantum-theoretical ideas, which was applied for over a decade—i.e., from 1913 to 1925—with ever-increasing skill and endurance to more and more complex atomic systems, did not rest on a consistent theoretical basis. Especially, the electrodynamics of James Clerk Maxwell—a theory which accounted so well for all phenomena of static and stationary electromagnetic phenomena and for electromagnetic waves and their propagation, reflection and diffraction—lost its validity when applied to the processes of creation and absorption of radiation in atoms.

Still one was able to reach a fair agreement between experimental and theoretical results in many examples, if one just stuck to the set of inconsistent rules implied in the Bohr–Sommerfeld theory of atomic structure. However, since 1922 a deeper analysis of the properties possessed by atoms with several electrons, or even by one-electron atoms under peculiar external conditions, had revealed increased failures of the theory. A more adequate quantum-theoretical description of the scattering of light by atoms, properly named the 'dispersion-theoretic approach'—as inaugurated in 1924 by Niels Bohr and his collaborators in Copenhagen and Max Born and his associates in Göttingen—however, fostered some distant hopes of gradually resolving all the fundamental discrepancies existing between experiment and theory. The theory, or, more accurately, the rules of this dispersion-theoretic scheme demanded: (i) replacing certain *differential* expressions in the classical description of atoms and their interaction with radiation by *difference* expressions, and (ii) substituting in the latter, for any (classical) frequency, the observed (quantum) frequency of radiation. From the dispersion-theoretic scheme, which the members of the Bohr school called occasionally 'the sharpened correspondence principle', there emerged in little more than a year the first consistent atomic theory.

The pioneering paper, in which the basic principles of the new theory were formulated, stemmed from Werner Heisenberg in Göttingen, a former student of Sommerfeld, Born and Bohr. In July 1925 he expounded emphatically that the main deficiency of the old, unsuccessful atomic theory (of Bohr and Sommerfeld) lay in the use of several concepts that did not correspond to observable quantities, these quantities, such as the orbital motion of electrons in atoms and molecules, should therefore be forbidden in future, and Heisenberg demanded that only observable quantities should occur in a consistent theory (Heisenberg, 1925c). In practice, he described periodic atomic systems by 'quantum-theoretical Fourier series,' whose amplitudes were the transition amplitudes between stationary states and whose frequencies were the corresponding atomic frequencies. These quantum-theoretical Fourier series represented the dynamical variables in the new theory; they obeyed equations of motion, whose structure resembled the corresponding equations in classical dynamics. A remarkable difference with the classical theory, however, showed up in the fact that products of two quantum-theoretical variables or Fourier series did not necessarily commute. Heisenberg further presented in his paper the solution of the simplest quantum systems, i.e., of the anharmonic oscillator and the rigid rotator, believing them capable of describing essentially the observed features of real atoms.

## The Göttingen Matrix Mechanics and Its Extensions

Heisenberg's colleagues in Göttingen, Born and Jordan, then obtained in the summer of 1925 a suitable mathematical scheme formulating the physical ideas of Heisenberg's 'quantum-theoretical reformulation' of the old atomic theory. They rewrote, in particular, the quantum-theoretical Fourier series as infinite Hermitean matrices and established within a matrix theory a description of atomic systems analogous to the Hamilton–Jacobi theory of classical systems (Born and Jordan, 1925b). The matrix mechanics provided a direct mathematical representation of the discreteness involved in the quantum concept.

Matrix mechanics allowed one to treat, in principle, any multiply-periodic system of atomic theory, provided one succeeded in finding suitable dynamical variables that could be described by Hermitean matrices (Born, Heisenberg and Jordan, 1926). As a special example, Wolfgang Pauli solved in October 1925 the hydrogen problem in matrix mechanics, i.e., he calculated the energy states of the hydrogen atom and reproduced Bohr's old result of 1913 in agreement with observation; at the same time he overcame a difficulty that had bothered the old atomic theory, because the matrix mechanical solution of the crossed-field problem did not lead to inconsistencies (Pauli, 1926a).

A little later than Born and Jordan, in the fall of 1925, Paul Adrien Maurice Dirac in Cambridge, also starting from Heisenberg's July paper, provided an even more general quantum-mechanical scheme than matrix mechanics (Dirac, 1925d). His theory, in which he represented the quantum-theoretical variables by what he called '$q$-numbers,' allowed one to take over the classical Hamiltonian theory more directly than was possible in matrix mechanics, thus offering the opportunity to deal with *any* kind of many-electron atoms and even with the relativistic problem of Compton scattering (Dirac, 1926a, b, c).

The new quantum mechanics, whether expressed in the matrix formulation of Born and Jordan or in the $q$-number formulation of Dirac, showed one characteristic feature: in its equations, say, the equations of motion, differences of products of dynamical variables appeared instead of differential expressions of classical mechanics. The very nature of the new quantum theory seemed therefore to rest, so the people in Göttingen, Copenhagen and Cambridge claimed, in the *algebraic* structure of its equations. However, in December 1925, Max Born and Norbert Wiener, who collaborated at the Massachusetts Institute of Technology, submitted a paper showing that the gap between the classical and the quantum mechanical treatment was not really so large as reflected in matrix or $q$-number theory: in their method of 'operator mechanics' not only the *differential* equation form of classical dynamics recurred; in certain cases, such as the uniform motion of a particle in a straight line, the quantum-theoretical problem also had the same solution as the corresponding classical problem (Born and Wiener, 1926a, b).

A similar consequence was derived even more explicitly in the 'field-like' formulation, which Cornelius Lanczos in Frankfurt proposed also in December 1925 (Lanczos, 1926a). He used explicitly the mathematical result already discovered twenty years earlier by David Hilbert in Göttingen, namely, the mathematical equivalence of the problem of solving integral equations of the Fredholm type on the one hand, and of obtaining the eigenvalues of an infinite matrix on the other hand (Hilbert, 1904a, b; 1905; 1906a, b; 1910). Lanczos reversed Hilbert's original procedure leading from the continuum problem (connected with integral equations) to the discrete problem (connected with matrices); thus he simply rewrote the matrix equations of Born and Jordan as linear integral equations. This rewriting implied the advantage, 'that all results of the theory [i.e., of quantum mechanics]... thus obtain a formulation which may appear to the physicist, who is used to working with analytical methods, more natural than the matrix formulation' (Lanczos, 1926a, p. 812). Lanczos also said: 'As far as the *interpretation* of the facts, that is, the real nature of the quanta, is concerned, however, one cannot exclude that the integral [equation] formulation is even superior to the matrix formulation; the reason is that it has the advantage of being immediately consistent with the field conception, indeed is even built on it, while the field concept is obviously far removed from the discontinuum formulation' (Lanczos, 1926a, pp. 812–813).

## Status of Quantum Mechanics in Early 1926

The discoveries and results of Heisenberg, Born, Jordan, Dirac, Pauli, Wiener and Lanczos, determined the picture of quantum mechanics during most of the first half of 1926. The theory seemed to be successful in describing the behaviour of atomic and molecular systems, especially their discrete energy states; in a very special example, namely that of the uniform motion of a particle in a straight line—as treated by Born and Wiener—it also showed itself to be capable of accounting for a continuous energy spectrum. Hence it possessed all the prerequisites of a scheme that could describe the phenomena in atomic physics completely.

Of course, the detailed application of quantum mechanics in one of the four equivalent formulations—that of Born, Heisenberg and Jordan (matrix mechanics), of Dirac ($q$-number theory), of Born and Wiener (operator mechanics), or Lanczos (integral equation representation)—to practical problems other than the most elementary ones already treated (i.e., the anharmonic oscillator, the rigid rotator, the hydrogen atom, and the motion of a particle in a straight line) had still to be shown; but promising progress was obtained, e.g., in treating the problem of diatomic molecules (Mensing, 1926a). Both in Europe and the United States (where, in particular, Born had imported quantum mechanics) many physicists were eagerly looking for problems that could be attacked within the new theory. One generally hoped to be able to cope with *all* problems of atomic theory in this way, provided one just tried hard enough and invented clever tricks, such as Pauli had employed in the case of the nonrelativistic hydrogen atom. Most physicists working on either side of the Atlantic could agree with the statement that quantum mechanics indeed *provided the solution of the pre-*1925 *difficulties of atomic and quantum theory.*

This general feeling notwithstanding, not all physicists interested in atomic theory were happy with or even liked the methods of calculation suggested by the Göttingen theoreticians or by Paul Dirac in Cambridge. There were some people who raised their voices against the new methods. An experimentalist, Bidhu Bhusan Ray, then working in Europe, even feared: 'If atomic physics had to progress on the lines of Born and Jordan, you will find very few people left in the atomic physics circle' (Ray to Niels Bohr, 6 March 1926). Also, physicists who in general approved of the results of Heisenberg, Born and the others, such as Arnold Sommerfeld, did not participate in the game of applying matrices or $q$-numbers. They perhaps stayed away and waited for more handy methods to arrive, although they would agree with the opinion that these methods would have to be as strange as those of matrix mechanics in order to overcome the fundamental difficulties of the old Bohr–Sommerfeld theory of atomic structure. However, hardly anybody expected, at least not within a short period of time, the arrival of another theory, which turned out to be as successful in removing the previous difficulties and whose methods were much more convenient for the normal theoretical physicist. Hence the surprise of the physics community was complete when Erwin Schrödinger, early in 1926, proposed an apparently totally different approach and right away computed, using the known methods of partial differential equations, the energy states of the hydrogen atom. Soon afterwards he also demonstrated the power of his wave mechanical theory by solving other problems of atomic physics with similar ease. Thus Ray's fears were removed; it now seemed that any theoretician, educated in the methods of classical theory, could again join the atomic physics circle.

## Schrödinger's Work in Physics Prior to 1926

There was more than just one surprise involved in the discovery of wave mechanics. First, how could, so very shortly after the development of the quantum mechanics in Göttingen and at the other places mentioned, there arise a *genuine alternative scheme* which had no connection at all to the original ideas of Heisenberg's 'quantum-theoretical reformulation' of July 1925? As a second surprise, one may note that the new theory came from Zurich, a place which had not particularly distinguished itself in atomic theory during the early 1920s, although some important contributions to quantum theory had originated there a while ago—e.g., Peter Debye had proposed in Zurich an important improvement in the theory of specific heats of solids (Debye, 1912b, c).

The third surprise consisted in the fact that it was Erwin Schrödinger who invented the new atomic theory. Having joined the University of Zurich in 1921, he was now thirty-eight years old, much older than most pioneers of quantum mechanics—Dirac, Heisenberg, Jordan and Pauli were all about 25 in 1925. Schrödinger was a reasonably well-known scientist of some accomplishment, from whose earlier work one might have had a good idea of what else could be expected from him. Since 1910 he had contributed to many experimental and theoretical topics; thus he had worked first on the explanation of the properties of dielectric substances, on radioactivity and electricity of the air, on problems of statistical physics, and later, on general relativity and colour theory. He had further written several good review papers and articles on various subjects, showing his ability to absorb, carefully analyze and organize the existing literature (i.e., mainly the work of other scientists). In short, Schrödinger possessed an established reputation among his colleagues; everything in the man's knowledge and scientific work was safe and solid; on certain questions of statistical mechanics, the theory of colour metric and colour vision, he could be considered a leading expert. But who in the scientific community would have expected of him a major breakthrough in quantum or atomic theory?

It would certainly be unfair and incorrect to say that Schrödinger was a stranger among the atomic physicists. He had dealt with certain problems of atomic and quantum theory since about 1912; he had lectured and written articles, for example, on the quantum theory of the specific heats (Schrödinger, 1917a, 1919b). After 1920 he had occasionally treated specific questions of the theory of atomic structure: a paper containing the idea of 'diving orbits' (Schrödinger, 1921a) had even found the approval of Niels Bohr, less so an investigation on the Doppler effect and the frequency condition for atoms, in which he had employed the light-quantum hypothesis (Schrödinger, 1922a). During the 1920s, Schrödinger had followed with interest the lively discussion of problems and difficulties in atomic theory, contributing a little bit to one question or the other. Being in Zurich, far away from places like Munich, Göttingen, Copenhagen or even Tübingen, where theoretical and experimental work on atomic properties was mainly performed, he had been more like a distant spectator, who had just occasionally raised his voice commenting on a specific point, such as the Bohr-Kramers-Slater theory of radiation (Schrödinger, 1924d).

## Schrödinger's Personality and Method of Work

Schrödinger had a rich, rather unique and complex personality. He showed a deep interest, besides the scientific topics, in art, literature and philosophy; and he occasionally dealt with those fields not only in a passive way, but by writing philosophical essays or even poetry. In physics, he worked on many different problems, experimental as well as theoretical, and mathematical problems as well as those dominated by data and requiring interpretation. In addition, he dared to stick his nose into the problems of other sciences, especially physiology and biology. He read widely, absorbing many facts in his mind, which he then had ready when he needed them for treating a problem in physics, or for making the connection between the sciences and humanities.

With this disposition, most of the things which Schrödinger did were somehow interconnected. Looking at a topic, a method or a result occurring in a particular paper, one cannot say whether it might not reappear or play a role in a problem, which is usually considered to be far away. This fact will be noticed in the various topics of physics treated by him. To give a special example, the detailed methods of tensor analysis worked out by Schrödinger, for the purpose of understanding Einstein's General Relativity Theory, proved to be very fruitful eight years later (i.e., in early 1926) for deriving the foundations of undulatory mechanics; or, as another example, one may notice that certain ideas about the statistical nature of physical laws, which Schrödinger obtained right at the beginning of his scientific career, influenced hiw views on atomic problems throughout his life. In order to lay bare the origins of Schrödinger's work on atomic theory, we have therefore to consider in some detail *all* his previous papers and analyze their contents carefully. Moreover, we have also to examine the influence exerted on him by other people, his teachers and friends, as well as the authors of books he read; and in doing so, we should not confine ourselves to representatives of physics or science alone. In a word, it is the whole educational background, especially the one existing in Vienna—where Schrödinger grew up and became a scientist—which ultimately contributed to the creation of his big theory: wave mechanics.

## Schrödinger and Wave Mechanics

A man of great culture and personal refinement, Schrödinger—after he had left Vienna—exchanged letters regularly with eminent colleagues: he wrote to Niels Bohr, Albert Einstein, Max Planck and Arnold Sommerfeld, and his Viennese countryman Wolfgang Pauli. In this correspondence he exhibited an intimate knowledge of what was going on at that time (i.e., in the 1920s) in atomic and quantum theory. He also published a handful of papers in that field, whose content was occasionally discussed in the letters. However, neither from the papers nor from the letters, it followed that he had clearly joined one of the existing schools, be it the Copenhagen or the Munich school; or that he had completely succumbed to the views which were held by Einstein or Planck in Berlin. Quite the contrary, Schrödinger sometimes supported a particular point of view, say, Einstein's concept of the light-quantum in 1922, but he was not afraid, two years later, of showing great sympathy for the radiation theory of Bohr, Kramers and Slater (which was thought to re-establish the full wave nature of atomic radiation). His attitude can be characterized on the whole as that of an independent observer, who kept himself at some distance from getting involved too much in heated discussions. He watched the situation carefully, but followed his own interests.

In all his work, Schrödinger tried to clarify the concepts thoroughly. Thus, at the moment when the last debates were conducted between Copenhagen, Göttingen and Munich, on sharpening the correspondence principle for solving problems of atomic structure, he became engaged in a fundamental discussion with Max Planck concerning the problem of seemingly lesser interest for atomic physics, namely the statistics of identical particles. The discussion of that question, in which Einstein also got involved, helped to bring Schrödinger on the path to wave mechanics.

In 1926, when he provided his greatest contribution to physics, Erwin Schrödinger was close to forty years old, being about the same age as Max Planck when he discovered quantum theory in 1900. Both men, who shared an interest in many topics of theoretical physics, already looked back on a long career in research and teaching at the time of their greatest contributions. Perhaps one could say that Planck in 1900 had a greater reputation among his colleagues—due to his earlier pioneering work on the second law of thermodynamics and on electrolytic dissociation—than did Schrödinger in early 1926. Schrödinger was probably hardly known outside the theoretical physics circles in Austria, Germany or Switzerland; in addition, some of his specialities, such as the work on physiological optics, did not interest people working on modern topics. Therefore, the publication of his work on wave mechanics signalled on the international scene the sudden appearance of a new star in atomic theory. These contributions seemed to come from nowhere, the only previous indication or forewarning being provided by the widely neglected work of another outsider, the Frenchman Louis de Broglie. The generality and maturity of Schrödinger's new theory, its elegance and perfection from the very beginning, made an enormous and lasting impression on the whole scientific community.

Wave mechanics immediately presented itself as a real competitor of the Göttingen-Cambridge quantum mechanics. After the identity of the results obtained in both theories and their, at least formal, mathematical equivalence was proved—first by Schrödinger himself and later, independently of him, by Pauli in Hamburg and Copenhagen, and Carl Eckart in Pasadena, California—wave mechanics soon won friends among quantum physicists, not only on the Continent, but in England and the United States as well. Schrödinger's scheme had already scored its first triumphs in the summer of 1926, for example, with Heisenberg's solution of the helium problem (Heisenberg, 1926c) or Dirac's derivation of the electron statistics (Dirac, 1926f).

The interpretation developed by Schrödinger at about the same time stressed the continuum features of the wave mechanical approach, in contrast to the discrete features exhibited by the matrix approach. Such an interpretation pleased some physicists, who had previously been at odds with the course that atomic theory had taken since 1920; on the other hand, it bewildered others. A substantial debate on the physical content of quantum and wave mechanics ensued during the following year until the case was settled and a really unified quantum mechanics was completed, applicable to all problems of atomic and quantum physics.

## Contents of this Volume

The present volume of *The Historical Development of Quantum Theory* (Volume 5) tells in detail the story of how wave mechanics was discovered, its background preparations—rooted in the past—and its first successes. The person Erwin Schrödinger, his background and previous accomplishments, plays a unique and dominant role in the preliminary steps, the genesis and the applications of the theory, hence this volume is largely a Schrödinger volume.

In Chapter I, we deal with the geographical, sociological, cultural and scientific background of Schrödinger. First, of course, the city of Vienna, his birthplace and the capital of a great European empire, simultaneously one of the centres of Western civilization, will be described. Here the young Schrödinger grew up as a boy and pupil; he received his education as a physicist, trained by competent teachers and surrounded by congenial fellow students and colleagues. Besides the sciences, he also took a deep interest in the philosophical, literary and musical life of Vienna during the first two decades of the twentieth century. Schrödinger's early scientific work reflects much of the heritage of the great Austrian theoretical physicist Ludwig Boltzmann, who died just before Erwin began his university studies. Such topics as the kinetic theory of solids, fluctuation phenomena and atmospheric electricity, fell into the tradition of Viennese physics in those days. World War I interrupted Schrödinger's professional career: he had first to perform duty as an artillery officer at the Austro-Italian front; during the second half of the war, he returned to Vienna to teach meteorology at a nearby military school. After 1916 he became acquainted with Albert Einstein's General Relativity Theory and wrote a few papers on it. Before he left Vienna in 1920 to assume positions at several German universities and finally at the University of Zurich, he also became engaged in problems of physiological optics; since he continued to work on these problems in Zurich, we shall discuss their results, however, in the next chapter.

We shall start Chapter II with a concise review of certain developments in quantum theory, which lie somewhere outside the 'canonical' path that the development had taken under the guidance of Niels Bohr, Arnold Sommerfeld and Max Born, described in the previous Volumes 2, 3 and 4. These developments include in particular the long-time discussions on the nature of light and radiation, initiated by Einstein's light-quantum hypothesis of 1905. Also the attempts to arrive, during the first twenty-five years of quantum theory, at a consistent statistics of atoms, electrons and radiation, which sometimes led to keen proposals for treating material particles and electromagnetic radiation on the same, unified footing, belonged to the less standard endeavours, at least after 1913, when the investigation of atomic structure began to dominate the attention of the quantum physicists. A particular idea, that of matter waves by the French physicist Louis de Broglie, must be considered more than anything else as the nucleus of wave mechanics.

Since the fall of 1921, Schrödinger had settled in the Swiss city of Zurich, whose peculiar scientific and cultural background will also be outlined in Chapter II. Then we report on the problems of atomic and quantum theory, on which Schrödinger worked up to 1925, including a discussion of his papers on colour theory and physiological optics. The topic, which brought Schrödinger on the route to wave mechanics, however, was the problem of obtaining a consistent quantum-statistical description of identical microscopic particles. With an analysis of his published and unpublished work on this subject we shall conclude Chapter II—and simultaneously Part 1 of this volume.

The discovery of wave mechanics proceeded in two distinct steps. The first resulted in setting up a wave equation for the hydrogen atom, and in exploring its solution. The second consisted in developing a systematic scheme, analogous to the Hamilton-Jacobi theory in classical dynamics, which provided the tool to 'derive' from the classical equations describing atomic systems the corresponding equations in undulatory mechanics. With these two fundamental steps Schrödinger established in late 1925 and early 1926 a new quantum theory of atoms, which he appropriately called undulatory or wave mechanics. The detailed description and analysis of the genesis of wave mechanics will be given in Chapter III of this volume.

The general undulatory theory of atoms, including the wave equations and their solution within the mathematical theory of eigenfunctions and eigenvalues, could be applied immediately by Schrödinger to a series of problems of atomic theory, yielding rather satisfactory results when compared to experiment, as we shall show in the beginning of Chapter IV. The conceptual ideas of underlying wave mechanics and the initial successes of the new scheme, attracted the attention and caused the positive reaction of several eminent colleagues, including Albert Einstein, Hendrik Lorentz and Max Planck. In the first, appreciative responses, however, some serious physical questions were also raised; for example: What properties of atomic systems could actually be described by the wave theory? Could undulatory mechanics, apparently a scheme completely different from the Göttingen quantum mechanics, be a consistent theory of atomic structure and repeat all the successes of Heisenberg and his followers? If so, was perhaps wave mechanics the superior theory?

Schrödinger showed very soon—i.e., still before the spring of 1926—that wave *and* quantum mechanics yielded, when applied to the problem of calculating the energy states of a given atomic system, *identical* results. His demonstration of equivalence of the two theories was soon followed by independent ones, which Wolfgang Pauli and Carl Eckart provided. These and other physicists began to solve further problems of atomic theory by wave mechanical methods, obtaining spectacular successes, e.g., the correct energy states of the helium atom (Heisenberg), the consistent treatment of the scattering of electrons by atoms (Born), and the derivation of the two quantum-theoretical statistics of Satyendra Nath Bose and Enrico Fermi, respectively (Dirac). We shall conclude Chapter IV with an outline of Schrödinger's attempts to supply his undulatory theory with a physical interpretation, which appeared to be radically different from that given to the previous theories of atomic structure—i.e., to the old Bohr-Sommerfeld theory and the Göttingen-Cambridge quantum mechanics. We shall not enter here, however, into the full debate that followed Schrödinger's interpretation, as this will be very much one of the themes of the next volume in this series.

# Chapter I
## Schrödinger in Vienna (1887–1920)

### Introduction

The history of quantum physics is connected particularly with several places where important events or developments occurred. Thus the Universities of Cambridge, Copenhagen, Göttingen and Munich acted as the principal centres of atomic theory in the years after 1913: there the quantum theory of atomic structure was created, extended and ultimately transformed into what one then called (in 1925) quantum mechanics and it is still called so. Two large cities, however, both capitals of large empires, also contributed considerable shares to the development of quantum theory. In the first place, it was Berlin, the capital of Prussia and of the post-1870 (Second) German *Reich*, where quantum theory was born and where for decades more experts of the new field of physics assembled than at any other place in the rest of the scientific world. In Vienna, until 1918 the capital of the Austro-Hungarian Empire, only a few contributions to quantum theory emerged; still Vienna must be mentioned in the history of quantum physics, because it gave birth to two of the leading pioneers, namely Wolfgang Pauli and Erwin Schrödinger.

Pauli and Schrödinger shared many qualities and experiences. Both received their primary and secondary education in Vienna: both became typical intellectual representatives determined by the rich culture and tradition of their native city; they were interested alike in art, literature and philosophy. However, their later lives and scientific careers developed quite differently. Pauli, after graduating from the *Gymnasium* in 1918, found Vienna to be a 'spiritual desert' ('*geistige Einöde*,' Enz, 1973, p. 767); he left his home and started university studies with Arnold Sommerfeld in Munich. Although he regularly returned to Vienna to spend, e.g., extended Christmas vacations with his parents, he continued his studies and sought a professional career in Göttingen, Hamburg and Copenhagen, settling finally in Zurich as a highly reputed professor of theoretical physics at the *Eidgenössische Technische Hochschule* (E.T.H.). The last connections with Vienna broke down when his father, the physiologist Wolfgang Joseph Pauli came to live in Zurich after his retirement.

Schrödinger, on the other hand, stayed in Vienna for a much longer time. He received his university education there, obtained his doctorate and *Habilitation* from the University of Vienna, and then continued for many years as a *Privatdozent* and assistant at Franz Exner's Institute. Only the worsening political situation, in which Austria found itself after total defeat at the end of World War I and the disintegration of the Austro-Hungarian Empire, forced him to seek the continuation of his professional career away from Vienna. In spite of becoming a distinguished scientist and professor abroad, occupying university positions in Zurich (1921–1927) and Berlin (1927–1933) of a kind which his country could not or did not offer him, he never cut his ties to Vienna and Austria. He even returned at a difficult and insecure time, namely in 1936, to take up a position at the University of Graz. Two years later, when Austria was united with Germany under the Government of the National Socialists, he had to leave again. Finally, after his retirement in 1955 from the Institute for Advanced Studies in Dublin, Ireland, he accepted a chair especially created for him at the University of Vienna. So at the end of his life the circle closed: Schrödinger returned to his hometown, where he also died a few years later in 1961.

Schrödinger loved Vienna. While he lived there, he participated fully in its cultural life. He went to concerts and theatre. He even wrote poems and philosophical essays in Vienna and at other places. In all his writings and lectures, the purely scientific ones as well as the popular ones, he displayed the wealth of erudition he owed to Vienna. The Viennese background has therefore to be considered as being of crucial importance for Schrödinger's life. Also it was the special scientific atmosphere of the Austrian capital, which provided the topics of his early scientific work. Since the experiences gathered in this early work later on proved to be important for the genesis of wave mechanics, we shall present, in Section I.1 a description of the Viennese background, especially the development of Austrian and Viennese science before Schrödinger entered university.

In Section I.2, the biography of the young Schrödinger will be given: we shall report about the family, the boyhood and the studies of Erwin at the University of Vienna; we shall sketch a picture of his teachers and comrades and consider the various influences on him as a student.

Schrödinger's early scientific work, apart from the apprentice work on the doctoral dissertation, will be considered in Section I.3. Schrödinger, as an assistant at Franz Exner's experimental institute, dealt with several subjects, both experimental and theoretical (e.g., with the kinetic theory of magnetism and of dielectric substances, with atmospheric electricity, penetrating radiation, and with the lattice structure of solids).

The outbreak and events of World War I, which form the background of Section I.4, influenced Schrödinger's life and career decisively. The Viennese science, having just experienced a high point in the last two pre-war years (e.g., the discovery of cosmic radiation in 1912 and the *Naturforscherversammlung* in 1913) was similarly affected, as most of its representatives were called to bear arms. Many European physicists were killed in action or died as a consequence of the war. The physics community of Vienna was hit badly by the death of Friedrich Hasenöhrl. Schrödinger, on the other hand, survived his first involvement as an artillery officer; he spent the last two years of the war at an officers' school near Vienna.

Although Schrödinger could not assume his duties at the University of Vienna until the war was over, he did work on various scientific topics during World War I, especially on questions of statistical fluctuations, the general relativity theory and quantum theory (Section I.5). When Schrödinger finally left Vienna, within a year of the end of the war, he had been introduced to the most advanced topics of theoretical physics.

## I.1 The Viennese Background

The sketch of the specific Viennese situation, culture and scientific life and its historical development, which we shall outline now, cannot of course, be considered as being complete. What we wish to do is just to call attention to certain features that seem to us characteristic of the place where Schrödinger grew up and lived through his formative stages.

### An *Aperçu* of Austrian and Viennese History[1]

It is impossible to understand the Viennese background without paying attention to the history and tradition of the country, or rather countries and states, of which Vienna had been the capital for centuries. Traces of the city go back to Neolithic and Bronze Ages, and it was already a Celtic settlement since the fourth century B.C. After the Romans took over (in the year 16 B.C.), they created the province of Pannonia (in A.D. 10) and founded Carnutum and later (around A.D. 80) Vindobona, a military city at the Limes, the northern border of the Roman Empire. Several civilian settlements lay in the region of Vienna, an open country having no insurmountable or easily defensible borders. Still the situation at the most important European East-West connection (between the Rhine Valley and the Black Sea) favoured commerce and traffic. As a consequence of this open and crucial geographic situation, foreign tribes, coming especially from Asia, regularly crossed the country; they often caused danger to the population, but contributed in an essential way to the special culture and civilization which can still be felt in today's Vienna.

The political situation became stabilized when the so-called *Ostmark* (*Ostarici*), the region around Vienna, came under the rule of the Babenberg margraves and dukes. From 1276 to 1918, for a period of 642 years the Habsburg family held sway over the country; the chief of the house was, for most of the time until 1806, elected head of the (German) Holy Roman Empire.[2] A principal

[1] For the history of Austria we have consulted: *Brockhaus Enzyklopädie*, seventeenth edition of *Der Große Brockhaus*, Volumes *1-23*, Wiesbaden: F. A. Brockhaus, 1966-1976; *Encyclopaedia Britannica*, Volumes *1-23*, Chicago-London-Toronto-Geneva: Encyclopaedia Britannica, Inc. (William Benton), 1962.

[2] The German King Rudolph I of Habsburg won Austria in the Battle on the Marchfeld from the Bohemian King Ottokar II, who had seized the country in 1246 after the death of the last Babenberg duke.

feature of the Habsburg rule consisted in the steady expansion of Austria, essentially by marriages and contracts, until finally Emperor Charles V also reigned over Spain and its new colonies in America; after his retirement in 1555 his brother Ferdinand inherited the Austrian *Erblande* (Austria, Carynthia, Carnobia, *Windische Mark*, Gorizia, Tyrol), the Bohemian states (Bohemia, Moravia, Silesia, Lusatia) and Hungary, while his son Philipp II got Spain, the Netherlands and the colonies.

The history of Austria and the associated states in the following centuries was dominated by three important features. The first consisted in the defence against the Turkish expansion, which after the second siege of Vienna in 1683 resulted in Austria expanding to the east, to Transylvania (1699), and parts of Walachia, Serbia and Bosnia (until 1718). The second essential aspect concerned Western politics, which led to wars with France over the leadership in Western Europe, especially over the succession to the throne in Habsburg Spain (War of Spanish Succession: 1701-1714). The third feature was the religious stiuation. In the great Reformation period the Habsburg emperors had remained faithful to the Catholic Church, and in the following religious wars of the sixteenth and seventeenth centuries they defended the old religion and strictly kept and restored it in their countries.

The enlarged, Catholic Austria and its widely scattered territories—to which, after the loss of Silesia, later in the eighteenth century the southern parts of Poland were added—essentially survived the tempestuous times of the French Revolution and the Napoleonic Wars. However, in 1806 Emperor Franz II renounced his title (for the Holy Roman Empire) and assumed the title of an Emperor of Austria. The German Empire, on the other hand, was not re-established after the Congress of Vienna (1814-1815); instead a German Federation (*Deutscher Bund*) was created, in which Austria played a leading role.

The Habsburg countries after 1815 consisted of the former Habsburg *Erblande*, except the most western parts (in the Alsace and the Netherlands); they included Vorarlberg, Tyrol, Salzburg (formerly a Bishop's state), Upper and Lower Austria, Styria and Carynthia as German-speaking parts; Bohemia, Moravia, the southern parts of Poland (Galicia, later Cracow and its surroundings) as Slavic-speaking parts; Hungary; the (mostly Slavic) Danubian and Dalmation provinces; the Kingdom of Lombardy-Venetia (ruled by a viceroy) and Tuscany and Modena (ruled by Habsburg archdukes) in Italy. During the first part of the nineteenth century the Austrian Emperor and his chief advisor, Clemens Lothar von Metternich, tried to restore the pre-French Revolutionary tradition as much as possible by establishing a bureaucratic, conservative administration and by suppressing any liberal and national movement within and outside the Empire through strict censorship. The revolutionary events in 1848 forced Metternich to resign. Although the liberal and nationalistic upheavals in Vienna, Hungary, Bohemia and Italy were suppressed by military force, a different Austria was taken over by the young Emperor Franz Joseph.

Franz Joseph was born on 18 August 1830 and received a severe education in a clerical atmosphere. Hard work from early morning until late at night, and deep respect for religious and political traditions, characterized him all his life. Having ascended the throne (on the resignation of his uncle Ferdinand), he soon established a strictly autocratic regime, based on uniforms (which he loved) and a unified administration in all parts of his diverse Empire; he indeed attempted to turn Austria and the associated countries into a centralized, essentially German state, exerting patriarchical and bureaucratic control over the governed peoples. With respect to external politics the young Austrian Emperor first developed a great ambition: he dreamt of re-establishing some kind of Holy Roman Empire based on church and sword. Bad fortune in political and military actions—such as the Crimean War (1854), the Italian War (1859) and the Austro-Prussian War (1866)—resulting in heavy political and territorial losses (especially in Italy: Lombardy, Venice) forced Franz Joseph to change his policy; he now sought peace wherever he could and gave up enforcing an absolute rule over his territories, approaching constitutional measures whenever necessary. In 1867 Hungary was granted certain constitutional rights—such as the rights of the individuals, an impartial judiciary, and freedom of belief and education. The same rights were given to the German parts of Austria, which now formed with Hungary the so-called *Austro-Hungarian Dual Monarchy.*

At the turn to the twentieth century, Austria was a highly complex system of complex states. It comprised many peoples of different nationalities, including Germans, Hungarians, Czechs, Solvacs, Poles, Slovenes, Croats and Italians, to name just the major ones.[3] This caused nationality problems, which were enhanced by the problems of political parties having various national programmes.[4] Further the entire large Empire, like the other European countries and states, underwent at that time rapid economic and industrial changes, bringing in quick succession boom and crisis. On top of this difficult situation in Austria only one integrating figure existed: the aging Emperor Franz Joseph, who grew into the role which he had always wished for himself, namely to be the father of his peoples. Personally isolated, he increasingly became separated from the actual administration of the Empire and lost his political influence and power. He constituted an ancient relic in a state system which appeared to have no place in the twentieth century. And still, in this very old-fashioned state system many developments emerged which would determine the life and politics of Europe and beyond after the Austrian Empire had ceased to exist.

Vienna, for centuries the capital of Austria, was not always the most important city of the Habsburg countries. Its roots went back about as far as those of the Hungarian city of Budapest, but certainly for a while the Bohemian capital of

[3] More nations were included when further parts of the Balkan countries, formerly under Turkish, rule were acquired; e.g., in 1878 Austria took over the administration of Bosnia and Herzegovina.

[4] For example, the German liberals, who came from the middle class, were anti-clerical and opposed to military expenditure in general. Franz Joseph occasionally used the conservative aristocracy in Austria *and* the Slavic parts in order to balance their influence.

Prague superseded Vienna as the finest large city—not to mention places like Cracow, Florence, Milan, or Venice, which also belonged to Austria for some time. Nevertheless, Vienna possessed several advantages that made it the natural Austrian capital, especially its geographical situation in the historical core territory, the *Ostmark*, at the most important junction of the connection from west to east (along the Danube) and from north to south (North Sea to Adriatic). The crusades strengthened the economic and political role of Vienna, which flourished due to viticulture and trade from the Middle East to Europe. Having been called a *city* since 1137, Vienna obtained the city charter (*Stadtrecht*) on 18 May 1221 and became the residence of the Habsburgs in 1276. Duke Rudolph IV founded the University of Vienna in 1365, the second one in the Holy Roman Empire (after Prague in 1348) north of the Alps. From 1439 to 1806 the city served, with short interruptions, as the hometown of the Holy Roman Emperors from the Habsburg house, its population increasing from 60,000 inhabitants in the fifteenth century to 175,000 in 1750 and then rapidly to several hundred thousand in the nineteenth century.

In late medieval times the citizens of Vienna acquired independence from their Habsburg rulers. While a majority of them became Protestants in the period of Reformation, from the middle of the seventeenth century the city was turned back fully to the Catholic faith. The world-wide relations of the Habsburg emperors attracted people from quite a few nations: members of many noble families moved to the Austrian capital, together with merchants and artists from western and eastern countries. During the nineteenth century, after the removal of most traditional feudal and social restrictions, industrialization and the growing wealth of the city attracted people from all over the Habsburg Empire, especially where the increasing population could no longer find employment in agriculture: Vienna doubled the number of its inhabitants from 1840 (population 440,000) to 1873 (900,000), and again from 1873 to 1910 (2,031,500), an increase which caused great difficulties in the housing situation.[5]

The architectural scene of Vienna, with the medieval inner parts around the *Hofburg* and St. Stephen's Cathedral, the great additions of the *Baroque* period (*Hofburg*, the palaces of Prince Eugene and of the nobility, the churches of St. Peter and Charles, and the imperial castle of *Schönbrunn*), and finally the representative architecture of the *Ring Streets* (monumental copies of the *Medieval*, *Renaissance* and *Baroque* styles), reflect in an enhanced manner the historical tradition of Austria. The same may be said, even more rightfully, about Viennese culture, especially the arts. Having roots in medieval times, it began really only in the seventeenth century to develop typical characteristics showing

[5] Although the medieval fortification was extended after the Turkish siege in 1683 by including the grounds around the major estates of noblemen, the city grew largely in size only after 1850 when the old walls were removed. At that time the *Ring Streets* were created, and 34 suburbs were included and reorganized in new districts. Finally, in 1892 the outer districts (XI to XIX) were established.

up first in architecture and the decorative arts, and then shifting over to music and finally to literature.[6]

## The Viennese Culture[7]

The wars against the Ottoman Empire following the siege of Vienna in 1683 created some prosperity in Austria. It was important then that the church and the nobility not only had available well-known architects from Italy, but could employ three great Austrian masters: Jakob Prandtauer (1660–1726), Johann Bernhard Fischer von Erlach (1656–1723) and Lukas von Hildebrandt (1686–1745). While the Tyrolian Prandtauer erected the most visible signs on the hills along the Danube (e.g., St. Florian and Melk), the works of the other two turned Vienna into a genuinely imperial city. Thus Fischer von Erlach was asked to construct, beginning in 1700, the huge castle of *Schönbrunn* in the south of the city, to extend the *Hofburg* further (e.g., the Residence Library from 1721) and to build the *Schönborn* (1699) and *Schwarzenberg* palaces (1697–1715) and others, as well as the Charles' Church.[8] Lukas von Hildebrandt built the two richly structured and ornate palaces of Prince Eugene, the Lower and Upper Belvedere (1714; 1720 and the following years), the churches of the Piaristes (1698) and St. Peter (1702) and many palaces of the nobility in Vienna.[9] In the late eighteenth century a less monumental style took over in the Viennese architecture, fashioned after French *Rococo* and classical tradition.

Baroque buildings demanded as a rule heavy decoration with sculptures and paintings. In the late seventeenth and throughout the eighteenth century, Austria and, in particular, Vienna, possessed an impressive number of masters in these fields: e.g., the sculptors Thomas Schwanthaler (1634–1707), Johann Meinrad Guggenbichler (1649–1723) and Georg Raffael Donner (1693–1741), or the painters Johann Michael Rottmayer (1654–1730), Daniel Gran (1694–1757), Paul Troger (1698–1762), Martin Johann Schmidt (1718–1801) and Franz Anton Maulbertsch (1724–1798).[10]

[6] We may remind ourselves that Vienna, the Babenberg Residence, was one of the centres of German poetry, the so-called *Minnesang*; thus Walther von der Vogelweide (ca. 1170–1230), an outstanding *Minnesänger*, learned his art there, But this, like the Gothic St. Stephen's Cathedral, remained within the fine German standards.

[7] For cultural development in general, as well as for the personalities involved, we have consulted the encyclopedias cited in footnote 1. Particular aspects of architecture in Vienna have been found in: *Wien* (*Dehio-Handbuch*: *Die Kunstdenkmäler Österreichs*), fifth edition (elaborated by Anton Macku and Erwin Neumann), Vienna-Munich: Verlag von Anton Schroll & Co., 1960. In dealing with aspects of the political and cultural situation at the turn of the century, the book of Allan Janik and Stephen Toulmin, *Wittgenstein's Vienna*, New York: Simon & Schuster, 1973, has been useful.

[8] After his death many of Fischer von Erlach's projects were completed by his son, Joseph Emanuel Fischer von Erlach (1693–1742), who also added the *Hofreitschule* to the *Hofburg*.

[9] Hildebrandt also exported the great Viennese style to Salzburg (Mirabelle Castle, 1721–1727) and other German states (Würzburg *Residenz*, Pommersfelden Castle).

[10] This tradition in decorative arts and painting seemed to be interrupted in the nineteenth century, but was revived in the early twentieth century.

The poetic tradition started later than those in architecture, sculpture and painting.[11] Even in the great period of German poetry at the end of the eighteenth century, represented by the names of Klopstock, Lessing, Goethe or Schiller, the voice of Austria remained silent. Only in the beginning of the nineteenth century did the country start to make important contributions to German literature. Franz Grillparzer (1791-1872), Anastasius Grün (1806-1876) and Nikolaus Lenau (1802-1850) worked in Vienna.[12] The Viennese playwrights Ferdinand Raimund (1790-1836) and Johann Nestroy (1801-1864) exerted a great impact on the public; Raimund, through fantastic and imaginative plays, in which he tried to merge the previous popular Vienna *Lokalposse* with highbrow classical drama, Nestroy through his bitter humour and frequent criticism of high sentiments or institutions, especially the existing censorship. Both Nestroy and Raimund were influenced largely by the Viennese atmosphere and tradition: Vienna provided the colourful world of different, popular characters, the knowledge of picturesque people, such as Turks, Greeks and Levantines, who figured in many of their plays.

The people of Vienna loved the theatre. The simple public, as well as the learned citizen and the nobleman, went to the *Vorstadttheater* (suburban theatre), where mainly popular plays were performed—often involving the Viennese *Hanswurst*, a person full of natural, at times malicious, wit and always prepared to do stupid things.[13] The Viennese public also knew operas since Emperor Ferdinand III (1637-1657) had brought an Italian opera company from Mantua, the home of his wife, to the Austrian capital.[14] Besides the traditional Italian opera, a *Nationalsingspiel*—i.e., an institution where popular plays containing extended musical pieces, like songs and couplets, were regularly performed—was established by Emperor Joseph II (1765-1790). Also the great Viennese masters Haydn and Mozart wrote *Singspiele.*

[11] We have already mentioned the medieval *Minnesang* period. During the sixteenth and seventeenth centuries one may just register the presence of two representatives of German literature in Vienna, both of whom came from abroad. In 1497 Conrad Celtis (born in 1459 in Wipfeld near Schweinfurt) obtained a chair for humanistic studies at the University of Vienna which he occupied until his death in 1508; besides writings poems, epigrams and plays, he devised a theory of poetry and a programme for national education. About two hundred years later the Capucine monk Abraham a Santa Clara (or Ulrich Megerle from Kronstetten in Schwaben, 1672-1709), preacher at the Imperial court in Vienna, published collections of sermons in which he combined satirical and folkloristic elements with a powerful language.

Finally, one might refer, in this context, to two events which happened in regions that had belonged to Austria for centuries. In 1401 Johannes of Tepl (or of Saaz) in Bohemia wrote one of the earliest pieces of New-High German prose. And in the seventeenth century members of the Silesian School of Poetry played a leading role in reforming German literature.

[12] The best known among these, the Viennese Grillparzer, wrote dramatic plays continuing the tradition of Johann Wolfgang von Goethe and Friedrich von Schiller.

[13] There were three suburban theatres, namely, the *Theater an der Josefstadt*, *Theater an der Leopoldstadt* and the *Theater an der Wien.*

[14] Ferdinand's successors, Leopold I (1658-1705) and Charles VI (1711-1740), had continued the tradition, bringing several Italian composers (e.g., Marc Antonio Cesti) and librettists (like the famous Pietro Metastasio) to Vienna.

In the late eighteenth and early nineteenth centuries Joseph Haydn (1732-1809), Wolfgang Amadeus Mozart (1756-1791) and Ludwig van Beethoven (1770-1827), formed the superb school of classical music, creating immortal works of vocal and instrumental music—operas, symphonies, pieces for single instruments and for groups of instruments, church music, etc. Still these three outstanding masters by no means exhausted the Viennese musical life of their times; numerous German and Italian composers lived and worked in the city before and after 1800, among them in particular the native Franz Schubert (1797-1828), the creator of hundreds of *Lieder.* Later in the nineteenth century the Hungarian Franz Liszt (1811-1886), Johannes Brahms of Hamburg (1833-1897) and the Upper Austrian Anton Bruckner (1824-1896) preserved and augmented the Viennese tradition of music.

The classical Viennese masters also acted as pioneers for the profession of an artist. In particular, Beethoven, although being assisted by friends and pupils of the nobility, became *the* example of a liberated artist working only on his own initiative. The liberation from the traditional patronage of church and court demanded, in the first place, the establishment of an enlarged, emancipated public willing to pay a fee for enjoying works of art. In the nineteenth century the majority of this public came from the middle-class bourgeoisie, say merchants, state and city officials, who had received a higher education and who earned sufficient salary. The middle-class citizens in post-Napoleonic times, the so-called *Biedermeier*—they existed also in other German states at the same time—were in general politically obedient and reliable, but socially ambitious. The Viennese bourgeois liked to live and let others live; he loved cultural festivities and presentations and public and military parades. Above all, he loved music and dancing and admired the creators of the Viennese *Waltz*, especially his countrymen, the Viennese Joseph Lanner (1801-1843), Johann Strauss, the father (1804-1897) and the son (1825-1899).[15] The latter also created a new tradition in Vienna by adopting that form of comic opera, which especially Jacques Offenbach had developed in Paris after 1850, the so-called *Operette.*[16]

The gay and joyous world of waltz represented just one aspect of life in Vienna, characterizing—at least in a glorifying retrospective—the first half of the nineteenth century. In fact, it provided the citizens an escape from the limitations set by traditional regulations and taboos. One of these was the strict censorship

[15] While the senior Strauss introduced the waltz to the Imperial Court and exported it beyond the Austrian borders, his son made it world famous on many tours all over Europe, including Russia and England.

[16] Johann Strauss, Jr. contributed distinguished examples of the Viennese *Operette* (e.g., the '*Fledermaus*'). In comparison with the French examples, his works were more *gemütlich* and less critical. Besides Strauss, Franz von Suppé (1819-1897), Karl Millöcker (1842-1899) and Karl Zeller (1842-1898) also perfected the Viennese *Operette*, while Franz Lehar (1870-1948) carried the tradition further into the twentieth century.

exerted by the authorities over the political and literary activities.[17] The 1848 Revolution removed many of the previous restrictions. A free press developed gradually (1848: '*Presse*'; 1864: '*Neue Freie Presse*'; 1867: '*Neues Wiener Tageblatt*'; 1871: '*Deutsche Zeitung*'), which represented the prevailing liberal movement.[18] However, the awakening of wider social groups or classes, and the rise of national conflicts within the Austrian Empire created new, more refined methods of censorship, exerted now on behalf of the parliament.

Towards the end of the nineteenth century Vienna was—like Berlin, London, New York or Paris—a world metropolis of commerce, politics and, in particular, of culture. The bearers of the new wealth no longer came from the old nobility but from the rich bourgeoisie; the bearers of culture were mainly bourgeois middle-class people and members of a population group that had obtained importance and influence just very recently. Since Babenberg times Jews had lived in Vienna, enjoying a certain security but having few privileges, primarily because anybody who was not a Catholic did not count much in Austria. After the 1848 Revolution, however, their situation gradually improved in the city; many Jews from all parts of Austria, especially Bohemia, moved to Vienna to enjoy the new rights and step upwards on the social ladder. The Jews soon contributed more than their share to the economics and culture of the city; as far as the arts were concerned, the Jews excelled particularly in music and literature, less so in decorative arts and architecture.[19] For example, the leading Austrian musician around the turn of the century, Gustav Mahler (1860-1911) was a Jew from Kalischt in Bohemia. He passed on the great heritage of Viennese music to the younger generation of the twentieth century, a generation more revolutionary than himself and represented by the names of Arnold Schönberg (1874-1951), Anton von Webern (1883-1945) and Alban Berg (1885-1935).[20]

In the Viennese architecture of the late nineteenth and early twentieth centuries three names stood out, those of Otto Wagner (1841-1918), Joseph Hoffmann

[17] For centuries the Catholic Church—whose distinguished representatives also acted often as chief advisors to the emperor—took part in controlling public opinion. When the spirit of the Enlightenment reached the Habsburg states in the middle of the eighteenth century, church control was replaced by state control. Thus, for example, after 1750 the texts of all plays performed on stages in Vienna, especially the *Vorstadttheater*, had to be approved by the police in advance. In the post-Napoleonic period (1815-1848) this censorship was sharpened by the Metternich system: writers of classic dramas, like Franz Grillparzer, suffered from censorship as much as the playwright Johann Nestroy, who loved to express, in extempore verses political opinions not agreeing with those of the government.

[18] Later on, other political movements also published their newspapers, e.g., the Christian-Social '*Deutsches Volksblatt*' (1888) and the Socialist '*Arbeiterzeitung*' (1889).

[19] In 1910, five per cent of the Viennese population was of Jewish descent. By the way, the rise of the Jews occurred not only in Vienna, but also in Budapest and Prague and other large cities. To give an example: the writers Max Brod, Franz Kafka and Franz Werfel, all born in Prague between 1883 and 1890, made important contributions to German literature.

[20] Schönberg, von Webern and Berg, were all born in Vienna; they created and established a new style in music, connected with the use of the twelve-tone (*Zwölfton*) technique. The senior Schönberg, who acted as the teacher of von Webern and Berg, later went to Berlin (1925-1933) and Los Angeles (1933-1951), passing some of the Viennese tradition of music on to students in the rest of the world.

(1870-1956) and Adolf Loos (1870-1933). The senior-most among them, Wagner, turned in 1893 to a functional style (*Nutzstil*), in which the function of the buildings, the materials used for their construction (including glass and steel) and other constructive elements were emphasized.[21] Hoffmann taught at the *Kunstgewerbeschule* and founded the *Wiener Werkstätte*, a school for modern decorative design. His Moravian countryman Loos—who received his education in Reichenberg and Dresden and obtained important impressions from a stay in the United States of America (1893-1896)—rather avoided decoration altogether in his buildings, which displayed simple geometric forms and the natural structure of the materials used. He would influence generations of architects in Austria and Europe and create a truly international style.

Like the architects the Viennese painters at the turn of the century also had to rid themselves of copying historic styles.[22] The eldest among them, Gustav Klimt (1862-1918), helped to found the *Wiener Secession*, an association of younger artists propogating a new modern style, the Vienna *Jugendstil*, and became its first director (1895-1905).[23] Egon Schiele (1890-1918), a student of Klimt, and Oskar Kokoschka (1886-1980) later achieved international reputation.[24]

Important and first rate as the Viennese contributions were to architecture and fine arts, those to literature should be rated higher still. Around the turn of the century, and for decades afterwards, the Austrian capital generated and housed a large number of poets and writers, whose fame went far beyond the Austrian borders.[25] A few years before 1900 a circle of artists, named *Jung Wien*, assembled regularly in the *Café Griensteidl*. The senior, guiding members of this group were

[21] As a professor of the Vienna *Kunstakademie*, Wagner influenced many students, among them Joseph Maria Olbrich (1867-1908)—who later went to Darmstadt and became cofounder of the *Deutscher Werkbund*—and Hoffmann.

[22] For the architects in Vienna the style to be abandoned was that of the *Ring Street* buildings. In the case of fine arts one had to overcome the fashion, in the second half of the nineteenth century, of huge paintings representing historical events, often in historical styles. The painter Hans Makart (1840-1884) from Salzburg, since 1879 the dominant figure of the Vienna Academy, was a main protagonist of that fashion.

[23] Klimt's works are characterized by an ornamental two-dimensionality of high colouristic power and refined decorative effects.

[24] Schiele portrayed, in particular, naked persons, giving them a thorough psychological characterization. Kokoschka, who like Schiele studied at the *Kunstgewerbeschule*, later went to Dresden and became a master of the German expressionistic movement; he travelled throughout Europe and North Africa and lived into his nineties at Lake Geneva. Especially known are his portraits, exhibiting a subtle psychological analysis of the portrayed persons, his paintings of large cities, and great, wide landscapes.

[25] It should be mentioned here that the tradition, established in the first half of the nineteenth century by writers and poets such as Grillparzer, Raimund and Nestroy—to whom one may add the novelist Adalbert Stifter (1805-1868)—had been interrupted later. Only after decades of economic and industrial progress—and occasional financial crashes like the one in 1873—was the Viennese literary tradition resumed. (See, e.g., the book on *Wittgenstein's Vienna*, referred to in Footnote 7.)

Hermann Bahr (1863–1934), born in Linz,[26] and the Viennese Arthur Schnitzler (1862–1931).[27] It included Richard Beer-Hoffman (1866–1945),[28] Hugo von Hofmannsthal (1874–1929),[29] and Stefan Zweig (1881–1942).[30] Viennese literature perhaps reached its highest level, though not appreciated at that time, in the novels of Robert Musil (1880–1942) and Joseph Roth (1894–1939); their writings, apart from being masterpieces of German prose, also conserved for later generations a lively impression of pre-World War I Austria.

No story of Viennese literature can be called complete without the mention of the critic, essayist and poet Karl Kraus (1874–1936). Born in Gitschin, Bohemia, he came to Vienna very early in his life and became not just a brilliant writer—whose essays revived the critical spirit of Nestroy—but a Viennese institution. A fighter in political—he attacked Austrian liberalism, journalism and militarism, and, though being a Jew himself, the Jewish nationalism—and social affairs—he pleaded for the equality of sexes, an improved criminal law, and against moral hypocrisy—Kraus left a lasting impression on the young intellectuals of his time.[31]

In Austrian literature one notices the relatively modest role played by philosophical topics, which reflects perhaps in some way the absence, in Vienna and Austria, of original philosophers in the nineteenth century—this situation being quite different from that existing in Germany. On the other hand, probably due to the dominance of social problems in the city and the country, Austria produced at the end of the nineteenth century a variety of socio-political schemes and ideologies, which become important for the world in the following decades. There was the evolutionary programme of Social Democracy of Victor Adler (1852–1918), the Christian-Social movement of Karl Lueger (1844–1910);[32] the pan-

[26] Bahr, who stemmed from the naturalistic school which dominated German literature during the second half of the nineteenth century, changed his style after the 1890s—leaning towards impressionism, symbolism and expressionism—as well as his social and political opinions. He was a prolific writer of successful plays, short novels, essays and critiques.

[27] Schnitzler began as a physician before turning to write plays (because of his interest in the phenomena of dreams and hypnosis). He used, in particular, the psychoanalytic methods developed in Vienna at that time.

[28] Beer-Hoffman, a Jew like Schnitzler, Hofmannsthal and Zweig, wrote novels and plays that were strongly influenced by Jewish religious traditions; the latter were not shared by Hofmannsthal and Zweig.

[29] Hofmannsthal, who studied law and philosophy, began to publish novels and plays as a teenager. After going through an inner crisis, he continued with dramatic plays in which he applied the philosophy of Friedrich Nietzsche and the psychoanalysis of Sigmund Freud. In collaboration with the Munich composer Richard Strauss he created a new type of opera by combining music and poetic drama; both became instrumental in founding the Salzburg Festival.

[30] Zweig used the psychoanalytic method in his widely-read novels and biographies of historical personalities.

[31] Kraus influenced such different personalities as the German poet and playwright Bert Brecht, and the Viennese philosopher and logician Ludwig Wittgenstein.

[32] Lueger was one of the founders of the Christian Social Movement (1888), whose programme demanded electoral reforms in favour of the lower classes, and the execution of large projects for the benefit of the entire city public—such as canal construction, enlarged recreation areas and modernization of city services. He served as mayor of Vienna from 1897 to 1910 and helped to carry out parts of the programme himself.

Germanic, anti-Semitic movement of Georg Ritter von Schönerer (1842–1924);[33] and the Zionist movement of Theodor Herzl (1860–1904).[34] Finally, Austria created at about the same time a philosophical tradition of the highest quality and greatest future significance. This was contained in the ideas of the physicists Ernst Mach and Ludwig Boltzmann, of the logician Ludwig Wittgenstein, and of the physicians and psychoanalysts Sigmund Freud and Alfred Adler. We shall deal with most of these personalities and discuss their work a little later.

## Austrian Science up to the Middle of the Nineteenth Century[35]

In contrast to the great tradition in arts and literature, not much is known about Austrian science before the middle of the nineteenth century. It appears to be rather young, especially when compared with that of other European countries, which is quite unexpected because of the early foundation of the University of Vienna (in 1365). Indeed, less than a hundred years later, mathematics and astronomy in Vienna were represented by two important personalities, Georg Peurbach and Regiomontanus. Peurbach (or Purbach), who was born on 30 May 1423 in Peuerbach, Upper Austria, studied in Vienna and lectured there from 1453 onwards. He revived the Greek and Arab trigonometric methods, calculated the occurrence of conspicuous astronomical events (constellations, occultations) and carried out stellar observations. While Peurbach, who died on 8 April 1461 in Vienna, may be considered as the first native Austrian scientist, his still more famous student Johann Müller (1436–1476) came from Königsberg near Nürnberg in Germany. He called himself (in Italy) Johann de Montereggio, which he later latinized to Regiomontanus; he began to study in Vienna in 1452 and assisted his teacher in working on Ptolemy's astronomy. After further studies in Italy he returned to Vienna, spent some time in Buda in Hungary and settled in 1471 in Nürnberg, where he established the first European observatory.[36]

[33] Schönerer, a member of the Austrian parliament since 1873, opposed both clericalism and liberalism; he rejected reason and progress and attacked the very structure of the Habsburg Empire, pleading for a romantic German Austria. In practical politics, he introduced violence and street-fighting, both in Vienna and in Bohemia. His fellow countryman, Adolf Hitler, would be influenced greatly by Schönerer's programme and methods.

[34] Herzl was correspondent of the *Neue Freie Presse* in Paris during the Dreyfus trial. As a reaction to growing anti-Semitism in Europe, and especially Austria, he founded the Zionist movement (1896, publication of the book '*Der Judenstaat*'; 1897, first Zionist World Congress in Basle); he propagated the idea of a Jewish nation, describing its future, in the homeland 'Israel', in a novel ('*Altneuland*,' 1902).

[35] For the development of science in Austria, we have taken some information, especially on the biographies of the personalities involved, from two encyclopaedias, *Brockhaus Enzyklopädie* and *Encyclopaedia Britannica* (Footnote 1). We have further used the *Dictionary of Scientific Biography*, Volumes I–XVI, New York: Charles Scribner's Sons, 1971–1980; *World Who's Who in Science. A Biographical Dictionary of Notable Scientists from Antiquity to Present* (A component volume of the Marquis Biographical Library) (Allan A. Debus, editor), Chicago: Marquis, Who's Who, Inc., 1968.

[36] Regiomontanus made extensive observations, e.g., of a great comet which is probably identical with Halley's comet. He was assassinated on 6 July 1476 during a visit to Rome in connection with the calendar reform.

After such an encouraging start, however, the mathematical and astonomical sciences in the Vienna of the following centuries did not produce personalities of comparable stature. On the other hand, three different cities in the Austrian states, namely Prague in Bohemia, Graz in Styria and Linz in Upper Austria, housed for about three decades during the late sixteenth and early seventeenth centuries two foreign astronomers of the highest reputation, the Dane Tycho Brahe (1546-1601) and Johannes Kepler (1571-1630) from Württemberg. Brahe had studied in Germany (Leipzig, Rostock, Augsburg) and built observatories in his homeland (e.g., on Uraniaborg on the island of Hven), before he assumed in 1599 the position of Court Astronomer to Emperor Rudolph II in Prague. He has been recognized as the most accurate observer before the introduction of the telescope; his instruments in Prague were also used by his disciples, especially by his successor Kepler. The latter arrived, after studies in Tübingen (where he had become a partisan of the Copernican system), in Graz to accept the position of mathematics teacher at the Protestant *Gymnasium* in 1594.[37] On being forced to leave Graz due to the counter-Reformation, he accepted Brahe's offer to come to Prague. There he published his main contributions to optics and astronomy: *Ad Vitellionem Paralipomena* (1604), in which he proposed to use light rays; *Astronomia Nova* (1609), in which he concluded that all orbits of planets were elliptical; and *Dioptrice* (1611), in which he gave the design of the astronomical telescope. Upon the death of Rudolph II in 1612, Kepler moved to Linz, assuming the position of mathematician to the *Stände* of Upper Austria. In Linz he completed the five books *De Harmonice Mundi* (1619), in which he expounded the third Keplerian law; and the *Tabulae Rudolphinae* (1627), containing the data on planetary orbits and fixed stars.[38]

In 1618 the Thirty Years' War began in Prague; it would devastate many parts of the Holy Roman Empire and Austria. The successors of Rudolph II, as well as private and official patrons, were far less interested in astronomy and science then, being completely absorbed in military, political and economic troubles. The war did settle the religious quarrels in cental Europe, especially in Austria where the Catholic faith was firmly re-established, and this fact may provide another reason for the interruption of scientific tradition, especially in astronomy. Astronomy had at that time led to a serious problem with the authorities of the Catholic Church. The mathematician Galileo Galilei of Florence (1564-1642), who had advocated the system of Copernicus, had been forced after a clerical trial in Rome to withdraw from these views (1633) and was sentenced to indefinite house arrest. This verdict not only hit Galileo personally but his scientific methods

[37] In Graz, Kepler published a calendar with astronomical and astrological predictions, as well as his important early work *Mysterium Cosmographicum* (1596), in which he presented speculations on the relationships between planetary orbits. Kepler's main claim was that he could fit five regular solids between the spheres of the six planets.

[38] It should be mentioned in this context that most of the observations stemmed from Tycho Brahe, who, by the way, had also developed a third planetary system—the so-called Tychonic system—constituting a compromise between the system of Ptolemy and Copernicus (that all planets except the earth move around the sun which itself describes an orbit around the earth).

as well; besides the Copernican astronomy the entire new physics, with its mathematical description of natural phenomena, also became suspect in the eyes of the Catholic Church. As a consequence Italy lost its eminence in science, and a similar situation occurred in other Catholic countries except in France.[39] The centre of scientific activity shifted to northern Protestant countries, notably Holland and England, where personalities like Christiaan Huygens (1629-1695), Antoon van Leeuwenhoek (1632-1723), Robert Boyle (1627-1691), Robert Hooke (1635-1703) and Isaac Newton (1642-1727) started a new age in science.[40]

Before Austria joined this scientific development, considerable changes, especially in the educational system had to be undertaken. Only after 1750 did the Enlightenment movement enter Austria; Emperor Joseph II (1765-1790) ordered important reforms, especially in the internal administration. Thus, for example the school system, which had previously been in the hands of the Church and local city authorities, was put under the authority of the central government. The *Studien-Hof-Kommission* represented the highest administrative supervisory body, including universities, in the Austrian countries, furnishing direct control of the church over science.[41]

The increasing centralization of administration and internal politics, which was enhanced after the establishment of an Austrian Empire in 1806, created a concentration of cultural and scientific activity in Vienna, the capital of the new empire. Centres like Prague, formerly an important city that had at times superseded Vienna, fell back in this respect behind it. Still this does not mean that everybody who later won an international reputation, worked in Vienna. Indeed, the first such Austrian scientist of the nineteenth century, the mathematician Farkas (Wolfgang) Bolyai came from a small place in the Transylvanian part of Hungary (now Rumania). Born on 9 February 1775 near Nagyszeven (or Hermannstadt), and having studied at the University of Göttingen (where he became a friend of Carl Friedrich Gauss), the older Bolyai taught from 1804 to 1853 at the evangelical reformed college of Morosvásárhely (now Sibiu, Rumania). In his main publication, *Tentamen*, a book of two volumes (published in 1832 and

[39] Italy had played a leading role in European science for many centuries, especially since the Renaissance. The last representatives of the great epoch were Galileo's pupils Evangelista Torricelli (1608-1647) and Vincentio Viviani (1622-1703). More than one hundred years later, when the Enlightenment movement reached Italy, science began to rise again with the discoveries by Luigi Galvani (1737-1798) and Alessandro Volta (1745-1827) in the field of electricity.

[40] The situation was different in France—in contrast to most other Catholic countries (like the Italian states and Spain)—because there a separation between church and state authorities had already been established at the end of the sixteenth century. In France, science was soon recognized to be a state affair, although men of the church contributed a considerable share.

[41] One hundred years later, after 1848, a Ministry of Public Education and a permanent *Reformkommission* were installed to take care of schools and universities. It would take some time before the last reminders of the trial of Galileo disappeared from Austria. One might perhaps consider it a consequence of the Catholic tradition in Austria that nearly 350 years after the trial it was an Austrian church leader, Cardinal König, who tried hard to have this matter reconsidered in an ecclesiastical court.

1833, respectively), he provided an introduction to the elements of pure mathematics, presenting himself as a kind of precursor of Gottlob Frege, Moritz Pasch and Georg Cantor.[42]

Volume I contained the *Appendix Scientiam Spati Absolute Veram Exhibens* (Appendix Explaining the Absolutely True Science of Space) of his son Janós (Johann) Bolyai. He was born on 15 December 1802 at Koloszvár (Klausenburg, now Cluj in Rumania) and educated at his father's college (1815–1818) and at the Imperial Engineering Academy in Vienna (1818–1822), where he prepared for service in the Army Engineering Corps (1823–1833). In 1823 he began to work on a mathematical paper, which dealt with constructing a geometry without the use of Euclid's axiom of parallels. Upon Gauss' approval, the paper was included in his father's book; however, Gauss' statement that he had earlier reached (but not published) similar conclusions prevented the younger Bolyai from further publication, until he noticed the work of the Russian Nicolai Lobatchevski (1793–1856) on non-Euclidean geometry. Farkas and Janós Bolyai died on 20 November 1856 and 27 January 1860, respectively, both in Marosvásárhely.

The work of the two Bolyais, important as it proved to be for the future, can hardly be considered as representing the official academic activities of the Austrian Empire; it was not only performed at a remote place, but the authors also did not have relations with any member of the scientific community in Austria. And yet, during their lifetimes there arose an Austrian science, starting practically from zero and finally arriving at a level which could well be compared to West European standards. The roots and causes of this rapid development in the first six or seven decades of the nineteenth century were manifold. First, Austria did not then cut itself off from the advances of science in the western countries, especially the neighbouring German states, but also France and England. Second, the Austrian Empire possessed rich natural resources, a great variety of minerals and even coal, which at that time of quick industrial development had to be exploited at an increasing rate.[43] The Metternich regime following the Napoleonic wars (1815–1848), although little progressive in other matters, strongly supported the industrial development. In particular, it fostered those sciences, which could help to raise industry and technology to a higher level, such as geology, mineralogy, mathematics and physics. Not surprisingly, we find among the earliest

[42] The full title of the book is: '*Tentamen Juventutem Studiosam in Elementa Matheseos Purae, Elementaris ac Sublimioris, Methodo Intuitiva, Evidentis Que Huie Propria, Introducendi, cum Appendice Triplici*' ('*An Attempt to Introduce Studious Youth Into the Elements of Pure Mathematics, by an Intuitive Method and Appropriate Evidence, With a Threefold Appendix*'). Volume I dealt with geometry and Volume II with arithmetic, algebra and analysis.

[43] In this respect we remind ourselves of the fact that in prehistoric and early historic times the salt mines of the Untersberg, near Salzburg, or the iron mines in Styria contributed important economic goods. In the late medieval times until the seventeenth century the gold and silver mines in Tyrol, Bohemia and Moravia covered even the most ambitious expenses of the Holy Roman Emperors.

Austrian scientists of international reputation the Viennese geologist and mineralogist Wilhelm Karl Ritter von Haidinger.[44]

The mathematician Joseph Petzval did not originate from Vienna, but was born on 6 January 1807 in Szepesbéla, Hungary, and had been professor at the University of Budapest (1832-1836) before he obtained a chair for higher mathematics at the University of Vienna in 1836. Besides working on various mathematical topics, he showed a great interest in physics, especially in geometrical optics; for example, he devised in 1840, by calculation, an achromatic objective of two pairs of lenses, which was duly constructed in practice by F. Voigtländer in the same year.[45] Being a brilliant personality, though not easily accessible to students—among them Ludwig Boltzmann—Petzval exerted an important influence on the scientific community in Vienna, where he lived to an advanced age (he died on 17 September 1891). His colleague, the physicist Andreas von Ettinghausen was born in Heidelberg on 25 November 1795 and he studied in Vienna, at the university and the Bombardier School; he spent nearly all his scientific career in Vienna, becoming in 1821 professor of higher mathematics and in 1835 professor of physics at the University of Vienna. On the premature death of Christian Doppler in 1853, von Ettinghausen was appointed director of the physics institute at the University of Vienna, a post which he filled with great vigour until becoming ill in 1862. He died on 25 May 1878 in Vienna.

Ettinghausen contributed to various fields in mathematics (analysis, algebra, and differential geometry) and in physics (mechanics, geometrical and wave optics, and electromagnetism), and even to engineering—e.g., in 1837 he constructed a magnetoelectric engine; he became a pioneer in the field of mathematical physics by combining his skills in mathematics and physics. Together with Andreas Baumgartner he founded and edited the journal *Zeitschrift für Physik und Mathematik* (10 volumes, 1826-1832). His talents were highly appreciated in Vienna where he occupied, besides his university professorship, the position of a director of mathematical studies at the Engineering Academy (from 1848) and a chair for higher engineering science at the *Polytechnikum* from (1852). He was also a co-founder in 1847 of the *Kaiserliche Akademie der Wissenschaften.*

Andreas von Ettinghausen became the first of a sequence of eminent scientists from his family; his son Constantin von Ettinghausen (1826-1897) was a renowned paleontologist, and the latter's nephew, Albert von Ettinghausen (1850-1932), again a noted physicist (and discoverer of thermoelectric effects). Such sequences of related scientific talents certainly existed in other countries as well; however, they played a more pronounced role in the young Austrian development in

[44] Haidinger, who was born on 5 February 1795, became in 1840 *Sektionsrat* in the Ministry of Agriculture and Mining and later director of the newly founded *Geologische Reichsanstalt* in Vienna (1849-1866). Under his leadership a geological map of the Austrian Empire was prepared. He died in Dornbach near Vienna on 19 March 1871.

[45] The main scientific work of Petzval may be his *Bericht über die Ergebnisse einiger dioptrischer Untersuchungen* (Report on the Results of Some Dioptric Researches) of 1843.

science. There was, for example, the Tschermak family: the first, Gustav Tschermak, Edler von Seysenegg (1836–1927) being a reputed mineralogist and professor at the University of Vienna; one of his sons, the botanist Erich Tschermak (1871–1962), would rediscover the Mendelian laws of inheritance; the other, Armin Tschermak (1870–1952), would become professor of physiology at the Universities of Halle, Vienna and Prague.[46] Like the family relations the teacher-student relationship also played perhaps a more decisive role in Vienna than elsewhere in Europe. Thus we may note an uninterrupted sequence of teacher and student on the physics chair in Vienna, from Andreas von Ettinghausen over Joseph Stefan, Ludwig Boltzmann, Fritz Hasenöhrl to Hans and Walter Thirring.

The 1848 Revolution considerably changed political and social life and had important consequences on cultural affairs. However, it affected only little the continuity of the relatively young scientific life. The reason is that the favourable attitude towards scientific activity, which had existed already in the Metternich era, was transferred without interruption into later times. One personality even represented that continuity perfectly, namely Andreas Baumgartner. Born on 23 November 1793 in Friedberg, Bohemia, he had studied at the University of Vienna (graduating in 1810) and was appointed professor of physics there in 1823. Ten years later he became director of the imperial porcelain factory, in 1842 he had advanced to the position of a director of the tobacco factories, and in 1847 finally to head the Austrian railroad construction. Now, in 1848, he became Minister of Public Works, and in 1851 Minister of Commerce and Finance (until 1855). Baumgartner, since 1851 also President of the *Kaiserliche Akademie der Wissenschaften*, may not be remembered as a great physicist of his times; but he deserves an honourable place in the history of Austrian science because of his popular writings on physics and his restless efforts for the welfare of science and technology.[47] Men like him were just indispensable in keeping the torch of science and technology burning in the reign of Franz Joseph, who himself showed only a weak interest in these fields.[48] With official and personal support many new chairs in science were established at the old universities (Budapest, Graz, Innsbruck, Prague and Vienna) and at the new ones (like Zagreb, since 1874) and at the new technical universities, e.g., in Prague (emerging from the *Ständisches Polytechnisches Institut*, founded in 1806; in 1869 split into a German and

[46] Father–son relationships often showed up in Vienna, both working in the same field, and still do; one may think of the two Thirrings, i.e., Hans Thirring (1888–1976) and his son Walter (born in 1927). Both were appointed in direct succession to the same chair, the father in 1921, the son in 1959.

[47] Baumgartner also supported the writer Adalbert Stifter, who had studied physics as well and later instructed the sons of Prince Metternich in scientific subjects. Baumgartner died on 30 July 1865 in Vienna.

[48] As we have mentioned, Franz Joseph came to rule the Austrian Empire when he was still very young. Being deeply absorbed in political and military actions, he had no time for arts and sciences although he liked the theatre. This point can be illustrated by a story. When Wilhelm von Hartel, the Minister of Education, approached the emperor with the plan to bring back Ludwig Boltzmann, then professor in Leipzig, to Vienna, Franz Joseph expressed unhappiness because Boltzmann had left the country once already, in 1890, to go to Munich. Then von Hartel won the emperor's approval of his plan by comparing Boltzmann to the famous dancer Fanny Elssler (1810–1884), who had also performed frequently on foreign stages, and of whom Franz Joseph had nevertheless been very fond.

a Czech technical university), or in Vienna (emerging in 1872 from the former *Zentralanstalt für Handel und Gewerbe*, founded in 1815, later the *Polytechnisches Institut*), or in Graz (previous institution founded in 1827).

## The Foundation of Physics in Vienna

As we have noted, physics was considered an important field of study in Vienna and Austria generally. After the revolution of 1848 this favourable opinion about physics grew further; it manifested itself in the foundation of a *Physikalisches Institut* at the University of Vienna. The new institute at the *Erdberg* was well equipped for the purpose of experimental training of future *Gymnasium* teachers.[49] It represented not only the first such state institution in Austria, but also preceded most similar state and university institutions in Europe.[50] The director of the institute was also appointed a full or ordinary professor of experimental physics at the Imperial University of Vienna, again something new in Austrian history. Johann Christian Doppler became the first occupant of the chair in 1850; he was perhaps the earliest physicist of the Austrian Empire who would acquire international fame. Christian Doppler was born in Salzburg on 29 November 1803 and educated in Vienna at the Polytechnic Institute (1822–1825) and at Salzburg, where his mathematical abilities were soon recognized. From 1829 to 1833 he served as an assistant in mathematics in Vienna; in 1835 he obtained the position of a professor of mathematics and accounting at the State Secondary School in Prague, from where he moved in 1841 to a professorship at the State Technical Academy in the same city, and in 1847 to a professorship of mathematics, physics and mechanics at the Mining Academy in Schemnitz, Bohemia. Doppler's growing scientific reputation brought him the directorship of the new physics institute in Vienna, which he helped to plan, establish and organize. However, he died a few years afterwards, on 17 March 1853 in Venice.

Doppler contributed papers on topics of mathematics and electricity since the 1830s. His scientific fame rests, however, on the discovery of what was later called 'Doppler's principle', an equation relating the observed frequency of a propagating wave to the relative motion of the observer with respect to the source of vibration. Doppler gave the first formulation of this principle in a paper, entitled '*Ueber das farbige Licht der Doppelsterne und einiger anderer Gestirne des Himmels*' ('On the Coloured Light of Double Stars and Several Other Celestial Stellar Objects,' Doppler, 1842) and presented on 25 May 1842 to the *Königliche Böhmische Gesellschaft der Wissenschaften* in Prague, whose associate member he

[49] Information about the early years of the Vienna *Physikalisches Institut* has been obtained from the following sources: Ludwig Boltzmann's talk on 'Josef Stefan,' delivered at the inaugration ceremony of the Stefan monument on 8 December 1895 (Boltzmann, 1905, pp. 92–103); Boltzmann's obituary of Loschmidt, presented before the *Chemisch-Physikalische Gesellschaft* in Vienna on 29 October 1895 (Boltzmann, 1905, pp. 228–240); Boltzmann's obituary of Loschmidt, delivered on 5 November 1895 at the University of Vienna (Boltzmann, 1905, pp. 240–252); Franz Exner's article on the centenary of Loschmidt's birth, published in *Die Naturwissenschaften* (Exner, 1921).

[50] It seems that the Physical Institute at the University of Leipzig, founded in 1835, was the first such institution in Europe.

had became in 1840 (full member in 1843). Several years later he generalized the formula in order to take into account the motion of both the observer and the source (Doppler, 1846). Doppler noted that his principle applied not only to the propagation of light waves—he believed that all stars emitted mainly visible light of different frequencies which combined to give them a white appearance, hence the observed coloured light of the double and the variable stars pointed to a velocity shift in the emitted frequencies—but also to that of sound waves. The acoustical effect was first confirmed in 1845 by the Dutch meteorologist Christoph Hendrik Didericus Buys Ballot (1817-1890), who had a locomotive drag an open car with several trumpeters. The French physicist Armand Hippolyte Fizeau (1819-1896), probably without knowing Doppler's work, suggested in 1848 the use of the effect for observing shifts in the frequency of stellar spectral lines. Doppler's insufficient justification of his principle—the Bohemian mathematician and theologian Bernhard Bolzano (1781-1848), for example, provided the extension to transverse vibrations of light—met quite some opposition; thus Buys Ballot criticized the unsound assumption (i.e., that all stars emit *in toto* white light) entering into the astronomical applications, and his Viennese colleague, the mathematician Joseph Petzval denied in 1852 the existence of the Doppler effect altogether. Later the Swedish physicist and astronomer Anders Jonas Ångström (1814-1874) failed to observe the optical shift. The controversy continued for decades after Doppler's death. In Vienna, for example, the young Ernst Mach became involved in the debate and he demonstrated the reality of the acoustical Doppler effect in 1860.

Upon Christian Doppler's early death, his Vienna colleague Andreas von Ettinghausen took over the directorship of the Physics Institute. When he became ill in 1862, an acting head of the institute had to be appointed to direct the work and Josef Stefan, since 1858 *Privatdozent* in mathematical physics at the University of Vienna, was chosen. He came from St. Peter near Klagenfurt, where he was born on 24 March 1835 and also received his school education. From 1853 to 1857 he studied at the University of Vienna; then he became a teacher at a *Realschule* in Vienna, besides working on water flow through tubes in the laboratory of the physiologist Carl Ludwig. In 1863 he was promoted to a full professorship of higher mathematics and physics at the University of Vienna and three years later he officially obtained the directorship of the Physics Institute and the chair for experimental physics. During his career Stefan occupied important positions at the University of Vienna—he served as dean of the philosophical faculty, 1869-1870, rector of the university, 1876-1877—and at the Imperial Academy, whose member he had been elected in 1865 (corresponding member since 1860)—secretary of the mathematics-science section in 1875 and vice-president from 1885 until his death. The Austrian Empire honoured him with the title of a *Geheimrat* (Privy Councillor), and several foreign academies with fellowships. To the international public Stefan became known as President of the Scientific Commision of the International Electricity Exhibition (1883) and of the International Conference on Musical Pitch (1885), both held in Vienna. He died on 7 January 1893 in Vienna.

'Stefan was a universal scientist and he treated all fields of physics with equal care,' remarked his student Ludwig Boltzmann later ('*Stefan war universell und behandelte alle Kapitel der Physik mit gleicher Liebe,*' Boltzmann, 1905, p. 101). Thus he worked successfully both on experimental and theoretical topics, especially in the field of the kinetic theory of gases, on problems of thermodynamics (evaporation, surface tension, ice formation, solutions), on heat conduction and diffusion in gases and liquids, and on acoustics. Of the many important results of his investigations, we want to mention just three. First, he analyzed the possible description of the mutual interaction of two electric currents demonstrating that not only the existing theory of André Marie Ampère (1775-1836) and the alternative theory of Hermann Günther Graßmann (1809-1877) represented the observed situation, but also several other different theoretical conceptions. As one of the earliest Continental supporters of the continuum field concept in electrodynamics, proposed by the English physicist Michael Faraday (1791-1867) and of the systematic theory developed on its basis by the Scottish physicist James Clerk Maxwell (1831-1879)—the other being Hermann von Helmholtz (1821-1894) in Berlin—he then showed its superiority over the earlier theories by applying it successfully to many problems, e.g., to those involving alternating currents. For the second of his principal fields of work, the kinetic theory of gases, he found much evidence. Especially, he constructed a device called the '*Diathermometer*,' which allowed him to measure heat conduction in gases: it consisted of two concentric copper cylinders such that the gas was confined between them, while the inner cylinder was filled with air and served as a gas thermometer to determine the temperature on the mantle of the cylinder. His measurements with the *Diathermometer* confirmed the predictions from kinetic gas theory; and the same did other experiments of Stefan on the diffusion of gases. He also performed several calculations in kinetic theory; thereby he obtained values for the diffusion and viscosity coefficients of gases as well as their dependence on the (absolute) temperature, which agreed with the calculations of Maxwell and his colleague Joseph Loschmidt. Perhaps the greatest future implications had a formula, which Stefan presented in 1879 to describe the energy of heat radiation from a body: he claimed that this energy is proportional to the fourth power of the absolute temperature of the emitting body (Stefan, 1879). This relation was confirmed theoretically five years later by his student Ludwig Boltzmann, who combined arguments from thermodynamics and Maxwell's theory of electrodynamics to reproduce Stefan's relation (Boltzmann, 1884b). The Stefan-Boltzmann law, as this relation is often called, only applied exactly to the heat radiation emitted from a completely black body; it became one of the milestones on the path to quantum theory.

Josef Stefan, who was an equally brilliant experimentalist and theoretician, trained many students at his institute, of whom Ludwig Boltzmann and Fritz Hasenöhrl became best known. Thus he fulfilled the tasks connected with the foundation of the institute perfectly. He had, however, the good fortune to have for some time a friend and collaborator of equal stature in his laboratory, Johann Joseph Loschmidt. Loschmidt was born in Putschirn near Carlsbad on 15 March

1821, the son of a poor peasant. His school education was supported by the Catholic clergy, and he entered the University of Prague in 1839 to study classical philology and philosophy. Franz Exner, professor of philosophy, not only gave him an assistant's position (as *Vorleser*), but also directed his attention to a particular problem, namely to carry further the attempt of the Göttingen philosopher Johann Friedrich Herbart (1776–1841) applying mathematical methods to philosophical and psychological problems. Loschmidt did not succeed in this task but, as a consequence of his involvement, his interest shifted to mathematics and science. When Exner became professor of philosophy at the University in Vienna in 1841, Loschmidt followed him there and completed his studies in 1843. Upon graduation he failed to obtain a teaching position and turned to earning his living in other ways. Thus he worked for three years in the laboratory of Anton Schrötter (1802–1875), professor of chemistry at the Vienna *Polytechnikum*, where he discovered, together with his friend Benedikt Margulies, a process for converting sodium nitrate into potassium nitrate, which was needed for gunpowder. The two friends tried to exploit this discovery by establishing a chemical factory in Atzgersdorf near Vienna, but it went bankrupt in 1849. Until 1854 Loschmidt worked with little success in several other industrial firms; then he returned to Vienna, where in early 1856 he completed his qualifications as a teacher and the following September he obtained a position at a Vienna *Realschule*. Besides teaching (chemistry, physics and algebra) he performed some scientific research on topics in chemistry and crystallography. Subsequently he met Josef Stefan, who had just taken over the directorship of the *Physikalisches Institut* at the University of Vienna, and who invited him to work there. Loschmidt became *Privatdozent* at the University of Vienna in 1866, two years later extraordinary professor of physical chemistry and in 1872 full professor and director of the Physico-Chemical Laboratory (now the Second Physical Institute at the University of Vienna). He retired in 1891 and died on 8 July 1895 in Vienna.

Loschmidt published his first scientific work, the book *Chemische Studien I*, in 1861 at his own expense. In it he pioneered the graphical representation of double and triple bonds of polyvalent atoms by means of connecting lines, applying it to organic and inorganic compounds—e.g., to benzene, for which he proposed a ring-shaped chain of carbon atoms. Though it contained many suggestions, which were later fruitfully extended—such as the assumption of mutiple valences for chemical elements—his book exerted little influence. In a further book, entitled *Zur Constitution des Aethers* and published the following year, he tried to construct a mathematical-mechanical model of the ether, which he assumed to consist of ether particles that repelled each other. He achieved lasting fame, however, with an investigation of material particles: the first accurate estimate of the size of air molecules. He presented the result in a paper submitted to the *Kaiserliche Akademie der Wissenschaften* in 1865 (Loschmidt, 1865). In this paper Loschmidt derived from kinetic gas theory, based on the ideas of Rudolf Clausius (1822–1888) and Maxwell, an equation relating $N$, the number of molecules per unit volume, to $l$, the mean free path, and $s$, the molecular diameter, i.e.,

$$\frac{1}{N} = \frac{16}{3}\frac{\pi l s^2}{4}. \tag{1}$$

Since the mean free path $l$ had already been determined by Maxwell and Oskar Emil Meyer (1834–1909), and a further equation existed connecting $N$ and $s$, Loschmidt finally arrived at a value of less than $10^{-7}$ cm for the diameter of air molecules. He also calculated the number $N$—which was later often referred to as the 'Loschmidt number' (especially in the German scientific literature)—finding $N \sim 10^{20}$ molecules per cubic centimetre. These molecular values, especially the latter, were frequently improved later on, and it took decades until the presently accepted values were obtained; still Loschmidt's calculation for the first time gave a reasonable estimate of the molecular size. This achievement made a great impression on his colleagues, and Loschmidt was elected in 1867 a corresponding member of the *Kaiserliche Akademie der Wissenschaften* (full member in 1870). He continued to work on problems of kinetic gas theory; for example, he investigated the rates of diffusion of gases and found agreement with the theoretical predictions of Maxwell. In a theoretical study he was the first to apply the second law of thermodynamics to solutions and chemical compounds, becoming a forerunner of Josiah Willard Gibbs (1839–1903) in this field (Loschmidt, 1869). He also entered into the discussion of the mechanical foundation of the second law of thermodynamics; in particular, he proposed the 'reversibility argument' ('*Umkehreinwand*') in order to escape the idea of the so-called heat death, i.e., the steady growth of entropy in the universe by the transformation of ordered mechanical motion into heat (Loschmidt, 1876a, b; 1877a, b). He claimed that one cannot establish the second law of thermodynamics on a purely dynamical basis, because for all mechanical processes—which might lead to an increase of entropy—there must also exist reverse mechanical processes in which the entropy is decreased.

Joseph Loschmidt, like Christian Doppler, was scientifically an autodidact. His work was characterized by an outstanding clarity of mathematical methods and a masterly use of rather modest experimental means. These modest means often prevented him from achieving success in well-planned and ingenious investigations; for instance, he failed to discover such phenomena as the action of a magnetic field on an electric current, which was found by the American physicist Edwin Herbert Hall (1855–1938), or the effect of a magnetic field on polarized light, which the Scottish physicist John Kerr (1824–1907) would detect. 'In spite of this failure it is remarkable what Loschmidt achieved with his modest means,' recalled his student and successor Franz Exner (the son of Loschmidt's former philosophy professor in Prague and Vienna), 'and that was also recognized by the educational administration (*Unterrichtsverwaltung*); however, it was not used in Loschmidt's favour, but the situation was rather regarded as a proof of the fact how superfluously well and fully equipped the institutes were' (Exner,

1921, p. 179–180). Loschmidt—who was simple in his personal life and despite his work in industry had little interest in practical matters, living only for his scientific ideas—accepted such unfavourable treatment. When he had to retire at the age of seventy, he lived in a small apartment, together with his caring wife (Karoline Mayr, whom he had married in 1887) and a son who would die at the age of ten of scarlet fever. His former pupil Ludwig Boltzmann, who visited him shortly before his death, remarked with the deepest regret and scorn: 'So houses Vienna its great men' ('*So beherbergt Wien seine großen Männer,*' Boltzmann, 1905, p. 237).

With Doppler, Stefan and Loschmidt the early tradition of physics in Vienna was established. All three had started from practically nothing and had risen to the international level in physics by contributing important results to its various fields. In spite of this they received little recognition abroad. While one may understand the reason in the case of Doppler, who lived at a time when Austria had hardly any scientific connection with foreign countries, the situation had changed considerably for Stefan and Loschmidt. Thus Boltzmann remarked: 'Neither Stefan nor Loschmidt went, according to my knowledge, on a travel beyond the borders of [their] Austrian homeland. At any rate, they never visited a *Naturforscherversammlung* and did not establish closer personal relationships with foreign scientists. I cannot approve of that, for I believe that they could have achieved still more if they had closed themselves off less. At least they would have made their achievements known faster and therefore more fruitful' (Boltzmann, 1905, p. 102). Boltzmann, a student of both men, would neither accept their seclusion from the world ('*Weltabgeschlossenheit*') nor the modest recognition they received at home from the official public and administration. With Ludwig Boltzmann and Ernst Mach, another student of Stefan's, would begin a new epoch of Austrian science, marked by an enormous influence on international science as well as on the national public.

## Pioneers in Medicine and Psychoanalysis, Biology and Physiology

Apart from a brief episode connected with the visits of the Swiss physician Paracelsus (Theophrastus Bombastus von Hohenheim, 1490–1541),[51] the medical tradition of Austria goes back to the eighteenth century, when the medical faculty of the University of Vienna did acquire some reputation.[52] However, the first Austrian, who obtained a well-remembered place in the history of medicine, was born in Budapest on 1 July 1818: Ignaz Philipp Semmelweis. He studied in Budapest and Vienna, becoming in 1846 assistant at the first obstetric clinic in Vienna. On studying the reasons for the fatal illness of puerperal (childbed) fever, he found that it was due to infections and achieved a considerable drop

[51] Paracelsus stayed in Austrian countries several times during his youth, e.g., he studied mining processes and diseases in the Tyrol. He died in Salzburg.

[52] Emperor Joseph II took a personal interest in medicine.

in mortality rate by instituting a thorough washing of the hands by all doctors and clinical personnel. Since his chief, Johann Klein, failed to acknowledge this advance and prevented his promotion, Semmelweis left Vienna in 1849 and returned to Budapest; there he repeated the previous success and made a successful career as a physician and university professor. He died on 17 August 1865 at Döbling near Vienna.

While Semmelweis was driven away from Vienna, the Prussian Albert Christian Theodor Billroth, born on 26 April 1829 in Bergen on the Island of Rügen, was called to Vienna in 1867, where he created an international centre of surgery at the university.[53] Having a strong artistic bent—he was a lifelong friend of the composer Johannes Brahms—he fitted perfectly into the Viennese and Austrian surroundings. A member of the Austrian *Herrenhaus* since 1887, Billroth died on 6 February 1894 in Vienna.

Billroth and Semmelweis certainly enriched the medical knowledge of their times. The medical Vienna of the late nineteenth century is even more remembered because it gave birth to psychoanalysis, a new method of treating psychological defects and diseases. Its origin is connected with the names of Josef Breuer, Sigmund Freud and Alfred Adler. The senior-most among them, the Viennese Breuer (born on 15 January 1842, died on 20 June 1925) was originally interested in the physiology of the human breathing mechanism and its relation to the nervous system.[54] When practising as a physician in Vienna he discovered a method to relieve a girl from severe hysteria by hypnosis. He got together with the neurologist Sigmund Freud; together, they wrote a book entitled *Studien über Hysterie* (*Studies on Hysteria*), which appeared in 1895 and provided the starting point of what was later to be called psychoanalysis.

Freud was born on 6 May 1856 at Freiberg, Moravia. From the age of four he lived in Vienna and later studied philosophy, chemistry, physiology and anatomy at the university.[55] He worked for several years in the physiological laboratory of Ernst von Brücke and later in the psychiatric clinic under the direction of the brain anatomist T. H. Meynerth; and finally he studied mental diseases in Paris and Nancy, before opening a neurological practice in Vienna in 1886.[56] After the joint work with Breuer on the treatment of hysteria, Freud developed the psychoanalytical method decisively through the replacement of

[53] Billroth introduced many new surgical techniques in Vienna: resection of the esophagus (1872); complete excision of the larynx (1873); resection of the pylorus for cancer (1881); a large number of intestinal resections and enterorrhaphies (1878–1883).

[54] Breuer performed this work while serving as assistant to the physiologist Ewald Hering. Later he was family physician (*Hausarzt*) to some famous contemporaries, such as Ernst von Brücke, Billroth and Franz Exner. He had a happy and balanced character; for instance, he wrote in his *curriculum vitae* for the Vienna Academy of Sciences: 'If anything can prevent the envy of the gods, then my deep conviction is that I have been lucky far beyond my merits.' (Quoted in Heller, 1964, p. 22, footnote 1.)

[55] Freud also translated one volume of John Stuart Mill's *Collected Works* into German.

[56] The Paris neurologist Jean Charcot had confirmed Freud's intention to investigate hysteria from a psychological point of view.

hypnosis by the method of free association; he also studied his own unconscious mental processes by a thorough self-analysis. While investigating the field essentially alone for about a decade, Freud slowly gathered supporters of psychoanalysis—besides his countryman Adler, in particular the Swiss psychiatrists Eugen Bleuler (1857-1939) and Carl Gustav Jung (1875-1961). In 1908 they met at the First International Congress of Psychoanalysis in Salzburg, and two years later the International Psychoanalytic Association was founded, propagating the new science beyond Austrian and middle European borders.[57] In 1938 when the National Socialist Government of Germany took over Austria, the 82-year-old Freud would have to emigrate to London, where he died on 23 September 1939.[58]

Psychoanalysis, as developed by Freud, was a method of understanding the origin of neurosis and psychosis and perversion, as well as the normal mind. Freud believed in the biological origin of human behaviour, making instincts responsible for it; the recognition of the origin of human behaviour—sometimes termed 'depth psychology' should therefore provide the particular method of treating mental disorders also.[59] Alfred Adler, who was born on 17 February 1870 in a suburb of Vienna and had joined Freud's circle in 1902, later established his own psychoanalytic school.[60] Adler also founded in 1919 the first child guidance clinics. From 1925 he visited the United States regularly, settling there in 1935; he died on a lecture tour in Aberdeen, Scotland, on 28 May 1937.

While psychoanalysis may be considered as a typical, new Viennese product, the first great Austrian contribution to a much older science, biology, came from the north of the Habsburg Empire.[61] On 22 July 1822 Johann (Gregor) Mendel

[57] In 1909 Freud and Jung were invited to the United States to deliver a series of lectures on psychoanalysis at Clark University in Worcester, Massachusetts.

[58] By 1926 psychoanalytic societies had been founded in Vienna, Berlin, Budapest, London, Switzerland, the Netherlands, Moscow, Calcutta, and the United States. The founder of the movement, Sigmund Freud, never obtained a full professorship in Vienna, although he held the personal title of professor at the university since 1902. He received several honours, including the Goethe prize of the city of Frankfurt-am-Main in 1930, and a fellowship of the Royal Society of London in 1936.

[59] The principal method of Freud's psychoanalysis consisted in the study of anomalous behaviour and of the dreams of the patient.

[60] Adler claimed that all important problems of the individual have a social origin; that the individual is a unique self-consistent unity unconsciously creating its own life style by making use of the objective biological and social condition. A neurotically disposed person, therefore, should be characterized by increased feelings of inferiority, underdeveloped social interest and an exaggerated goal of superiority presenting itself as open aggression.

[61] The Viennese character of psychoanalysis shows itself in another aspect. The pioneers of the field were, as a rule, deeply interested not only in scientific but also in cultural matters. Freud, for instance, interpreted works of art and literature: he wrote on Michelangelo's 'Moses,' or Dostoevsky's novels; and he applied psychoanalysis to problems of anthropology and religion (his last book, published in 1935, was *Moses and Monotheism*). Jung, on the other hand, tried to illuminate the creative processes in the artist.

The psychoanalytic method exerted an enormous influence on writers and poets, from the Vienna circle of Hugo von Hofmannsthal and Arthur Schnitzler, to the north German Thoman Mann, to the English playwright Thomas Stearns Eliot, and to the Irish author James Joyce. Joyce's novel *Ulysses* (1922) has been considered to be a special application of Freud's *Traumdeutung*. A strong bond

was born in Heinzendorf, Austrian Silesia, just on the border between German and Czech areas. He studied philosophy, physics and mathematics (1840–1843) at the *Philosophische Lehranstalt* in Olmütz. In 1843 he entered the Augustinian monastery in Brünn (Brno), where he received the name Gregor. Free of financial worries, he then studied theology and—encouraged by his abbot—natural sciences and agriculture (1845–1848).[62] After becoming an ordained priest in 1847, Mendel first worked as a chaplain, later as a substitute teacher at the *Gymnasium* in Znaim (Znojmo), southern Moravia (1849–1851). Since he failed to pass the university examination for teachers in natural sciences (in Vienna, 1850), he was sent—on the recommendation of Andreas Baumgartner—to the University of Vienna to study physics, mathematics, natural history (including paleontology, botany and zoology) and chemistry (1851–1853).[63] Returning to Brünn in 1853 he taught physics and natural history at the *Technische Schule* of the city (1854–1868), although he never actually obtained the teacher's licence. In 1868 he was elected abbot of his monastery; he died on 6 January 1884 in Brünn, and was mourned as a popular and respected man of the church.[64]

Gregor Mendel's scientific fame rests mostly on the botanical experiments in the garden of his monastery in Brünn (1856–1863); these experiments consisted of cultivating plants and studying their characteristic features, especially the problem of inheritance.[65] Mendel was able to obtain in certain cases of hybridization (e.g., garden peas) definite results and explain them according to the rules of combinatorics.[66] His best known conclusion constituted the simplest example of what was later called 'Mendel's law of segregation.' Mendel presented it, together with more complex cases, at the meetings of 8 February and 8 March 1865 to the *Naturforschender Verein* in Brünn; but neither the presentation, nor the publication in the proceedings of the society (Mendel, 1866) made any impact

---

connects psychoanalysis with the surrealism in fine arts and literature; this is expressed as much in André Breton's *Manifeste du Surréalisme* (Paris, 1924), as in the works of prominent painters like Max Ernst or Salvador Dali.

[62] Mendel learned especially from Franz Matthaeus Klaćel (1808–1882), a member of his monastery and director of the experimental garden there, and from Franz Diebl (1770–1859) who gave lectures on agriculture, pomology and viticulture at *Philosophisches Institut* in Brünn.

[63] Mendel attended lecture courses on experimental physics (held by Christian Doppler), on physical apparatus (A. von Ettinghausen), on chemistry (F. J. Redtenbacher) and on plant physiology (F. Unger). He also became a member of the *Zoologisch-botanischer Verein* in Vienna, to which he contributed his first two short communications on the damage to plant cultures by insects.

[64] Mendel performed the mainly administrative duties of an abbot with great devotion and skill; he was a good diplomat and upright fighter on behalf of his monastery in the politically difficult time of Austrian liberalism. (For detailed information on Mendel's life see biography by Hans Iltis: Iltis, 1924.)

[65] Mendel had learned the methods of cultivating plants in his early youth (from his father), and he received a more scientific training in these matters from his Brünn and Viennese teachers (Klaćel, Diebl and Unger).

[66] Mendel learned the methods of combinations in Olmütz (from a course of Johann Fux). In Vienna both C. Doppler and A. von Ettinghausen knew combinatorics and might have taught it in their classes.

on the scientific community.[67] Later he also occasionally published on scientific problems, especially on meteorological ones.[68]

Mendel's researches on hybridization did not receive proper recognition during his lifetime and were forgotten afterwards.[69] In 1900, however, three botanists independently rediscovered the results of 1865: Carl Erich Correns (1864-1933), a student of Carl Wilhelm Nägeli, whom Mendel had tried unsuccessfully to convince of his results; Hugo de Vries (1848-1935), then professor at the University of Amsterdam; and Erich Tschermak, Edler von Seysenegg.[70] Erich Tschermak, born on 15 November 1871 in Vienna, would establish a scientific tradition which could have been started several decades earlier; he also obtained the first chair of plant breeding at the *Hochschule für Bodenkultur* in Vienna (1909).[71]

It seems to have been a characteristic feature of Austrian and Viennese scientists to work on topics that connected different disciplines: Breuer, Freud and Adler combined in their psychoanalysis methods from medicine and psychology; Mendel introduced the mathematical methods of combinatorics into biology in order to create the science of heredity. Keeping this liking for interdisciplinary topics in mind, one might not be surprised to find many Austrian scientists contributing to the field connecting medicine with physics and chemistry, which received particular attention during the nineteenth century and was called physiology.[72]

[67] Mendel had been an active member of the Brünn *Verein* since its foundation a few years earlier. The *Verhandlungen*, containing his paper, were sent to 134 scientific institutions including those in the United States; and the author personally sent copies to two famous experts in hybridization, Anton Kerner (1831-1898) in Vienna and Carl Wilhelm Nägeli (1817-1891) in Munich.

[68] Besides plant hybridization Mendel investigated, for example, bee culture and meteorological problems from 1856. From 1863 to 1869 he published systematic reports on weather observations in Brünn and Moravia, and he used graphical and statistical methods. Later he confined himself to write on occasional, exceptional phenomena (e.g., the 1870 tornado in Brünn). He also supported the first weather forecasts in central Europe, appearing in 1877 for the benefit of farmers in Moravia.

[69] Several reasons may be given for this neglect. First, Mendel's experiments were subtle and required great skill and luck (in selecting the proper characteristics). Second, in the last third of the nineteenth century Charles Darwin's ideas about the origin of species and their evolution by natural selection gained increasing support in the scientific community; now in the ideas of biological evolution there seemed to be no place for Mendel's law of segregation, which demanded the *constancy* of hereditary 'elements'.

[70] Correns also searched through the literature and rediscovered—as did de Vries and Tschermak—the pioneering work of Mendel. De Vries further discovered the sudden appearance of new forms among a display in his cultures of *Oenothera lamarckiana* (American evening primrose), which he attributed to the process of 'mutation' of heredity characteristics; thus he removed the main obstacle that had previously prevented the simultaneous validity of Mendel's laws of inheritance and Darwin's theory of evolution of species.

[71] Tschermak died on 11 October 1962 in Vienna. In this century the Viennese tradition in biology was brilliantly continued by Karl Ritter von Frisch (1886-1981) and Konrad Lorenz (born in 1903).

[72] In historic times, physicians had frequently been interested in various aspects of science. Thus in the development of science, since the Renaissance, men of medicine contributed a large share: in physics one may remember the investigations of William Gilbert (1544-1603), the physician of Queen Elizabeth I of England, contained in his book *De Magnete* (London, 1600), dealing with the action of magnets and magnetic bodies and with electrical attraction; or the researches of the Italian anatomist Luigi Galvani (1737-1798) on the reaction of frog's legs if brought into contact with metals, which led to the development of electrodynamics. While these results exclusively served physics, a notable

Perhaps the first representative of Austrian physiology, who received a world-wide reputation, was Johannes (Jan) Evangelista Purkinje (Purkyne), born on 17 December 1787 in Libochowitz (Libochovice) near Prague and from 1850 professor of physiology in Prague. He performed research on a wide variety of topics in physiology, histology and embryology, obtaining many original and valuable results; in the field of optical vision he found the so-called Purkinje images (i.e., the reflections on the surfaces of the cornea and lens of the eye) and the Purkinje phenomenon (i.e., the change in the brightness of blue and red in the dark); in microscopy he discovered the Purkinje cells in the cerebellar cortex, the Purkinje fibres and network (formed by the large heart muscle cells beneath the endocardium) and the nucleus of the ovum; he introduced the notion of protoplasm to describe the actually living substance in the cells of animals and plants.

When Purkinje was a young scientist, the famous poet and scientist Johann Wolfgang von Goethe admired and welcomed his contributions; when he died on 28 July 1869 in Prague, physiology was an established science in Austria, though with the assistance of German imports like Ernst Wilhelm von Brücke (1819-1892), Carl Friedrich Wilhelm Ludwig (1816-1895) and Ewald Hering (1834-1918). Brücke was called in 1849 from Königsberg to a chair for physiology at the University of Vienna. He studied, among other topics, the properties of protoplasm and of blood and investigated stimulated motions (*Reizbewegungen*). In Vienna he wrote books on the physiology of spelling sounds (*Sprachlaute*)—thus becoming one of the founders of modern phonetics—on the physiology of colour vision, and on the theory of fine arts. Carl Ludwig arrived in Vienna in 1855, having been before a professor in Marburg and Zurich. He worked on all topics of physiology—which he planned to turn into a physico-chemical science

---

change occurred in the nineteenth century. On the one hand, in medical science there occurred an increasing specialization of methods. On the other hand, the connection between medicine and science was considerably deepened. For example, the discovery that microscopically small bacteria caused human diseases—established by the French chemist and microbiologist Louis Pasteur (1822-1895) and the German physician Robert Koch (1843-1910)—provided another fundamental link between biology and medicine. But the role of medicine as a partner of science had already been fixed earlier in the name of an important German society (founded in 1822 in Leipzig on the instigation of the biologist Lorenz Oken), the *Gesellschaft Deutscher Naturforscher und Ärzte* (Association of German Scientists and Physicians). In the meetings of this association a substantial portion of the progress in science and medicine, achieved in Austria, Germany and Switzerland, was presented during the nineteenth and twentieth centuries.

In the nineteenth century man tried to reduce *all* phenomena in nature to a mechanical origin; this also happened in medicine and biology, where one tried to explain all functions of living systems on a physico-chemical basis. Thus it was not a complete accident that physicians established one of the fundamental physical principles, which also played a crucial role in the physical description of biological phenomena, namely the law of energy conservation (Robert Mayer, Hermann von Helmholtz). That law and the discovery of a great number of electrical phenomena and chemical reactions led to important advances in the science of physiology. In Germany Johannes Müller (1801-1856), who pioneered the investigations of nerves and senses, was sometimes called the father of this science. His student, Hermann von Helmholtz (1821-1894), who held chairs of physiology and physics at the universities of Königsberg, Bonn, Heidelberg and Berlin, became the outstanding representative of physiology in the nineteenth century.

of biological organisms. In his investigations on the hydrodynamics of blood vessels he temporarily received help from the physicist Josef Stefan. He left Vienna in 1865 for Leipzig. The same year Ewald Hering came from Saxony as professor of physiology at the medico-surgical *Josephs-Akademie* (1865-1870).[73] His research fields included the physiology of smelling, visual perception and colour vision of the human eye. He developed, for example, the six-colour theory, based on the antagonistic pairs of colours: black-white, blue-yellow, and red-green, opposing the usual Young-Helmholtz three-colour theory; also in the theory of visual perception he disagreed with the great Hermann von Helmholtz by proposing a 'nativistic' theory of space perception as opposed to the latter's 'empiristic' theory.

A year before Hering left Vienna to accept Purkinje's chair in Prague, Wolfgang Joseph Pauli was born in the Bohemian capital (on 11 September 1869). He began his scientific career at the University of Vienna in 1898 as *Privatdozent* in physiology, being promoted in 1907 to an extraordinary professorship and finally obtaining in 1922 the chair for *biologisch-physikalische Chemie* and the directorship of the *Institut für medizinische Kolloidchemie.* Pauli is remembered for his pioneering studies of the connection between electrical, mechanical and chemical processes in muscles on the one hand, and for establishing relations between physiological problems and a new field in chemistry, called colloid chemistry, on the other. Indeed, he can be counted, together with his Leipzig colleagues Wolfgang Ostwald (1883-1943)—son of the physico-chemist Wilhelm Ostwald—and the Göttingen professor Richard Zsigmondy (1865-1929), who originated from Vienna, as the founder of colloid chemistry. While Pauli had to retire from his chair in 1938 and join his son, the physicist Wolfgang Pauli Junior in Zurich, another Austrian physiologist, namely Armin Eduard Peter Gustav von Tschermak-Seysenegg (1870-1952)—the elder brother of the botanist and rediscoverer of Mendel's laws—who held from 1913 to 1939 a chair at the German University of Prague, found himself after World War II exiled to Regensburg.

## Physiology, Physics and Cognition: Ernst Mach[74]

Physiologists played a crucial role in the life of Ernst Mach, perhaps the best-known physicist of Austria at the end of the nineteenth century. During his university studies he read the books on physiology of Hermann von Helmholtz (1821-1894) and Emil Heinrich du Bois-Raymond (1818-1896), and he attended the lectures of Ernst Wilhelm von Brücke (1819-1892). One of his early research problems was to construct a practical instrument to accurately measure pulse beats and record the changes of blood pressure (1862). Later, in Prague, he met

[73] From Vienna, Hering proceeded to Prague (1870-1895) and finally to Leipzig.

[74] We owe much to the books by K. D. Heller (1964) and John T. Blackmore (1972) for information about the life and work of Ernst Mach.

Purkinje and co-operated with Ewald Hering, Purkinje's successor in the physiology chair. Among Mach's closest supporters was Wolfgang Pauli Senior, who asked him to be godfather to his son, Wolfgang Pauli Junior. Still Mach did not become a physiologist but remained a physicist; and besides working successfully on various topics of physics he concerned himself with what we now call science theory, that is, the systematic methods used in establishing results and knowledge in physics. The particular views he obtained left a deep, occasionally sweeping impression on his contemporaries and the following generation.

Ernst Waldfried Joseph Wentzel Mach was born on 18 February 1838 in Turas-Chirlitz (Chralice-Tuřany) near Brünn (Brno) in Moravia, the son of the *Gymnasium* teacher Johann Mach and his wife Josephine Wentzel Langhaus. In 1852 the family moved to Untersiebenbrunn, 30 km east of Vienna, where Johann Mach ran a farm, in which he tried to improve upon methods of agricultural production (orchards, silkworm cultivation). Ernst Mach received instruction at home from his father. From 1853 he attended the *Piaristen-Gymnasium* in Kremsier (Kroměříž), Moravia.[75] In 1855 he went to Vienna to study mathematics, physics and philosophy at the university.[76] He received his doctorate in January 1860 (with a dissertation on electrical charge and induction). Deciding on a university career, he became *Privatdozent* at the University of Vienna a year later. The early lecture courses already indicated the direction of his later scientific work; in the winter semester 1861/62 he taught: '*Physik für Mediziner*' (physics for medical students, 3 hours per week), '*Methoden der physikalischen Forschung*' (methods of physical research, 1 hour), and '*Höhere Physiologische Physik*' (advanced physiological physics). The first lecture course also gave rise to his first book publication, the *Compendium der Physik für Mediziner* (*Compendium of Physics for Medical Students*, Vienna 1863). In the summer semester 1862 Mach presented a one-hour course on '*Principien der Mechanik und der mechanischen Physik in ihrer historischen Entwicklung*' (Principles of mechanics and of mechanical physics in their historical development), which laid the foundation of one of his principal books, the *Mechanik in ihrer Entwicklung, historisch-kritisch dargestellt* (Mach, 1883).[77] When in 1862 Andreas von Ettinghausen fell ill, Mach was considered among the candidates for the acting director of the Physics Institute; however, the older and more experienced Stefan obtained the position in 1863. A year later Mach received the call to a chair of mathematics at the University of Graz; he accepted and taught mathematics and physics during the

[75] At the *Gymnasium*, Mach was especially impressed by his teacher in natural history, F. X. Wessely; Wessely introduced him to the ideas of Jean Baptiste Lamarck (1744–1829) on biological evolution and to the cosmogony of Immanuel Kant (1724–1804) and Pierre Simon de Laplace (1749–1847).

[76] Mach's university teachers included Andreas von Ettinghausen and Joseph Petzval.

[77] Mach's later lecture courses in Vienna included one on psychophysics and another on '*Akustik als physikalische Grundlage der Musiktheorie*' ('Acoustics as the Physical Foundation of Musical Theory'), which he delivered in the winter semester 1863/64 on the basis of Helmholtz' researches.

following three years, being appointed full professor of physics in early 1866. In 1867 he then moved to Prague as professor of experimental physics at the famous *Karls-Universität.* The following 28 years, spent in the Bohemian capital, meant the peak of Mach's life and career. He enjoyed a happy family life amidst a growing number of children.[78] He worked in a suitably equipped laboratory and produced and completed most of his important work in physics, psychophysics and methodology of physics—altogether he published over 100 scientific papers. He also ascended the highest academic position, becoming *Rector* of the University of Prague twice, first in the academic year 1879/1880 and again in 1883/1884—then for the German university, because the institution had been split into a German and a Czech university the year before. When he returned in 1895 to Vienna to accept a chair of philosophy at the university, he had risen to world fame as a physicist and as a philosopher of science.

Through the variety and breadth of his scientific work, Ernst Mach belonged to the small circle of universal scientists in the nineteenth century, among men like Alexander von Humboldt (1769–1859) and Hermann von Helmholtz. His investigations extended over many fields of experimental physics, especially optics, acoustics, gas dynamics, to physiology and psychophysics—i.e., the physics underlying psychology—history of physics, science theory and philosophical aspects of physics and science generally, and last but not least physics teaching. He included numerous original results in his seventeen books, the first being published in 1863 and the last posthumously edited in 1921 by his son Ludwig Mach.[79] This variety showed up at a rather early stage and was caused by unfavourable conditions. Mach, the trained mathematician and physicist, had little opportunity to use the equipment of a physical laboratory. Also, as a young *Privatdozent*, who was not blessed with a rich purse from his father, Mach had to accept all kinds of extra jobs—like tutoring in mathematics or teaching at the

[78] Shortly before leaving Graz, Mach married Ludovica (Louise) Marussig. The Machs had four sons and a daughter: Ludwig (born 1868), Carolina (born 1873), Heinrich (born 1874), Felix (born 1879) and Victor (born 1881).

[79] Mach's books, in chronological order of publication, are (see Hiebert, 1973): *Compendium der Physik für Mediciner* (Vienna, 1863); *Einleitung in die Helmoltzsche Musiktheorie. Populär für Musiker dargestellt* (Graz, 1866); *Die Geschichte und die Wurzel des Satzes von der Erhaltung der Arbeit* (Prague, 1872); *Optisch-akustische Versuche. Die spectrale und stroboskopische Untersuchung tönender Körper* (Prague, 1873); *Grundlinien der Lehre von der Bewegungsempfindungen* (Leipzig, 1875); *Die Mechanik in ihrer Entwicklung, historisch-kritisch dargestellt* (Leipzig, 1883); *Beiträge zur Analyse der Empfindungen* (Jena, 1886), second, revised and enlarged edition as *Die Analyse der Empfindungen und das Verhältnis des Physischen zum Psychischen* (Jena, 1900); *Grundriss der Naturlehre für die unteren Classen der Mittelschulen*, written with Johann Odstrčil (Prague, 1887); *Grundriss der Physik für die höheren Schulen der Deutschen Reiches* (Leipzig, 1890); *Leitfaden der Physik für Studienrende*, written with Gustav Jaumann (Prague–Vienna–Leipzig, 1891); *Popular Scientific Lectures*, translated by Thomas J. McCormack (La Salle, Illinois, 1895); *Die Principien der Wärmelehre. Historisch-kritisch entwickelt* (Leipzig, 1896); *Erkenntnis und Irrtum. Skizzen zur Psychologie der Forschung* (Leipzig, 1905); *Space and Geometry in the Light of Physiological, Psychological and Physical Inquiry* (La Salle, Illinois, 1906); *Kultur und Mechanik* (Stuttgart, 1915); *Die Leitgedanken meiner naturwissenschaftlichen Erkenntnislehre und ihre Aufnahme durch die Zeitgenossen. Sinnliche Elemente und naturwissenschaftliche Begriffe* (Leipzig, 1919); *Die Principien der physikalischen Optik. Historisch und erkenntnispsychologisch entwickelt* (Leipzig, 1921).

elementary school—to earn his living.[80] The difficult financial situation proved to be advantageous for the future. First, he learned to present popular lectures drawing big audiences—for example, he taught a physics course for medical students. Second, he got to know a lot about new topics, like Helmholtz' acoustical theory or psychology, by lecturing on them.[81] Third, he was forced to figure out inexpensive laboratory experiments. As he later recalled: 'After my university studies I lacked, for bad or good luck, the means for undertaking physical researches, hence I was forced to concern myself with the field of physiology of senses. There, where I could observe my own sensations and at the same time also the conditions of the surroundings, I arrived at, I believe, a natural view of the world, free of speculative-metaphysical additions.' (Quoted in Heller, 1964, p. 13.) He was allowed to carry out the first physiological investigations at *Garnisonsspital*, in the laboratory of Dr. Duschek; a little later he received a grant from the Vienna Academy to continue these researches. Mach would never cease to be interested in physiological or psychological topics, even after getting a well-equipped physics laboratory.

Although the young Mach soon turned away from purely physical problems, his very first investigation—shortly after receiving the doctorate—was devoted to one. Andreas von Ettinghausen, his professor, wished him to decide the dispute in favour or against the Doppler effect, and Mach constructed in 1860 a simple apparatus, which allowed him to prove the existence of the acoustical effect.[82] Only after settling in Prague seven years later and having established his laboratory at the *Obstmarkt* (next to the *Carolinum*, i.e., the oldest part of Prague University) did he return to the problems of experimental physics. He then first designed instruments and equipment, which was constructed subsequently by his talented mechanic Franz Hajek for class demonstrations.[83] With the new equipment Mach

[80] In a letter of 1861 to his parents Mach reported that a renowned occulist, a Dr. Jäger, planned to write a book, 'which he wishes to embroider with mathematical considerations' ('*das er mit mathematischen Betrachtungen zu verbrämen wünscht*'). He continued: 'Since he himself is not a master of mathematics, he approached [Andreas] Baumgartner, the president of the Academy [of Sciences], requesting him to recommend somebody who would be able to take over the work. Baumgartner answered that he knew of nobody, who was simultaneously a mathematician and a physiologist, except for a Dr. Mach.' (see Herneck, 1961, pp. 455–456). In spite of the honourable recommendation by Baumgartner, Mach did not wish to accept the task.

[81] In his lectures on psychology, Mach discussed Gustav Theodor Fechner's recent book *Elemente der Psychophysik* (*Elements of Psychophysics*, 1860).

[82] The apparatus was later used in the laboratories of Central Europe for many years.

Mach did not succeed in proving the optical effect experimentally. Still he remained a firm defender of the Doppler effect in all wave phenomena throughout his life. For example, when the astronomer Johann Heinrich von Mädler in Dorpat, Estonia, denied its existence in the 1870s, Mach answered him by republishing his earlier articles on the subject and by writing a new one in 1878. Further, he arranged in the same year a public demonstration: he got his students and colleagues in Prague to listen to whistles emitted by approaching and departing trains, respectively; then he had them sign a prepared statement confirming the existence of the acoustical effect.

[83] Mach invented, for example, a special pendulum to demonstrate the dependence of its period on acceleration, and a polarization apparatus for light with a rotating analyzer. Another instrument was the *Wellenmaschine* (wave machine) of Mach and Hajek, in which propagating or standing longitudinal and transverse water waves could be produced. The apparatus was frequently used in other places.

performed a large number of experiments on refraction, interference and polarization of light, and on optical spectra. He was especially interested in all kinds of wave motion, whether of an acoustical, electrical or optical nature. After 1875 he also concerned himself with the mechanical effects of spark discharges and studied their propagation within solids or on surfaces. These early observations of shock waves yielded some unexpected results, which could later be explained by a new set of investigations.

While visiting the First International Electrical Exhibition of 1881 in Paris, Mach heard a lecture by the Belgian artillerist Louis Melsens, in which the speaker expounded a theory explaining the effect of the explosive impact on condensed air. On his return to Prague, Mach decided to set up an experiment to test this theory; for this purpose he devised and perfected, assisted by his collaborators, optical and photographic techniques for the study of flying projectiles, sound waves, spark waves and shock waves. Especially in 1884 he succeeded in developing an apparatus and a method suitable for obtaining clear photographs of shock waves created by moving projectiles.[84] Since Mach failed to get shock waves from pistol bullets—because they propagated with subsonic speed in air—he requested in early 1866 the help of P. Salcher, professor at the Austrian Naval Academy at Fiume (near Rijeka). Salcher and his collaborators carried out the experiment which Mach had proposed; on 10 July 1886 Mach submitted a short note to the Vienna Academy of Sciences, accompanied by the first two successful photographs of shock waves arising from supersonic projectiles (Mach, 1886). The previous year Mach had published a paper in the *Sitzungsberichte* of the Academy, giving the theory of the observed phenomena and introducing for the first time what was later called (by J. Ackeret in 1928) 'Mach's number' (Mach and Wentzel, 1885). In 1887 a detailed joint paper with Salcher appeared, in which the results of the Fiume experiments plus their theoretical explanation were given (Mach and Salcher, 1887). Through these investigations Mach advanced to the rank of founder of supersonic gas dynamics.[85]

The researches on shock waves certainly constituted Mach's most important contribution. In theoretical physics he provided an equally important contribution by his historical-critical studies of mechanics. Mach had already lectured on that topic in 1862, as we have mentioned earlier. Ten years later he published the content of a talk, presented on 15 November 1871 before the *Königlich-böhmische Gesellschaft der Wissenschaften* in Prague, as a book, *Die Geschichte und die Wurzel des Satzes von der Erhaltung der Arbeit* (*The History and the Root of the Theorem of Conservation of Work:* Mach, 1872). In it he claimed that energy

[84] For that purpose Mach used an idea of the physicist August Toepler (1836-1912), who had made density differences in a gas visible by his *Schlierenmethode* (striation method).

[85] Salcher constructed the first supersonic blown-down wind tunnel in a torpedo factory in Fiume, in which air travelling at supersonic speed moved past a stationary projectile. Mach's eldest son, Ludwig, from the 1880s his principal assistant, would improve the observation methods; e.g., he adapted the interferometric method of the French physicist Jules Célestin Jamin (1818-1886) for the purpose of studying shock waves (Mach-Zehnder interferometer, 1891).

conservation had been known since the earliest times, and was nothing less than a specific form of the law of causality. He further opposed the view, held by many contemporaries, that mechanics was the basis of all physics; in his opinion, optics, acoustics, electricity, magnetism or the theory of heat were as fundamental as mechanics, and the historical priority of the latter had to be considered accidental. Finally, he analyzed carefully the concepts used in mechanics, arriving at the conclusion that the introduction of the mass as a measure of matter was a purely metaphysical step—mass should rather be a quantity derived from the observable acceleration of objects—and that the concepts of absolute motion, absolute space and absolute time were of metaphysical origin as well. Mach continued and deepened the analysis of the historical development of mechanics and the criticism of its concepts in his book *Die Mechanik in ihrer Entwicklung historisch-kritisch dargestellt* (*The Development of Mechanics, Presented in a Historical and Critical Perspective*, Mach, 1883; English translation, 1893). In four long chapters the author discussed systematically the various lines of development in the science of mechanics; in Chapter Five he established the relation of mechanics to the other fields of physics and to philosophy.[86] The most original and important ideas of Mach were contained in Chapter Two, where he analyzed and criticized Isaac Newton's foundation of mechanics, and in Chapter Four, where he described the role of 'economy of thought' in physics and science.

The main criticism, which Mach directed against Newton's *Principia*, concerned the concepts of mass, absolute space, absolute time and absolute motion. He had already discussed the problem of mass in his Prague lecture of November 1871, referred to above; now he also enlarged on other concepts. After presenting Newton's concepts, their definitions and applications, he concluded: 'The question whether a motion is in *itself* uniform is senseless. With as little justice, also, we may speak of an "absolute time" . . . , This absolute time can be measured by comparison with no motion, it has therefore neither a practical nor a scientific value . . . . It is a metaphysical conception' (Mach, 1883, English translation, p. 273). And: 'No one is competent to predicate things about absolute space and time motion; they are pure things of thought, purely mental constructs, they cannot be produced in experience . . . . All our principles of mechanics are, as we have shown in detail, experimental knowledge concerning the relative positions and motions of bodies' (Mach, 1883, English translation, p. 280). Mach demonstrated the non-existence of absolute motion especially when discussing Newton's

[86]The contents of Mach's *Mechanik* are: *Erstes Kapitel*: *Entwicklung der Prinzipe der Statik* (Chapter One: The Development of the Principles of Statics); *Zweites Kapitel*: *Entwicklung der Prinzipe der Dynamik* (Chapter Two: The Development of the Principles of Dynamics); *Drittes Kapitel*: *Weitere Verwendung der Prinzipien und die deduktive Entwicklung der Mechanik* (Chapter Three: The Extended Application of the Principles and the Deductive Development of Mechanics); *Viertes Kapitel*: *Die formelle Entwicklung der Mechanik* (Chapter Four: The Formal Development of Mechanics); *Fünftes Kapitel*: *Beziehungen der Mechanik zu anderen Wissenschaften* (Chapter Five: The Relations of Mechanics to Other Sciences).

We have used for quotation the English translation, *The Science of Mechanics*, La Salle, Illinois, 1893.

famous 'bucket' experiment; in contrast to the claim that this experiment proved an absolute rotational motion, he argued:

> Newton's experiment with the rotating vessel of water simply informs us that the relative rotation of the water with respect to the sides of the vessel produces *no* noticeable centrifugal forces, but that such forces *are* produced by its relative rotation with respect to the mass of the earth and the other celestial bodies. No one is competent to say how the experiment would turn out if the sides of the vessel increased in thickness and mass till they were ultimately several leagues thick .... Relatively, not considering the unknown and neglected medium of space, the motions of the universe are the same whether we adopt the Ptolemaic or the Copernican mode of view. Both views are, indeed, equally *correct*, only the latter is more simple and more practical. (Mach, 1883, English translation, p. 284)

Mach's second important contribution to theoretical physics was one concerning the method. He had already obtained the basic idea, economy of thought, in the 1860s. As he wrote later: 'Through my interaction in 1864 with the political economist, E. Herrmann, who, according to his own speciality, thought to trace out the economical element in every kind of occupation, I became accustomed to designating the intellectual activity of an investigator as economical' (Mach, 1919, p. 3).[87] Mach connected this idea with the theory of biological evolution of Charles Darwin, whose *Origin of Species by Means of Natural Selection* (1859) impressed him deeply. He first applied the principle of economy of thought in a lecture in 1868, dealing with the form of liquids; especially he compared the principle of the minimum surface of liquids and the mercantile principle of a tailor working with the greatest saving of material. 'Should science be ashamed of such a principle?' he asked, continuing: 'Is science itself anything more than—a business? Is not its task to acquire with the least possible work, with the least possible thought, the greatest possible part of eternal truth?' (from 'The Form of Liquids', in Mach, 1895; reprint 1943, p. 16). In his *Mechanik*, Chapter Four, he carried the idea further; for example, he drew attention to the fact that algebraic and analytic relations in mathematics do replace economically in an innumerable amount of counting processes; in physics a similar economy of thought is reached by introducing the force function, or the moment of inertia, not to mention the minimum principles.

The principle of economy of thought served not just as a useful tool to describe the historical development of mechanics; it formed an essential part of Mach's philosophy or theory of science. Other crucial elements of this philosophy emerged from his dealing with problems of physiology and psychology. As we have mentioned earlier, the young Mach worked in Vienna essentially on experimental problems of physiology. Continuing this engagement in Graz, he discovered, e.g., the phenomenon of what was later called 'the Mach bands' (1865).[88] In Graz

[87] The translation of the quotation has been taken over essentially from Blackmore, 1972, p. 25. Emanuel Hartmann (1839-1902) was *Privatdozent* for *Nationalökonomie* at the University of Graz.

[88] A Mach band is observed when a spatial distribution of light contains a sharp change in illumination at some point. For example, if one takes a white colour wheel with an irregular black nick and lets it spin, then one observes two colour bands. Mach claimed that this observation was merely subjective and reflected a neurological inhibition of the human eye.

and Prague he also continued his earlier interest in psychology, which he had begun in Vienna with a study of Gustav Theodor Fechner's book *Elemente des Psychophysik* (1860); especially he had put, in the summer of 1860, the so-called Weber–Fechner law to test, finding from an experiment involving two pendula that the sensation intensity was directly proportional to the stimulus intensity rather than to its logarithm, as the law claimed.[89] Gradually he developed his own views, in which he put together more closely the fields of physiology and psychology to form a nearly unified science, with only the differences remaining between the specific methods of approach.

Mach may have been put on the path of uniting physiology and psychology by two major experimental findings of his own. We have already mentioned the first one, the so-called 'Mach bands,' which were a physiological as much as a physical discovery. The other result arose from investigating the human sense of motion. Mach started out in the 1860s by refuting an idea of Hermann von Helmholtz, who had explained a certain observation by a man travelling in a train rounding a curve by 'unconscious interference.' Mach, however, found such an explanation unsatisfactory and, after carrying out an experiment with a whirling chair in Prague, suggested a different solution: acceleration is felt as motion by a human observer, because the liquid in the semicircular canal of the ear's labyrinth is pressed by physical interia against the receptors of the ampullae.[90]

Mach also disagreed with Helmholtz on other topics of physiology and psychology. For example, he raised several objections to Helmholtz' theory of audition, and against his adoption of Thomas Young's three-colour theory of human vision, and against the latter's attempt to relate the reception of colours to particular nerve fibres. With respect to colour theory, Mach proposed a new scheme of six colours, which was later modified by Ewald Hering, his colleague in Prague. In Mach's opinion colour sensations had to be considered as basic sensations or 'elements,' similar to the sensations of space, time, sound, temperature and

[89] The Leipzig anatomist and physiologist Ernst Heinrich Weber (1795–1878), the elder brother of the physicist Wilhelm Weber, had tried to determine the sensibility of human senses quantitatively, notably the sensibility of the ear for sound. His observations were put into mathematical form by his colleague Gustav Theodor Fechner, i.e., in the equation, sensation intensity = constant × logarithm of stimulus intensity. Fechner based his entire system of psychophysics on this law.

In spite of disproving the Weber–Fechner law, Mach remained impressed by Fechner's ideas. For instance, he stuck for some time to the latter's distinction of psychology (as the science of determining the connection of presentation of physical phenomena or facts), physics (as the science for discovering the laws of connection of sensations or perceptions) and psychophysics (as the science establishing the laws and presentations). Mach still continued to do this in his lecture and book on *Die Geschichte and die Wurzel des Satzes von Erhaltung der Arbeit* (Mach, 1872). He also exchanged letters with Fechner, met him personally and planned to dedicate a book on the analysis of sensations to him. However, Fechner reacted rather negatively to Mach's ideas; hence Mach postponed writing and publishing the planned book until 1900 (see Footnote 79).

[90] Mach published the result first in a paper submitted to the Vienna Academy of Sciences on 6 November 1873, having only the slight priority of one week over Joseph Breuer and of six weeks over the Scottish chemist Crum Brown (1838–1922)—who arrived at the same conclusion independently by using different methods. Mach discussed the problem again in a wider context in his book *Grundlinien der Lehre von den Bewegungsempfindungen* (Leipzig, 1875).

pressure. He expounded a psychology, based on such small units—for which he synonymously used the name 'elements or atoms'—in his book '*Beiträge zur Analyse der Empfindungen*' (Jena, 1886).[91] For the purpose of analysis he organized the elements of sensation into three groups: the *äußere* (external) elements characterize human bodies and belong to, even define, the discipline of physics; the *innere* (internal) elements characterize human bodies and define the discipline of physiology; and the *innerste* (innermost) elements characterize our *ego* (memory, volitions) and define the discipline of psychology. It is not the subject matter or the facts which distinguish the three disciplines, but rather the different methods of investigation. In principle all sensation can be studied from the point of view of physics, or of physiology, or of psychology. Thus the sensation of space has a physiological and a physical aspect: the physiological space constitutes a pure sensation; the physical space also represents a functional dependence.[92]

Mach's ideas concerning sensations involved two noticeable features which emerged from his biological views. First, he was convinced that the experiences collected as a child, exert an important, if not decisive, influence on the later activity of a human being.[93] He combined this conviction with his unshakable belief in the theory of inheritance of acquired characteristics, which had been propagated by the French biologist Jean Baptiste de Lamarck (1744-1829). Mach argued that the behaviour of children, when they constructed toys, gave insights into the techniques of making tools by primitive people—he would expand on this particular point later in the book *Kultur und Mechanik* (Stuttgart, 1915). Second, he early adopted the opinion that the bodily and mental natures of men and animals are essentially the same and that most of the differences are quantitative rather than qualitative ones. That is, Mach believed that man was superior to animals only through the richness of his intellectual life, his broader range of

[91] By introducing small psychological units, Mach came close to the then influential Leipzig school of psychology of Wilhelm Wundt (1832-1920). Wundt, like Mach, concerned himself with the contents of consciousness; both of them represented the so-called 'content psychology'. On the other hand, Franz Brentano (1838-1917), since 1874 professor of philosphy at the University of Vienna, represented a different scheme of psychology, the so-called 'act psychology,' in which the psychical processes or acts play a fundamental role.

In later times Mach would often have to respond to the criticism of students and partisans of Brentano, e.g., of the philospher Edmund Husserl (1859-1938)—see the letter of Husserl to Mach, dated 18 June 1901 (quoted in Heller, 1964, pp. 61-64).

Finally, we would like to add that Brentano did influence two younger Austrian psychologists, namely Christian von Ehrenfels (1859-1932) and Max Wertheimer (1880-1943), who founded the so-called *Gestalt* psychology.

[92] A functional dependence was considered by Mach to be the most economical description of the relation between sensations in physics. Besides the concepts of physiological and physical space, he further introduced that of metric space; the latter represented a purely mathematical object, which could have arbitrarily many dimensions (especially more than three!).

[93] These experiences, by the way, included (in Mach's opinion) dreams and hallucinations, which he did not consider as sensations that are different in principle from conscious psychological sensations. The importance of childhood experiences and of subconscious sensations reappeared later in the ideas of Sigmund Freud.

interests, his ability to reach biological goals by more indirect and subtler means, and his capacity to make use of the experience of his fellow creatures.[94]

The concept of sensation was the unifying concept of Mach's physical, physiological and psychological views. Sensations also played a crucial role in his approach to physics. On the other hand, he attributed comparatively modest importance to theory in physics and in science generally, stating rather early: 'The object of science is the connection of phenomena; but the theories are like dry dreams which fall away when they have ceased to be the lungs of the tree of science' (Mach, 1872, p. 46; English translation, p. 74). His contemporary, the physicist Ludwig Boltzmann, expressed a totally different opinion on the value of theory, when he said in 1890: 'I hold the view that the task of theory is the construction of a picture of the external world that exists merely in ourselves, and which has to serve as a guiding star to all our thoughts and experiments' (Farewell address at the University of Graz, published in Boltzmann, 1905, p. 77; English translation in Broda, 1973, p. 22). Boltzmann also disagreed on other points with his colleague, as for example, on the question whether atoms existed or not; he thus became the antithesis of Mach, as we shall show below.

## A Giant in Theoretical Physics: Ludwig Boltzmann

Ludwig Eduard Boltzmann was born on 20 February 1844 in Vienna, the eldest son of Ludwig Boltzmann, a civil servant in the imperial internal revenue service (*Kaiserlich-königlicher Cameral-Concipist*), and his wife Maria Pauernfeind, the daughter of a merchant in Salzburg.[95] He grew up in Wels and Linz, where his father was soon transferred, and received his primary education from a private tutor at home and from the *Mittelschule* (*Gymnasium*) in Linz. He stood out as a very good, industrious pupil, nearly always at the top of his class. He also took a deep interest in nature (collecting butterflies and plants) and in music and literature.[96] Already at the age of fifteen he had lost his father to tuberculosis.

[94] As a result of his conviction that man and animals showed only quantitative differences, Mach opposed any treatment of animals that was not allowed with human beings, e.g., vivisection. The opposition to vivisection prevented him from studying the anatomic peculiarity of animals (which might have helped him to obtain more satisfactory results in certain physiological and psychological problems).

[95] Boltzmann's grandfather, the watchmaker (*Uhrenfabrikant*) Ludwig Gottfried Boltzmann came to Vienna from Berlin; his great-grandfather Samuel Ludwig Boltzmann had moved from Königsberg, East Prussia, to Berlin.

Ludwig and Maria Boltzmann had two other children: Albert, two years younger than Ludwig Eduard, died when a secondary school boy; and a daughter Hedwig. Detailed dates of the life of Boltzmann can be found in a paper presented by his grandson Dieter Flamm to the International Symposium on '100 Years Boltzmann Equation', Vienna, 4–8 September 1972 (Flamm, 1973). We have obtained further information about Boltzmann's life, work and views from Engelbert Broda's and Martin J. Klein's papers presented at the same time (Broda, 1973; M. J. Klein, 1973), and from Broda's book on Boltzmann (Broda, 1955).

[96] Boltzmann took some piano lessons from Anton Bruckner, the great master and composer in Linz. Continuing later with this interest in music, he used to play chamber music with friends and with his son Arthur at home; especially, he loved to play on the piano the symphonies of his favourite composer, Ludwig van Beethoven, in the adapted version of Franz Liszt.

In 1863 Boltzmann returned to Vienna in order to study mathematics and physics at the university. Josef Stefan, then acting director of the physics institute, became his main teacher in physics; in mathematics he learned most from Joseph Petzval, a highly original and stylistically elegant teacher and author who, in addition, showed strong interest in combining mathematical theory and practical application.[97] The lively atmosphere at the Physics Institute with the small laboratory in a house lying in the third district of the city, *Erdbergstraße* 15, and the personal closeness of the professor to his students left a lasting impression on Boltzmann. Many years later he recalled:

> *Erdberg* remained through all my life a symbol of honest and inspired experimental work. When I managed to bring some life into the Physics Institute in Graz, I called it jokingly *Klein-Erdberg*. I didn't mean that the space was small, it was probably twice as big as Stefan's institute; but I had not yet achieved the spirit of *Erdberg*. Still, in Munich, when the young graduate students came and did not know what to work on, I thought: how different we were in *Erdberg*. Today there is nice experimental equipment, and people are looking for ideas on how to use it. We always had enough ideas, our only worry was the experimental apparatus. (Boltzmann, 1905, pp. 100-101; English translation in Flamm, 1973, p. 7).

Thus properly guided by Petzval and Stefan, Boltzmann embarked upon scientific work; already in 1865 he published his first paper dealing with a problem of electrodynamics in the *Sitzungsberichte* of the Vienna Academy of Sciences (Boltzmann, 1865). He obtained his doctorate in 1866 and his *Habilitation* at the University of Vienna in the following year. At the Physics Institute his colleague Joseph Loschmidt soon became his friend. In 1869, then only 25 years old, Boltzmann was appointed *Ordinarius* of mathematical physics at the University of Graz. Besides teaching and performing theoretical research there, he seized the opportunity of doing experimental work in the laboratory of August Toepler, his colleague. During the years from 1869 to 1871 he paid visits of several months to Heidelberg (as guest of the chemist Robert Wilhelm Bunsen and the mathematician Leo Königsberger) and Berlin (as guest of Hermann von Helmholtz). From 1873 to 1876 he returned to Vienna as professor of mathematics at the university. When Toepler left Graz in 1876, Boltzmann, after a competition with Ernst Mach, succeeded him. He spent the following 14 years, the happiest in his life, as director of the Physics Institute.[98] In this period he became dean

[97] We have discussed Petzval's life and work earlier in this section.

[98] In 1876 Boltzmann married Henriette von Aigentler, a mathematics and physics student at Graz, who was 10 years younger than himself. In those days it was very unusual for a girl to study at a university; thus in her second semester the philosophical faculty in Graz passed a rule excluding female students. She was able, however, to make a petition to the Austrian Minister of Education, a former colleague of her father at the Graz court of justice, whereupon the rule was suspended. However, she did not continue her studies after becoming engaged to Boltzmann.

The Boltzmanns had five children, two sons and three daughters: Ludwig (1877-1889), Henriette, Ida, Arthur (1881-1952) and Elsa (1891-1965). Elsa was married to the physicist Ludwig Flamm (see Section I.5).

of the faculty (in 1878), *Regierungsrat* (in 1881), member of the Vienna Academy of Sciences (1885), *Rector* of his university (1887) and *Wirklicher Hofrat* (1889). In 1888 he turned down the honourable offer to succeed Gustav Robert Kirchhoff (1824-1887) as professor of theoretical physics at the University of Berlin, hesitating to exchange the lovely surroundings and conveniences of a southern provincial city with the hectic, official atmosphere of the German imperial capital.[99] The death of his eldest son Ludwig from appendicitis in 1889, however, depressed Boltzmann deeply; in 1890 he left Graz and Austria for the Bavarian capital of Munich, to accept the chair of theoretical physics at the university. He felt rather well among his colleagues there and drew students from all over the world.[100] Only the fact that the Bavarian government did not, in those days, pay a pension to university professors, and the offer to succeed his beloved teacher Josef Stefan as professor of physics at the University of Vienna, brought him back four years later to his hometown.

Boltzmann devoted his first publication to a problem of electrodynamics, which arose from the lectures of Stefan on the theory of electric current. Stefan had derived a strange-looking result: the resistance of a spherical conducting surface, when the endpoints of a diameter are connected to a battery, is equal to that of an infinite plane (of the same material) if measured at two points that are separated by the same distance as the diameter. He then asked Boltzmann to study the distribution of the electric potential on the sphere's surface and to check whether the result for the resistance also held if arbitrary points were taken on the surface. Boltzmann proved that it was indeed so and wrote his first research paper containing the derivation (Boltzmann, 1865).[101] While this apprentice's work remained within the old electromagnetic theory of André Marie Ampère (1775-1836), he also learned the most recent one. He recalled in particular: 'When I came—still being a student at the university—in closer contact with Stefan, the

[99] Boltzmann bought a farmhouse near Oberkroisbach in the vicinity of Graz, where he lived with his family. While here he enlarged his butterfly collection and herbarium. However, Boltzmann already knew Berlin from a previous visit. He highly respected Hermann von Helmholtz, then (from 1887) president of the *Physikalisch-Technische Reichsanstalt.* It is reported that, after negotiations in the ministry, Frau Helmholtz said at a dinner with his colleagues: 'Professor Boltzmann, I am afraid that you will not feel comfortable here in Berlin.' (See Broda, 1955, p. 6, and Flamm, 1973, p. 11.)

[100] Boltzmann especially enjoyed the weekly meetings at the *Hofbräuhaus*, where he met colleagues from the university and the *Technische Hochschule*, such as the mathematicians Walther von Dyck (1856-1934) and Alfred Pringsheim (1850-1941), the physicists Eugen von Lommel (1837-1899) and Leonhardt Sohncke (1822-1897), the chemist Adolf von Baeyer (1835-1917), the astronomer Hugo von Seeliger (1849-1924) and the refrigeration engineer Carl von Linde (1842-1937).

[101] Boltzmann's first paper was entitled '*Über die Bewegung der Elektrizität in krummen Fläschen*' ('On the Motion of Electricity on Curved Surfaces') and published, as we have mentioned earlier, in the *Sitzungsberichte der Kaiserlichen Akademie der Wissenschaften.*

The scientific papers of Ludwig Boltzmann have been collected and reprinted in three volumes of *Wissenschaftliche Abhandlungen* (Fritz Hasenöhrl, editor) (Boltzmann, 1909a, b, c). A new edition of Boltzmann's papers and books is now in the course of publication: *Ludwig Boltzmann Gesamtausgabe* (Roman U. Sexl, editor), Graz: *Akademische Druck- und Verlagsanstalt*, Braunschweig/Wiesbaden: Friedrich Vieweg & Sohn. The following have so far appeared: *Vorlesungen über Gastheorie* (Volume 1, 1981); *Vorlesungen über Maxwells Theorie der Elektricität und des Lichtes* (Volume 2, 1982).

first thing that he did was to hand me copies of Maxwell's papers [on electrodynamics]; and, since I did not know a word of English at that time, he gave me an English grammar book; I had a dictionary I had inherited from my father' (Speech at the unveiling of Stefan's monument, 8 December 1895, published in Boltzmann, 1905, p. 96). Thus Stefan transferred his interest in the theory of electrodynamics of James Clerk Maxwell (1831–1879), which had only recently been published (Maxwell, 1861a b; 1862a, b; 1865), to his student, and this would bear fruit in the future.[102]

A few years later, as a guest in Helmholtz' physics laboratory in Berlin, Boltzmann investigated experimentally the change in the capacity of condensers due to the introduction of insulating material between the plates. He found that the measured value of the dielectric constant $\varepsilon$ of insulating materials always satisfied the equation

$$n = \sqrt{\varepsilon\mu}, \tag{2}$$

with $n$ denoting the index of refraction and $\mu$ the coefficient of magnetic induction (or magnetic susceptibility) of the same substance. This was the relation which also followed from Maxwell's theory, of which Boltzmann provided the first experimental proof.[103] In 1880 he again worked on problems of electrodynamics, this time discussing theories of the Hall effect and of electrostriction.[104] Boltzmann searched eagerly for a crucial test of Maxwell's theory for many years. The success fell to Helmholtz' former student Heinrich Hertz (1857–1894), who found electromagnetic waves in 1888 in his Bonn laboratory (Hertz, 1888a). Boltzmann, who was very impressed by the result, repeated the experiment of Hertz before a large audience in Graz, adding to it an interference experiment—of the type that Augustin Fresnel (1780–1827) had carried out many decades earlier with optical waves (Boltzmann, 1890).

After Hertz' discovery of electromagnetic waves, Boltzmann became an ardent partisan and promoter of Maxwell's electrodynamic theory in Austria and Germany. Thus he presented regular lecture courses on the subject in Graz, Munich

[102] Boltzmann's concern with Maxwell's electromagnetic theory has been discussed by Walter Kaiser, in the introduction to the new edition of *Vorlesungen über Maxwells Theorie der Electricität und des Lichtes* (Kaiser, 1982).

[103] Boltzmann continued the Berlin experiments in Graz, partly with his colleague August Toepler, and then published several papers on the results during the following years, the first in the fall of 1872 (Boltzmann, 1872b).

[104] Boltzmann pointed out in particular that the phenomenon observed the year before by the American Edwin Herbert Hall (1855–1935) might serve to determine the 'absolute value of velocity of electricity in an electric current' (Boltzmann, 1880a, p. 12). In the two papers, which he wrote on electrostriction (i.e., on the elastic deformation of a dielectric substance created by the action of an electric field), Boltzmann investigated the possibility whether it was possible to decide between Maxwell's theory and the previous potential theory (Boltzmann, 1880c, d). He returned to the Hall effect six years later and derived some consequences from a theory of the effect based on Maxwell's electrodynamics. Albert von Ettinghausen and Walther Nernst, then working at Boltzmann's institute, confirmed experimentally some of these consequences.

and Vienna.[105] He made these lectures available to a wider audience by publishing them as a two-part book, *Vorlesungen über Maxwells Theorie der Electricität und des Lichtes* (Boltzmann, 189lb, 1893). He further discussed specific problems of the theory at the *Naturforscherversammlung* at Halle (Boltzmann, 1892a). Those aspects of the theory especially interested him, which Maxwell had called 'dynamical illustrations' (Maxwell, 1865, p. 467) and discussed *in extenso* in his *Treatise on Electricity and Magnetism* (Maxwell, 1873). Boltzmann constructed, for example, a theoretical model of the ether having such mechanical properties that Maxwell's equations could be derived from it (Boltzmann, 1892b).[106] He also illustrated more specific results of the theory by mechanical models.[107] One should not, however, interpret these dynamical illustrations as meaning a purely mechanical picture of the electrodynamical phenomena. Boltzmann used them rather as a suitable tool for theoretical description. In the same, very general, sense he also considered the derivative of Maxwell's equations from a Hamiltonian principle as a dynamical illustration.[108] Although Boltzmann's *Vorlesungen* did spread the modern theory of electrodynamics in Central Europe, they were soon replaced as a textbook by August Föppl's *Einführung in die Maxwellsche Theorie der Electricität* (Föppl, 1894). This elementary, introductory text made systematic use of the vector calculus, a practical mathematical method, which became indispensable for the further progress of electrodynamics, especially electron theory.[109] Strangely enough, Boltzmann did not participate in the rapid development of electron theory, which was so close in spirit to his own work on the kinetic theory of gases.

In the same year as he took his doctorate, Boltzmann published his first paper on the kinetic theory of gases, entitled '*Über die mechanische Bedeutung des zweiten Haupsatzes der Wärmetheorie*' ('On the Mechanical Significance of the Second Law of Thermodynamics,' Boltzmann, 1866). Also his last scientific paper,

[105] Boltzmann lectured, for instance, in the summer semester of 1890 at the University of Graz on '*Elektromagnetische Theorie des Lichtes*' (Electromagnetic Theory of Light), and in the summer semester of 1891 in Munich on '*Maxwellsche Electricitätstheorie*'; and again on the same topic in the winter semesters of 1891/1892, 1892/1893, and in the summer semester of 1893. (See Kaiser, 1982, p. 17*.)

[106] Many years earlier, in a more popular demonstration of Maxwell's electrodynamic theory, Boltzmann explained it by referring to certain hydrodynamical properties of the ether (Boltzmann, 1873).

[107] Boltzmann gave, for example, in the first part of his *Vorlesungen*, the explicit construction of mechanical devices, which consisted of complicated systems of turning wheels and gears and which were supposed to describe the phenomena occurring in electrical circuits involving condensers and coils (Boltzmann, 1891b; Lecture Six and Figures 14 and 15).

[108] Boltzmann did not particularly like the presentation, which Heinrich Hertz gave of Maxwell's theory in 1890, when the latter assumed Maxwell's equations as given postulates from the very beginning (Hertz, 1890a). He criticized this procedure in Lecture Two of Part II of his *Vorlesungen* (Boltzmann, 1893, pp. 13-15).

[109] Föppl's *Einführung* would be extended after 10 years by a second part on electron theory, written by the theoretician Max Abraham (1875-1922) of Göttingen (Abraham, 1905). The Föppl-Abraham volumes became the standard textbook in German scientific literature during the following decades.

a review article on '*Kinetische Theorie des Materie*' ('Kinetic Theory of Matter'), which he wrote jointly with his student Joseph Nabl for the *Encyklopädie der mathematischen Wissenschaften* (Boltzmann and Nabl, 1907), dealt with the same topic. One can rightly state that Boltzmann devoted most of his scientific activity, more than half of his scientific papers and two books (*Vorlesungen über Gastheorie*, *Teil I*, Boltzmann, 1896d; *Teil II*, Boltzmann, 1898), to this topic. Indeed, his work on kinetic theory shows so many facets that we have to organize it into four subtopics:

(i) the dynamical interpretation and derivation of the second law of thermodynamics;
(ii) the study and extension of Maxwell's distribution law;
(iii) the transport equation and the $H$-theorem; and
(iv) the foundation of statistical mechanics.[110]

In his first, above-mentioned paper on kinetic theory, Boltzmann tried 'to give a purely analytical, completely general proof of the second law of thermodynamics, as well as to find the corresponding theorem in mechanics' (Boltzmann, 1866, p. 195). He derived, from a general formulation of the principle of least action, a purely mechanical expression—involving the kinetic energy, $E_{\text{kin}}$, of molecules and a period ($\tau$) of the molecular configuration (for technical reasons he had to assume periodicity of motion)—i.e.,

$$\sum \ln(E_{\text{kin}}\tau)^2 + \text{const.}, \tag{3}$$

which appeared to possess the properties of the thermodynamic entropy $S$.[111]

Although Boltzmann later returned to the problem of giving the second law of thermodynamics a purely mechanical foundation (see. e.g., Boltzmann, 1884c), he soon noticed that extra aspects had to be considered as well. He especially studied Maxwell's papers on the velocity distribution of molecules in a gas of given temperature, which involved probability arguments (Maxwell, 1860, 1867). As a first step he generalized the distribution law by including the action of a potential $V$ on the molecules, finding the new distribution formula

$$f(v) = \text{const. exp}\left\{ -h\left(\frac{mv^2}{2} + V\right)\right\}, \tag{4}$$

where $mv^2/2$ is the kinetic energy of a molecule having velocity $v$, and $h$ denotes a quantity inversely proportional to the absolute temperature (Boltzmann, 1868).

[110] Detailed analyses of Boltzmann's work on kinetic gas theory have been published by Stephen G. Brush in his article on Boltzmann for the *Dictionary of Scientific Biography* (Brush, 1970), and in his introduction to the English translation of Boltzmann's *Vorlesungen über Gastheorie* (Brush, 1964); and by Martin J. Klein in a paper for the 1972 Boltzmann Symposium in Vienna (M. J. Klein, 1973). (See also Brush's introduction to Volume 1 of the new Boltzmann edition, referred to in Footnote 101: Brush, 1981.) We have discussed certain aspects of Boltzmann's contribution to kinetic theory in Volume 1, especially in the Introduction and in Chapter I.

[111] Five years later Rudolf Clausius (1822-1888) published a paper, in which he arrived, on the basis of a slightly different reasoning, at practically the same conclusion (Clausius, 1871), upon which Boltzmann did not hesitate to point out his priority (Boltzmann, 1871).

Expression (4) did not represent the most general situation in kinetic gas theory, because the molecules could not be considered as pointlike objects without internal degrees of freedom. Boltzmann undertook the necessary generalization in 1872, in his long memoir '*Weitere Studien über das Wärmegleichgewicht unter Gasmolekülen*' ('Further Studies on the Thermal Equilibrium Among Gas Molecules,' Boltzmann, 1872a). The main content of this paper consisted of the investigation of the collision process between gas molecules, in the derivation (Section I) and the application (Section III) of the equation describing these processes. Boltzmann found that the distribution function $f(x, t)$, in dependence of the kinetic energy ($x$) of the molecules and the time ($t$), obeyed the partial differential equation

$$\frac{\partial f(x,t)}{\partial t} = \int_0^\infty dx' \int_0^{x+x'} \left[ \frac{f(\xi,t)}{\sqrt{\xi}} \frac{f(x+x'-\xi,t)}{\sqrt{x+x'-\xi}} - \frac{f(x,t)}{\sqrt{x}} \frac{f(x',t)}{\sqrt{x'}} \right] \cdot \sqrt{x x'} \cdot \psi(x, x', \xi)\, d\xi, \tag{5}$$

where $\psi(x, x', \xi)$ depends on the nature of the interaction force between two molecules. The author immediately applied Eq. (5)—now called 'Boltzmann's collision equation'—to describe the irreversible processes of diffusion, friction and heat conduction in gases. In later papers he established on its basis a general theory of friction (Boltzmann, 1880b; 1881a, b) and diffusion in gases (Boltzmann, 1882, 1883a).[112]

The collision equation became a common and powerful tool in the physics of irreversible transport phenomena in the twentieth century.[113] Boltzmann, however, also used it for a different, more fundamental purpose, namely to prove the second law of thermodynamics. In particular, he defined an expression $E$,

$$E = \int_0^\infty f(x,t) \left\{ \ln\left[ \frac{f(x,t)}{\sqrt{x}} \right] - 1 \right\} dx \tag{6}$$

[112] In the simplest case of a one-dimensional motion in the $z$-direction the transport equation reduced to the equation,

$$\frac{\partial f}{\partial t} + \dot{z}\frac{\partial f}{\partial z} + \frac{F_z}{m}\frac{\partial F}{\partial \dot{z}} = \text{collision term},$$

where $F_z$ denotes the force in the $z$-direction (of diffusion, friction, or temperature difference) acting on a molecule with mass $m$. The collision term on the right-hand side consists of an integral of the expression $(ff_1 - f'f_1')$ over the variables of the relative motion of the two molecules before and after collision (described by the distribution functions, $f, f_1$ and $f', f_1'$, respectively). Maxwell, for instance, used the above equation to calculate the pressure in rarefied gases under the influence of heat conduction (Maxwell, 1880).

An exact solution of the collision equation could only be obtained in the case when the molecules interacted with each other through inverse fifth-power forces (i.e., the force was proportional to $r^{-5}$, with $r$ the distance between the molecules). For other power laws of force Boltzmann developed complicated approximations. A systematic treatment of collision equation with arbitrary forces was given by Sydney Chapman and David Enskog, on the basis of David Hilbert's theory of integral equations. (We have discussed the latter theory in Volume 3, Section 1.2.)

[113] On the one-hundreth anniversary of the equation an international symposium was held in Vienna. The contributed papers were published as a book, *The Boltzmann Equation. Theory and Applications* (E.G.D. Cohen and W. Thirring, editors). Vienna and New York: Springer-Verlag, 1973.

(see Boltzmann, 1872a, Section I, Eq. (17)), and showed that it possessed the properties of a negative entropy: thus $-E$ always increased monotonically with time $t$, if the distribution function $f(x, t)$ deviated from the Maxwellian distribution ($\sqrt{x}\exp(-hx)$, with $x$ denoting the kinetic energy); it stayed constant if the latter distribution was assumed. This result—i.e., $-dE/dt \geq 0$—was later called the '$H$-theorem' (see Boltzmann, 1895b).[114] While it made Boltzmann very happy, since it was identical with the second law of thermodynamics, his friend and colleague Joseph Loschmidt raised serious objections (Loschmidt, 1876a, b; 1877a, b). Loschmidt was bothered by the apparently inevitable consequence from the second law that was discussed frequently at that time, i.e., the 'heat death of the universe' or all energy becomes dissipated into irregular heat motion. He claimed, however, that the 'heat death' should not be considered as following necessarily from mechanical laws; especially, that Boltzmann's derivation of $dE/dT < 0$ could not be correct, since the collisions between molecules were *reversible*, not irreversible processes.[115] Boltzmann replied immediately to this criticism; he pointed out: 'Loschmidt's theorem [i.e., the theorem stating that to each entropy-increasing process between molecules there had to exist a reverse process decreasing the entropy by the same amount] seems to be of great importance, since it demonstrates how closely the second law is related to the probability calculus' (Boltzmann, 1877a; Reprint, p. 121). In a following, long paper entitled '*Über die Beziehung zwischen dem zweiten Hauptsatze der mechanischen Wärmetheorie und der Wahrscheinlichkeit respektive den Sätzen über das Wärmegleichgewicht*' ('On the Relation Between the Second Law of the Mechanical Theory of Heat and Probability, and the Theorems Concerning Thermal Equilibirum, Respectively,' Boltzmann, 1877b), he introduced the fundamental relation between the entropy $S$ and what he called the 'measure of probability' ('*Permutationsmaß*') $\Omega$ as[116]

$$S + \text{const.} = \int \frac{dQ}{T} = \tfrac{2}{3}\Omega. \tag{7}$$

Boltzmann's 1877 memoir containing the statistical definition of the thermodynamical entropy of a gas marked the beginning of what was later baptized—

[114] The original notation $E$ for the negative entropy was simply replaced by the notation $H$.

[115] Loschmidt also attacked the generalized distribution law, Eq. (4), of Boltzmann; he claimed that it failed to describe the behaviour of gas molecules in the presence of a gravitational field.

[116] $\Omega$ denotes, up to factors and constant additional terms, the logarithm of the 'permutability' ('*Permutabilität*'), the latter being the number of permutations of the molecules in a given state of the gas.

To derive Eq. (7), Boltzmann took a given volume of gas containing a given number of molecules and having a given total energy. The permutability then is the number of possibilities, in which the thus given state (the macroscopic state) of the gas can be obtained from distributions of individual molecules (the microscopic state). In the later notation (introduced by Planck in 1900), the right-hand side of Eq. (7) would be replaced by $k \ln W$, where $W$ denotes the number of possible molecular configurations (microscopic states) corresponding to a given macroscopic state of the gas (characterized by total energy, volume, temperature, and number of molecules)—or another system—and $k$ is Boltzmann's constant, as introduced by Planck (1900f; see our discussion in Volume 1, Section I.2).

by the American theoretical physicist Josiah Willard Gibbs (1839–1903)—'Statistical Mechanics.' Equation (7) stated, in particular, that purely mechanical concepts did not suffice to describe the thermodynamical behaviour of macroscopic bodies, consisting of a large number of molecules, but one also had to refer to statistical considerations.[117] Boltzmann later tried to deepen the foundation of the new theory by introducing the concept of '*Ergoden*'—meaning a collection (ensemble) of similar systems (of gas molecules) having the same energy but different initial conditions (Boltzmann, 1884c, 1887). Ergodic ensembles satisfied the second law of thermodynamics.[118] The ensemble concept in kinetic theory—whose beginnings may be traced back to an earlier paper of Maxwell—was especially developed by Gibbs in his *Elementary Principles in Statistical Mechanics* (Gibbs, 1902), the book which, together with Boltzmann's research papers, established statistical mechanics as a scientific theory. In 1905 the mathematician Felix Klein would request Boltzmann to write the review article on the new field for the *Encyklopädie der mathematischen Wissenshaften.* After Boltzmann's death the task would be passed on to his student Paul Ehrenfest, who would elaborate on it for many years (P. and T. Ehrenfest, 1911).

## Do Atoms Really Exist: Boltzmann or Mach?

In 1894 Ludwig Boltzmann obtained the chair of physics at the University of Vienna, held previously by Joseph Stefan. The return to his hometown created some difficulties for Boltzmann; being fifty years old he was no longer used to living in the big city, which he had left eighteen years previously, and he feared that he would not find there as nice a group of friends and colleagues as he had enjoyed in Graz and Munich. Finally, he did not feel in the best of health: thus in 1873 at the Vienna World's Fair, when his Cracow colleague Zygmunt Wróblewski had asked Boltzmann to write a book on gas theory, he had refused to do so, excusing himself by saying that he 'did not know anyway how soon the eyes might fail' (Boltzmann, 1896d, p. V). Yet, he soon sat down to follow Wróblewski's advice: in Vienna he completed writing the two volumes of his *Vorlesungen über Gastheorie* (1896d, 1898). The first book or *Part One* contained an exposition of the general topics of gas theory; Boltzmann attempted especially 'to make clearly comprehensible the path-breaking works of Clausius and Maxwell' (Boltzmann, 1896d, p. V). While working in 1896 on *Part Two*, however,

[117] The first use of a statistical consideration in gas theory goes back to Maxwell, who, in his above-quoted paper on the velocity distribution of molecules, introduced the error distribution of Carl Friedrich Gauss (Maxwell, 1860). In the 1860s Maxwell suggested that the second law of thermodynamics did not strictly hold (Maxwell to Tait, 11 December 1867). William Thomson, Lord Kelvin (1824–1907) remarked at that time that the second law would be violated by the reversibility of the motion of the molecules, whereupon Maxwell concluded that the 'Second Law of Thermodynamics has only a statistical certainty' (see Knott, 1911, p. 214; also Maxwell's textbook *Theory of Heat*: Maxwell, 1871, p. 308).

[118] In agreement with the essential use of the non-dynamical probability concept, it was not possible to associate ergodic systems with a single trajectory in the multidimensional phase space (built up from the co-ordinates and the momenta of the individual molecules).

he noticed that he had to go beyond a pure presentation of the results of the theory, because 'just at the time the attacks on the gas theory began to increase' (Boltzmann, 1898, p. V); the author even realized a 'hostile mood' (*feindliche Stimmung*,' Boltzmann, 1898, p. VI) against it. What had actually happened in this respect between 1894 and 1896?

In the fall of 1894 Boltzmann visited the Oxford meeting of the British Association for the Advancement of Science. The mathematician George Hartley Bryan (1864-1928) of Cambridge had prepared a report for the meeting on 'The Laws of Distribution of Energy and Their Limitations' (Bryan, 1894); and Boltzmann eagerly participated in the discussion of Bryan's report, submitting short comments on it (Boltzmann, 1894a, b). At Oxford Boltzmann enjoyed the discussions immensely, and also the support of the kinetic theory by the British physicists, recalling later 'the unforgettable meeting of the British Association at Oxford' (Boltzmann, 1896d, p. VI). He also participated in the post-Oxford debate on the subject, which was carried on into the following year in the British scientific journal *Nature*.[119]

A year later, at the *Naturforscherversammlung* of September 1895 in Lübeck, Boltzmann wished to stimulate a similar debate in Germany—as in Britain—on the issues of kinetic gas theory. Therefore he wrote in advance to the renowned physico-chemist Wilhelm Ostwald in Leipzig: 'At the *Naturforscherversammlung* Professor Helm will report on *Energetik*. I would like, if possible, to provoke a debate *à la* British Association, mainly for the purpose of learning myself. For this, it is first of all necessary that the main representatives of this direction [i.e., *Energetik*] be present. I need not mention how much especially your presence would please me' (Boltzmann to Ostwald, 1 June 1895, quoted in Körber, 1961, pp. 21-22). Ostwald (1853-1932), since 1887 professor of chemistry at the University of Leipzig, had indeed been responsible for the concept of *Energetik*; i.e., he had since 1891 tried to establish science on the concept of energy alone, discarding in particular the hypothesis of an atomic structure of matter.[120] His views had won him ardent supporters, among them the French physicist Pierre Duhem (1861-1916) and Georg Helm (1851-1923), *Ordinarius* for analytic geometry, analytic mechanics and mathematical physics at the *Technische Hochschule* of Dresden.

In Lübeck the first public encounter between *Energetik* and atomism occurred. Helm presented his report, '*Über den derzeitigen Zustand der Energetik*' ('On the Present Status of *Energetik*,' Helm, 1895), and Boltzmann raised critical questions.

[119] One difficulty discussed at that time presented the explanation of the observed specific heats of gases. The other issue in debate was the derivation of the second law of thermodynamics from reversible mechanical processes. Boltzmann, in his letters to *Nature*, seized the opportunity to reply in detail to the critical questions asked by Edward P. Culverwell and S. H. Burbury and to clarify the meaning of his *H*-theorem (Boltzmann, 1895b, c, d).

[120] In 1891 and 1892 Ostwald published two articles, bearing the title '*Studien zur Energetik*' ('Studies on Energetics,' Ostwald, 1891) and '*Grundlinien der allgemeinen Energetik*' ('Outlines of the General Energetics,' Ostwald, 1892a). A little later he stated: 'In fact, energy constitutes the only reality in nature, and matter is not actually a *carrier* of energy but its representation' (Ostwald, 1892b, p. 771).

Arnold Sommerfeld, one of the younger participants, recalled nearly fifty years later:

> The review on *Energetik* was presented by Helm of Dresden. Wilhelm Ostwald stood behind him, and behind him stood the philosophy of science of the absent Ernst Mach. The opponent was Boltzmann, assisted by Felix Klein. The fight between Boltzmann and Ostwald resembled, externally and internally, the fight of a bull with a supple fencer. But this time the bull defeated the matador, in spite of all the fencing skill of the latter. Boltzmann's arguments broke through. We, then young mathematicians, all took the side of Boltzmann; it was quite obvious to us that one could not derive from a single energy equation the equations of motion, not even in the case of a single mass point—to say nothing about a system of arbitrarily many degrees of freedom. (Sommerfeld, 1944, p. 25)

Ostwald's plea against 'scientific materialism' did not win the support of the younger scientists.[121] Immediately after the Lübeck meeting Boltzmann wrote further papers, in which he attacked Ostwald's and Helm's views (Boltzmann, 1895e; 1896a, c); and soon he was assisted in this task by Max Planck (1895b).[122] However, Boltzmann did not really enjoy Planck's support, for at about the same time Planck's assistant Ernst Zermelo (1871–1953) made a serious objection, the 'recurrence argument' ('*Wiederkehreinwand*'), against the $H$-theorem (Zermelo, 1896). Zermelo claimed, in particular, that the mechanical laws forced any complex system to return after a characteristic period of time to a given state, hence no mechanical system should be an ergodic one. Boltzmann, a little angry, replied that Zermelo's argument did not hold in a statistical theory, and further that the recurrence period for systems consisting of many gas molecules would be extremely large (Boltzmann, 1896b; 1897b). However, at home in Vienna he faced a really serious enemy of atomism, namely Ernst Mach.

Although Mach, in his early research, had not been unsympathetic to the hypothesis of an atomic structure of matter, he had changed this attitude before leaving Vienna in 1864.[123] In Graz, and even more in Prague, he had developed into a staunch anti-atomist. For example, in 1872 he had written: 'If then, we are astonished at the discovery that heat is motion, we are astonished at something, which has never been discovered' (Mach, 1872, English translation, p. 47).[124] In

[121] In Lübeck, Ostwald presented a talk with the title '*Überwindung des wissenschaftlichen Materialismus*' ('The Overcoming of Scientific Materialism,' Ostwald, 1895), in which he identified atomism with scientific materialism.

[122] We have discussed Planck's papers against *Energetik* in Volume 1, Section I.1.

[123] It has been suspected that Mach's unsuccessful attempt to understand the motion of liquids in terms of molecules, and his failure to identify experimentally chemical elements through their spectral lines, had converted him into an opponent of atomism. (See Blackmore, 1972, p. 21; also Hiebert, 1970.) Another reason which may have contributed to his giving up the atomic hypothesis was the growing opposition to Helmholtz, who supported it, and the fact that the atomist Stefan was preferred over him (Mach) in the Vienna physics chair.

[124] Mach shared the opinion of Robert Mayer, the discoverer of energy conservation, that one should not interpret the first law of thermodynamics as necessarily following from mechanical theories, or as a mathematical formula based on the dynamics of atoms and molecules. In the preface to the second edition of *Die Geschichte und die Wurzel des Satzes von der Erhaltung der Arbeit* of 1909 he wrote: 'Any metaphysical or any other one-sided mechanistic interpretation of physics has been avoided.'

later years Mach had not retired from this opinion, but rather sharpened it; and he had continued to have bad relations with Stefan, the head of the Viennese school of atomism. When Boltzmann succeeded Stefan, the situation did not improve for Mach. Still he (Mach) had friends in Vienna—among them Theodor Gomperz, the historian of Greek philosophy, and the mineralogist Gustav Tschermak. They managed to bring him back to Vienna and get him a chair of philosophy with the additonal assignment for history and theory of exact sciences ('*Geschichte und Theorie der exakten Wissenschaften*').[125]

Mach's lectures, which dealt with all topics of physics (especially its history), physiology and psychology (on which he had already written papers and books, or planned to write them), attracted a large audience: students and senior people came from all faculties.[126] At the University of Vienna he won several supporters of his views on scientific and philosophical questions: especially Anton Lampa (1868-1938), since 1891 *Privatdozent* for experimental physics, the two philosophers Wilhelm Jerusalem (1854-1923)—*Privatdozent* since 1891—and Heinrich Gomperz (1873-1942)—son of Theodor Gomperz and *Privatdozent* after 1900—and the physiologist Wolfgang Joseph Pauli. He influenced not only students of physics and philosophy and people interested in these topics, but writers and poets as well—such as Hermann Bahr, Hugo von Hofmannsthal, Arthur Schnitzler and Robert Musil[127]—and even politicians like the Austrian Social Democrat Victor Adler.[128] His fame reached far beyond the borders of Austria, even to the German speaking countries.[129] However, times did not remain lucky for Mach. In July 1898, while on a railway train, he suffered a stroke, which permanently paralyzed the right half of his body. Although he never gave up work—he began to use a typewriter with his left hand—and continued to lecture for a while, he finally had to retire from his university position in the spring of 1901.

The philosophical attitudes of the influential Mach bothered Ludwig Boltzmann quite a bit. In two articles, which he published in *Annalen der Physik*, he emphasized the indispensability of the atomic hypothesis in science, attacking his opponent at the same time (Boltzmann, 1897a, e). Mach's statement, 'I don't

[125] Franz Brentano formerly held this chair, but had resigned it several years previously.

[126] In 1896 he published two books, the *Populäre-Wissenschaftliche Vorlesungen* (Leipzig, 1896)—an extension of the English book *Popular-Scientific Lectures* (Chicago, 1895)—and *Die Prinzipien der Wärmelehre* (Leipzig, 1896). Earlier books received new editions as well.

[127] Musil wrote a thesis on the understanding of Mach's philosophical ideas, with which he obtained his doctorate at the University of Berlin in 1908. His great novel, *Der Mann ohne Eigenschaften*, published in three volumes (1931, 1933, 1943), reflects the influence of Mach's views, as do his other writings.

[128] Adler's son, Friedrich Adler (1897-1960), studied chemistry, physics and philosopy at the University of Zurich, becoming a *Privatdozent* there in 1907. In Zurich Mach also won strong supporters at the E.T.H.; some partisans of the philosopher Richard Avenarius (1843-1896), who had been professor at the E.T.H. from 1877 and whose ideas showed some overlap with Mach's, now became Mach's allies, among them Joseph Petzold (1862-1929). Students in Zurich founded a Mach circle, of which Albert Einstein was a member.

[129] Mach gained followers particularly in France and America.

believe that atoms exist,' shocked him deeply.[130] He began to study more thoroughly the philosophical aspects underlying that statement. Thus he wrote in an article, bearing the philosophical title '*Über die Frage nach der objektiven Existenz der Vorgänge in der unbelebten Natur*' ('On the Problem of the Objective Existence of Processes in Inanimate Nature,' Boltzmann, 1897c), in which he analyzed with Machian seriousness the methods applied by physicists in order to obtain a description of phenomena in nature; he concluded in particular that one might give similar arguments for the existence of atoms, as Mach had previously suggested when he wanted to prove the existence of thoughts and feelings of people different from himself.

Boltzmann continued the crusade for atoms in Leipzig, where he accepted in 1900 an appointment as professor of theoretical physics; there he fought against Wilhelm Ostwald, who had arranged the position for him.[131] The stay in Saxony did not last long: in the fall of 1902 Boltzmann returned to his old chair in Vienna, which had remained vacant. In addition he was entrusted with Mach's former philosophical chair, which now was associated with the special condition to teach '*Methode und allgemeine Theorie der Wissenschaften*' ('Method and General Theory of Science'). Boltzmann enjoyed the latter appointment as a triumph over his scientific opponent Mach.[132] His lecture courses on 'natural philosophy [philosophy of science] were immensely popular,' as Ludwig Flamm recalled (Flamm, 1944, p. 30).[133] Still, the second return to Vienna did not mean a pure triumph: the establishment and authorities would not easily forgive his previous going away to Leipzig.[134] Above all, however, his health worsened: his eyesight failed to such an extent that he had to employ somebody to read to him, and he suffered at night from strong asthma attacks—probably he also had *angina pectoris.* Being overworked, he frequently had headaches. Even during his last great journey, a visit to California—which he enjoyed very much and described in a humorous way in a popular article, the '*Reise eines deutschen Professors ins Eldorado*' ('Journey of a German Professor into Eldorado'; see Boltzmann, 1905)—Boltzmann's personal condition did not improve. On 6 September 1906,

[130] Boltzmann mentioned in a later lecture that Mach made this statement in a debate on atomism at one of the 1897 meetings of the Vienna Academy of Sciences (Boltzmann, 1903; Reprint, p. 339).

[131] It should be mentioned that, in spite of their disagreement on atomism, Boltzmann and Ostwald had great respect for each other. It could even be said that they were friends.

[132] As in the case of Ostwald, Boltzmann also had great personal respect for Mach. (See, for example, the statements in his inaugural lecture Boltzmann, 1903.)

[133] Only a few of Boltzmann's philosophical lectures have become available in print: e.g., an essay criticizing the content of a lecture of Wilhelm Ostwald in Vienna (Boltzmann mocked the latter's application of thermodynamic formulae—or even formulae from *Energetik*—describing quantitatively what is human luck and well-being; see Boltzmann, 1905, pp. 364–378); or another essay criticizing the philosophy of Arthur Schopenhauer (Boltzmann, 1905, pp. 384–402).

[134] Boltzmann had to wait until 1904 for his re-election as a full member of the Imperial Academy of Sciences; and the budget for his *Physikalisches Institut* was severely cut down.

while on vacation in Duino near Trieste, he committed suicide during an attack of depression.[135]

Ernst Mach, semi-paralyzed, outlived Boltzmann by ten years. Although his health gradually weakened with increasing age—due to rheumatism (in 1903), neuralgia (in 1906) and prostate infection (in 1912)—his willpower sustained him. In 1905 he published his Vienna lectures as the book *Erkenntnis und Irrtum* (Mach, 1905); he revised and enlarged his older books, and he began to work on a new book, *Die Prinzipien der physikalischen Optik* (*The Principles of Physical Optics*), whose first part he finished in 1913 (Mach, 1921). Besides writing more than a dozen new articles on philosophy, popular science and his own experimental investigations, he composed the volume *Kultur und Mechanik* (*Culture and Mechanics*, Leipzig 1915), the last book of Mach that appeared in his lifetime. In May 1913 he left Vienna and moved to the home of his son Ludwig near Haar, close to Munich.

There he recovered a little, began to walk again and to receive visitors, e.g., Joseph Petzold and Hugo Dingler (1881–1954)[136] A pacifist and anti-nationalist throughout his life, Mach died on 19 February 1916 at his son's home.

The importance of Ludwig Boltzmann and Ernst Mach and their influence on their contemporaries, especially in Austria and Germany, cannot be overestimated. In the years around 1900 their opposing views in science and science theory dominated the specific debates in wide circles: philosophy of science (in Germany: *Naturphilosophie*), after having been suspect in the eyes of the scientific community during most parts of the nineteenth century, experienced a great renaissance. The discussions were certainly most intense in Vienna, where pro-Mach and pro-Boltzmann professors and students fought their battles. To many people the arguments of *both*, Mach *and* Boltzmann, seemed to be equally persuasive; thus they simultaneously adhered to Mach and Boltzmann. The physicists among them essentially believed in Boltzmann's physical theories, including atomism, and in Mach's methods of science theory.[137] Farther away from Vienna one generally did not share this typical Austrian compromise. For

[135] Occasionally Boltzmann's suicide has been partly attributed to his despair that atomic theory, to which he had essentially devoted his scientific career, was not universally accepted by the physics community. (For quotations, see Broda, 1955, pp. 26–27.) On the other hand, the health situation of the 62 year-old Boltzmann appears to have been bad enough to drive a sensible man to death: he suffered from melancholia and severe migraine headaches. In any case, we see no reason to refer *too much* in this respect to the debate on atomism; after all, Boltzmann paid hardly any attention to the decisive progress in proving the existence of atoms, like that obtained by Albert Einstein and Marian von Smoluchowski in the explanation of Brownian motion. (We have discussed this development in Volume 1, Section I.3.)

[136] Petzold and Dingler were the most important partisans of Mach's ideas in Germany. In 1911 they started a pro-Mach movement by preparing a public manifesto, which was signed by Albert Einstein, David Hilbert, Felix Klein, Georg Helm and Sigmund Freud. In 1912, in Berlin, Petzold and Dingler founded the *Gesellschaft für positivistische Philosophie*, which published its own journal (*Zeitschrift für positivistische Philosophie*) from 1913.

[137] One of these Mach and Boltzmann partisans was the physicist Heinrich Mache (1876–1954), who studied at the University of Vienna from 1894; another was the physicist Philipp Frank (1884–1966), a later graduate (1907) of the University of Vienna.

example, Max Planck of Berlin—in his lecture on '*Die Einheit des physikalischen Weltbildes*' ('The Unity of the Physical World Picture') on 8 December 1908 at Leyden—directed a harsh attack against 'a school of science philosophy which just at the moment, under the leadership of Ernst Mach, is very much in fashion, in particular in scientific circles' (Planck, 1909, p. 73). He argued that Mach's thesis, according to which 'all science is ultimately nothing but an economical fitting to our sensations, enforced on us by the struggle for survival' (Planck, 1909, p. 73), was disproved by the development of physics in the past decades, a development represented by personalities like Rudolf Clausius and Ludwig Boltzmann. The actual progress in physics and science, Planck claimed, consisted first of all in increasing unification of different branches of science; Boltzmann's statistical mechanics, for example, had resulted in uniting thermodynamics and mechanics; and Mach's opposition to atomic theory, the basis of statistical mechanics, would prevent—if adopted—the establishment of a unified picture of the world. Planck ended his talk at Leyden by pointing to the crucial test, which allowed one 'to distinguish false prophets from true ones: By their *fruits* shall ye know them' ('*die falschen Propheten von den wahren scheiden*: *An ihren Früchten sollt Ihr sie erkennen*,' Planck, 1909, p. 75).

Planck's statements did not remain unchallenged: a debate arose, in which partisans of Mach and Mach himself protested against unjustified simplifications.[138] With respect to atomism, Mach repeated his rejection. 'If belief in the reality of atoms is so crucial for you [i.e., for physicists],' he wrote, 'then I renounce the physical way of thinking; I will not be a professional physicist, and I hand back my scientific reputation. In short: thank you so much for the community of believers, but for me freedom of thought comes first' (Mach, 1910; Reprint, p. 603). Evidently this antiatomistic credo could hardly be accepted in 1910, when the atomic structure of matter had been established beyond doubt, not least by students of Mach's opponent Boltzmann.[139] Thus physics, even in Vienna, was dominated in the early years of the twentieth century by Boltzmann and not by Mach.[140] Of course, the physicists outside Vienna and Austria were

[138] Several physicists, including Friedrich Adler, Joseph Petzold, Philipp Frank and Albert Einstein, wrote papers and letters, in which they argued that Planck had misunderstood Mach's views. Mach himself answered, in detail, in an article entitled '*Leitgedanken meiner naturwissenschaftlichen Erkenntnislehre und ihre Aufnahme durch Zeitgenossen*' ('The Leading Thoughts of My Science theory and Their Reception by contemporaries,' Mach, 1910). He protested against Planck's 'polemical last remarks, their unusual form, and complete ignorance of the matter in dispute' (Mach, 1910; Reprint, p. 603). Planck immediately replied to Mach's article without, however, revising his earlier conclusions (Planck, 1910c).

[139] Boltzmann's students included the physico-chemists Svante Arrhenius (1859-1927) and Walther Nernst (1864-1941) (both came to his Graz institute in the mid-1880s); in Vienna they included Paul Ehrenfest (1880-1933), Fritz Hasenöhrl (1874-1915), Lise Meitner (1878-1968) and Stefan Meyer (1872-1949). Planck had meant with his 'by their *fruits* shall ye know them' this particularly fruitful and flourishing school of Boltzmann. Mach's disciples, on the other hand, were fewer in number and they gained a more modest reputation. The most prominent ones were Gustav Jaumann (1863-1924) and Georg Alexander Pick (1859-1929); both had been students in Prague and later professors of physics in Brünn and Prague, respectively.

[140] In Vienna Anton Lampa was called a partisan of Mach. Mach's influence in physics existed for some time mainly in Prague.

also, in general, supporters of Boltzmann's views; besides Arrhenius, Nernst, Planck and Sommerfeld, nearly the entire generation of physicists, chemists and mathematicians fell into this camp.[141] Apart from his anti-atomistic views, Mach enjoyed the support of a great physicist: Albert Einstein read his books carefully, especially *Die Mechanik* (Mach, 1883); and Mach's critique of Newton's concepts of mechanics offered therein helped to formulate the concepts of the new, relativistic mechanics. 'It was Mach who, in his *History of Mechanics* [1883] shook this dogmatic faith [in Newtonian concepts]; this book exercised a profound influence on me in this regard while I was a student,' Einstein wrote after nearly half a century (Einstein, 1949, p. 21). He further admitted then that he was even 'influenced very greatly' by 'Mach's epistemological position,' which he later found to be untenable because 'he [Mach] did not place in the correct light the essentially constructive and speculative nature of thought and more specifically of scientific thought' (Einstein, 1949, p. 21).[142] Independently of any change of judgement about Mach, certainly Einstein's statement, 'that those who consider themselves opponents of Mach, hardly know how much of Mach's views they have so-to-speak sucked with mother's milk' (Einstein, 1916a, p. 102), remained unaffected.

## I.2 The Young Erwin Schrödinger, His Youth and Study at the University of Vienna

In an address to the Physics and Mathematics Section of the Prussian Academy of Sciences, presented on 18 February 1932, the speaker discussed 'the question of how far the spirit of a cultural epoch is also expressed in exact science.' He stated in particular:

> Our civilization forms an organic whole. Those fortunate individuals who can devote their lives to the profession of scientific research are not merely botanists or physicists or chemists, as the case may be. They are men and they are children of their age. The scientist cannot shuffle off his mundane coil when he enters his laboratory or ascends the rostrum in his lecture hall. In the morning his leading interest in class

[141] The effect and support of the Göttingen mathematician Felix Klein, who had the atomistic view of matter presented in the *Encyklopädie der mathematischen Wissenschaften*, should not be forgotten in this context.

[142] As a consequence of this anti-speculative attitude Mach not only condemned the kinetic theory of matter but ultimately also the relativity theory. Thus he stated in the preface to *Die Prinzipien der physikalischen Optik*, dated July 1913: 'I must disclaim to be a forerunner of the relativists as I personally reject the atomistic doctrine of the present-day school or church. The reasons why, and the extent to which, I respect the present-day relativity theory—which I find to be growing more and more dogmatic—together with the particular reasons which have led me to such a view—considerations based on the physiology of the senses, epistemological doubts, and above all the insight resulting from my experiments—must remain to be treated in a sequel' (Mach, 1921, p. viii). For details of the Mach-Einstein relationship we refer to the book of Blackmore, especially Chapters 16 and 17 (Blackmore, 1972, pp. 247-287).

> or in the laboratory may be his research, but what was he doing the afternoon and evening before? He attends public meetings just as others do or he reads about them in the press. He cannot and does not wish to escape discussion of the mass of ideas that are constantly thronging into the foreground of public interest, especially in our day. Some scientists are lovers of music, some read novels and poetry, some frequent theatres. Some will be interested in painting and sculpture. And if one should believe that he could really escape the influence of the cinema, because he does not care for it, he is surely mistaken. For he cannot even walk along the street without paying attention to the pictures of cinema stars and advertisement tableaux. In short, we are all members of our cultural environment. (Schrödinger, 1932; English translation, pp. 98–99)[143]

The speaker, Erwin Schrödinger, was then close to forty-five years of age and at the height of his fame. For nearly five years he had occupied one of the most prestigious positions in the academic world, the chair of theoretical physics at the University of Berlin as the successor of Max Planck, who presided over the meeting of the Prussian Academy.

## Schrödinger's Family and School Education

Erwin Schrödinger was born in Vienna on 12 August 1887, the son of Rudolf Schrödinger who owned a linoleum business inherited from the family.[144] Although Rudolf Schrödinger ran his business successfully nearly until he died in December 1919, and thus secured a good income for his family, he did not have a great interest in it. 'He possessed an unusually wide education,' his son reported later, explaining: 'After studying chemistry he occupied himself for years intensively with Italian painting, besides being involved in drawing and etching landscapes. Then, finally, he turned to use the plant-box and the microscope, which resulted in a series of papers on the philogeny of plants' (Schrödinger, 1935, p. 86).

Rudolf Schrödinger had married the second of three daughters of his former teacher, Alexander Bauer, professor of chemistry at the *Technische Hochschule* in Vienna. Schrödinger described his mother as follows: 'My mother was very nice, with a cheerful character; she was of poor health and helpless towards life, but also unassuming' (Schrödinger, 1935, p. 86). However, she cared restlessly for her only child, who would always retain for her a tender memory and great

[143] See the announcement in *Sitzungsberichte der Preussischen Akademie der Wissenschaften* 1932, No. VI, p. 45, on the meeting of 18 February 1932. The text of the elaborated address was published, together with that of an earlier lecture—entitled "*Über Indeterminismus in der Physik*' and presented on 16 June 1931 at a philosophy congress in Berlin—in a book (Schrödinger, 1932).

[144] The Schrödinger family originated from the *Oberpfalz* in Bavaria and had settled in the Austrian capital for three or four generations. We obtained this and other information from Erwin Schrödinger's autobiographical sketch in *Les Prix Nobel en 1933* (Schrödinger, 1935).

Rudolf Schrödinger published scientific papers in botany in the *Abhandlungen* and in the *Verhandlungen der zoologisch-botanischen Gesellschaft in Wien.* (See the biographical sketch by Armin Herman, 1963.)

esteem. Her mother was British, and young Erwin visited England and his grandmother when he was a boy.[145] He learned English very early in life, and would later master this language nearly as well as his native German. Nevertheless, the decisive influence on Erwin's education was exerted by his father. 'For the growing son he was a friend, a teacher and an untiring discussion partner and a court of appeal for everything which might interest me,' recalled Schrödinger (Schrödinger, 1935, p. 86). Much later, after World War I, when Erwin wanted to marry and doubted whether he could support a family in an academic position, he asked his father: 'Couldn't you take me into your business?' Upon which Rudolf Schrödinger replied: 'No, my dear, you won't go into this business. I don't want you to do this. You stay at the university and you go on [with your academic career].'[146]

The young Schrödinger did not go to the elementary school; instead he was taught at home by a *Volkschullehrer* until the eleventh year. From 1898 to 1906 he attended the *Öffentliches Akademisches Gymnasium* in Vienna.[147] This *Gymnasium* was, of course, a humanistic one, concentrating heavily on the classical languages, Greek and Latin; however, science was also presented, at least partly, on a good level.[148] 'I was a good pupil, regardless of the subject; I loved mathematics and physics, but equally well the rigorous logic of the ancient grammars,' he once summarized his secondary school education, adding that he 'hated only memorizing "accidental" historical and biographical data and facts' (Schrödinger, 1935, p. 86). He further 'loved the German poets and writers, especially the dramatists, but hated the school-like dissection of their works' (Schrödinger, 1935, p. 86). A classmate, R. Heydner, recalled that Schrödinger always stood first ('Primus') in the *Gymnasium*, mentioning in particular:

> I cannot recall a single occasion that our *Primus* had ever failed to answer a question. We all knew that he really grasped everything already during the class hours, that he was no "plodder", no place hunter .... In physics and mathematics especially Schrödinger developed a perceptive faculty enabling him to pick up and work through all subjects of instruction instantaneously and even immediately, even without further study at home. Our Professor Neumann who taught the two fields during the last three *Gymnasium* classes could, after presenting a topic, instantaneously call Schrödinger to the blackboard and give him problems, which the latter solved with playful ease .... It was certain that Schrödinger spent his spare time in the afternoon preferably to study other subjects, not to work hard on the topics taught [at school]. In particular, he devoted himself to the study of the English

[145] See AHQP Interview with Mrs. Schrödinger, 5 April 1963 (AHQP, Tape 60a, p. 2).

[146] See AHQP Interview with Mrs. Schrödinger, pp. 1-2.

[147] See the autobiographical note, entitled '*Lebenslauf*' and dated 2 July 1938, AHQP Microfilm, No. 39, Section 1.

[148] Schrödinger later stated: 'To scientific disiplines fewer hours [than to the classical languages] were devoted; however, the instruction was excellent' (Schrödinger, 1935, p. 86). Among the notebooks of his school days, which he kept, were eight issues on physics containing the topics taught in the last two gymnasium classes.

> language, which was not taught—like French—in the Austrian *Gymnasiums*. Besides he did a lot of sports; he was an enthusiastic hiker and alpinist. His best friend was [Tonio] Rella, later [from 1932] professor of mathematics at the *Technische Hochschule* [in Vienna], who was fatally hit by a stray Russian bullet on the last day of war in April 1945. (Heydner, quoted in Hermann, 1963, pp. 174–175)[149]

Another remarkable quality of the student Schrödinger was his quickness and presence of mind. Thus Heydner remembered that his classmate worked secretly on some other topic during a school hour; then he reported: 'Suddenly Professor Haberl asked him [Schrödinger] something from the Greek Bible—as quick as lightening the student of the *Maturaklasse* adapted himself and gave a correct answer to the question' (Heydner in: Hermann, 1963, p. 175).

The love for drama made the student Schrödinger 'an addict of the famous *Burgtheater* in Vienna' (Heitler, 1961, p. 221). He especially adored Franz Grillparzer's plays.[150] He would keep his interest in literature to the end of his life, even publishing a book of poems of his own (Schrödinger, 1949). Being a talented linguist, moreover, he would speak 'an entire host of foreign languages ('*eine ganze Reihe von Fremdsprachen*'), presenting 'his lectures as he liked or found necessary in German, English, French or Spanish'; and he would 'translate Homer from the original [Greek] into English, or old-Provençal poems into German' (Thirring, 1947, p. 109). He inherited another artistic love from his father. As his friend and former fellow student Hans Thirring remarked on the occasion of his sixtieth birthday: 'He really possesses an understanding for old and modern painting, and enjoys himself during his leisure time with working on sculptures' (Thirring, 1947, p. 109). In contrast to his talents for literature and sculpturing, he showed little interest in music. 'No, he was not musical at all,' stated his widow, explaining: 'He could tell you whether it is good music or bad music, but he was not interested at all. His mother was very, very musical. . . . [She] wanted him to learn some instrument. A teacher suggested he should sing a tune which he [the teacher] was playing on the piano. He [i.e., Schrödinger] said: "I am not a piano, I can't sing!"' (Mrs. Schrödinger, AHQP Interview, pp. 2–3).

What Schrödinger enjoyed immensely at school was, besides mathematics and physics, the extended learning of the Greek and Latin languages. It provided him the access to classical literature and culture of Antiquity, especially to the philosophy of the Greeks. The earliest documentary evidence of this interest is a notebook, entitled '*Griechische Praeparationen*', which Schrödinger worked out during his last *Gymnasium* class (grade) and which contains a short, concise account of Greek philosophy from Thales of Miletus to Plato.[151] Schrödinger would return to the subject of the philosophy of the ancient Greeks again and

[149] Rella was born on 24 March 1888 in Brünn; he died on 8 April 1945 when the Soviet Army took Vienna.

[150] Schrödinger saved and collected, in a file, the theatre programmes of the plays he went to, and he wrote detailed notes on them. (See Hermann, 1963, p. 175.)

[151] The above-mentioned notebook has been filed on AHQP Microfilm, No. 30, Section 1.

again in his life, being attracted by its intimate and basic relation to the origin of science in Europe. Thus, for example, he would devote his *Shearman Lectures* delivered in May 1948 at University College, London, to demonstrate the continuation of the Greek philosophical traditions in modern science, including relativity and atomic theory. At the beginning of these lectures, which he published in an extended form as the book *Nature and the Greeks* (Schrödinger, 1954), he would state 'the motives for returning to ancient thoughts' as: 'That in passing the time with narratives about ancient Greek thinkers and with comments on their views I was *not* just following a recently acquired hobby of mine; that did not mean, from the professional point of view [of a theoretical physicist], a waste of time which ought to be relegated to the hours of leisure; that it was justified by the hope of some gain in understanding modern science and thus *inter alia* also modern physics' (Schrödinger, 1954, p. 1).

Schrödinger developed, as he remarked himself (and this has been testified by a classmate), during his *Gymnasium* years a particular liking for mathematics and physics. Consequently, when he finished school and obtained the *Maturat* in the summer of 1906, he decided to study these fields at the University of Vienna. Why, one may ask, did he not follow more closely the particular interests of his father, which lay in chemistry and botany? Although no straightforward answer exists to this question, some reasons can be given. First, the *Gymnasium* curriculum did not leave much time for science in general, and for chemistry and botany in particular. Second, according to his own testimony, Schrödinger was also attracted by the logical order represented in the grammar of the ancient languages; the same logical order existed *a fortiori* in mathematics and physics, while it could be discovered less in the more empirical and descriptive sciences of chemistry and biology. Third, the interest in the Greek philosophers, especially those of the Ionic school, opened the path to mathematics and physics rather than to the other sciences.

## Student at the University of Vienna

In the winter semester 1906/1907 Erwin Schrödinger enrolled in the University of Vienna, as a student of 'mainly physics and mathematics, and [as minors] of chemistry and astronomy' (*Lebenslauf*, see Footnote 147). To a large extent the topics he studied, and the professors whose lectures he attended, can be derived from Schrödinger's notebooks, kept in the *Nachlaß*. These include notebooks on the following lecture courses[152]: Viktor von Lang's course on experimental physics (*Experimentalphysik II*; see *Phys. Zs.* **7**, p. 686) and Franz Mertens' course on differential calculus (*Differential- und Integralrechnung*, *Phys. Zs.* **7**, p. 687) in the winter semester 1906/1907, Schrödinger's first semester at the university; Gustav

[152] The titles of Schrödinger's notebooks are given on AHQP Microfilm, No. 39, Section 1. We have compared these titles with those of the lecture courses, announced in the *Vorlesungsverzeichnis* of the University of Vienna, or in the one of the *Technische Hochschule* of Vienna, which were published in *Physikalische Zeitschrift* (Volumes 7-11 of the years 1906-1910).

von Escherich's course on probability theory (*Wahrscheinlichkeitstheorie, Phys. Zs.* **8**, p. 271), Gustav Kohn's course on synthetic geometry of space (*Synthetische Geometrie, Phys. Zs.* **8**, p. 271), plus the *Proseminar* course of the same lectures on spherical trignometry (*Übungen in mathemetischen Proseminar, Phys. Zs.* **8**, p. 271) in the summer semester of 1907; Emanuel Czuber's course on differential geometry[153], Escherich's course on function theory (*Einleitung in die Funktionentheorie, Phys. Zs.* **8**, p. 719), Kohn's course on analytic gemetry (*Analytische Geometrie, Phys. Zs.* **8**, p. 719), Mertens' course on algebra (*Algebra, Phys. Zs.* **8**. p. 719), Julius Ferdinand Hann's on meteorology (*Meteorologie, Phys. Zs.* **8**, p. 719), and Josef von Hepperger's course on spherical astronomy (*Sphärische Astronomie, Phys. Zs.* **8**, p. 719) in the winter semester 1907/1908; Mertens' courses on algebra (*Algebra, Fortesetzung*; *Phys. Zs.* **9**, p. 287) and on probability calculus (*Wahrscheinlichkeitsrechnung, Phys. Zs.* **9**, p. 287), Lothar von Schrutka's course on algebra (*Ausgewählte Kapital der höheren Algebra, Phys. Zs.* **9**, p. 287), and von Hepperger's course on the three-body problem (*Über das Problem der drei Körper, Phys. Zs.* **9**, p. 287) in the summer semester 1908; Anton Lampa's course on acoustics (*Akustik, Phys. Zs.* **9**, p. 695), Zdenko Hanns Skraup's course on general and inorganic chemistry (*Experimentalchemie I: Anorganische Chemie, Phys. Zs.* **9**, p. 695), Kohn's course on continuous groups (*Kontinuierliche Gruppen, Phys. Zs.* **9**, p. 695) and Wilhelm Wirtinger's course on function theory (*Funktionentheorie, Phys. Zs.* **9**, p. 695) in the winter semester 1908/1909; Skraup's course on organic chemistry (*Experimentalchemie II: Organische Chemie, Phys. Zs.* **10**, p. 294), and Wirtinger's course on function theory (*Funktionentheorie, Fortsetzung, Phys. Zs.* **10**, p. 295) in the summer semester 1909; Fritz Hasenöhrl's course on heat theory (*Wärmelehre, Phys. Zs.* **10**, p. 734), Escherich's course on definite integrals (*Bestimmte Integrale und Variationsrechnung, Phys. Zs.* **10**, p. 735), Joseph Nabl's course.on screw geometry (*Schraubengeometrie*)[154], Wirtinger's course on differential equations (*Differentialgleichungen, Phys. Zs.* **10**, p. 735) in the winter semester 1909/1910; Hasenöhrl's course on optics (*Optik, Phys. Zs.* **11**, p. 374), Kohn's course on algebraic curves (*Algebraische Kurven, Phys. Zs.* **11**, p. 375), and Wirtinger's course on mathematical statistics (*Mathematische Statistik, Phys. Zs.* **11**, p. 375), in the summer semester 1910, Schrödinger's last semester.[155]

[153] This course is not listed in the *Vorlesungsverzeichnis* for the winter semester 1907/1908, either of the university or the *Technische Hochschule* of Vienna, where Czuber was professor. However, he must have presented it at one of these institutions.

[154] Nabl's lecture course was not announced in the *Vorlesungsverzeichnis* reproduced in *Physikalische Zeitschrift 10* (1909).

[155] The corresponding notebooks, whose titles have been reproduced on AHQP Microfilm, No. 39, Section 1, are: Lang, *Experimentalphysik, Winter* 1906/1907 (four notebooks); Kohn, *Synthetische Geometrie, Winter* 1906/1907 (four notebooks); Mertens, *Infinitesimalrechnung, Winter* 1906/1907 (seven notebooks); Mertens, *Sphärische Geometrie*, undated; Escherich, *Wahrscheinlichkeitsrechnung, Sommer 1907*; Kohn, *Synthetische Geometrie des Raumes, Sommer 1907*; Mertens, *Integralrechnung, Sommer 1907* (two notebooks); Mertens, *Mathematisches Proseminer, sphärische Trigonometrie, Sommer 1907*; Czuber, *Differentialgeometrie, Winter 1907/1908* (two notebooks); Escherich, *Funktionentheorie, Winter 1907/1908* (three notebooks); Kohn, *Analytische Geometrie, Winter 1907/1908* (two notebooks); Mertens, *Algebra, Winter 1907/1908, Sommer 1908* (seven notebooks); Hann,

These notebooks certainly do not cover all the lecture courses which Schrödinger attended in his student years. He must have attended others as well, especially in physics: for example, several laboratory courses, or the courses of Hasenöhrl on topics other than heat theory and optics.[156] If we include those, the picture of Schrödinger's studies becomes fairly complete, agreeing with what one would expect from a dutiful student (and former *Primus* at school), who wishes to obtain a complete and thorough education in physics. The names of the lecturers are also fairly consistent with what Schrödinger recalled in 1938, when he stated explicitly: 'His teachers were, among others, Fritz Hasenöhrl (theoretical physics), Viktor von Lang and Franz Exner (experimental physics), Franz Mertens and Wilhelm Wirtinger (mathematics)' (*Lebenslauf*, cited in Footnote 147, p. 1). One may perhaps be surprised that no other notebooks have been kept on experimental physics, except the four on von Lang's course in the winter semester of 1906/1907; one especially misses those on the lectures of Franz Exner. The main reason for the absence of the latter is probably that Exner mainly directed the laboratory courses on experimental physics at the University of Vienna, which Schrödinger, of course, did attend.[157]

One of the mathematicians, whom Schrödinger did not forget, was Franz Mertens. As we have noticed above, he took several courses from him—on calculus, on algebra, on probability theory and on number theory. Mertens

---

*Meteorologie*, undated; Hepperger, *Sphärische Astronomie, Winter 1907/1908* (four notebooks); Mertens, *Wahrscheinlichkeitsrechnung, Sommer 1908*; Schruttka (!), *Algebra, Sommer 1908*; Hepperger, *Das Problem der drei Körper. Astrophysik, Sommer 1908* (two notebooks); Lampa, *Akustik, Winter 1908/1909* (two notebooks); Skraup, *Allgemeine und anorganische Chemie*, undated (five notebooks); Kohn, *Kontinuierliche Gruppen*, undated; Mertens, *Zahlentheorie, Winter 1908/1909* (four notebooks); Wirtinger, *Funktionentheorie, Winter 1908/1909* (four notebooks); *Skraup, Organische Chemie, Sommer 1909* (four notebooks); Wirtinger, *Funktionentheorie, Sommer 1909* (two notebooks); Hasenöhrl, *Wärmelehre, Winter 1909/1910* (two notebooks); Escherich, *Bestimmte Integrale, Winter 1909/1910* (two notebooks); Nabl, *Schraubengeometrie, Winter 1909/1910* (two notebooks); Wirtinger, *Differentialgleichungen, Winter 1909/1910* (three notebooks); Hasenöhrl, *Optik, Sommer 1910*; Kohn, *Algebraische Kurven, Sommer 1910*; Wirtinger, *Mathematische Statistik, Sommer 1910*. In addition the notebooks, Kohn, *Invarianten, Sommer 1910*, and [Philipp] Frank, *Integralgleichungen, Winter 1911/1912*, also contain notes on university courses attended by Schrödinger, while the other notebooks listed (*Höhere Mathematik, Darstellende Geometrie*, and Neumann, *Physik, vii* and *viii Kl*) contain notes of the *Gymnasium* years, or of courses taken by Schrödinger in his early period of studies. (For example, he might have attended the course of Emanuel Czuber, entitled *Grundlehren der höheren Mathematik* and presented in the winter semester 1906/1907 at the *Technische Hochschule*, Vienna—see *Phys. Zs.* **7**, p. 687—and the course of E. Müller, entitled *Darstellende Geometrie* and also presented in the winter semester 1906/1907 at the *Technische Hochschule—see Phys. Zs.* **7**, p. 687—in order to improve on his mathematical knowledge. The lecture courses on mathematical topics at the *Technische Hochschule* were on a more elementary level than those given at the University of Vienna.)

[156] Hasenöhrl lectured in the winter semester 1907/1908 on mechanics, in the summer semester 1908 on the mechanics of continua, in the winter semester 1908/1909 on electrodynamics, and in the winter semester 1909/1910 on heat theory II.

[157] For example, in the *Vorlesungsverzeichnis* we find under the name of Exner the *Physikalisches Praktikum* (winter semester 1907/1908—see *Phys. Zs.* **8**, p. 718—and summer semester 1908—see *Phys. Zs.* **9**, p. 286) or the *Physikalische Übungen für Vorgeschrittene* (winter semester 1908/1909—see *Phys. Zs.* **9**, p. 695).

obtained some reputation by a result in number theory, i.e., a hypothesis (later called Mertens' hypothesis) on the distribution of prime numbers.[158] The other mathematician mentioned by Schrödinger is even better known. Wilhelm Wirtinger was born on 19 July 1865 in Ybbs on the Danube and had studied at the University of Vienna, obtaining his doctorate in 1887 and—after post-doctoral studies in Berlin and Göttingen (with Felix Klein)—his *Habilitation* in 1890, also at the University of Vienna. Five years later he had been promoted *Extraordinarius*, then the same year he had been called to a mathematics chair at the University of Innsbruck, from where he returned to Vienna in 1905. Wirtinger's fame stemmed from a book on the theory of the theta-function based on the ideas of Bernhard Riemann and Felix Klein (1896), which developed from his Göttingen *Preisschrift*. He published further papers on various topics of geometry, algebra, number theory, theory of invariants, statistics, relativity theory and other problems of mathematical physics—e.g., on the theory of the rainbow—remaining active nearly to the end of his life.[159] Schrödinger attended several of Wirtinger's advanced lecture courses on function theory, differential equations and mathematical statistics; he also kept contact with his former teacher later on.[160]

Besides Mertens and Wirtinger, Schrödinger also took mathematical courses from Gustav Kohn (who died in 1922)—on geometry and group theory—Gustav von Escherich (1849-1935)—on probability theory, function theory and determined integrals—and Lothar von Schrutka, Edler von Reichenstamm (1881-1945).[161]

In his three minors (*Nebenfäscher*), astronomy, meteorology and chemistry, Schrödinger attended the courses of Joseph von Hepperger, Julius Hann and Zdenko Skraup. Von Hepperger had been *Extraordinarius* for astronomy and geology at the University of Graz before being appointed full professor at the University of Vienna in 1901; in 1909; in 1909 he also became director of the university observatory.[162] Even better known internationally were the other two

[158] Mertens' hypothesis stated that a function (the so-called Mertens' function), $M(x)$, which is the sum $\sum_{n \leqq x} \mu(n)$—with $\mu(n)$ the number of prime numbers up to $n$—is of the order of $O(\sqrt{x})$. Mertens retired from his chair at the University of Vienna in 1911 and died on 5 March 1927.

[159] Wirtinger died on 15 January 1945 in Ybbs. (Information on the biography and work of Wirtinger may be found in his obituary by Constantin Carathéodory: Carathéodory, 1948.)

[160] In Schrödinger's *Nachlaß* there exist two letters from Wirtinger to Schrödinger, dated 9 and 10 June 1926; they deal with a mathematical problem arising from wave mechanics, on which Schrödinger had asked Wirtinger's advice. (These letters have been filed on AHQP Microfilm, No. 41, Section 11.)

[161] When von Schrutka became *Privatdozent* at the University of Vienna in 1907 (later, in 1925, professor of mathematics at the *Technische Hochschule* of Vienna), Kohn and von Escherich were already senior professors. Kohn was occupied (at the time of Schrödinger's studies) in writing, together with Gino Loria of the University of Genoa, the article on '*Spezielle ebene algebraische Kurven*' ('Special Planar Algebraic Curves') for Volume III/2 of *Encyklopädie der mathematischen Wissenschaften* (completed in May 1908).

[162] Like Kohn, von Hepperger contributed an article to the *Encylopädie der mathematischen Wissenschaften*, namely '*Bahnbestimmung der Doppelsterne und Satelliten*' ('Orbit Determination of Double Stars and Satellites') in Volume VI/2, Part 1 (completed in December 1910).

senior professors. Of these Julius Ferdinand Hann was born on 23 March 1939 in *Schloß Haus*, near Linz; in 1877 he became professor of meteorology at the University of Vienna and director of the *Zentralanstalt für Meteorologie* (Central Institute for Meteorology).[163] After a short stay at the University of Graz from 1897 to 1900, he returned to Vienna. Because of his scientific publications and organizational work at the *Zentralanstalt*, Hann was recognized as the head of the outstanding Austrian school of meteorology in his time.[164] Zdenko Skraup, on the other hand, was born on 3 March 1850 in Prague; he had served since 1881 as professor of chemistry at the *Handelsakademie*, Vienna, then gone to Graz and finally, in 1906, was called to the University of Vienna. Skraup, who died on 10 September 1910 (in Vienna), obtained a good scientific reputation through his research in organic chemistry, notably on the properties of chinoline and alkaloids. In spite of his family relationship with chemistry—via his grandfather Alexander Bauer—Schrödinger would never in his later career show particular concern for chemical topics. However, he would be attracted, especially during the Vienna period, to problems of or connected with meteorology—such as atmospheric electricity.[165]

The central field of Schrödinger's university studies was undoubtedly physics, and his main teachers in this field were Professors Franz Exner, Fritz Hasenöhrl and Viktor von Lang. The most senior among them, von Lang (1838-1921), had been associated with the *Physikalisches Institut* of the University of Vienna nearly since its creation; he had been appointed in 1865 to the chair of Andreas von Ettinghausen, second director of the institute. However, since Josef Stefan had already taken over the direction of the institute, von Lang had 'no rooms of his own, no assistants and only extremely poor apparatus' (Benndorf, 1927, p. 398). This situation had improved slightly in 1875, when the *Physikalishes Institut* was moved to a new location in the *Türkenstraße* No. 3; there the small *Physikalisches Cabinet*, which had been conceded a little earlier to von Lang, and the *Chemisch-Physikalisches Institut* of Joseph Loschmidt, had obtained rooms for the next 37 years in a rented house. Hans Benndorf, who had been a student of Franz Exner

[163] The *Zentralstalt für Meteorologie und Erdmagnetismus* was founded in Vienna in 1851, its task being to establish and conduct observations of the entire Austrian Empire, ranging from Lombardia to Galicia and Vorarlberg to Transylvania. Later, due to the loss of northern Italy and to the autonomy of the Hungarian parts of the empire, the region of control shrank but not the reputation of the Vienna *Zentralanstalt*. In 1904 it was renamed *Zentralanstalt für Meteorlogie und Geodynamik* (Central Institute for Meteorology and Geodynamics), which implied an extension of its tasks. Apart from the interest of the Austrian government in meteorology there were also public and private interests; these resulted in 1865 in the creation of the *Österreichische Gesellschaft für Meteorlogie* (Austrian Society for Meteorology). The Austrian efforts received international recognition; thus the first international meteorological congress was held at Vienna in 1873. (We obtained this information from an article written on the 75th anniversary of the *Zentralanstalt*, written by Felix M. Exner, nephew of Franz Exner, then director: F. M. Exner, 1926.)

[164] Hann was co-founder and long-time editor of the *Meteorologische Zeitschrift*, the journal of the *Österreichische Gesellschaft für Meteorologie*. He died on 1 October 1921 in Vienna.

[165] Note also Schrödinger's teaching of meteorology at the *Flack-Offiziersschule* in Wiener Neustadt during World War I, which we shall discuss in Section I.4.

in the *Türkenstraße*, gave (several decades later) a lively description of Loschmidt's and later Exner's institute, which may be immediately taken over in general for the *Physikalisches Institut* and the *Physikalisches Cabinet.* He wrote:

> The institute consisted of three rooms plus an entrance room in each of the second, third and fourth floors [in America: third, fourth and fifth floors] lying in the court wing of the rented house in the *Türkenstraße.* By removing the wall between two of the rooms a lecture hall was created. It was not furnished with benches nor did it ascend to the lecturer's desk. Hence the students sat on chairs and had to take notes on their knees. The floor was a completely old, inlaid one through which gaping slits ran, in which even today uncountable amounts of mercury might rest. Each step made the entire room shake, hence every bit of sensitive apparatus had to be placed on a wall console fixed to the main walls of the house, but even the latter trembled when a strong wind blew outside or a truck passed by in the street. (Benndorf, 1927, p. 402)

We have reproduced here the details of Benndorf's report, because the situation in the physical institutes had not changed much when Schrödinger studied and worked there. Even the equipment may not have considerably improved over what Benndorf recalled, namely:

> When I became assistant to Exner in 1893, our most valuable instrument was an old mirror galvanometer with a ring magnet *à la* Plath; whenever it was made more astatic, we often noticed, after a few hours, a mysterious swinging forwards and backwards of the magnet. As was found out later, this always occurred when in the neighbouring house during the ironing of the laundry the iron was moved forwards and backwards. I just mention all this in order to give an impression of the difficult conditions under which experimental physics had to be carried out in Vienna at that time; one should know what immense amounts of work had to be wasted, then due to these insufficient conditions, before one is able to properly evaluate the results achieved. Some of the older colleagues might perhaps still remember with fright our institute on the occasion of the *Naturforscherversammlung* in 1894; Exner had then fled [from Vienna] because he was ashamed of his institute. (Benndorf, 1927, p. 402)

Again, what Benndorf said about Exner's institute, more or less also applied to the institutes of Stefan and von Lang. Of course, Exner and Stefan's successor Ludwig Boltzmann tried to improve the situation, but the big change came later, in 1913. Then the successor of von Lang—who retired in 1909—Ernst Lecher, could work in a more suitable and better equipped institute.[166]

Schrödinger attended the introductory courses on experimental physics taught by Viktor von Lang. His physics education, however, was influenced more decisively by a student of von Lang, Franz Exner, and a student of the latter, Fritz Hasenöhrl. Both were excellent representatives of their fields, namely Exner of experimental physics and Hasenöhrl of theoretical physics; and they continued, both in teaching and research, the meanwhile world-famous tradition of the Vienna school of physics.

[166] We shall describe the new physical institutes of the University of Vienna in Section I.3.

## Franz Exner and His School

Franz Serafin Exner was born in Vienna on 24 March 1849, where his father Franz Exner, formerly professor of philosophy at the University of Prague, had moved to in the previous year in order to assist Count Leo Thun in carrying out the reform of the Austrian educational system.[167] Franz, his three elder brothers and his sister lost their father very early and soon afterwards their mother; however, good friends of the family—like the philologist Hermann Bonitz, who had also been active in the university reform, and the physicist Christian Doppler—took care of their education.[168] After attending the *Gymnasium* in Vienna (1860–1867), where he was an average pupil with a strong interest in science, Franz Exner enrolled in the fall of 1867 at the University of Vienna to study physics. At that time Josef Stefan, Joseph Loschmidt and Viktor von Lang taught physics. Of these von Lang gave Exner his first instructions in doing his own research, but Loschmidt—whom he knew as a visitor in his father's home (see Exner, 1921, p. 178)—also left a strong impression upon him. While he had to perform military service during his second year of study, Franz Exner left Vienna in the third year for Zurich.[169] August Kundt (1839–1894) became his teacher there; and, after Kundt left in the spring of 1870 to assume a chair in Würzburg, Exner turned, during the following summer semester, mainly to studying chemistry with Johannes Wisclicennus (1835–1903). When the Franco-German War broke out in the fall of 1870, he went back to Vienna and completed his doctorate in July 1871. In the following fall he joined Kundt in Würzburg as a second, unpaid assistant.[170] In the spring of 1872 Kundt moved to Strasbourg, to the new German *Reichsuniversität*, taking Exner with him as his first assistant. Returning to the University of Vienna in November 1873, Exner got his *Habilitation* in the summer of 1874 and an assistantship with von Lang in the fall of 1874.[171] Simultaneously, he assumed the position of a lecturer at the *Hochschule*

[167] This reform was initiated before the 1848 Revolution in Vienna; by it the censorship of the Austrian universities was removed. Franz Exner contributed essential ideas to the reform. (See Benndorf, 1927, p. 397.)

[168] All the sons of Franz Exner pursued an academic career. Adolf Exner, the eldest son, became professor of Roman law in Zurich and later in Vienna—succeeding the famous Rudolf Ihering; Karl Exner became professor of theoretical physics in Innsbruck, while Siegmund Exner succeeded Ernst Brücke in the physiology chair at the University of Vienna. Finally, Marie von Frisch, *née* Exner, conducted a correspondence with the Swiss Poet and writer Gottfried Keller.

[169] In Zurich his brother Adolf was already a professor at the university.

[170] Kundt's main assistant in Würzburg was Wilhelm Conrad Röntgen, whom he had already converted in Zurich to studying physics.

[171] The Strasbourg period was scientifically very interesting for Exner, who communicated not only with Kundt and Röntgen, but also with Emil Warburg (1846–1931), then extraordinary professor in physics, and the mathematician Elwin Bruno Christoffel (1829–1900). However, he suffered from the absence of a rich cultural life in the city. (See Benndorf, 1927, p. 400.) This had been rather different in Zurich, where he had had via his brother access to eminent intellectuals, including Gottfried Keller and the architect Gottfried Semper. (See Benndorf, 1927, p. 399.)

*für Bodenkultur*, which not only provided extra money but also the welcome opportunity to lecture before a large audience. His professional career proceeded in 1879, when he was appointed *Extraordinarius*.[172] In 1885 he was elected corresponding member of the Imperial Academy of Sciences, and in 1891 he finally succeeded Joseph Loschmidt as professor of physics and director of the *Chemisch-physikalisches Institut* in the *Türkenstraße*.

The new *Ordinarius* developed a great and fruitful activity, both in research and in teaching. He installed new laboratory courses for advanced students, for physics teacher's candidates and for medical students. In particular he created his own, very successful school of experimental physics. Hans Benndorf, having been one of the students himself, characterized the Exner school as follows:

> As the shoots of grass push out after a warm rain in spring, after the installation of Exner [as full professor] the young physicists also pushed forth from the bottom. I just mention from that first period A. Lampa, J. Tuma, H. Benndorf, M. von Smoluchowski, St[efan] Meyer, E. von Schweidler, E. Haschek, F. Hasenöhrl, A. Szarvassi, H. Mache, V. Conrad, his nephew F. M. Exner—the later meteorologist—F. von Lerch and F. Ehrenhaft; then later followed E. Lohr, W. Schmidt—the meteorologist—F. Aigner, V. F. Hess, K. W. F. Kohlrausch, L. Flamm, Karl Przibram and E. Schrödinger. All these persons mentioned have become university professors; and perhaps nothing characterizes the efficacy of Exner better than the fact that all chairs for experimental physics in Austria are presently [in 1927] occupied by former assistants of Exner, with the exception of Exner's own chair, which is held by G. Jäger, a student of Stefan. This happened although Exner never moved a finger to establish his students anywhere. Many other good students have found positions, partly in industry and partly as secondary school teachers. (Benndorf, 1927, p. 402)

All this was achieved in the *Türkenstraße* with the restricted rooms and the limited equipment. Very late, in 1913, the situation improved considerably: then Exner's institute, which had been renamed in 1905 as the *Zweites Physikalisches Institut*—with Boltzmann's being the *Erstes Physikalisches Institut*—received adequate space within the new physics institutes. But the fortunate situation did not last long; in the following year World War I would reduce all scientific activities considerably. Soon after the end of the war Franz Exner retired from his chair; he died in Vienna 6 years later, on 15 November 1926.[173]

[172] In Vienna, Exner remained interested in cultural matters. Thus he went in 1876 on an extended trip to Greece, Asia Minor and Istanbul, following the guide of the ancient Greek author Pausanias. He frequently visited Italy to enjoy art and the southern landscape and to recover from personal hardships.

Exner married in the spring of 1877 the Viennese Auguste Bach; they had two daughters. but Auguste Exner died after a few happy years of marriage. His second wife was Friedericke Schuh, a friend from his youth, who cared for Exner with great devotion for the rest of his life.

[173] In completing Franz Exner's *curriculum vitae*, we should add that in the fall of 1907 he became *Rector Magnificus* of the University of Vienna.

In comparison with his importance as a powerful and extremely successful teacher, Exner's scientific work might not be remembered as much. Still, in his time, he was a skilful and gifted experimentalist who achieved many interesting results. Exner devoted his earliest publications to caloric researches, e.g., the determination of the temperature at which water assumes its highest density. For the *Habilitation* he wrote a thesis, entitled '*Über die Diffusion durch Flüssigkeitslamellen*' ('On the Diffusion through Lamellae of Fluids,' 1874). From 1877 to 1894 he concentrated on studies of problems of electrochemistry and developed the chemical interpretation of Galvanic processes. Later he shifted his interests to a wide range of topics in meteorology, spectroscopy and radioactivity. Exner became in particular a pioneer in the field of atmospheric electricity; he initiated and carried out systematic measurements of the electric field in the atmosphere under different weather conditions and at different locations.[174] Stimulated by the idea of determining the chemical composition of the meteorites contained in the large Vienna collection, Exner turned his attention to spectral analytic investigations: from 1895 onwards he developed, with his student Eduard Haschek (1875-1947), a method for fast measurement of wavelengths by enlarging photoplates with spectra on a white screen.[175] From 1902 to nearly until his death he investigated the consequences of the Young-Helmholtz three-colour theory, attempting to establish further its experimental foundation and defending it against attacks from other researchers.[176]

From the above summary we recognize Exner's great talents as an organizer for systematic studies in many fields; the detailed work he often passed on to his students, who would partly achieve more important results than their inspiring teacher. Still all these activities did not exhaust the work of Franz Exner. We must not forget his book *Vorlesungen über die physikalischen Grundlagen der Naturwissenschaften* (*Lectures on the Physical Foundations of Sciences:* Exner, 1919), which he wrote during World War I. This book, which reflected the wide horizon of the physicist Exner extending far beyond his professional views to philosophy and the development of civilization and culture, would exert a crucial influence on his student Erwin Schrödinger, as we shall discuss in Chapter II. After his retirement Exner began to work on a big treatise, to which he gave the title '*Vom Chaos zur Jetztzeit*' ('From the Chaos until the Present Time'); it would describe the history of the earth and of life on this planet, from the initial chaotic state to the following geological epochs, as well as the development of biological organisms, and finally the history of man and his culture from the beginning up

[174] Exner also suggested a theoretical explanation of the electric field existing in the atmosphere, which was later disproved by his students. We shall discuss the subject of atmospheric electricity in Section I.3.

[175] Exner and Haschek collected and reproduced about 100,000 lines in the ultraviolet region, measured to an accuracy of 0.01 Å units, in four volumes of wavelength tables (Exner and Haschek, 1902, 1904).

[176] We shall discuss the problem of colour theory in connection with the work of Erwin Schrödinger in the following chapter (Section II.3).

to World War I.[177] Exner conceived his new book in the spirit of Alexander von Humboldt's *Kosmos*, a work which he had long admired. He did not, however, proceed very far with this ambitious project.

As we have mentioned, Exner was a great teacher and a natural inspirer of his students. His institute formed a centre of intellectual exchange. Thus Hans Benndorf reported:

> A circle of similarly inclined friends, we surrounded our esteemed and beloved teacher. Late in the afternoon we assembled for tea around "Daddy." Then each of us had to report about his work and activity; there were no secrets and no priority claims, because everything belonged to everybody through an intensive exchange of ideas. Exner regarded the fear of intellectual theft as an indication of poverty. There we talked about God and the world; very frequently we discussed seriously, quarrelling about scientific problems. One could very well quarrel with him [Exner], because he admitted any contradiction, as long as it was to the point. He never let us feel his superiority, and he always behaved like a youngster among youngsters. (Benndorf, 1927, p. 408)

With these qualities it cannot be considered a miracle that Exner's students profited enormously from their teacher, and not only in scientific matters. Many of them tried later to continue the style of physical research and discussion. Thus the spirit of Exner lived for a long time in his students.

## Fritz Hasenöhrl and Theoretical Physics

Many of Exner's superb qualities were shared by one of his most outstanding students: Fritz Hasenöhrl. He was born on 30 November 1874, the son of Dr. Victor Hasenöhrl and his wife Gabriele, *née* Freiin Pidoll zu Quintenbach. His father was a state official, who made a name for himself through scientific publications[178]; his mother came from a family whose members had served Austria for generations in military affairs. The young Hasenöhrl was foreseen to choose a military career; thus he entered the *Theresianische Akademie* in 1884, a *Gymnasium* usually attended by officer candidates, as a boarding student.[179] He did very well at school, having a special preference for one of his teachers, namely Alois Höfler (1853–1922), a former student of the philosopher Franz Brentano.[180] Höfler would also submit the pupil Hasenöhrl's first scientific paper on a mathematical problem to the journal *Österreichische Mittelschule* for publication

[177] Since his youth Exner had also become interested in all aspects of human culture; we have mentioned his visits to Italy and Greece, and his participation in the intellectual circle in Zurich. Exner loved music very much; although he did not play any instrument himself, renowned musicians frequently met at his home to perform chamber music.

[178] Victor Hasenöhrl had the official titles of a *Regierungsrat, Hof- und Gerichtsadvokat.* (See Stefan Meyer's obituary of Hasenöhrl: Meyer, 1915, p. 430.)

[179] The director of the *Gymnasium* was Michael von Pidoll, an uncle of Fritz Hasenöhrl.

[180] In 1903 Höfler obtained a professorship at the University of Prague and in 1910 a chair at the University of Vienna. He became a pioneer of theoretical *Gestalt* psychology.

(Hasenöhrl, 1892). Hasenöhrl finished the last class in 1892 with distinction, being honoured as the best student of his school by the golden *Kaiserpreis-Medaille.* He then decided not to enter military service, however, and enrolled, in the fall of 1892, at the University of Vienna to study mathematics and physics. Again he excelled as a brilliant student; thus he soon published another scientific paper, arising from a problem given in the mathematics *Seminar* of Professor Leopold Gegenbauer, in the *Anzeiger* of the Imperial Academy of Sciences (Hasenöhrl, 1894). He interrupted his studies in the year 1894/95 in order to serve a *Freiwilligenjahr* (voluntary year) with the Austrian cavalry. On his return to Vienna, he came under the influence of Ludwig Boltzmann who had just succeeded Josef Stefan as director of the Physics Institute. Boltzmann converted Hasenöhrl's interests to problems of theoretical physics. Before long, from a seminar problem arose a paper on a mechanical analogue of the action of induction in electrical circuits (Hasenöhrl, 1896b). In spite of his early preference in his publications for mathematical and theoretical topics, Hasenöhrl selected experimental physics as a dissertation topic: under the supervision of Franz Exner he investigated in particular the temperature dependence of the dielectric constants of fluids (Hasenöhrl, 1896a).

On receiving his doctorate in 1897 he continued to perform both experimental and theoretical research.[181] Ludwig Boltzmann recommended Hasenöhrl to Heike Kamerlingh Onnes (1853-1926) for the position of an assistant at the Leyden low temperature laboratory; thus he spent a year, from the fall of 1898 to the fall of 1899, in Holland, obtaining valuable scientific experience and becoming friends with Kamerlingh Onnes and the theoretician Hendrik Antoon Lorentz (1853-1928). At that time Hasenöhrl was one of the few Austrians who went abroad early in their scientific careers.[182] Among his fellow students only Marian von Smoluchowski (a little earlier) and Friedrich von Lerch (a little later) shared this experience. We do not wish to argue that Hasenöhrl's other friends and colleagues, who remained at home—like Hans Benndorf, Eduard Haschek, Gustav Jäger, Heinrich Mache, Stefan Meyer and Egon von Schweidler—, made less progress in their scientific careers. But the fact remains, and it was deeply deplored by the great Boltzmann, that young Austrian scientists in general avoided leaving home and breathing an international scientific air. Back in Vienna Hasenöhrl got his *Habilitation* at the end of the year 1899.

The young Hasenöhrl soon won a reputation as an excellent teacher. Stefan Meyer would say later: 'The special care which he devoted to the preparation of his lectures assembled around him a host of faithful auditors (among others, [Paul] Ehrenfest, [Felix] Ehrenhaft and [Gustav] Herglotz); and it also characterized his entire later teaching' (Meyer, 1915, p. 431). His research in theoretical physics also won Hasenöhrl the recognition of the scientific community in Austria and abroad. Quite justly he obtained in 1905 an extraordinary professorship for

[181] For instance, he studied further the experimental topic of his doctoral dissertation.

[182] Previously, his teacher Franz Exner had done so.

'*allgemeine und technische Physik*' ('general and technical physics') at the *Technische Hochschule*, Vienna, where the senior Gustav Jäger held a full professorship in the same field. Two years later he even succeeded Ludwig Boltzmann in the theoretical chair at the University of Vienna.[183]

Hasenöhrl fully justified the expectations placed in him: he became a worthy and effective propagator of Boltzmann's ideas, and he even surpassed his former teacher in creating a flourishing school of theoretical physics in Vienna. Thus he counted among his students Ludwig Flamm, Karl Herzfeld, Friedrich Kottler, Erwin Schrödinger and Hans Thirring. But Hasenöhrl was also a wonderful human being, whom Stefan Meyer recalled as 'always cheerful and merry on social occasions, full of original humour and, if at times becoming sarcastic, never hurting, always helpful and filled with natural kindness' (Meyer, 1915, p. 432).[184] An active member of the *Akademischer Verein deutscher Mathematiker*, he especially enjoyed—being a skilled skier and mountaineer—its excursions to the Alps.[185]

Hasenöhrl's scientific *Oeuvre* is small, compared to that of his teachers Exner and Boltzmann.[186] At the beginning of his career—after the two mathematical papers mentioned earlier—experimental problems dominated his work, especially on the temperature dependence of the dielectric constants of fluids and solids. The main result of these investigations, which he carried out at Exner's institute in Vienna and at Kamerlingh Onnes' laboratory in Leyden—see, e.g., the paper on the dielectric constant of fluid oxygen (Hasenöhrl, 1899a)—was that the formula of Rudolf Clausius and Ottaviano Mosotti could be confirmed over a wide range of temperature for a variety of different substances. After his first theoretical paper on a mechanical analogue of a problem of electrical circuits (Hasenöhrl, 1896b), we find an interruption in Hasenöhrl's output in theory. However, his *Habilitation* thesis was devoted to a problem of potential theory (Hasenöhrl, 1899b); and 3 years later Hasenöhrl started publishing a series of theoretical papers devoted to problems of electromagnetic wave theory, which established his reputation as a theoretical physicist in the scientific world. In 1904 he presented three contributions to the Vienna Academy of Sciences, all dealing with moving bodies and the radiation emitted and absorbed by them (Hasenöhrl, 1904b, c, d). The last of these papers (Hasenöhrl, 1904d, see also the summary of Hasenöhrl, 1904e) won him the *Haitingerpreis* of the Academy for the year 1905, which was justified with the following words: 'He has derived

[183] Hasenöhrl also carefully edited the scientific papers of Boltzmann in three volumes (Boltzmann, 1909a, b, c).

[184] Kamerlingh Onnes had already acknowledged the amiability and social qualities of his young Austrian collaborator, recalling many wonderful hours with Hasenöhrl and his wife Ella, *née* Brückner—whom Hasenöhrl had married in the spring of 1899. Hasenöhrl's marriage turned out to be very harmonious: they had a son and a daughter. (See Meyer, 1915, pp. 430–431.)

[185] Hasenöhrl also took a great interest in questions of geology.

[186] Stefan Meyer added a fairly complete list of Hasenöhrl's scientific papers to the latter's obituary, containing 38 items.

in it the very remarkable law that the pressure forces occurring in the cavities of a body, which is completely opaque with respect to radiation and performs a [uniform] motion relative to the ether of light, result in an apparent increase of the mass of the body under consideration' (Meyer, 1915, p. 431).

Hasenöhrl's result was indeed remarkable and unexpected.[187] However, little more than a year later, in the summer of 1905, Albert Einstein arrived at a similar conclusion, based on quite different arguments. Then he stated quite generally: 'The mass of a body is a measure for the energy content of the latter; if the energy alters by an amount $L$, the mass will change in the same sense by $L/9 \times 10^{20}$ [i.e., $L/c^2$, $c$ being the velocity of light *in vacuo*], provided the energy is measured in ergs and the mass in grams' (Einstein, 1905e, p. 641). This is the famous energy-mass relation, which Einstein—like Hasenöhrl's result—derived only in an approximation, neglecting terms of higher order than $v^2/c^2$, with $v$ the velocity of the body. Hasenöhrl did not get the general equivalence of mass and energy but a more restricted result: namely the equivalence between what was then called the 'electromagnetic mass' of the body and its energy.[188] In spite of this restriction and the pre-relativistic language he used—he talked about an 'apparent' increase of mass rather than of the 'real' increase, as did Einstein in 1905—Hasenöhrl might be counted among the founding fathers of Special Relativity Theory.[189]

Hasenöhrl followed these truly pioneering communications with further papers on the theory of radiation in moving cavities (1907a) and on the electrodynamics of moving systems (1907b, 1908). Then he participated in the discussion on the blackbody radiation; in particular, he investigated the possibilty of obtaining Planck's law (Planck, 1900f) from the laws of electron theory and classical statistical mechanics (Hasenöhrl, 1909, 1910). He thus became engaged in the most difficult theoretical problem that had arisen during the first decade of the twentieth century. On account of his deep and original contributions to theoretical physics he was elected corresponding member of the *Kaiserliche Akademie der Wissenschaften* in 1911; he was further invited to present one of the main reports on the foundation of kinetic theory at the 83rd *Naturforscherversammlung*, Karlsruhe, in September 1911 (Hasenöhrl, 1911). He also attended the first two Solvay Conferences in Physics in 1911 and 1913, being one of the few selected participants.

[187] Still the result had already been obtained more than two decades earlier by Joseph John Thomson (1858-1940) for a charged spherical conductor (Thomson, 1881).

[188] Further Hasenöhrl, like Thomson before him, obtained the relation $\Delta m \cdot c^2 = \frac{4}{3}$ times the electromagnetic energy. In 1900 Henri Poincaré (1854-1912) suggested that electromagnetic energy might possess mass density equal to $(1/c^2)$ times energy density (Poincaré, 1900).

[189] Philipp Lenard would later even claim that Hasenöhrl's result of 1904 contained all the useful information needed to describe the motion of fast electrons, and that Relativity Theory constituted nothing but an unnecessary, wrong theoretical extension. (See Lenard and Becker, 1927, especially pp. 401-402 and footnote 11 on p. 402.)

## Felix Ehrenhaft and Other Viennese Experimentalists: The Search for Subelectrons

When Schrödinger began to study in the winter semester of 1906/07, Ludwig Boltzmann had just committed suicide.[190] Some of his students had left the University of Vienna, like Paul Ehrenfest and Lise Meitner[191]; or, like Philipp Frank and Ludwig Flamm, they completed their studies with other professors. Still Boltzmann's influence remained very much alive. As Schrödinger recalled nearly a quarter of a century later:

> The old Vienna institute, which had just mourned the tragic loss of Ludwig Boltzmann, where Fritz Hasenöhrl and Franz Exner worked and where many other students of Boltzmann came and went, provided me with a direct insight into the ideas of that powerful mind. His world of ideas may be called my first love in science. No other personality has ever since thus enraptured me or will do so in future. (Schrödinger, 1929b, p. C)

Boltzmann's world of ideas, which caught the imagination of the young Schrödinger, implied of course the central point of the atomic structure of matter. Fritz Hasenöhrl, Boltzmann's former student, certainly advocated the atomic hypothesis firmly. Franz Exner, the friend and successor of Joseph Loschmidt, also supported the hypothesis, although one may not find among his investigations an important study concerned with it. However, his student Marian von Smoluchowski (1872-1917) filled the gap perfectly with his thorough and detailed experimental and theoretical studies of the phenomenon of Brownian motion (Smoluchowski, 1906), which he—like Einstein independently a little earlier (Einstein, 1905c, 1906b)—interpreted as an effect of fluctuations on the basis of the kinetic theory of matter.[192] Einstein and Smoluchowski thus provided the decisive work for demonstrating the existence of atoms.[193] As Sommerfeld wrote in 1944 (on the occasion of the hundredth anniversary of Boltzmann's birth): 'It

[190] Boltzmann delivered his last lecture course in theoretical physics in the winter semester of 1905/1906. He did not feel very well, as Ludwig Flamm reported many years later: 'I myself, as a student, was able to hear the last lecture [course] which Boltzmann held in theoretical physics; it was in the winter semester of 1905/1906. A nervous complaint [headaches] prevented him from continuing his teaching activity. Together with another student I took and passed my oral examination in his villa in Währing. On leaving, after the examination was over, we heard from the front hall his heartrending groans' (Flamm, 1944, p. 30; English translation in Blackmore, 1972, p. 212).

[191] Ehrenfest moved with his Russian-born wife Tatyana to Göttingen and later to St. Petersburg (see Volume 1, Section II.3). Lise Meitner (1878-1968) served, after receiving her doctorate first, as a teacher in a girls' secondary school in Vienna, then she went on to Berlin to pursue a scientific career there.

[192] Felix M. Exner, nephew and student of Franz Exner, also contributed an experimental investigation on Brownian motion.

[193] In 1904 Smoluchowski and Einstein had already been the first to consider fluctuation effects in the kinetic theory of matter; Smoluchowski estimated density fluctuations in gases (Smoluchowski, 1904), while Einstein dealt with energy fluctuations in blackbody radiation (Einstein, 1904). We have discussed the role of fluctuations in establishing a proof of the atomic hypothesis (see Volume 1, Section I.3).

is a pity that he [Boltzmann] was no longer able put together the final proof of atomism by a theoretical explanation of Brownian motion, which was done shortly before his death .... This explanation also persuaded [Wilhelm] Ostwald, as he told me from time to time, to accept atomism' (Sommerfeld, 1944, p. 26). That is, in the first decade of the twentieth century Boltzmann's main opponent at the Lübeck *Naturforscherversammlung* not only acknowledged his conversion to atomism, but also Boltzmann's genius as 'a man who was superior to us all in intelligence, and in the clarity of his science' (quoted in Feyerabend, 1967, p. 334).[194] Thus the last pillar of antiatomism to be persuaded, was Ernst Mach; and in his, at least partial, conversion another student of Exner's played a role, as we shall report below.

It might be of interest to call attention to the fact that one of Exner's students, Erwin Lohr (1880–1951), turned against atomism after having joined the Mach supporter Gustav Jaumann in Brünn in October 1905.[195] But he cannot be called a representative of the large host of physicists that emerged from Exner's institute; indeed, none of the many other, later prominent, students there followed his example. The senior-most among them, Hans Benndorf, had already left Vienna before Schrödinger began to study, accepting in 1904 a professorship at the University of Graz.[196] Also Heinrich Mache was in Graz from 1906 in the position of an *Extraordinarius*; however, two years later he returned to Vienna to take over Hasenöhrl's former extraordinary professorship at the *Technische Hochschule*.[197] Three other students and collaborators of Exner, all being about the same age, were around during most of Schrödinger's student time: Felix Ehren-

[194] In his biography of Ernst Mach, John Blackmore argued that the story of Ostwald's acceptance of the atomic hypothesis was more complicated than might seem from the above-quoted remarks; Ostwald really did not give up *Energetik* when he agreed to accept the existence of atoms in the physical description. (See Blackmore, 1972, p. 217.)

[195] Lohr, after studying physics, mathematics and philosophy at the universities of Graz (1899–1901) and Vienna (1902–1904), obtained his doctorate with a dissertation on the electrical conductivity of sodium. Then he spent a year with Joseph John Thomson at the Cavendish Laboratory in Cambridge (1904/1905). In Brünn he received his *Habilitation* in 1908; he was promoted to *Extraordinarius* (in 1919) and a full professor (in 1924). In April 1945 he had to leave Brünn and retire to an estate in the Tennengebirge near Salzburg, where he died on 23 October 1951.

In the scientific community Lohr became known as the one who supported and extended Jaumann's electrodynamical theory. Jaumann did not deny the existence of stable configurations of mass and electricity, such as atoms and electrons; however, he and Lohr tried to derive such configurations on the basis of a continuum theory. Jaumann and Lohr also collaborated later on the thermodynamics of irreversible processes and on a continuum theory of X-ray diffraction. (See Lohr's obituary by Clemens Schaefer: Schaefer, 1952.)

[196] Schrödinger would become his colleague in Graz much later, i.e., in 1936.

Hans Benndorf was born on 13 December 1870 in Zurich. He studied physics with Franz Exner at the University of Vienna, where he received his doctorate (in 1896) and his *Habilitation* (in 1899). In 1904 he was appointed *Extraordinarius* at the University of Graz; in 1910 he was promoted to a full professorship. He retired in 1936, but took over the Graz Institute again in May 1945. Two years later he finished his academic career. He died in Graz on 11 February 1953.

[197] Heinrich Mache was born on 27 April 1876 in Prague and had studied physics and astronomy since 1894 with Boltzmann, Exner, Mach and von Hepperger. In 1901 he became *Privatdozent* at the University of Vienna. Later, in 1911, he would become *Ordinarius* at the *Technische Hochschule*, Vienna, and stay there until his retirement in 1946. He died in Vienna on 1 September 1954. (See his obituary by Ludwig Flamm: Flamm, 1955.)

haft, Friedrich von Lerch and Karl Przibram. Of these the first one became perhaps the best-known scientist in the international community, because of his experimental skill and of his stubborn, lonely stand for the existence of subelectronic charges. Since these investigations fell into the period when Schrödinger began scientific research, and were much debated at that time, it may be worth while to give some details here.

Felix Ehrenhaft, who was born on 28 April 1879 in Vienna, studied at the university there. After his doctorate (1903) he had been Viktor von Lang's assistant and obtained his *Habilitation* in late 1905.[198] Ehrenhaft's early researches were devoted to the physical properties, especially the optical behaviour of metallic colloids (Ehrenhaft, 1903).[199] They provided him with a natural entry into the investigations on Brownian motion. In 1907 he succeeded, for example, in confirming the theory of Einstein and Smoluchowski; that is, he compared his experimentally determined values for the average square of the position shifts, which microscopic particles suspended in gases undergo, with the formula of Smoluchowski, finding agreement in the order of magnitude (Ehrenhaft, 1907). The researches on Brownian motion won Ehrenhaft the *Lieben* prize of the Vienna Academy of Sciences; they also led him more or less continuously to that research topic, which would occupy him for the rest of his scientific career, notably the problem of determining the smallest electric charge existing in nature. In a paper, submitted in March 1909 to the Vienna Academy, he first suggested the possibility of observing the motion of the individual, charged microscopic metal particles in gases (Ehrenhaft, 1909a). He reported in the same paper some preliminary results, which he further described in a following note, submitted to *Physikalische Zeitschrift* (Ehrenhaft, 1909b). He stated then: 'To sum up, we conclude *the existence of large ions, having mass* $10^{-12}$[g] *and bearing charges of both signs. Their mechanical mass can be determined* by microscopic observation of their *falling motion in the gravitational field of the earth*, or by direct microscopic measurement [of their size]; and their *electric charge can be determined by a microscopic measurement of their mobility in an electrostatic field. Their charge comes out to be identical with that of a single-valued ion*, for which the values $e = \{4.46 \times 10^{-10}, 4.51 \times 10^{-10}, 4.68 \times 10^{-10}$ electrostatic units$\}$ were obtained' (Ehrenhaft, 1909b, p. 310).[200] Ehrenhaft immediately planned to refine his

[198] In 1912 Ehrenhaft was promoted to an extraordinary professorship, and in 1920 to a full professorship at the University of Vienna. After the incorporation of Austria into the *Deutsches Reich* in 1938, he lost the Viennese chair. He first went to England and later emigrated to the United States, becoming a U.S. citizen. In 1947 he returned to Vienna as a guest professor; he took over the direction of the *Erstes Physikalisches Institut*. He died on 4 March 1952 in Vienna. (For details, see his obituary by Hans Thirring: Thirring, 1952.)

[199] Ehrenhaft's topic fitted perfectly into the Viennese scientific scene, where one pioneer of colloid science—Wolfgang Joseph Pauli—was active, and the other—Richard Zsigmondy—had originated. Ehrenhaft in his 1903 paper, to which we have referred above, quoted some work of Zsigmondy. (See Ehrenhaft, 1903, p. 510, footnote.)

[200] Ehrenhaft explicitly noted the agreement of his value for the elementary charge with the one derived by Max Planck from blackbody radiation (Planck, 1900f), and with half of the value of the charge that Ernest Rutherford and Hans Geiger had found for alpha-particles (Rutherford and Geiger, 1908).

observational methods by introducing microphotographic techniques using ultraviolet light and also by extending his investigations to observe the motion of microscopic particles suspended in fluids. The new experiments yielded an unexpected result.

In a detailed paper, presented in May 1910 to the Vienna Academy, Ehrenhaft arrived at the following conclusion: 'The author would like to tend to the opinion that . . . the *charge measurements carried out for single particles seem to show the existence of electric charges in nature smaller than the charge of the single-valued hydrogen atom or electron, which have been thought to be indivisible*' (Ehrenhaft, 1910a, p. 861).

The claim that subelectron charges occurred in nature caused considerable unrest among physicists, who learned about it from Ehrenhaft's article in the widely read *Physikalische Zeitschrift*, which appeared in the issue of 15 July (Ehrenhaft, 1910b). The same issue contained, immediately behind that paper, a short note from Karl Przibram, who confirmed the conclusions of his Viennese colleague (Przibram, 1910a).[201] At the 82nd *Naturforscherversammlung* in Königsberg, during the afternoon meeting of the physics section on 19 September 1910, Ehrenhaft delivered a detailed report on the status of the elementary charge problem. He argued, in particular, that only the observation of *single* particles would allow one to decide the question 'whether all [microscopic] particles carry the same charge, or whether *the particles emitted from radioactive substances do not hint as well at some fluctuation of the charge transported by single particles*' (Ehrenhaft, 1910c, p. 942).[202] After describing his own method in detail, as well as the results, he emphasized the conclusion:

> Presently the author believes that in nature there have to exist, besides the already known charges—which are close to the values known for single and multi-valued ions—still other values of the charge between these; these must be assumed to show up in nature, as well as quanta of electricity which are just a fraction of the electron's charge, although the latter has been assumed to be indivisible. According to the present state of his investigation it appears [to the author] that an indivisible atom of electricity, having the order of magnitude of $10^{-10}$ absolute electrostatic units, cannot be assumed to exist in nature. (Ehrenhaft, 1910c, p. 948)

Ehrenhaft's presentation at the *Naturforscherversammlung* was followed by a lively and extended discussion, in which the speakers (including Arnold Sommerfeld, Edgar Meyer, Max Born and Max Planck) suggested possible sources of

[201] Przibram had observed, in a slightly different set-up, the fall of fog droplets (formed by the action of phosphorus in moist air) in a homogeneous electric field of variable strength. In an addendum to the proofs the author remarked, 'that, in agreement with the conclusions of Ehrenhaft, the errors of observation do not provide a sufficient explanation for the values [of the electric charge] which have been found to lie between the accumulation points [i.e., integral multiples of the charge of the electron or the positive hydrogen ion]' (Przibram, 1910a, p. 632).

[202] Ehrenhaft pointed out earlier in his talk, that the previous experimental determinations of the elementary charge—from J. J. Thomson's before 1900 to Harold A. Wilson's in 1903 to Ernest Rutherford and Hans Geiger's in 1908—had in common the fact that the desired value of the elementary charge was obtained as the average over many particles.

error which might have led to apparent fractional charges.[203] This debate on the validity of Ehrenhaft's conclusion went on for years, with numerous physicists participating, especially in Austria. For example, Karl Przibram started systematic investigations, searching for all possible corrections and sources of error in the experiments of Ehrenhaft and himself on the determination of the elementary charge.[204] He, as well as Anton Mayer—a student of Benndorf in Graz—removed *some, but not all*, of the discrepancies between Ehrenhaft's results of 1910 and the assumption that the electron's charge was the smallest one existing in nature and that only integral multiples of it occurred (Mayer, 1912). The most powerful experimentalist, who competed in the problem of determining the elementary charge, was the American Robert Andrews Millikan (1868-1953). He developed the very same experimental methods as Ehrenhaft, simultaneously with and independently of the latter, in 1909. When he gave the first complete account of his results in June 1913, he reported only an accurate value for the elementary charge without referring to any observed deviations from it. However, in a later paper, entitled 'The Existence of a Subelectron,' he explicitly dealt with the problem, arriving at the conclusion that 'all of Professor Ehrenhaft's results are easily explained on the assumption of incorrect assumptions as to the density and sphericity of his particles, but even if these assumptions are correct, his results have yet no bearing on the existence of a subelectron' (Millikan, 1916d, p. 625).[205] Six years later Millikan would receive the Nobel prize in physics 'for his work on the elementary charge of electricity and on the photoelectric effect' (*Nobel Lectures*: *Physics 1922-1941*, Amsterdam, Elsevier, 1965, p. 49). His

[203] Sommerfeld said that, if the microscopic particles observed by Ehrenhaft did not have spherical but elliptical shape, the charges derived (on the assumption of spherical shape) would turn out to be smaller than the real elementary charge; Heinrich Siedentopf supported this explanation, but in a note added in proof to his paper he admitted that, after having seen Ehrenhaft's experiment, the particles did indeed possess an approximately spherical shape. Meyer mentioned the possibility that the particles could change their charge during their falling down; they might be neutralized, and therefore their observed average motion suggested a smaller average charge. Born expressed the opinion that the Brownian motion might have a decisive influence, while the experimentalist Walther Kaufmann pointed to fluctuations in the density of the metallic particles as a possible source of error. Finally, Planck questioned the applicability of Gabriel Stokes' formula to microscopic particles; Ehrenhaft immediately replied that his own and Przibram's investigations so far were consistent with Stokes' law.

[204] Przibram presented reports to the Vienna Academy of Sciences from time to time (e.g., Przibram, 1910b, 1912a).

[205] Millikan repeated very much the arguments already presented at the *Naturforscherversammlung* in Königsberg. Especially, he claimed that Ehrenhaft's subelectronic charges were caused by 'the existence of a multiple relationship between the charges carried by a given particle in conjunction with my direct proof, extended in this paper, that the apparent value of [the elementary charge] $e$ is not in general even with mercury droplets a function of the radius of the drop on which it is found' (Millikan, 1916d, p. 625). Millikan further accused Ehrenhaft of having copied his experimental set-up, an accusation which is difficult to justify. Gerald Holton has treated the Ehrenhaft-Millikan debate in detail (see 'Subelectrons, presuppositions, and the Millikan-Ehrenhaft dispute' in G. Holton: *The Scientific Imagination: Case Studies*, Cambridge University Press, Cambridge, U.K., New York, U.S.A., 1978, pp. 25-83).

opponent Ehrenhaft, however, would continue to fight for subelectrons for years to come.[206]

As we have mentioned, besides Ehrenhaft, another Viennese physicist became involved in the problem of the elementary quantum of charge, namely Karl Przibram.[207] He was born on 21 December 1878 and studied physics and mathematics at the University of Graz, obtaining his doctorate in 1901. He became acquainted with investigating electric charges especially while staying with Joseph John Thomson at the Cavendish Laboratory in Cambridge (1901–1903); and he continued experimental researches on gas discharges at the University of Vienna, where he became *Privatdozent* (1905). In 1907 he studied the condensation of alcohol vapours in air ionized by X-rays, measuring the electric charges of the condensated droplets and finding values between 1.4 and $6.2 \times 10^{-10}$ electrostatic units (Przibram, 1907). This work then started a series of investigations on the elementary charge, which extended over the subsequent years. In late 1912 he submitted a paper dealing with the Brownian motion of non-spherical particles; especially, he extended the formulae for the spatial and angular fluctuations of spherical particles in Einstein's and Smoluchowski's theory to non-spherical particles and tested the formulae thus obtained in experiments with dead bacillae (Przibram, 1912a).[208] He also turned his interests to problems of radioactivity, becoming associated with the *Institut für Radiumforschung* since its foundation.[209]

Friedrich von Lerch, who was born on 30 May 1878 in Pressburg (now Bratislawa, Czechoslovakia) and studied at the universities of Prague and Vienna, approached the subject of radioactivity even earlier than Przibam. It happened thus: After receiving his doctorate in 1901, he went to Göttingen and became private assistant to Walther Nernst (1903/1904). While working there on various subjects, he was stimulated by Nernst to study radioactive phenomena: thus he started to analyze the radioactive deposit of thorium—what one then called the

[206] As late as 1938, in a review of the problem, Ehrenhaft stated: 'The author believes that he has succeeded in showing, in this review on the development of the problem of charges carried by single microscopic or submicroscopic spheres, *that in every check charges always occur which are smaller than the electron's, and that in particular also the direct microscopic determination of the size of the particles has provided a renewed proof for the correctness of the prerequisites of the* [author's] *method, and also for the real existence of charges smaller than that of the electron*' (Ehrenhaft, 1938, p. 685).

[207] Przibram's further investigations of the problem, however, did not fully support those of his colleague Ehrenhaft. Thus Millikan reported that Przibram wrote to him in 1912 that 'although his [Przibram's] work in this field had "commenced with such a grievous mistake [as assuming the existence of subelectronic charges in 1910] it was now [in 1912] in good agreement" with mine [i.e., Millikan's preliminary results on the value of the elementary charge]' (Millikan, 1916d, p. 604).

[208] Przibram improved his approach in a second paper, finding rather good agreement between experiment and theory (Przibram, 1913).

[209] Przibram concerned himself especially with the luminescence phenomena and colour changes caused by the influence of radioactive rays on substances.

Przibram obtained in 1927 an extraordinary professorship at the University of Vienna. He lost this position in 1938 and emigrated to Brussels (1940). In 1946 he returned to Vienna and took over the *Zweites Physikalisches Institut* of the university (1947–1950). He retired in 1951 and died on 10 October 1973 in Vienna. (For further information see the article by Berta Karlik and Erich Schmid on Przibram's eightieth birthday: Karlik and Schmid, 1959.)

'induced' activity of thorium—developing a suitable experimental method by chemically dissolving the deposit on different metals in acids (Lerch, 1903). After his return to Vienna in October 1904—von Lerch became assistant to Franz Exner—he continued to do research on radioactivity, succeeding, for example, in 1905 in separating the thorium deposit into two substances—then called Th A and Th B (today renamed Th B and Th C)—and determining their half-lives (Lerch, 1905).[210] Von Lerch was on the way to becoming an expert on radioactivity; unfortunately, however, he was called to an extraordinary professorship at the University of Innsbruck in 1908 (full professor, 1914-1946), where he devoted himself primarily to teaching and the education of physics students. (He died on 19 December 1947.) In Vienna, on the other hand, radioactivity had already found a first rate treatment in the work of Stefan Meyer and Egon von Schweidler.

## Two Pioneers of Radioactivity: Stefan Meyer and Egon von Schweidler

Stefan Meyer and Egon von Schweidler were about the same age, being separated by less than a year. The slightly senior Meyer was born on 27 April 1872, the son of a wealthy Jewish bourgeois family.[211] He attended the *Gymnasium* in Vienna and in Horn, Lower Austria, graduating in 1892; after a voluntary year of military service with the artillery, he enrolled at the University of Vienna to study physics and chemistry. There he met Egon von Schweidler, who was born on 10 February 1873 in Vienna as the son of the *Hof-* and *Gerichtsadvokat* Emil Ritter von Schweidler, and who went, immediately after finishing the *Schottengymnasium* in 1890, to the university to study physics and mathematics. Von Schweidler got his doctorate in 1895 and soon became assistant to Franz Exner, while Meyer obtained his doctorate in 1896 and joined Ludwig Boltzmann as assistant at the *Physikalisches Institut.* Thus the scientific careers of Meyer and von Schweidler began practically at the same time and at the same place, i.e., in the *Türkenstraße.*

The situation in physics at that time seemed to be rather settled after very rapid progress in the nineteenth century. Hans Benndorf, a witness from those days, reported:

> When in the year 1888 Henrich Hertz succeeded in creating electromagnetic waves, the final stone seemed to have been laid on the splendid edifice of physics. All processes in the inorganic world known at that time seemed to be intelligible and understood in principle; it was thought to be extremely improbable to discover really essentially new phenomena. We young physicists were worrying then, whether

[210] Later in 1905 he also separated the radium deposit by the same method (Lerch, 1906).

[211] Stefan Meyer's father, Gotthelf Karl Meyer—a doctor of law—was a descendant of Dr. Josef Magnus Österreicher, who had been General Medicus to the Holy Roman Emperors Joseph II, Leopold II and Franz II. He was a man of high cultural interests; he passionately collected paintings. His wife Klara, *née* Goldschmidt, was the sister of Victor Mordechai Goldschmidt (1853-1933), professor of mineralogy in Heidelberg. The Meyer family were Christians. (See the talk in memory of Meyer by Hans Benndorf: Benndorf, 1951.)

we could still find a sufficiently large, uncultivated area [in physics] in which to apply our scientific ambition. (Benndorf, 1951, p. 156)

However, that fear soon turned out to be completely unsubstantiated, and Benndorf continued:

A new development of science began, with considerable speed, which mankind had not experienced before. Discovery followed upon discovery: 1895, X-rays; 1896, H. Becquerel's uranium rays (discovery of radioactivity); April 1898, P. and M. Curie's polonium; December 1898, radium; transmutation of chemical elements; constitution of atoms from elementary parts; constitution of matter from atoms; and a hundred other things, of which one had no idea at the end of the [nineteenth] century. (Benndorf, 1951, p. 156)

The initial item of this development, the discovery of X-rays by Wilhelm Conrad Röntgen (1845-1925) in Würzburg on 8 November 1895, soon became known in Vienna. Within a week Franz Exner received a letter from Röntgen, and on 4 January 1896 he spoke about X-rays to a group of Vienna professors. Ernst Mach then suggested a method of obtaining a three-dimensional picture of an X-rayed object, and the spectroscopists Josef Maria Eder and Eduard Valenta published in February 1896 the first X-ray atlas based on this suggestion.[212] An even greater impact than that of X-rays was felt in Austria by the discovery in Paris of radioactivity by Henri Becquerel (1852-1908), which in turn had been stimulated by Röntgen's discovery and had been announced in March 1896 (Becquerel, 1896a, b). The main reason was that the substance in which Becquerel found the new radiation, uranium, then existed in large amounts in *St. Joachimsthal* in Bohemia. In the old, practically exhausted silver, copper and lead mines of that place uranium pitchblende had been mined since 1853, and was used for producing uranium colours. When Pierre (1859-1906) and Marie Curie (1867-1934) became interested in closely investigating the phenomena of radioactivity they requested, in September 1898, the president of the *Kaiserliche Akademie der Wissenschaften* in Vienna, the geologist Eduard Suess (1831-1914), for support in obtaining residues of the *St. Joachimsthal* uranium production. Suess helped immediately, and the Curies obtained 100 kg free of charge to continue their experiments. These experiments yielded the discovery of a new element, radium (in December 1898). Up to 1906 more than 26 tons of the ore was sent to Paris at the lowest possible price.[213] Pierre and Marie Curie used the uranium residues to produce especially the strongly radioactive element radium, which played a crucial role in research at that time.

[212] We have taken this information from John Blackmore's biography of Ernst Mach (Blackmore, 1972, p. 162).

[213] See the detailed report on the relationship of Pierre Curie to the Vienna Academy of Sciences in Stefan Meyer's article on the Vienna *Institut für Radiumforschung* (Meyer, 1950, pp. 7-9).

Still, Stefan Meyer in Vienna did not prepare his first radium probe from the Austrian raw material himself, but got it from Friedrich Giesel (1852–1927), professor of chemistry at the *Technische Hochschule* in Braunschweig, whom he met at the 71st *Naturforscherversammlung* in September 1899 in Munich. On this occasion Giesel demonstrated his radium probes, and Benndorf reported later:

> Stefan Meyer, who concerned himself at Boltzmann's institute with determining quantitatively the magnetic properties of the chemical elements, also wished to determine those of radium; hence he requested Giesel to lend him one of his [Giesel's] probes. Since Giesel most kindly fulfilled this request, then for the first time an actually weak radium probe, which was at that time a rather strong one, arrived in Austria. Thus the Austrian physicists obtained the opportunity to participate in revealing the mysteries of radioactive substances. (Benndorf, 1951, pp. 156–157)

For his systematic investigation of the magnetic properties of chemical elements Stefan Meyer had many substances available in 1899, which had formerly been used by Franz Exner and Eduard Haschek for spectroscopic studies.[214] After receiving Giesel's radium probe—which contained 2 g radium chloride—and a weaker radium-polonium probe from Pierre Curie, he immediately measured, together with Egon von Schweidler, the respective susceptibilities of radium and polonium; both authors submitted their results in a short note to *Physikalische Zeitschrift* in November 1899 (Meyer and von Schweidler, 1899a). In the same note Meyer and von Schweidler wrote: 'The discharging effect of radium on electrically charged bodies turned out to be very much weakened by the presence of a magnetic field' (Meyer and von Schweidler, 1899a, p. 90). The same effect had earlier been observed by Julius Elster (1854–1920) and Hans Geitel (1855–1923) in Wolfenbüttel; however, they had observed the situation in a partial vacuum (Elster and Geitel, 1899a). Meyer and von Schweidler continued their analysis during the following week and concluded (in a second note) that, 'in agreement with some results, which we have meanwhile learned through letters from Giesel, Elster and Geitel,' the radium source emitted rays which 'thus behaved completely analogous to cathode rays' (Meyer and von Schweidler, 1899b, pp. 113–114).

Their observation about the magnetic deflection of the rays emitted by radium—the first Austrian contribution to the brand new science of radioactivity—was soon followed by further pioneering findings, e.g., on the absorption of radium rays by different substances from aluminium to lead (Meyer and von Schweidler, 1900), Meyer and von Schweidler soon moved into the front row of experts on radioactivity, together with the Curies in Paris, Giesel in Braunschweig, Elster and Geitel in Wolfenbüttel and Ernest Rutherford (1871–1937) in Montreal. Indeed, the pair of researchers Meyer and von Schweidler obtained a similar

[214] In March 1899 Meyer published a paper containing values for the susceptibility of the elements from beryllium to thorium (Meyer, 1899).

reputation as the pair Elster-Geitel.[215] Among the prominent results of their joint research was the experimental proof that RaF (a decay product of radium and the element polonium—discovered by the Curies) was identical and in agreement with Rutherford's suggestion (Meyer and von Schweidler, 1906).

Occasionally the two friends split their investigation, hence some important results became associated with one of the two names. Thus it was Egon von Schweidler, who submitted a paper, entitled '*Über Schwankungen der radioaktiven Umwandlung*' ('On Fluctuations in the Radioactive Transmutation') to the first International Congress for the Study of Radiology and Ionization, held from 11 to 16 September 1905 during the *World's Fair* in Liège, Belgium (von Schweidler, 1905). In this paper the author interpreted, in just three pages, the decay process of radioactive substances in a statistical way. Assuming that each atom of the radioactive substance possesses a probability of decaying—which is large when the decay time of the substance is small—he concluded that, if $Z$ denoted the average number of atoms decaying in a given period of time, there always occurred deviations from $Z$, whose fluctuation (i.e., the square root of the average of the quadratic deviation) was inversely proportional to $\sqrt{Z}$.

The first experimental test of von Schweidler's formula was given a little later by K. W. Fritz Kohlrausch, then a student in Franz Exner's institute (Kohlrausch, 1906a). He and others, for example, Hans Geiger (1882-1945) in Manchester, found fluctuations (Geiger, 1908). Later von Schweidler applied his statistical consideration to suggest an experiment which might decide the nature of $\gamma$-rays; i.e., whether they were impulses in the sense of classical electrodynamics, or corpuscular rays as assumed at that time by William Henry Bragg (1862-1942), or light-quanta in the sense proposed by Albert Einstein (von Schweidler, 1910a, b).[216] Norman Campbell (1880-1949) of Leeds, however, raised objections against the validity of the test suggested by von Schweidler; especially, he argued that no decision was possible if the ionization measured was created not by the $\gamma$-rays themselves but by secondary rays (Campbell, 1910). Campbell's criticism also extended to the earlier observations of fluctuations in radioactive decay by Kohlrausch and Geiger. Only eight years later another student in Vienna, Miss

[215] The pioneers of radioactivity showed, despite their active competition, a friendly relationship, which was expressed in mutual support on the necessary radioactive probes—as we have seen in the case of Giesel and Meyer. Hans Benndorf recalled the following story: 'The pair of researchers (Elster and Geitel) performed, at the same time as St. Meyer and E. Schweidler in Vienna, experiments with radioactive substances. Both pairs of researchers steadily exchanged letters about their investigations. One day a happy letter from Elster and Geitel arrived [in Vienna], stating that they had found the deflection of Becquerel's rays in a magnetic field, and that they soon wanted to report this fact in Berlin. Now, however, Meyer and Schweidler had also discovered the deflection of these rays in a magnetic field; from the letter of Elster and Geitel they were able to conclude that the others must have made an error, because they had indicated the deflection of the rays in the wrong direction. What did Meyer and Schweidler do then? They sent a telegram stating that the Wolfenbüttel people must have made a mistake; thus they prevented this embarrassing error of Elster and Geitel from becoming public. Some others may have omitted this, possibly rejoicing about the fact that a scientific competitor had failed' (Benndorf, 1951, p. 161).

[216] We have reported about this fluctuation discussion in radiation theory in Volume 1, Section II.5.

E. Bormann would establish what Kohlrausch, in a review paper, called 'the first *incontestable* proof of the existence of von Schweidler's assumption on the statistical decay of atoms' (Kohlrausch, 1926, p. 211).[217] But even before this final proof, the correctness of the statistical assumption could hardly be doubted by the experts in radioactivity; and von Schweidler would be called by David Hilbert simply 'the father of fluctuations' ('*Vater der Schwankungen*').[218]

While Schweidler contributed several very important results to the understanding of radioactive processes, Stefan Meyer—who joined Exner's institute in 1907—carried out extended, systematic investigations on natural radioactivity. For example, he went (together with Heinrich Mache) to *St. Joachimsthal* for this purpose. As Meyer noted:

> We intended to study the radioactivity of the waters at different depths. It was in winter, and we were not suitably equipped at all. In the morning we were supposed to be brought through the deep snow higher up to the entrance of the [uranium] mine. *Bergrat* J. Stěp looked at our clothing and shoes, and shook his head. Then he said, 'I know what to do! I'll get dogs placed at your feet!' Thus it was done. Early in the morning the sledges came, each with a big dog for use as a hot water bottle. In the mine I then caused a stir, because I suddenly had to sneeze. The miners threw reproachful glances at me. I could not guess that there was a rumour that the ghosts of the mountain might be offended by such a disturbance. Apparently they did not mind this, and we could take probes of water from different depths. The result [of our measurement] was that the radioactivity increased with increasing depth. The radon concentrations in the water at the bottom surpassed several times those of the Gastein sources. These measurements, afterwards continued by J. Stěp, caused the establishment of a *Kurhaus* and of cure facilities in *St. Joachimsthal.* (Meyer, 1950, pp. 11-12)[219]

[217] Since Schrödinger contributed essentially to this proof, we shall discuss it in some detail in Section I.5.

[218] In planning the *Wolfskehlwoche* on fluctuation phenomena—probably the one in which Marian von Smoluchowski presented his '*Drei Vorträge über Diffusion, Brownsche Molekularbewegung und Koagulation von Kolloidteilchen*' ('Three Lectures on Diffusion, Brownian Molecular Motion, and Coagulation of Colloid Particles'), 20 to 22 June 1916—Hilbert had invited von Schweidler 'in the most honourable way, because the Göttingen people did not want to miss in their discussions [at the *Wolfskehlwoche*] the critique of the "father of the fluctuation phenomena"' (Benndorf, 1951, p. 300).

[219] In his reminiscences of the prehistory of the Vienna *Institut für Radiumforschung*, from which we have quoted above, Meyer also gave a colourful description of the radioactivity boom in Austria in the case of *St. Joachimsthal.* He wrote: 'The little mining town of *St. Joachimsthal* became a health resort only after the discovery of radioactivity. It is situated in a rather steep valley. At the bottom was the uranium colour plant. In the upper parts stood the big hotels, "*Zur Stadt Wien*" and "*Zur Stadt Dresden*". When we (Mache and St. Meyer) arrived there in 1905, we found the place very much in the spirit of the age. Several "radium roasts," "radium soaps," "radium cigars," "radium cigar tips," "radium cookies," etc., were available. One of the owners had called his inn "*Radium-Gaststätte*" ("Radium Inn"). This annoyed the competing owner; he therefore wanted to call his inn "*Zur Emanation*" and ordered this name to be printed on the cigar tips, which he intended to distribute for advertising purposes. The words, however, were not known to the printer; hence he printed instead the more often used word "*Zur Imitation*"' (Meyer, 1950, p. 11).

Mache and Meyer published the results of their radioactivity measurements (of the water from various Austrian mineral sources) in several reports, submitted to the Vienna Academy of Sciences. They wrote a review paper summarizing the results, entitled '*Über die Radioaktivität östereichischer Thermen*' ('On the Radioactivity of Austrian Thermal Springs'), for the *Physikalische Zeitschrift* (Mache and Meyer, 1905). The springs included the famous waters of Bad Gastein, Karlsbad, Marienbad, Franzenbad (the last three being situated in Bohemia) and Baden near Vienna. Their investigations also led to 'the suggestion to establish special facilities and new sanatoriums, for example, in *St. Joachimsthal* itself,' since 'the curative power of certain thermal springs was associated with its content of radium emanation' (Meyer, 1920, pp. 1-2).

## The *Institut für Radiumforschung* in Vienna

The possible importance of natural radioactivity for the economy of the country and the health of its people, and the fact that Austria possessed particularly rich resources of radioactive minerals, provided an increasing interest for the Imperial administration and for the public in research on radioactivity. In 1904 the Vienna Academy of Sciences, under the president Eduard Suess (1831-1914), ordered 'the processing of 10,000 kg of residues of the uranium production [in *St. Joachimsthal*] in the *Auer-Welsbachschen Gasglühlampenfabrik*, Atzgersdorf, near Vienna, by L. Haitinger and C. Ulrich' (Meyer, 1920, p. 1). This processing took three years (from 1904 to 1907) and 'was performed in constant and close collaboration with the Vienna *Physikalisches Universitätsinstitut*, where St. Meyer and E. Schweidler controlled the progress or guided it occasionally (partly) in the right direction, by measuring the radioactivity of all enriched probes and of all intermediate and side products of the process' (Meyer, 1950, p. 10).[220] The work resulted in the gain of 4 g of radium chloride, which constituted a great treasure at that time (worth several millions of Austrian *Kronen*).[221] Now the problem was to establish a suitable institute for research in radioactivity, as the old university institute in the *Türkenstraße* clearly could not do so. Then, on 2 August 1908 the Vienna Academy received a letter, signed by Dr. Karl Kupelwieser, which began:

> The fear that my native land of Austria might miss something in mastering scientifically one of the greatest treasures given to it by nature, namely the mineral uranium-pitchblende, has concerned me since the mysterious emanation of its

[220] Meyer recalled that besides Suess, Exner and Boltzmann were also interested in radioactivity. Meyer knew Suess through his acquaintance with the latter's son Erhard Suess; hence 'from the very beginning he was called on by the three of them [i.e., Suess, Exner and Boltzmann] for information about this special subject [i.e., radioactivity], in particular for the exchanges with Paris and England' (Meyer, 1950, p. 4).

[221] One may recall here the fact that Ernest Rutherford in Manchester received 300 mg radium from the Austrian Academy of Sciences in 1908; with it he performed his fundamental studies on the nature of the atomic nucleus and on the transmutation of chemical elements in the following decade.

> product, the "radium," became known. I wish, as far as my powers go, to prevent my native land incurring the disgrace that others will take away a task which nature so-to-speak has assigned to it as a privilege. (Meyer, 1920, p. 2)

Kupelwieser therefore offered to donate an amount of 500,000 *Kronen* to the academy, for the purpose of constructing a building for the physical and chemical studies of radium, but with the following conditions: (1) The Austrian government should provide suitable grounds next to the already-planned future physical institute at a cheap rate; and (2) the Vienna Academy of Sciences should take over the direction of the new radium institute and be supplied by the government with the necessary amount of radioactive substances. Kupelwieser's offer was gladly accepted, and as early as 1909 construction began. On 28 October 1910 the opening ceremony of the *Institut für Radiumforschung* was celebrated, with the happy donor remaining in the background. As Benndorf wrote:

> That the beautiful institute really stood complete after two years, was mainly due to the efforts of Stefan Meyer. Exner himself had no experience of radioactive research; he might also have been too old for prosecuting all these tasks and obtaining all the information necessary for planning such an institute. Therefore he had to search for a collaborator. As such primarily his assistants E. Schweidler and St. Meyer qualified; both of them had already won reputations by their work in activity. Exner chose St. Meyer, who appeared to be the most suitable on account of his strong interest, his vitality, his unselfishness and his great joy in work and working power; time has confirmed that Exner had made an excellent choice. (Benndorf, 1951, pp. 158-159)

Since no institute of the kind envisaged existed at that time, Meyer had to begin from scratch. Benndorf wrote:

> First of all, together with the architect, he had to devise the plans for the institute. For this purpose everything possible had to be carefully taken into account. At that time, of course, many properties of radioactive substances had not yet been studied sufficiently; the danger of contamination had to be avoided, and one had to think of the possibility of explosions and many other things. (Benndorf, 1951, p. 159)

Meyer solved his tasks perfectly. The *Institut für Radiumforschung*, located at the Boltzmann-Gasse between the later constructed *Physikalisches Institut* and the *II. Chemisches Institut*, had a front length of 21.7 m and a depth of 15 m. It consisted of a basement (*Kellergeschoß*) embracing various installations for power supply, etc., a ground floor with two large chemical laboratories and extra workshops, as well as an apartment for the mechanic of the institute; a second floor with five offices (*Arbeitszimmer*) of different sizes, and a small library and dark-rooms for photographic work; a third floor again with five offices, a large room for instruments (*Sammlung*) and a small room housing gauge apparatus for $\alpha$- and $\gamma$-radiation; on the fourth floor a huge, movable electromagnet was installed, as well as five offices plus a mechanical workshop.[222] The institute

[222] About half of the donation money was spent on the construction of the building, general installation and furniture, the remainder on physical and chemical apparatus. (See Meyer, 1920, pp. 5-7.)

possessed no lecture or seminar rooms; however, after the completion of the *Physikalisches Institut*—about which we shall report in Section I.4—the lecture hall of the latter could be used easily, since a passage between the respective fourth floors of both institutes was established.

Although Franz Exner figured officially as the director of the *Institut für Radiumforschung* until 1920, Stefan Meyer had to run it from the very beginning of its existence.[223] It was he who selected the assistants and supervised their work, as well as that of collaborators and guests. For his first assistant he chose Victor (Viktor) Franz Hess; later the chemist Friedrich Adolf Paneth served as second assistant, then Karl Friedrich Herzfeld and Karl Przibram were added.[224] Besides the assistants of the institute, guests from the University of Vienna and the *Technische Hochschule*, like Fritz Kohlrausch, Heinrich Mache and Egon von Schweidler, carried out investigations during the early years. Finally, scientists from other Austrian universities, and even people from abroad, received the opportunity to work at the *Radiuminstitut*—as it was called in short—especially when they had already acquired some experience with physical and chemical work on radioactivity.[225]

The working programme of the *Radiuminstitut* was a mixed one. An important portion consisted of routine measurements, such as the determination of the strength of radioactive probes. At the Brussels Conference for Radiology in September 1910 Stefan Meyer was elected Secretary of the International Radium Standard Commission (with Ernest Rutherford as President). The Commission then decided to have radium standard probes made in Paris (by Marie Curie) and in Vienna. Otto Hönigschmid (1878-1945), since 1911 professor of chemistry at the *Technische Hochschule* in Prague and regular guest at the *Radiuminstitut*, prepared three probes of great purity of about 10, 31, and 40 mg radium chloride. At the Paris meeting of the commission in 1912 (25-28 March) the Paris probe of 21.99 mg radium chloride was declared the primary standard, and the Vienna probe of 31.17 mg radium chloride the substitute standard.[226] Meyer and his collaborators now used the pure probes of Hönigschmid to measure accurately

[223] Benndorf noted the curious fact, 'that F. Exner did not re-enter the building after the opening ceremony until ten years later, i.e., after he had retired and wanted to assume his "*Austragsstüberl*" (retirement office) in a room located on the third floor of the *Radiuminstitut*' (Benndorf, 1951, p. 159).

[224] The personnel expenses of the institute were provided by the Austrian government, which also took over the running expenses, such as water and electricity costs. (See Meyer, 1920, p. 7.)

[225] Among the foreign visitors we may name the American Samuel Colville Lind (1879-1965), who had received his doctorate at the University of Leipzig and spent some time afterwards in Paris; he investigated at the *Radiuminstitut* the effect of $\alpha$-rays on the formation of ozone. Another visitor was Robert William Lawson (1890-1960), who came on a grant from Newcastle-upon-Tyne and tried to isolate the end-product of the radioactive decay of uranium (in collaboration with Arthur Holmes from Imperial College, London); later he measured the rate of emission of $\alpha$-particles from radium.

[226] During the following years Hönigschmid performed precision determinations of the atomic weight of radium (1911/1912, obtaining 226.0), of uranium (1913/1914, obtaining 238.1) and of Ra G (1914, obtaining 206.05).

important constants connected with radioactive decay: thus Meyer and Hess obtained the heat developed by radium and the amount of $\gamma$-radiation emitted in the decay process (Meyer and Hess, 1912). Radioactive substances different from radium were also studied in Vienna; for example, Meyer and Paneth re-examined the intensity of $\alpha$-radiation emitted from uranium (Meyer and Paneth, 1912), and Meyer determined the lifetime of the same substance (Meyer, 1913). The detailed absorption of the different types of radiation, especially of $\alpha$- and $\gamma$-rays, by a variety of substances represented another major task of the *Radiuminstitut*, which also devoted in those days considerable interest to the understanding of physical phenomena. Finally, Victor Hess soon turned his attention to investigating the origin of penetrating $\gamma$-radiation in the atmosphere. This investigation would lead him to discover the existence of 'cosmic radiation.'[227] Thus, within a very short period, Meyer's institute developed a distinguished research programme, which established for it a high reputation in the world of science.[228]

## Relativity and Quantum Theory in Vienna

Compared to such ample activities in the field of what was called in those days 'radiology,' the Austrian physicists seemed to pay little attention to two other important subjects, which came up at about the same time, namely relativity and quantum theory. If viewed from a distance, this fact appears to be rather surprising, since it was two Austrian physicists, Mach and Boltzmann, who provided essential steps on the route to both theories.

In the previous section we have presented in some detail Ernst Mach's analysis of the fundamental concepts on which, two centuries earlier, Isaac Newton had based the entire system of physics. Especially, his criticism of absolute space, absolute time and absolute motion, displayed in detail in the book *Die Mechanik in ihrer Entwicklung* (Mach, 1883), provided the young Albert Einstein with important hints towards the new kinematics of special relativity.[229] Nevertheless the publication of the pioneering paper '*Zur Elektrodynamik bewegter Körper*'

[227] We shall discuss atmospheric radioactivity and the discovery of cosmic rays in some detail in the following sections.

[228] The acting head of the institute, Stefan Meyer, received the *Lieben-Preis* of the Austrian Academy (1913) for his pioneering work in radioactivity. In 1920 he was officially appointed director of his institute; in 1921 he was elected corresponding member of the Austrian Academy of Sciences (in 1932 full member). He remained in this position until his forced retirement in 1938; then he lived with his wife and daughter in Bad Ischl. After World War II he returned to Vienna and the Radium Institute, again assuming its directorship (until 1949). He died on 29 December 1949 in Bad Ischl.

[229] The influence of Mach's criticism of Newton's mechanical concepts on Einstein was confirmed by Einstein himself in his autobiographical notes.

It may also be worth noting in this context, that Ludwig Boltzmann, whose books were eagerly studied by Einstein, exerted some influence on the creation of relativity theory. In a recent investigation Siegfried Wagner (1982) pointed out some ideas, especially with respect to the concept of co-ordinate systems, which Einstein might have taken over from Boltzmann's *Vorlesungen über die Prinzipe der Mechanik* (Volumes I and II, Leipzig: J. A. Barth, 1897, 1904).

('On the Electrodynamics of Moving Bodies,' Einstein, 1905d) and of the following investigations on the subject by Einstein, Max Planck, Hermann Minkowski and others, did not stimulate great interest in relativity theory on the part of the Viennese physicists. It is true that Ernst Mach was then partly paralyzed and was also working on general science theory and its relation to psychology rather than on purely physical problems. But the 'official' theoretician in Vienna after Boltzmann's death, Fritz Hasenöhrl, should have reacted more positively to the new theories, especially because he himself was concerned with the electrodynamics of moving bodies and their interaction with electromagnetic radiation since 1902.[230] In the years after 1905 Hasenöhrl turned to what he called the 'thermodynamics of moving bodies' (Hasenöhrl, 1907a, b; 1908), arriving at results similar to those that Planck found at the same time on the basis of Einstein's theory (Planck, 1908).

The main reason why the importance of Einstein's work on relativity theory was not noticed particularly in the Austrian capital, must therefore be seen in the fact that Hasenöhrl, the natural candidate for a full appreciation, had already embarked *before* 1905 on a programme of his own, attacking similar problems as Einstein would in special relativity theory. Hasenöhrl did study the papers of his colleagues, notably those of Hendrik Lorentz and Max Abraham on the theory of the electron and the radiation emitted from moving bodies. But even as late as 1907, when he replied to a criticism raised by Max Planck's doctoral student Kurd von Mosengeil against his treatment of radiation in moving bodies (Hasenöhrl, 1904e, 1905), he claimed that these problems had not yet obtained a satisfactory solution. He wrote in particular:

> The foundation of Lorentz' theory must at present be considered just a hypothesis. As is well known, another theory also exists, which is notably advocated by M. Abraham. Hence it is certainly of interest that Mr. von Mosengeil [(Mosengeil, 1907)] has treated the problem of cavity radiation from a different point of view. Which of us will be right [i.e., whether Hasenöhrl, who used Lorentz' theory explicitly or von Mosengeil who avoided it], is a question to be decided by the further development of the science. (Hasenöhrl, 1907a, p. 792)

While Hasenöhrl remained cautious and open towards the further development of the problems of electron and radiation theory and related questions, Philipp Frank, who was born in Vienna on 20 March 1884 and received his doctorate at the University of Vienna in 1907, became an ardent partisan of Einstein's theory of special relativity. In 1909 he published his first paper on the subject, describing 'the position of relativity theory within the system of mechanics and electrodynamics' (Frank, 1909). A few months later he submitted a second paper, coauthored with Hermann Rothe and submitted like the first one to the Vienna Academy of Sciences; in it they tried to establish mechanics on the basis of the invariance against general linear transformations arriving at a theory which slightly extended relativistic mechanics (Frank and Rothe, 1910).

[230] See our discussion of Hasenöhrl's work earlier in this section.

Before these papers of Frank appeared in print, the community of Austrian physicists had the opportunity to learn about various aspects of relativity theory at first hand. At the 81st *Naturforscherversammlung*, held from 21 to 25 September 1909 at Salzburg, three talks were connected with relativity theory. Thus Max Born spoke '*Über die Dynamik des Elektrons in der Kinematik des Relativitätsprinzips*' ('On the Dynamics of the Electron in the Kinematics of the Relativity Principle,' Born, 1909c), and Arnold Sommerfeld '*Über die Zusammensetzung der Geschwindigkeiten in der Relativitätstheorie*' ('On the Addition of Velocities in Relativity Theory,' Sommerfeld, 1909). The greatest interest was certainly aroused by the first public appearance of Albert Einstein, who presented at the beginning of the afternoon session of 21 September a long and thorough talk, entitled '*Über die Entwicklung unserer Anschauungen über das Wesen und die Konstitution der Strahlung*' ('On the Development of our Views on the Nature and Constitution of Radiation,' Einstein, 1909b). Ernst Mach, who was unable to attend the Salzburg meeting, received a report about Einstein's address by the mathematician Wilhelm Wirtinger. Wirtinger wrote:

> A. Einstein has indeed held a lecture here [in Salzburg]. Unfortunately, I could not understand the first part in physical terms and therefore I left. I was told, that in what followed he developed the idea of light as a quantity, but I wasn't able to get anything more precise on it. On the other hand, your colleague Anton Lampa was here, and he may be able to give you a more exact account. (Wirtinger to Mach, 5 October 1909; quoted in Blackmore, 1972, p. 262)

Einstein, unlike his colleagues Born and Sommerfeld, did not restrict himself to pure questions of relativity theory—indeed, he gave only in the first part of his Salzburg lecture a rather general, fairly elementary account of the physical ideas that had led to special relativity theory—but devoted himself mainly to the problem of the nature of light, i.e., to the apparent conflict between the concepts of radiation in classical electrodynamics and in quantum theory. This topic was certainly even more difficult to understand than the questions of pure relativity theory presented by Born and Sommerfeld. In addition, Max Planck, the father of quantum theory and chairman of the discussion, immediately raised objections against Einstein's hypothesis of the light-quantum. The physicists, who were present at the meeting, got the impression that this question was far from being settled. Nevertheless, the physicists, and especially those in Austria, got a lasting impression of relativity theory and its creator at the Salzburg *Naturforscherversammlung.* Less than two years later Einstein would find himself in the chair for theoretical physics at the German University of Prague. When he left this position in August 1912, his Vienna propagandist Philipp Frank succeeded him.[231]

[231] Frank was promoted in 1917 to a full professorship and stayed in Prague until 1938. Then he went to the United States, occupying various positions until 1948, when he founded the Institute for the Unity of Science in Cambridge, Massachusetts. He died on 21 July 1966 in Cambridge, Massachusetts.

As Mach could be connected with the foundations of relativity theory, so Boltzmann was connected with those of quantum theory. In particular, Max Planck based the theoretical derivation of his blackbody radiation law on Boltzmann's statistical interpretation of entropy (Planck, 1900f, 1901a).[232] He sent a copy of his paper—probably the one published in *Annalen der Physik* (Planck, 1901a)—to Boltzmann. As Planck recalled in his Nobel lecture: 'It brought me much-valued satisfaction for many disappointments when Ludwig Boltzmann, in the letter acknowledging the receipt of my essay, expressed his interest and basic agreement with the train of thought expounded in it' (Planck, 1920, p. 14; English translation, p. 412).[233] Still Boltzmann did not, after 1900, concern himself further with the problem of radiation theory, leaving it to his student Paul Ehrenfest (1880–1933). Ehrenfest analyzed, in a paper submitted in November 1905 to the Vienna Academy of Sciences, the physical prerequisites of Planck's theory of irreversible radiation processes (Ehrenfest, 1905). Ehrenfest then concluded, in particular, that Planck's introduction of energy-quanta of radiation 'lacked an analogy in Boltzmann's theory [i.e., in the statistical mechanics as developed by Boltzmann]' (Ehrenfest, 1905, p. 1313). He proposed, therefore, to study more deeply the true foundations of Planck's radiation theory, a task which he undertook in a second paper that was received by *Physikalische Zeitschrift* in July 1906 (Ehrenfest, 1906b). Before composing this article, he had seen an advance copy of Planck's book *Vorlesungen über die Theorie der Wärmestrahlung* (Planck, 1906), which the author had sent to Vienna.[234] Ehrenfest's second paper on radiation theory should be considered as the official point of view of Boltzmann's institute in 1906. There was no further comment from Vienna, because Boltzmann died a few months later and Ehrenfest—although he would continue to publish papers dealing with radiation and quantum theory—went soon afterwards to Göttingen and finally to St. Petersburg and Leyden.[235] Several years passed before quantum theory received new attention in the Austrian capital, and then it was a candidate for a *Habilitation* in philosophy, not in physics, who connected Planck's quantum of action with the size of the hydrogen atom.

In early March 1910 the Vienna Academy of Sciences received a paper, entitled '*Über die elektrodynamische Bedeutung des Planckschen Strahlungsgesetzes und über eine neue Bestimmung des elektrischen Elementarquantums und der Dimensionen der Wasserstoffatoms*' ('On the Electrodynamical Significance of

[232] We have discussed the role of Boltzmann's work on statistical mechanics in the genesis of quantum theory in some detail in Volume 1, Prologue and Section I.1.

[233] In the English translation we have corrected the phrase 'in the letter returning my essay' by 'in the letter acknowledging the receipt of my essay'. Planck stated in German: '*in dem Briefe, in dem er die Zusendung meines Aufsatzes beantwortete.*'

[234] This advance copy might have been the one presented to Ludwig Boltzmann. On the other hand, Ehrenfest exchanged letters at that time with Planck on radiation theory; hence Ehrenfest might have obtained an extra copy.

[235] We have presented the background of Ehrenfest and his contributions to quantum theory in some detail in Volume 1, Sections I.4 and II.5.

Planck's Radiation Law and on a New Determination of the Electrical Elementary Quantum and of the Dimensions of the Hydrogen Atom') and authored by one Arthur Erich Haas (Haas, 1910a).[236] Haas, who was born on 30 April 1884 in Brünn, the son of the lawyer Dr. Gustav Haas, had come to Vienna after attending the German *Gymnasium* of his hometown, enrolling in the University of Vienna from the winter semester 1902/03 to the winter semester 1903/04.[237] He then continued his studies at the University of Göttingen (from the summer semester 1904 to the winter semester 1905/06), where he heard lecture courses in mathematics, physics, chemistry, Sanskrit and on the Romantic period.[238] After obtaining his doctorate with a thesis on '*Antike Lichttheorien*' ('Theories of Light in Antiquity') in October 1906, he concerned himself with topics of the history of physics during the following years.[239] Thus he presented, for example, a talk on '*Historische Analyse des Energie-Prinzips*' ('Historical Analysis of the Energy Principle') at the annual meeting of the *Deutsche Gesellschaft für Geschichte der Medizin und Naturwissenschaften*, held in September 1908 together with the 80th *Naturforscherversammlung* in Cologne.[240] He had also written an essay on the same topic, hoping to get it acknowledged for his *Habilitation* (Hass, 1909). The philosophical faculty of the University of Vienna, however, did not accept it immediately but required the candidate to complete a paper on a purely physical topic first.[241] The disappointed Haas had then successfully completed a study of law, begun in the summer semester 1908, with a *Rechtshistorische Staatsprüfung* in October 1909, before turning to physics again and selecting a problem related to quantum theory as a suitable topic.

In his Salzburg lecture, mentioned earlier, Einstein had discussed the theoretical foundations of Planck's blackbody radiation law, arriving at the conclusion that one might possibly be able to explain the 'quantum problem' by suitably generalizing the equations of (classical) electrodynamics, referring to his

[236] In the title of the paper, Haas' first name is spelt as 'Artur' (see Haas, 1910a, p. 119). The same spelling is also repeated in a later paper published in the *Sitzungsberichte* of the Vienna Academy of Sciences (Haas, 1911). In all other publications Haas signed with 'Arthur Haas'.

[237] A detailed biography of Arthur Haas has been given by Armin Hermann (1966).

[238] In Göttingen, Haas studied scientific topics with Felix Klein, Max Abraham, Ludwig Prandtl, Eduard Riecke, Hermann Theodor Simon, Johannes Stark, Woldemar Voigt and Walther Nernst.

[239] Haas had been an excellent pupil at the *Gymnasium* in all topics, especially history and languages. Being financially independent through support from his father, he thought of establishing for himself a reputation as a *Naturphilosoph* in succession to Ernst Mach.

[240] Haas' lecture won him the approval of Karl Sudhoff (1853–1938), founder of the abovementioned society and professor of the history of medicine in Leipzig. However, it did not attract much attention among the physicists; his name was not registered in the list of participants of the *Naturforscherversammlung* (see *Präsenzliste der Abteilung II: Physik* in *Physikalische Zeitschrift* **9**, issue No. 22 of 1 November 1908, pp. 730–731).

[241] It might be of interest to mention at this point that Haas had earlier submitted two short physics papers to *Physikalische Zeitschrift*, one '*Über ein Maßsystem, das die Längeneinheit und die Lichtgeschwindigkeit als Grundeinheit enthält*' ('On a System of Units Containing the Unit of Length and the Velocity of Light as Fundamental Units,' Haas, 1905) from Göttingen, and another on '*Die Beziehungen zwischen dem Newtonschen und dem Coulombschen Gesetze*' ('The Relations between Newton's and Coulomb's Law,' Haas, 1906) from Vienna.

earlier proposal to extend the fundamental equation of optics, namely,

$$\frac{1}{c^2}\frac{\partial^2 \phi}{\partial t^2} - \left(\frac{\partial^2 \phi}{\partial x^2} + \frac{\partial^2 \phi}{\partial y^2} + \frac{\partial^2 \phi}{\partial z^2}\right) = 0, \tag{8}$$

so as to include terms involving the quantity $e$, the charge of the electron (Einstein, 1909a).[242] Einstein claimed that this idea was promising, because the equation

$$h = e^2/c \tag{9}$$

held between Planck's constant and the constants $e$ and $c$ (velocity of light *in vacuo*), up to a dimensionless factor of the order of 1,000. It is doubtful that Haas attended Einstein's lecture, as he was just then preparing for his law examination.[243] But he found essentially the same idea discussed in the article on '*Theorie der Strahlung*' ('Theory of Radiation'), which Wilhelm Wien had written for the *Encyklopädie der mathematischen Wissenschaften* and which appeared in issue No. 2 of Volume V/3, dated 28 September 1909 (Wien, 1909).[244] Wien concluded his article with the remarks:

> Thus if Planck's theory of radiation in its present form cannot be considered as the definite solution of the problem, one may still conclude from it that Maxwell's theory of electrodynamics does not suffice to describe the phenomena of radiation; notably, one may perhaps succeed just with the assumption that Maxwell's theory no longer represents the phenomena in the interior of the atom. (Wien, 1909, p. 356)

Arthur Haas came upon Wien's article at the moment when he finished his law studies; it provided him with a suitable topic to work on in order to fulfil the demands of the Vienna physicists, a topic satisfying his own philosophical interests. He was especially attracted by a remark of Wilhelm Wien, who declared: 'For the moment I cannot follow the opinion expounded by Einstein [(Einstein, 1909a)], that the magnitude of the quantum of energy may be related to the elementary quantum of electricity; the reason is that the quantum of energy, if it possesses a physical explanation at all, can only be derived from a universal property of the atoms' (Wien, 1909, p. 356).[245] Now, when searching for a universal

[242] The new equation for the wave function $\phi$ should be non-linear and pass over into Eq. (8) in the classical limit, that is, in the limit in which the radiation law of Lord Rayleigh and James Hopwood Jeans described the blackbody radiation.

[243] The certificate (*Zeugnis*) of Haas' examination is dated 11 October 1909, i.e., twenty days after Einstein's lecture in Salzburg. (See Hermann, 1966, p. 9, footnote 1.)

Strangely enough, in his publications Haas never referred to Einstein's papers of 1909; this is all the more surprising, since Einstein had used dimensional arguments—as did Haas in his previous papers on physical topics. In addition, both Einstein and Haas published in the same journal, the *Physikalische Zeitschrift*, and one would expect that Haas regularly studied the issues of this journal. (However, see Footnote 245.)

[244] Haas quoted Wien's encyclopaedia article as the first reference. (See Haas, 1910a, footnote 1 on p. 119.)

[245] Perhaps this opinion of Wien prevented Haas from referring to Albert Einstein's idea of deriving the quantum of action from the quantum of electricity.

property of atoms, what better suitable property could be chosen, especially in Vienna, than the size of atoms.[246] The problem still remained to be solved for Haas of how to relate the size of the atoms to the quantum of action. For that purpose he had to refer, first, to an electrodynamical model of the atom, and second, he introduced a quantum-theoretical hypothesis.

As the atomic model Haas selected the one which Joseph John Thomson had proposed several years previously (Thomson, 1904a) and seemed to be the most preferred model in those days.[247] He learned the details of it from Thomson's book *Electricity and Matter*, the second German edition of which had just appeared (Thomson, 1909). As Haas noted: The Thomson atom 'consists of a uniformly dense sphere of *positive electricity*, in whose interior a number of *negative electrons* perform *circular orbits* around the centre of the sphere. The total amounts of positive and negative electricity are in all cases equal' (Haas, 1910a, p. 121). While in the hydrogen atom only one electron existed, Thomson suggested for atoms containing several electrons a grouping of the latter in the form of regular polygons. Haas, because of his *Gymnasium* background and his strong interest in Greek science and philosophy, was very attracted by this fact and commented in a later paper:

> The old favourite idea of Plato, that all qualitative differences of matter would originate basically just from *geometrical* differences, had been brought to honour again most recently by the English physicist J. J. Thomson. Modern physics has [thus] renewed this old idea in the form of the hypothesis that the atoms of all chemical elements consist of exactly the same building stones; and that the latter, however, are [in different chemical elements] present in different numbers and are assembled in different geometrical figures. (Haas, 1911, p. 1111)

In this first paper on the connection of Planck's constant with the size of the atom (Haas, 1910a), Haas, after a short description of Thomson's atom, first calculated the total energy $E$ of an electron (i.e., the sum of its potential and kinetic energies) moving on the surface of a uniformly charged sphere of radius

[246] As we have mentioned in the previous section, Joseph Loschmidt had obtained the first reliable value for the size of the atoms (Loschmidt, 1865).

[247] For example, Max Born—who had been a fellow student of Haas in Göttingen—published in the 2 December 1909 issue of *Physikalische Zeitschrift* his *Habilitation* lecture '*Über das Thomsonsche Atommodell*' ('On Thomson's Atomic Model,' Born, 1909d). In this lecture he started from the observation that the spectral lines emitted from atoms should follow from the model of the atom and added: 'Thomson's atomic model, about which I want to speak today, resembles a piano score made from the big symphony of the radiant atom. Even if it might, in many aspects, be rough and inadequate, it still provides some hints to understand this powerful music' (Born, 1909d, p. 1031).

Incidentally, Haas had attended, together with Born, the Göttingen Seminar on '*Ausgewählte Kapitel der Elastizitätstheorie*' ('Selected Topics from the Theory of Elasticity'), held during the winter semester of 1904/1905. He had volunteered then to talk about the topic '*Stabilität der elastischen Drähte und Bänder*' ('Stability of Elastic Wires and Bands'). Since he became sick, however, Born had to replace Haas as speaker. Born developed from this talk his doctoral thesis on '*Untersuchungen über die Stabilität der elastistischen Linie in Ebene and Raum unter verschiedenen Grenzbedingungen*' ('Investigations on the Stability of the Elastic Line in the Plane and in Space under Different Boundary conditions,' Born, 1906). See also Volume 1, Section III.2.

$a$, obtaining

$$E = \frac{e^2}{a} \tag{10}$$

if $e$ is the total charge of the sphere.[248] As a second step he made the assumption, that 'the energy-quantum (element) of the hydrogen atom is identical with its maximum energy' (Haas, 1910a, p. 125), hence

$$h\nu^* = \frac{e^2}{a}, \tag{11}$$

with $\nu^*$ a limiting frequency emitted by the atom.[249]

To obtain the physical significance of the constant $h$, Hass referred to a dimensional argument. Max Planck proposed in 1899 a system of 'natural' units in which the quantity $\sqrt{hc/f}$ (where $f$ denotes Newton's constant) provided a unit of mass (Planck, 1899, p. 480).[250] Haas, who liked dimensional arguments, now considered $C$, the ratio of gravitational forces to electric forces, exerted on each other by two unit masses (having the mass value $5.42 \times 10^{-5}$ g), both carrying a unit charge $e$, finding

$$C = \frac{h^2 c}{e^2} \tag{12}$$

or—by inserting Planck's values for $e$ and $h$—a value of the order of 1,000. He concluded: 'In any case, the suspicion is not far-fetched that this ratio might be either equal to the ratio of the masses, or to the ratio of the dimensions of atoms and electrons, respectively, because the constant $h$ should depend only on very general properties of the ether, of the ponderable matter and of the electrons' (Haas, 1910a, p. 128). The last suspicion could be substantiated, if one took the value of $\nu^*$, Eq. (11), as the frequency of the electron in its circular orbit of radius $a$ and assumed a balance between the centrifugal $(4\pi\nu^{*2}m_e a)$ and Coulomb

[248] Haas noticed that this energy value represented a maximum; that is, $E$ became smaller whenever the electron moved on a circle with a different radius (i.e., for $r < a$ and $r > a$).

[249] Haas discussed in Section 3 of his paper the possibility that in his theory fractions of the energy quantum $h\nu^*$ might exist. Especially he claimed that any visible radiation emitted by the sun was not capable of exciting an atomic resonator, contrary to experience. He wrote: 'In particular it might be advisable to give up the hypothesis of the indivisibility of energy-quanta; thus many difficulties in radiation theory might be removed, and many complicated assumptions caused by it might become superfluous' (Haas, 1910a, p. 126).

[250] Planck again presented this scheme of natural units in his book of 1906 (Planck, 1906, Section 159, p. 164).

forces ($e^2/a$) acting on the electron (mass $m_e$), then the relation[251]

$$h = 2\pi e\sqrt{m_e a} \tag{13}$$

followed for Planck's constant, which seemed to be fairly well satisfied.[252]

Arthur Haas, soon after submitting his first paper (Haas, 1910a), summarized his results in two short papers, one for the *Jahrbuch der Radioaktivität und Elektronik* (1910b) and the other for *Physikalische Zeitschrift* (1910c). The early response in Vienna to his work did not, however, encourage the author. He recalled:

> When I spoke in the Vienna *Chemisch-Physikalischen Gesellschaft* on my work, [Ernst] Lecher thought himself to be particularly witty when he called the whole thing (in the public discussion) a carnival joke ('*Faschingsscherz*'); and when Laurenz Müllner [the professor of philosophy] inquired about me from Hasenöhrl, the latter told him that I could not be taken seriously, because I naively mixed scientific subjects which could not possibly have anything to do with each other, such as quantum theory (a topic of the theory of heat) and spectroscopy (a topic of optics). Obviously I was most deeply discouraged. (Haas: *Autobiographie*, quoted in Hermann, 1966, p. 16)

The first to acknowledge Haas' work was the great Hendrik Lorentz, who referred to it in his fifth Wolfskehl lecture of 28 October 1910 at the University of Göttingen (Lorentz, 1910b). Although he called the whole thing 'a little bold' ('*etwas gewagt*,' Lorentz, 1910b, p. 1251) and mentioned several difficulties, especially concerning Haas' connection of the limiting frequency of the Balmer series and the maximum rotational frequency of the electron in the hydrogen atom, he ended with the fairly positive statement: 'Through the hypothesis of Haas the riddle of the energy quantum is connected with the problem concerning the nature and the action of positive electricity; and it may well be so that these different problems will be completely soluble only simultaneously' (Lorentz, 1910b, p. 1253). Lorentz' plea to give Haas' ideas a fair hearing would not be pushed aside by the Viennese physicists; they could no longer declare that the *Habilitation* candidate Haas was crazy. Still Hasenöhrl repeated his proposal, made a year previously, that Haas should write a further paper on a purely physical topic, one, however, which did not imply any unconventional hypothesis.

[251] He had already studied a comparison between the two forces in a previous paper on '*Die Beziehungen zwischen dem Newtonschen und dem Coulombschen Gesetze*' ('The Relations between Newton's and Coulomb's Law,' Haas, 1906).

[252] For the radius $a$ of the hydrogen atom Haas inserted the value $1.75 \times 10^{-8}$ cm and for $e$ several values existing in the available literature; thus he found $h$ assuming values between 6 and $10 \times 10^{-27}$ erg · sec, in reasonable agreement with Planck's value of $6.58 \times 10^{-27}$ erg · sec following from the blackbody radiation law. On the other hand for the frequency $\nu^*$, he derived values between 6 and $20 \times 10^{14}\ \mathrm{sec}^{-1}$, which again appeared to be reasonable, because $\nu^*$ should correspond to the series limit ($8.23 \times 10^{14}\ \mathrm{sec}^{-1}$) of the Balmer series.

Haas gave in and composed another article, again on Thomson's atomic model; this paper, entitled '*Über Gleichgewichtslagen von Elektronengruppen in einer äquivalenten Kugel von homogener positiver Elektrizität*' ('On Equilibrium Positions of Electron Groups in an Equivalent Sphere of Homogeneous Positive Electricity'), was properly communicated to the meeting of 28 July 1911 of the Vienna Academy of Sciences (Haas, 1911). With this work, based on conventional electrodynamical considerations, Haas obtained the necessary approval of the Vienna faculty; on 31 July 1912 he was duly appointed *Privatdozent* for history of physics.[253]

We have displayed here in some detail the interesting and fundamental contribution of Arthur Haas to quantum theory. Haas had to overcome great prejudicies in Vienna because he did not belong to the physics establishment—he was not the student of a respected physics professor; especially, his ideas did not fit at all into the Viennese atmosphere. Only gradually and reluctantly did the official physicists pick up certain aspects of Haas' atomic model, in Vienna first Arthur Schidlof, and later, in a very modified way, Fritz Hasenöhrl and Karl Herzfeld.

## Schrödinger as a Graduate Student

With such reservations of the official Viennese physicists against relativity and quantum theories, the student Schrödinger, in his classes, would not have heard much about these most recent, fundamental discoveries. Still he learned about some modern topics, for example, in Karl Przibram's lectures on '*Die neueren Anschauungen über die Konstitution der Materie*' ('The Recent Views Concerning the Constitution of Matter'), delivered in the winter semester 1908/09 and continued in the following summer semester of 1909.[254] Schrödinger kept two notebooks on these courses.[255] In the first semester Przibram presented the following topics: the kinetic theory of matter, even including such an advanced item as Boltzmann's collision equation and its application describing the diffusion of gases; the phenomena of electrolysis; the discovery of the electron through

[253] After about a year he accepted a call to an extraordinary professorship for the history of physics at the University of Leipzig; he moved in the spring of 1914 to Leipzig, where he soon took charge of a new volume of Poggendorff's *Biographisch-Literarisches Handwörterbuch der exakten Wissenschaften.* The outbreak of World War I interrupted Haas' career; in 1917 he came back from military service to his Leipzig chair. In 1921 he returned to Vienna as *Privatdozent* of physics at the university; two years later he was promoted to *Extraordinarius.* Several times he was invited to deliver a series of lectures abroad, e.g., in England at University College, London (1924), and in the United States (1927, 1931). In October 1935 he again went to America as a visiting professor at Bowdoin College, Brunswick, Maine. The following spring (1936) he was appointed professor of physics at the University of Notre Dame, Indiana. He died in Notre Dame on 20 February 1941.

[254] See the announcements in *Physikalische Zeitschrift* **9** (1908), p. 696, and **10** (1909), p. 294.

[255] Schrödinger's two notebooks on Przibram's lectures ('*Przibram: Konstitution der Materie, Winter 1908*') have been filed on AHQP Microfilm No. 39, Section 1. In the middle of the second notebook Schrödinger wrote '2. *Semester*' (i.e., indicating the beginning of the notes on the lectures of the summer semester of 1909).

Joseph John Thomson's experiments; the dynamical theories of the electron put forward by Max Abraham and Hendrik Antoon Lorentz a few years previously; a discussion of the experiment of the Americans Albert Abraham Michelson and Edward Williams Morley to determine the velocity of the earth with respect to the ether; an analysis of the motion of electrons in electric and magnetic fields, including a discussion of the fundamental experiments of J. J. Thomson, Walther Kaufmann and Philipp Lenard; a presentation of the existing experiments determining the electric charges of the electron, the hydrogen ion and the $\alpha$-particles (by Gabriel Stokes, Charles Thomson Rees Wilson, Karl Przibram and Joseph Nabl, Ernest Rutherford and Hans Geiger), as well as the experiments on the Loschmidt-Avogadro number (e.g., those of Jean Perrin); comments on the radius of the electron; a discussion of the phenomena connected with canal rays (especially the experiments of Eugen Goldstein and J. J. Thomson); an introduction to further types of radiation, e.g., the so-called $\delta$-rays and the rays emitted from radioactive substances; a discussion of the action of the different rays on matter, especially of what followed from the absorption experiments of William Henry Bragg, Pierre Curie, and Philipp Lenard; a review of the diffusion of ions (the experiments of John S. Townsend and others); and considerations about electric conduction in metals. In his second course in the summer of 1909 Przibram treated the (normal) Zeeman effect of spectral lines and its explanation in terms of Lorentz' electron theory; the anomalous Zeeman effect, including the rules of Carl Runge; the inverse Zeeman effect; the band spectra; the phenomena of radioactivity from 1897 to the present (i.e., summer 1909); the laws of radioactive decays, and the radioactive families.

Przibram's lectures covered an enormous variety of topics and results from the previous twenty years of physics, in particular all the spectacular experimental discoveries, with the exception of X-rays. Schrödinger learned about other results of recent research in the physical seminars in which he participated; for example, in the summer semester of 1909 Gustav Jäger of the *Technische Hochschule* presented a talk on '*Fortschritte der Gastheorie*' ('Advances in Gas Theory'). In the spring of 1908 the famous British chemist William Ramsey (1852-1916) came from London and delivered a lecture at the University of Vienna, entitled '*Neue Theorie der Chemischen Bindung*' ('A New Theory of Chemical Binding').[256] Based on the assumptions that 'every negatively charged body possesses more electrons than [it] had in the state of electrical neutrality, and every positively charged body had given away some of the electrons which it *possessed even in the state of neutrality*' (*Notebook*, quoted in Footnote 256, pp. 2-3), Ramsey explained the phenomena of chemical binding, of valences, of the formation of ions, and of the contact potential arising when different metals are connected. He noticed, in particular, that 'each atom seems to have eight valences with respect to the atom of electricity' ('jedes Atom *scheint bezüglich des Elektrizitäts-*

[256] A notebook of Schrödinger ('*Vortrag Ramsay, Frühjahr* 1908, *über seine neue Theorie der chemischen Verbindung und der Electricität*') filed on AHQP Microfilm No. 39, Section 1, contains 15 pages of notes in German.

*atoms 8-wertig zu sein*'), that is, the properties of the chemical elements exhibited a periodicity with the period eight; 'the number of positive valences of an atom indicates how many electrons it possesses in the free, neutral state, i.e., how many electrons it can give away to another atom while forming a compound with it'; 'the number of negative [valences indicates], how many [electrons] it can receive maximally' (*Notebook, loc. cit.* p. 8). Evidently, in his talk Ramsey presented an extremely modern physical explanation of the electrical and chemical properties of atoms and of their combination to form compounds.

Schrödinger was a dutiful, serious student at the University of Vienna and known as such to his fellow students. Thus Hans Thirring, his junior by less than a year, reported: 'In the winter semester of 1907/08, being a complete beginner, I used to sit in the library of the *Mathematisches Seminar.* When one day a blond student entered the room, my neighbour pushed me and said suddenly: "This is Schrödinger." ("*Das ist der Schrödinger*!") I had never heard this name before, but the respect with which it was said and the sight of the colleague [i.e., of Schrödinger] made such an impression on me that I, from the very first meeting, obtained the conviction which became confirmed in the course of the years: "This man is really somebody special."' (Thirring, 1947, p. 106). And Thirring continued:

> The acquaintance soon developed into a friendship in which Schrödinger was nearly always the giving partner. During the common studies in preparation for the examinations, in discussing those aspects of the lecture courses which were difficult to grasp, the friend [i.e., Schrödinger] always played the role of the elder brother, whose intellectual superiority was acknowledged without jealousy. Long before he scored his great success with the formulation of wave mechanics, his narrow circle of friends was convinced about the fact that something very important was expected from him. We saw in him quite clearly a fiery spirit at work, who laboured with that scholarly ardour [and quest for exploration] which breaks the narrow limits of the special discipline in order to proceed independently on new paths probing into nature. (Thirring, 1947, p. 106)

Not only were his fellow students full of admiration; Schrödinger enjoyed a likewise good reputation among his professors, as a story—which happened in the summer of 1910—shows. At that time Ernst Mach was interested in the content of a manuscript of Paul Gerber, dealing with the relationship between gravitation and electromagnetism.[257] Since Mach could not arrive at a clear opinion about it, he requested Gustav Jäger to study the manuscript. The latter, however, passed it on to the theoretician Fritz Hasenöhrl, who after briefly looking at it, declared it worthy of attention, upon which Jäger asked Philipp Frank's opinion. Frank

[257] The title of Gerber's manuscript was *Gravitation und Elektrizität* (Gravitation and Electricity). For the details of the story we refer to Blackmore, 1972, pp. 262-263 and p. 352, footnote 8.

Gerber, a *Gymnasium* teacher in Stargard, had earlier published a paper on gravitation, entitled '*Die Fortpflanzungsgeschwindigkeit der Gravitation*' ('The Velocity of Propagation of Gravitation,' Gerber, 1902). This paper was later, after Einstein had expounded his theory of general relativity, reprinted at Ernst Gehrcke's request in *Annalen der Physik* (Volume 52, 1917).

now found Gerber's ideas rather confused and the mathematical derivations unsatisfactory. Upon Mach's insistence the mathematician Wilhelm Wirtinger pursued the case further. After a while he reported the results to Mach, writing: 'I have given Gerber's paper to a young electron man [i.e., a man experienced in the problems of electron theory], who otherwise seems quite reasonable, and he offered to give me a detailed opinion in return' (Wirtinger to Mach, 28 July 1910). Schrödinger indeed studied the paper carefully and found, as Frank had previously done, that 'some of the essential points in Gerber's paper remain completely unclear'. Wirtinger summarized the young man's answer as follows: 'Dr. E. Schrödinger has now written that detailed letter, and what seems striking to me is his objection that the whole thing [i.e., Gerber's proposed relation between the phenomena of gravitation and electromagnetism] is quite different when another kind of radioactive material is taken under serious consideration' (Wirtinger to Mach, 28 July 1910, reported in: Blackmore, 1972, p. 262).

In spite of his reputation in theoretical questions, which followed from these early testimonies, Schrödinger did not consider himself a theoretical physicist yet. Indeed, he carried out his first scientific research at Franz Exner's institute; it was an experimental investigation, '*Über die Leitung der Elektrizität auf der Oberfläche von Isolatoren an feuchter Luft*' ('On the Conduction of Electricity on the Surface of Insulators in Humid Air'), and led to his first publication, a paper which was communicated to the meeting of 30 June 1910 of the Imperial Academy of Sciences and appeared later in the *Sitzungsberichte* of the academy (Schrödinger, 1910).

Schrödinger's experimental work at Exner's institute originated, in connection with the interest existing there, in the problem of atmospheric electricity.[258] During the period of Schrödinger's university studies, Egon von Schweidler, especially carried out measurements on that subject (see, e.g., Schweidler, 1906; 1909a, b); and it was he who supervised the work of the doctoral candidate Schrödinger, because the latter thanked von Schweidler in his paper for the 'kind support' ('*gütige Unterstützung*,' Schrödinger, 1910, p. 1221). The particular question investigated by Schrödinger played quite some role in the research on atmospheric electricity. Thus Hans Benndorf wrote in a later review article on that subject:

> One of the biggest difficulties in the measurements of atmospheric electricity constitutes the maintenance of the necessary insulation. For that purpose often the most perfect insulators like amber, *Ambroid*, quartz glass, sulphur, hard rubber (ebonite) and paraffin fail because of the fact that conducting surface films (water films, dust bridges, snow, frozen fog, snail and worm secretion) are being formed. (Benndorf, 1928, p. 319)

Now, this problem of insulation was given to the doctoral candidate Schrödinger. As he noted: 'Starting from the known fact that electrostatic experiments succeed badly in moist air, I have tried to investigate the influence of moisture

[258] We shall discuss the subject of atmospheric electricity in Vienna in the following section.

on those insulating materials which are most often used in laboratories' (Schrödinger, 1910, p. 1215). Thus he took rods of ebonite, glass, polished amber, sulphur and paraffin, wrapping their ends in tin foil and connecting one with an accumulator battery, the other with an electroscope. While in dry air the electroscope showed no indication, it was charged when under the influence of moisture the surfaces of the rods became conducting. Schrödinger measured in particular the speed of charging the electroscope and derived from the results the electrical resistance of different materials as a function of the humidity. He found finally that glass had to be considered the worst, and paraffin the best insulator for measurements of atmospheric electricity.

Schrödinger submitted this research on the conductivity of insulators in moist air as his doctoral thesis to the philosophical faculty of the University of Vienna. He was then duly admitted to the necessary examinations for obtaining his doctorate, i.e., the *Hauptrigorosum* in physics and mathematics and the *Nebenrigorosum* in philosophy; he passed those in May 1910. In the fall of the same year he took leave of absence from the university to perform military service for a year.

At about the time when Schrödinger completed his doctorate, he also witnessed the partial resolution of the old debate between Boltzmann and Mach about the reality of atoms. Of course, by that time Boltzmann was dead and Mach, semiparalyzed, could hardly participate in any public debate. However, the new discoveries of the first decade of the twentieth century, especially in radioactivity, turned the scale on the side of Ludwig Boltzmann. According to the recollections of Stefan Meyer, a demonstration of the light flashes, created by $\alpha$-particles from radioactive substances on a scintillation screen, brought Mach to exclaim: 'Now I believe in the existence of atoms' ('*Nun glaube ich an die Existenz der Atome*,' Meyer, 1950, p. 4). One should perhaps not overestimate the significance of Ernst Mach's statement as implying a deep and complete conversion to atomism. Still this statement, made more or less privately, reflects perfectly the thorough change in the philosophical discussion at Vienna, occurring in the years after Boltzmann's death. What had given rise to bitter fights earlier now mellowed into a relaxed and conciliatory atmosphere. Erwin Schrödinger would describe this transition decades later in a letter to Arthur Eddington. He stated in particular:

> I was born and educated in Vienna with E. Mach's teaching and personality still pervading the atmosphere. I was devoted to his writings which I read practically all . . . , maybe about the same time when we were initiated into the restricted theory of relativity. Just as strong or even stronger than Mach's was in this time in Vienna the after-effect of the great Boltzmann, whose splendid pupil and admirer Hasenöhrl had just taken Boltzmann's chair (so cruelly evacuated a year before, 1906). Both Boltzmann and Mach were, as you know, just as much interested in philosophy, more especially in epistemology, as they were in physics, in fact all their later writing was pervaded by the epistomological ("*erkenntnistheoretisch*") outlook. Their views were not the same. But fulfilled with great admiration for the candid and incorruptible struggle for truth in both of them, we did not consider them irreconcilable. Boltzmann's idea consisted in forming absolutely clear, almost naively clear and detailed "pictures"—mainly in order to be *quite* sure of avoiding contradictory

assumptions. Mach's ideal was the cautious synthesis of observational facts that can, if desired, be traced back till to the plain, crude sensual perception (pointer reading). He was most anxious not to contaminate this absolutely reliable timber with any other one of a more doubtful origin.

However, we decided for ourselves, that these were just different methods of attack and that one was quite permitted to follow one or the other provided one did not lose sight of the important principles that were more strongly emphasized by the followers of the other one, respectively. (Schrödinger to Eddington, 22 March 1940)[259]

The reading of Mach and Boltzmann left a lasting impression on Schrödinger. He would, for the rest of his life, be interested in epistemological questions; he would state his points of view with the same clarity and sincerity as the great Viennese physicist-philosophers; and he would defend them with the same eagerness and stubbornness. With respect to atomism he would adopt from the beginning the strict advocacy of Ludwig Boltzmann, trying himself to contribute further proofs for the existence of atoms through his own physical investigations. Some of these investigations he had already performed in the first period of his career as a scientist, in the years before World War I.

## I.3 The Early Scientific Work of Erwin Schrödinger

The first period of Schrödinger's scientific activity embraces less than 3 years, from the fall of 1911 to the summer of 1914. During this time he contributed ten papers on various topics of experimental and theoretical physics and completed a review article on dielectric substances for a physical handbook. His major fields included the study of dielectric properties of matter, atmospheric electricity (both experimental and theoretical), the theory of X-ray patterns (von Laue diagrams) and of dynamic point lattices. On account of his work at the physical institute of Franz Exner he obtained his *Habilitation* in January 1914, the first promotion in his scientific career.

### Assistant at the *II. Physikalisches Institut*

In October 1911 Schrödinger returned to Vienna from military service and obtained the position of a substitute assistant (*Aushilfsassistent*) at his old, i.e., Exner's physical institute. After Egon von Schweidler left in the summer of 1911 to accept the professorship in Innsbruck, Fritz Karl Wilhelm Kohlrausch became Exner's main assistant. A member of a famous family of physicists, K. W. Fritz Kohlrausch—as he would sign his name—was born on 6 July 1884 at Gestettenhof near Türnitz, Lower Austria.[260] He studied at the University of Vienna, obtaining

[259] Letter filed on AHQP Microfilm No. 37, Section 5.

[260] K. W. Fritz Kohlrausch's grandfather was Rudolf Kohlrausch (1809-1858), who performed with Wilhelm Weber the well-known determination of the velocity of light by comparing the units for the electric charge in the systems of the electrostatic and electromagnetic units, respectively. Another member of the family, Friedrich Wilhelm Kohlrausch (1840-1910), president of the *Physikalisch-Technische Reichsanstalt* in Berlin, was K. W. Fritz Kohlrausch's uncle.

his doctorate in 1907. The following year he joined Exner's institute as assistant. Kohlrausch's early research was devoted especially to experimental investigations of problems in radioactivity and atmospheric electricity. Thus, in a paper of 1906, he confirmed the existence of statistical fluctuations, which Schweidler had predicted would occur in radioactive processes, in the case of the $\alpha$-decay rates of polonium (Kohlrausch, 1906a). In August 1907 he went on a voyage to Puerto Rico, in order to observe the electrical properties of tropical rain (October 1907 to January 1908), and to determine the atmospheric electricity and radioactivity of the atmosphere above the Atlantic Ocean (Kohlrausch, 1909). Shortly before Schrödinger started his new position, Kohlrausch became *Privatdozent* for physics at the University of Vienna. At the same time he presented lecture courses at the Vienna *Musikakademie*, where he had become a *Dozent* for acoustics.

The experienced, senior experimentalist Kohlrausch was just the right person to organize, together with the more theoretically-minded Schrödinger, the physical laboratory courses at Exner's institute. He also stimulated some of Schrödinger's research, especially on topics of atmospheric electricity and radioactivity, and later in colour theory. Schrödinger recalled later on: 'In the following years [i.e., after 1911] I learned, as laboratory assistant of Franz Exner and in close collaboration with my friend K. W. F. Kohlrausch, what performing experiments meant—without ever learning to do it myself' (Schrödinger, 1935, p. 87). Both would become close friends for life, and at Kohlrausch's holiday place of Seeham near Salzburg Schrödinger would meet, in 1913 through Kohlrausch's mediation, his future wife, Annemarie Bertel. (See AHQP Interview with Mrs. Schrödinger, 5 April 1963, p. 2.)[261]

Schrödinger devoted only part of his publications to purely experimental topics; in fact, he submitted only one paper in which he reported about experimental observations made completely by himself (Schrödinger, 1913b). Still, in these days one cannot say that Schrödinger showed no interest in experiment; not only were all of his investigations closely related to experimental problems, he also became involved in them even without being able to publish a neat result. Thus his Viennese friend and colleague Hans Thirring remembered that: 'Already two decades before high-voltage arrays were used in atomic physics, Schrödinger one day carried a large influence machine into his office, in order to find out experimentally whether the penetration power of $\beta$-rays would be reduced if the source of radiation is charged to a high negative potential' (Thirring, 1947, p. 106). This particular experimental effort may be related to an investigation, which Schrödinger performed in 1914 with Fritz Kohlrausch and which led to the joint publication, '*Über die weiche ($\beta$) Sekundärstrahlung von $\gamma$-Strahlen*' ('On the Soft ($\beta$) Secondary Radiation from $\gamma$-Rays,' Kohlrausch and Schrödinger, 1914). Other experimental work would emerge from Schrödinger's interest in theoretical questions. Again Thirring reported: 'A few years later [i.e., after 1914] he attacked

[261] In 1920 Kohlrausch became *Ordinarius* for physics at *Technische Hochschule* in Graz, a position which he occupied until his death on 17th September 1953.

experimentally the then frequently discussed problem of "needle radiation" by studying the question of whether two bundles of rays, which are emitted from the same light source and whose respective directions make a large angle, can still interface' (Thirring, 1947, p. 106).[262] Indeed, even as late as 1925 Schrödinger, the meanwhile firmly installed professor of theoretical physics, would undertake simple experiments in physiological optics.[263] There remains no doubt that Schrödinger, when he began his scientific career at Exner's experimental institute, was not bound to become a pure theoretician. His own talents and abilities, however, soon indicated a preference for theoretical, even mathematical methods, as can already be seen from the analysis of his early publications.

## Schrödinger's Paper on the Kinetic Theory of Magnetism

The first publications of the new assistant for experimental laboratory courses dealt with calculations within the kinetic theory of matter and electron theory. Schrödinger had already thoroughly been acquainted with these subjects as a student.[264] In his paper '*Zur kinetischen Theorie des Magnetismus*' ('On the Kinetic Theory of Magnetism'), which was communicated to the meeting of 20 June 1912 of the Vienna Academy of Sciences, he now suggested how an additional diamagnetic contribution to atomic moments might arise, besides the one created by the electrons orbiting in the atoms (Schrödinger, 1912a).

The study of the magnetic properties of chemical elements already belonged to the working programme at Exner's institute before 1900. Especially, Stefan Meyer had—as we reported earlier (in Section I.2)—in a series of papers published since 1899, measured systematically the magnetic susceptibility of atoms and molecules of different chemical elements and compounds. Later on he concentrated upon the rare earth elements (e.g., Meyer, 1901, 1908). The latter substances turned out to be basically paramagnetic, while many other substances were found to be diamagnetic. Based on electron theory Paul Langevin developed a complete description of the 'complicated phenomena of magnetism and diamagnetism' in 1905 (Langevin, 1905a, p. 71). He explained the diamagnetism by referring to the fairly old idea—of André-Marie Ampère and Wilhelm Weber—that electrons (mass $m$, charge $-e$) move in atoms in orbits with the mean square radius $\overline{r^2}$; thus he obtained the constant negative susceptibility (for $N$ electrons)

$$\chi = -\frac{Ne^2}{4m}\overline{r^2} \tag{14}$$

independent of the temperature. However, he pointed to the fact that the metal

[262] We shall discuss the experiment in detail in Section I.5.

[263] We shall report about this investigation in Chapter II (Section II.4).

[264] Besides the lectures of Fritz Hasenöhrl on statistical mechanics and the courses of Karl Przibram on the constitution of matter—to which we referred in the last section—there exist in the Schrödinger *Nachlaß* two notebooks, entitled '*Hasenöhrls Seminar 1910/1911*'. These notes deal with problems of electron theory and molecular collisions. (See AHQP Microfilm No. 39, Section 1.)

bismuth, which was diamagnetic, did show a different behaviour—i.e., the absolute value of its susceptibility decreased linearly with falling temperature. In this context, he referred to a suggestion, made by Joseph John Thomson at the 1900 Paris Conference (Thomson, 1900), namely, that this might be due to 'the presence of free cathode ray particles in the metals which provide their conductivity, and whose mean free paths—particularly long in the case of bismuth—are curved by the external magnetic field in the sense corresponding to diamagnetism' (Langevin, 1905a, p. 90). In 1909 Kōtarō Honda at the Bosscha Laboratory in Berlin, and 2 years later Morris Owen at the same place (when measuring magnetic susceptibility), further found, in the case of most diamagnetic elements, a definite temperature dependence (Honda, 1910; Owen, 1912). This seemed to support Thomson's hypothesis.

In his notebook on the seminars in the summer of 1909 Schrödinger discussed in detail the motion of an electron in a magnetic field.[265] Now in 1912 he thought of applying his knowledge to determine whether Thomson's idea gave rise to the observed temperature-dependent diamagnetism. He based his calculation on Hendrik Lorentz' picture of the constitution of metals (Lorentz, 1905), according to which a metal consisted of two elementary objects, 'the relatively large and heavy metal atoms, and one type of electrons with charge $e$ and mass $m$, which move freely in the spaces between the atoms' (Schrödinger, 1912a, p. 1307).[266] Both elementary objects perform a motion according to the kinetic theory; due to the large mass one could, however, neglect the motion of the atoms, and due to the small size of the electrons only collisions between electrons and atoms played a role, yielding a mean free path $\lambda$ of the electrons. Schrödinger further assumed a Maxwellian velocity distribution of the electrons in the absence of any external forces and demonstrated—in Appendix 2 of his paper—that it was not changed when an arbitrary static magnetic field was applied. In a static magnetic field the originally straight orbits of the electron (between collisions) will be curved, thus they turn into pieces of screw lines, and the author therefore had to compute the magnetic moment of a macroscopic body arising from these motions of the electrons in it. He obtained, after integrating over the possible paths and over the velocity distribution of the electrons, indeed *a magnetic moment in the opposite direction* of the magnetic field and proportional to the field strength $H$; that is, it yielded a diamagnetic susceptibility $\chi_{\mathrm{f}}$,

$$\chi_{\mathrm{f}} = -\frac{1}{3}\frac{e^2}{m}\lambda^2 N(1 - 2\alpha^2), \tag{15}$$

with $\alpha = (\sqrt{3/2})[\lambda e H / m\sqrt{\overline{c^2}}]$ ($\overline{c^2}$ being the average square velocity of the electrons

[265] This notebook, entitled '*Physikalisches Seminar, Sommer 1909*,' has been filed on AHQP Microfilm No. 39, Section 1.

[266] Lorentz' model of electrical conductivity had also been used earlier by Austrian physicists, e.g., by Ernst Lecher in Prague (Lecher, 1906) and by Gustav Jäger in Vienna (Jäger, 1908).

in the Maxwellian distribution).[267] By inserting the values for $\lambda$ and $N$ for bismuth, lead, copper and silver from a table given by Eduard Riecke (1909) and the known value for $e/m$, and by comparing the theoretically obtained results with the experimentally observed ones by Honda and Owen for a temperature of 18°C, Schrödinger arrived at the conclusion: 'We see that $\chi_f$ varies between a fifth [for bismuth] and more than two-hundred-fold [for copper] of the observed value and assumes for the good conductors copper and silver an order of magnitude which is not reached by any of the diamagnetic substances we know' (Schrödinger, 1912a, p. 1317). In order to restore agreement with experiment Schrödinger proposed the explanation, 'that in these cases [of disagreement] the diamagnetism of the free electrons must be covered in part by paramagnetism' (Schrödinger, 1912a, p. 1318).

Schrödinger's early contribution to the theory of diamagnetic properties of metals constituted a nice application of the theoretical methods, which he had learned as a student, to physical ideas proposed by renowned senior physicists, such as Thomson, Lorentz or Langevin. As it would turn out in the course of the years, he did not provide the right solution to the problem posed, as it was too early to think of the solution of any of the problems involving metal electrons before the arrival of the correct quantum statistics for electrons.[268]

## Ambitious Speculations on the Kinetics of Dielectric Substances and on the Theory of Solids

Schrödinger continued, encouraged by the apparent sucess of his first paper on kinetic theory, with a more ambitious programme. On 17 October 1912, less than four months later, he presented a second paper to the Vienna Academy of Sciences, bearing the title '*Studien über Kinetik der Dielektrika, den Schmelzpunkt, Pyro- und Piezoelektrizität*' ('Studies on the Kinetics of Dielectric Substances, the Melting Point, Pyro- and Piezo-Electricity,' Schrödinger, 1912b). Schrödinger addressed the ultimate goal in the first sentences of the introduction, stating: 'While today the difference between the gaseous and fluid state, at least in principle, seems to have been solved and removed to a certain degree by Van der Waals' theory, the same does not hold for the transition from the fluid to the solid state. This latter process indeed seems to be much more mysterious than the first one mentioned' (Schrödinger, 1912b, p. 1937). The mystery mentioned showed up in the fact that the transition from fluid to solid was not necessarily

[267] Under normal conditions we have $\alpha \ll 1$. For a very large magnetic field strength $H$ and very low temperatures (corresponding to small values for $\overline{c^2}$), the second term on the right-hand side of Eq. (15) contributed noticeable 'paramagnetic' susceptibility, proportional to the square of the field strength. As an example, Schrödinger found, in the case of copper at 20° absolute temperature for a field strength of $5 \times 10^4$ gauss, a reduction in the diamagnetic susceptibility (first term in Eq. (15)) of 25 per cent.

[268] Nearly fifteen years later Wolfgang Pauli, Schrödinger's younger Viennese countryman would start the consistent explanation of the phenomena with his paper on the paramagnetism of metal electrons (Pauli, 1927a).

accompanied by a contraction in volume but rather by the occurrence of a regular crystal structure. Schrödinger continued: 'This can certainly be explained by general thermodynamics; atomistics so far, however, has not provided a proper description' (Schrödinger, 1912b, p. 1938). As a scientific descendent of Boltzmann he felt that this problem had to be attacked, and he believed he had in hand the key to a kinetic theory of solids.[269]

Schrödinger derived the stimulus for his proposed theory from an article of Peter Debye in Zurich on '*Einige Resultate einer kinetische Theorie der Isolatoren*' ('Some Results of a Kinetic Theory of Insulators'), published in a February issue of *Physikalische Zeitschrift* (Debye, 1912a). In it Debye had assumed that an extra term had to be introduced into the description of the effect of the electric field on a fluid substance; that is, the electric field $E$ not only caused the elastically bound electrons in the molecules of the fluid to shift their position (leading to a dielectric polarization $P$), but also acted on 'permanent dipoles of constant electric moment [$m$]' (Debye, 1912a, p. 37), existing in the interior of the insulating substances. He had thus obtained a temperature dependence of the polarization $P$ according to the equation

$$P = \left(E + \frac{P}{3}\right)\left(\frac{3a}{T} + 3b\right), \tag{16}$$

with

$$a = \frac{Nm^2}{3k}, \tag{17}$$

where $N$ denoted the number of molecules per $cm^3$ and $k$ the Boltzmann constant. Hence the dielectric constant $\varepsilon$ ( $= 1 + P/E$) satisfied the relation

$$\frac{\varepsilon - 1}{\varepsilon + 2} = \frac{a}{T} + b, \tag{18}$$

in which $b$ described the temperature-independent term provided by the polarization of the bound electrons—the only term considered so far—and $a/T$ the new, temperature-dependent part arising from the permanent dipoles.[270] Debye had further noticed that the dielectric constant $\varepsilon$ assumed a singular (infinite)

[269] Ludwig Boltzmann displayed in detail Van der Waals' theory of gaseous and fluid states in the second volume of *Vorlesungen über Gastheorie* (Boltzmann, 1898, Section I). He did not address there the problem of the solid state at all, nor had he published anything on it later (or earlier). However, in the Boltzmann *Nachlaß* a manuscript, entitled '*Versuche einer Theorie des festen Körpers vom Standpunkte der mechanischen Wärmelehre*' ('Attempts at a Theory of Solids from the Point of View of the Mechanical Theory of Heat') has been found. In this manuscript the author dealt with the law of Pierre Louis Dulong and Alexis Thérèse Petit for the specific heats of solids; he tried to derive it on the basis of the assumption that the solids consist of molecules performing thermal vibrations about their average positions. (See Fasol, 1982, pp. 86-87.)

[270] Experiments—Debye quoted, e.g., an investigation by the young Hasenöhrl (1896a)—had demanded such a temperature-dependent term.

value at a critical temperature $T_k$,

$$T_k = \frac{a}{1 - b} \tag{19}$$

and concluded: 'If the temperature $T$ is decreased below $T_k$, the detailed theory shows that a polarization may exist in the interior of the body even when an external field is absent; i.e., one would be able to observe then a residual dielectric polarization' (Debye, 1912a, p. 99). However, he had immediately dismissed this possibility as being unrealistic, since the fluids turned into solids at temperatures considerably above $T_k$, Eq. (19), and therefore 'the validity of the relation (18) ceases, since the molecules cannot rotate freely any longer' (Debye, 1912a, p. 99). Still, he had pointed out the analogy of the temperature $T_k$ for dielectric substances to the Curie temperature for ferromagnetic substances.

The last remark in Debye's paper interested Schrödinger a few months later. He suggested in particular the idea 'that a certain ... "critical" temperature following for each dielectric substance, below which spontaneous electrization must occur, and which is the exact analogue of the Curie point for ferromagnetic substances; that this critical temperature is identical with the melting temperature' (Schrödinger, 1912b, pp. 1938–1939). He knew the theory of ferromagnetism, which Pierre Weiss had developed by generalizing Paul Langevin's molecular-kinetic description of the magnetic properties of matter (Weiss, 19ᴗ7). If one took Debye's Eq. (16) as describing the electrical analogue of paramagnetism, then a generalized equation could be written down, namely,

$$P = 3b\left(E + \frac{P}{3}\right) + Nm\left(\frac{\cosh \alpha}{\sinh \alpha} - \frac{1}{\alpha}\right), \tag{20}$$

with

$$\alpha = \frac{m\left(E + \frac{P}{3}\right)}{kT} \tag{20a}$$

corresponding to Weiss' equation for para- *and* ferromagnetic substances.

As in the magnetic analogue this equation passed over, in the limit of small $\alpha$ or of high temperatures, into Debye's Eq. (16), which in Schrödinger's theory accounted only for fluids or for irregular, amorphous solid substances. Below the critical temperature $T_k$ the full Eq. (20) must be applied at least in the case of crystalline substances. A difficulty remained in so far as Schrödinger wished to identify $T_k$ with the melting temperature of solid crystals, while Eq. (19) provided too low values. In order to establish agreement with experiment, Schrödinger modified Eq. (20) by inserting everywhere for $E + \frac{1}{3}P$ the new expression $E + (\mu/3)P$. He obtained the dimensionless parameter $\mu$ by identifying the new critical temperature, $T_k^{\text{Schr.}}$, i.e.,

$$T_k^{\text{Schr.}} = \frac{\mu a}{1 - \mu b}, \tag{21}$$

with the experimental melting temperature of a given crystalline substance, finding $\mu$-values slightly above unity. He thus concluded, 'that the dielectric substances of Debye below the critical temperature are nothing but solid dielectric substances; that the solid substances owe the strong binding of their constituents to just this intense self-electrization, which arranges all molecular moments parallel to a far higher degree than can be achieved by a field applied to a fluid; and that thus the anisotropic structure exhibited by all real solids is created' (Schrödinger, 1912b, p. 1943).[271]

Schrödinger was not bothered by the observation that the usual solids did not show any observable, macroscopic electrization. Even for an 'elementary crystal,' in which all permanent moments have the same direction in the absence of an external electric field, one could not measure a gross electric moment, 'because this electrization remains constant as long as we do not change the temperature, and therefore will be fully compensated, as far as its external actions are concerned, by surface charges' (Schrödinger, 1912b, p. 1949). If, on the other hand, an external electric field were applied, the difference between the electrization caused by it and the spontaneous electrization would be observed. Schrödinger further explained the phenomena of piezo- and pyro-electricity—i.e., the creation of electric charges due to mechanical deformation and to temperature change of certain crystals, respectively—by studying in detail the cases of quartz and tourmaline, noticing a 'quite satisfactory agreement between theory and experiment' (Schrödinger, 1912b, p. 1970).

One may note one feature of the theory stressed by Schrödinger as being quite remarkable. He wrote explicitly: 'I want to put the main emphasis on the fact that the theory is able to predict correctly the order of magnitude of all these phenomena without any assumption about the absolute magnitude of the electric charges and moments of the molecules' (Schrödinger, 1912b, p. 1939). In this remark we again recognize the opposition of the official Viennese physicists against models of atoms or molecules, such as Arthur Haas had proposed in his treatment of the hydrogen atom; and in addition a cautious attitude with respect to the still unsettled question—at least in Vienna—about the elementary charge.[272]

[271] Schrödinger realized, of course, that in an anisotropic solid there existed more than one value for the parameter $\mu$; but at least one of the values had to be larger than unity and thus would determine the critical temperature $T_k^{\text{Schr.}}$, Eq. (21).

Schrödinger also contradicted Debye's argument, that in a solid free rotation of the molecular dipoles was not possible, by pointing to the fact that the same assumption (of free rotation) worked in the case of the ferromagnetic analogue.

[272] Schrödinger's specific opinion concerning detailed atomic models and their constants may be represented by a remark in a later paper (Schrödinger, 1913b). There he referred, as we shall discuss a little later, to a work of Peter Debye on the role of molecular dipoles on dispersion, and praised it with the words: 'With it a point in the dispersion formula, where usually vague electron constants are inserted, is clarified by introducing the macroscopically observable constant of internal friction; this must be welcomed for the same reasons as, a while ago, the introduction of the coefficient of elasticity in the quantum formula for the specific heats' (Schrödinger, 1913b, p. 1167).

Evidently, Schrödinger wished to remain on safe, sure ground, both with respect to his theoretical assumptions and hypotheses and with respect to a reliable comparison of theoretical predictions and experimental data.

The problem of establishing an equation of state for solids, notably on the basis of the kinetic theory of matter, occupied quite a few physicists in the early 1910s, among them Walther Nernst and Eduard Grüneisen in Berlin, Peter Debye and Simon Ratnowsky in Zurich, and Rudolf Ortvay in Munich. For example, Grüneisen (1877–1949) presented at the Karlsruhe *Naturforscherversammlung* in September 1911 quite detailed ideas in his talk '*Zur Theorie einatomiger fester Körper*' ('On the Theory of Monatomic Solid Bodies,' Grüneisen, 1911). He assumed, in particular, that the atoms in monatomic solids exert repulsive and attractive forces on each other—the latter being identical with the Van der Waals forces known from gas theory—and that the atoms perform an oscillatory temperature motion, possessing the energy $\int_0^T c_v \, dT$, where $c_v$ denotes the specific heat at constant volume.[273] He thus obtained the following equation of state

$$pv + \frac{A}{v} - m\frac{B}{v^m} = \frac{3m+2}{6} \int_0^T c_v \, dT, \tag{22}$$

where $p$ denotes the pressure and $v$ the volume. The term $A/v$ arose from the Van der Waals cohesive force, while the negative term on the left-hand side came from the repulsive force.[274] Grüneisen evaluated the expression on the right-hand side for the oscillation energy by inserting for $c_v$ either Einstein's quantum-theoretical formula (Einstein, 1906g) or Nernst and Frederick Alexander Lindemann's modified formula of 1911 (Nernst and Lindemann, 1911a). Equation (22) described a large number of experimental observations on metals so well that Nernst remarked in the discussion of Grüneisen's talk: 'You [i.e., the audience] have certainly got the impression that we have now proceeded far enough to be able to treat the solid state on the basis of a molecular theory; and it may also be hoped that we will soon have in hand a similarly complete theory of the solid state, as we have possessed for some time for the gaseous state' (Nernst in: Grüneisen, 1911, p. 1028).

Simon Ratnowsky pursued the same basic ideas as Grüneisen in a paper published in the *Verhandlungen der Deutschen Physikalischen Gesellschaft* of January 1913 (Ratnowsky, 1913).[275] The main point, which he modified, was the right-hand side of Eq. (22), where he used for $c_v$ the temperature function derived the year before by his teacher Debye (Debye, 1912b, c). Debye then extended the theory of solids by also admitting anharmonic oscillations of the atoms and molecules; he gave a detailed report on this in his lecture '*Zustandsgleichung und*

[273] Grüneisen based his approach on an earlier approach to the same problem by Gustav Mie (Mie, 1903). Mie, unlike physicists before and after him (including Schrödinger), started from the assumption 'to consider fluids and solids to be essentially in the same state' (Mie, 1903, p. 658).

[274] Grüneisen's integer $m$ was determined by the repulsive forces; the author assumed that at absolute zero temperature they contributed the energy term $B/(v_0)^m$, with $v_0$ the corresponding volume.

[275] We shall return in Chapter II (Section II.3) to the background and work of Simon Ratnowsky.

*Quantenhypothese*' ('Equation of State and Quantum Hypothesis') presented at the Wolfskehl week of April 1913 in Göttingen (Debye, 1914). The further solid state theory developed more or less on the lines suggested by Grüneisen and Debye, with essential contributions by Max Born. It bypassed completely Schrödinger's theory of late 1912 based on classical kinetic theory and the analogy to ferromagnetism.[276]

## The Review Article on Dielectrics and a Paper on Anomalous Dispersion

Schrödinger also treated the subject of the dielectric properties of matter in two further publications; one was a 75-page article on '*Dielektrizität*' ('Dielectrics') for the *Handbuch der Elektrizität und des Magnetismus* (Schrödinger, 1918e), the other was a short note on the theory of anomalous electric dispersion (Schrödinger, 1913a). The above-mentioned *Handbuch*, edited by the Munich professor for experimental physics Leo Graetz (1856–1941), was planned to extend over five volumes. While parts of Volumes I to IV appeared in print between 1912 and 1914, the first issues of Volume I, including the contribution of Schrödinger, were delayed until 1918. Among the authors of the first Volume, entitled '*Elektrizitätserzeugung und Elektrostatik*' ('Generation of Electricity and Electrostatics'), we find the editor Leo Graetz himself (with an article on '*Reibungselektrizität*' ['Frictional Electricity'], Eduard Riecke ('*Pyroelektrizität und Piezoelektrizität*' ['Pyro-electricity and Piezo-electricity']), Woldemar Voigt ('*Elektrooptik*' [Electro-optics']), and Egon von Schweidler ('*Die Anomalie der dielektrischen Erscheinungen*' ['The Anomaly of Dielectric Phenomena']). Schrödinger found himself as '*Fachgelehrter*' (expert) in good company with his former teacher Schweidler, or with his colleague Fritz Kohlrausch, who contributed to another volume of the same *Handbuch.*[277]

Schrödinger's article consisted of one theoretical part (I) and four primarily experimental parts (II–V). In Part I, after a detailed historical introduction to older theories, he discussed Maxwell's theory of dielectrics and then the electron theory of absorption and dispersion of electromagnetic radiation. In Part II, he presented the experimental methods of determining the static dielectric constant, in Part III those of the dynamical dielectric constant (which includes the refraction and absorption index). He devoted Part IV to the changes in the dielectric constant (occurring with a change in temperature, in the state of the substance, in the pressure, and in the electric field strength), and in Part V he considered the relationship between the dielectric constant and the chemical constitution and composition of the substances. Schrödinger presented the material in a clear and well organized form; he referred to an enormous amount of literature—i.e., to

[276] One may, in turn, ask why Schrödinger did not pay attention to Grüneisen's earlier work. A reason certainly was that the latter involved the quantum hypothesis, with which Schrödinger was not yet familiar. He might also have preferred an approach which did not involve detailed assumptions about molecular forces.

[277] See the title page of *Handbuch der Elektrizität und des Magnetismus*, Volume I.

several hundred papers including his own (Schrödinger, 1912b)—published up to the end of the year 1912. Certainly the article on dielectrics demonstrated the high diligence and skill of the young author and established for him a reputation as a writer of review articles for the future.

In his '*Notiz über die Theorie der anomalen Dispersion*' ('Note on the Theory of Anomalous Dispersion'), submitted in November 1913, Schrödinger changed to that topic, on which Egon von Schweidler wrote the article in Graetz' handbook (Schrödinger, 1913a).[278] With anomalous dispersion one usually denoted in optics the phenomena, caused by a refractive index very different from unity; this occurs for frequencies of the incident light close to the characteristic frequencies of the substance, i.e., the frequencies they strongly absorb and emit. Dielectrics, however, show anomalies not only in the optical range. Thus Schweidler investigated in a paper of 1907, entitled '*Studien über die Anomalien im Verhalten der Dielektrika*' ('Studies on the Anomalies in the Behaviour of Dielectrics'), which won him the *Baumgartner Prize* of the Austrian Academy of Sciences, the unusual properties of several substances under the action of a constant or a slowly varying electric field (Schweidler, 1907). What Schrödinger addressed in 1913 were not the effects treated earlier by Schweidler (which exhibited a certain analogy to ferromagnetism, including the occurrence of a hysterisis); he rather meant the empirical fact that the refractive index of various substances, e.g., water and alcohol, decreases from a rather high value for low frequencies to a value not much different from unity for optical frequencies. In order to explain such behaviour, Peter Debye had applied in a recent contribution, '*Zur Theorie der anomalen Dispersion im Gebiete der langwelligen elektrischen Strahlung*' ('The Theory of Anomalous Dispersion in the Region of Long-Wave Electromagnetic Radiation'), published in the issue of 30 August 1913 of the *Verhandlungen der Deutschen Physikalischen Gesellschaft*, his idea of the existence of rigid, constant electric moments in the interior of molecules (Debye, 1913c). He argued qualitatively that an electric field acting on a dielectric substance would align the axis of their permanent electrical moments within a short but finite time interval, thus providing a macroscopic electric moment which causes the high dielectric constant observed in a static field; in an alternating field, especially one with rapid oscillations, the molecules would be unable to align, and as a consequence only the bound electrons and not the permanent dipoles contribute to the dispersion phenomena. Debye had subsequently, based on the mathematical formulation of these considerations, obtained an expression for the complex refractive index $n$ as a function of the frequency $\nu$ of the electromagnetic radiation,

$$n^2 = \frac{\dfrac{\varepsilon_0}{\varepsilon_0 + 2} + 2\pi i \nu \tau \dfrac{\varepsilon_\infty}{\varepsilon_\infty + 2}}{\dfrac{1}{\varepsilon_0 + 2} + 2\pi i \nu \tau \dfrac{1}{\varepsilon_\infty + 2}}, \tag{23}$$

[278] We have already mentioned Schweidler's article on anomalous dispersion.

with $\varepsilon_0$ and $\varepsilon_\infty$ denoting the dielectric constants for very low (static) and very high (optical) frequencies, respectively, and $\tau$ representing a 'relaxation time' given by the expression $\rho/kT$ (with $\rho$ a quantity proportional to the coefficient of friction $\eta$ and the volume of molecules, i.e., $\rho = 8\pi\eta a^3$; $k$ Boltzmann's constant, and $T$ absolute temperature). The change of the refractive index from the high static value to the low optical value occurred around the frequency $(2\pi\tau)^{-1}$.[279] Debye concluded:

> Even though the results of the above discussion require quantitative improvements, we still believe we have obtained a sound theoretical treatment of the essential features of anomalous dispersion and absorption in the electric spectrum, both with respect to the dependence on the frequency of the alternating field and on the temperature. (Debye, 1913c, p. 793)

Schrödinger, who considered Debye's idea of permanent dipoles an important contribution to the kinetic theory of matter, analyzed in his paper of November 1913 the above-mentioned results of Debye on anomalous dispersion. He claimed, in particular, that the frequency dependence of the refractive index could only be explained, 'if the permanent dipoles provide a *new mathematical element* in the dispersion theory, which is not provided by the polarization electrons' (Schrödinger, 1913a, p. 1167). However, Schrödinger noted: 'Unfortunately, it turns out now that this hope failed. The expression, which Debye obtained for the complex refraction index, indeed agreed formally with a formula, which [Paul] Drude derived on the assumption of *one type of aperiodically damped electrons*' (Schrödinger, 1913a, pp. 1167-1168). Paul Drude had found, for the refractive index in that case, the equation

$$n^2 = \varepsilon_0 + \frac{\varepsilon_1}{1 + i\,.\,2\pi\nu a_1}, \tag{24}$$

with $\varepsilon_1$ and $a_1$ denoting electron constants (Drude, 1898). Schrödinger, after introducing the optical dielectric constant $\varepsilon_\infty\ (= \varepsilon_0 + \varepsilon_1)$ and identifying the expression $a_1[(\varepsilon_0 + 2)/(\varepsilon_\infty + 2)]$ with a relaxation time $\tau$, transformed Eq. (24) into the equivalent one of

$$n^2 = \frac{\dfrac{\varepsilon_\infty}{\varepsilon_\infty + 2} + 2\pi i\nu\tau\dfrac{\varepsilon_0}{\varepsilon_0 + 2}}{\dfrac{1}{\varepsilon_\infty + 2} + 2\pi i\nu\tau\dfrac{1}{\varepsilon_0 + 2}}, \tag{24'}$$

which formally agreed with Debye's equation, except for the exchange of $\varepsilon_0$ and $\varepsilon_\infty$. Since Eq. (24) did not describe the empirical situation correctly, Debye's Eq. (23) would not do either. Schrödinger also analyzed the temperature dependence, found experimentally for various fluids, and concluded, 'that besides the per-

[279] With a molecular radius of $10^{-8}$ cm the 'relaxation frequency' for water at 18°C would be $2.35 \times 10^{10}\ \text{sec}^{-1}$, corresponding to a wavelength of 1.28 cm.

manent dipoles other parts of the dielectric constant [formula] also follow the slow oscillations but not the rapid ones' (Schrödinger, 1913a, p. 1170). However, he did not propose an alternative to Debye's theory; instead, he suggested to investigate experimentally the situation further and to observe the details of the dispersion curves and their temperature dependence.[280]

## Research on Atmospheric Electricity in Vienna

During his years as an assistant Erwin Schrödinger also published two papers in the series '*Beiträge zur Kenntnis der atmosphärischen Elektrizität*' ('Contributions to the Understanding of Atmospheric Electricity'), which were distributed among many issues of the *Sitzungsberichte* of Vienna Academy of Sciences. The initiator of the series was the chief of his institute, Franz Exner, who himself contributed to the study of atmospheric electricity from the 1880s and transformed Vienna into one of the leading centres of research in that field far into the twentieth century.

The interest in atmospheric electricity originated in the early eighteenth century, when several authors, including the English clergyman, D. William Wall and an inhabitant of an old-people's home in London, Stephen Gray (1666/1667–1736), suspected the identity of electric sparks created artificially—say, by rubbing amber—and the lightning phenomena connected with thunderstorms.[281] The American printer, publisher, author, diplomat and scientist Benjamin Franklin later took up the subject by suggesting and performing experiments on thunderstorm electricity. The crucial experiment, however, had been made at Marly-La Ville, France, where—stimulated by Franklin's writings—the director of the Royal Botanical Gardens, Thomas François Dalibard (1703–1799) put up a large lightning conductor device; after a thunderstorm on 10 May 1752 electric sparks were drawn from the well-isolated pole of the device. The very same year the French astronomer Pierre Charles Lemonnier (1715–1799) discovered that even in fair weather a state of electricity persisted in the atmosphere.[282] Later observers improved the experimental methods; thus, e.g., Horace Bénédict de Saussure (1740–1799) of Geneva constructed a sort of electrometer allowing one to observe the annual variation of the fair weather electric field, and the Italian priest and

[280] Schrödinger was fairly convinced that the description of the anomalous dispersion was more complicated than Debye had assumed; he suggested, for example, considering besides the permanent dipoles one type of aperiodically damped electrons as Paul Drude had assumed—this would provide more constants to fit into experimental data than Debye's theory; he also thought it possible that 'a simple, generally valid, electric dispersion formula was excluded from the very beginning' (Schrödinger, 1913a, p. 1170).

[281] For the early history of atmospheric electricity we refer to Fritz Fraunberger's *Elektrizität im Barock* (Electricity in the Baroque Age, Fraunberger, 1964).

[282] The history of atmospheric electricity has been treated, for example, in review articles by E. von Schweidler (1918), H. Benndorf and V. F. Hess (1928), and in the article on 'Atmospheric Electricity' in *Encyclopaedia Britannica* **8**, Chicago-London-Toronto-Geneva: Encyclopaedia Britannica Inc., 1962, especially pp. 212–213.

physicist Giovanni Battista Beccaria (1716-1781) noticed that the charges registered in the atmosphere were always positive. In the nineteenth century further observations yielded the conclusion, 'that for fair weather at all places on earth, where measurements were performed, the [electric] potential increases with the rising height of the air; that the electric field strength is greater in winter than in summer; and that at most places there exists a daily change of the field, having two maxima within 24 hours' (Benndorf and Hess, 1928, p. 519). After 1859 William Thomson, later Lord Kelvin (1824-1907), raised the research on atmospheric electricity to a new scientific level by devising new precision instruments—such as reliable, even portable, electrometers to determine the electric charge, and an instrument to measure the electric field—providing the basis of an exact theory of observational methods, and by inaugurating a programme of continuous recordings at Kew Observatory in England. Interest in the topic continued, notably in Great Britain, the United States (where Henry Rowland became involved), Italy, Germany, and in particular Austria.

Benndorf and Hess stated in their review article:

> The research on atmospheric electricity received a mighty impetus from the activity of Franz Exner in Vienna, who—stimulated by Thomson's work—turned, in the middle of the eighties of the last century, to investigations of atmospheric electricity. Exner constructed a comfortably portable set of instruments, which could be used while travelling, to measure the electric field above the earth, and he himself carried out numerous observations at sea and in tropical regions. He was the first to emphasize strongly the necessity of establishing absolute measurements, normalized to the plane surface, of the 'fair-weather electricity,' in order to be able to derive interpretable laws [i.e., laws which could be physically interpreted]. Exner also succeeded to a high degree in getting his students interested in studies of atmospheric electricity; hence a great number of papers subsequently emerged from the Exner school; they appeared under the collective title '*Beiträge zur Kenntnis der Luftelektrizität*' ('Contributions to the Understanding of Atmospheric Electricity') in the *Sitzungsberichte* of the Vienna Academy, and they quite properly document Exner's influence on the research in atmospheric electricity. (Benndorf and Hess, 1928, p. 520)

Moreover, as von Schweidler noted: 'He [i.e., Exner] was also the first to arrange balloon observations, which ... proved the existence of *positive* charges in the atmosphere' (Schweidler, 1918, p. 236).

Exner's simple explanation of the earth's electric field, based on the assumption that negative electric charges exist on the surface of the earth and that fluctuations of field strength are caused by negative charges carried by water vapour occurring in the atmosphere, had to be modified considerably in the light of later discoveries. First, the fact that air is not a perfect insulator, concluded already in 1785 by Charles Auguste Coulomb (1736-1806), received new attention more than one hundred years later (by W. Linss in 1887), upon which physicists eagerly started to study conduction properties of the atmosphere. In Wolfenbüttel especially Julius Elster and Hans Geitel became interested in air-electricity research from

Exner's work, and constructed 'a device with which the dispersion of electricity in the free atmosphere might be determined in an unobjectionable manner' (Elster and Geitel, 1899b, p. 11).[283] They then proved beyond doubt that even clean and dust-free air conducts [electricity]; and, based on observations at different locations, they arrived at the conclusion 'that atmospheric air is ionized to a certain degree' (Elster and Geitel, 1900, p. 247) and that all empirical findings 'may be easily interpreted on the basis of ionic theory' (Elster and Geitel, 1900, p. 246). The question now arose as to how to investigate the distribution of ions in the atmosphere and how to search for their origin. Again Elster and Geitel came upon an important clue to answer the last question, when they detected the presence of radioactive substances in the atmosphere (Elster and Geitel, 1901). A little later they proved the existence of radioactive substances 'in that part of the atmosphere, which lies *below* the earth's surface and which is in permanent exchange by diffusion with its overground mass' (Elster and Geitel, 1902, p. 577). Thus, they established the fact that the radioactivity of the earth's surface, especially the gaseous emanation observed as the decay products of many radioactive minerals, must be considered as the primary source of atmospheric electricity.

The very important results obtained by Elster and Geitel in Wolfenbüttel encouraged work at many other places in Germany.[284] Meanwhile in England, especially at Joseph John Thomson's Cavendish Laboratory, measurements of the electrical properties of the atmosphere were also carried out and theoretical ideas were put forward to explain the observed phenomena.[285] However, Vienna still retained its reputation as one of the main centres for research into atmospheric electricity, with Exner's students performing widely appreciated investigations. Thus, for example, Heinrich Mache suggested in 1903 a new method for determining the velocities of positive and negative ions (Mache, 1903)—simultaneously

[283] Julius Elster was born in Blankenburg on 24 December 1854. He studied science, especially physics, at the universities of Berlin and Heidelberg (1875–1879), obtaining his doctorate in 1879 from Heidelberg. After performing his year of military service he became a teacher at the Wolfenbüttel *Gymnasium* in 1881. In spite of offers of a better scientific career he remained in school service until illness forced him to retire in October 1919. He died in Wolfenbüttel on 8 April 1920. Since the 1880s Elster published over 100 scientific papers, most of them jointly with Hans Geitel, on a wide range of topics in ionic conduction in gases, the photoelectric effect, radioactivity and atmospheric electricity. Elster and Geitel entered the latter field in 1885 with a study of thunderstorm electricity. They became pioneers in these fields. (For further information on Elster's life see the commemoration speech of Emil Wiechert: Wiechert, 1921.)

Hans Geitel was born on 6 July 1855 in Braunschweig and studied at the universities of Heidelberg and Berlin (1875–1879). In the fall of 1879 he joined the staff of the Wolfenbüttel *Gymnasium.* He died on 15 August 1923 in Wolfenbüttel. Geitel's scientific work is closely, nearly inseparably, connected with that of his friend Elster. (They knew each other since school days; later Geitel lived in Elster's house until the death of Elster's wife in 1920). Further information on Geitel's life may be obtained from two obituaries: Hoppe, 1923, and Pohl, 1924.

[284] In Munich, for example, H. Ebert from the *Technische Hochschule* became interested in investigating atmospheric electricity, and in Göttingen notably Eduard Riecke, Woldemar Voigt, and Emil Wiechert.

[285] One of the most active persons in Cambridge, who became deeply involved in atmospheric electricity was Charles Thomson Rees Wilson, the discoverer of the cloud chamber. (See Volume 4, Section II.1.)

and independently of Hans Gerdien (1877–1951) of the University of Göttingen (Gerdien, 1903). Mache and Egon von Schweidler systematically measured the ionic conductivity of the atmosphere as a function of varying meteorological conditions at different geographical locations (e.g., Mache and Schweidler, 1905; Schweidler, 1906, 1912). Two problems received particular attention at Exner's institute: What was the absolute amount of radioactive induction in the atmosphere (i.e., of induced radioactivity or decay products of radium emanation), and did this amount suffice to explain the observed phenomena and properties of atmospheric electricity?

The determination of radioactivity contained in the atmosphere was continued in many places after Elster and Geitel's discovery, e.g., by Ernest Rutherford and S. J. Allan in Canada. In 1905 H. Gerdien criticized the experimental methods used hitherto, as they neither took care of the specific velocities of the carriers of radioactive emanation, nor of the general current situations in the atmosphere (Gerdien, 1905). In Vienna, Heinrich Mache and Travis Rimmer then suggested an idea which avoided such errors (Mache and Rimmer, 1906). Fritz Kohlrausch, also at Exner's institute, devised a new method for the absolute determination of the radioactive emanation content of the atmosphere (Kohlrausch, 1906b). He described his new apparatus as follows:

> A tube of 100 cm length and 16 cm diameter, which is mounted horizontally and can be closed at both ends by lids, sat on the neck of an aluminum-foil electrometer *á la* Exner–Elster–Geitel. On the foil support of the latter a T-piece could be mounted in such a manner that its 30-cm-long horizontal part was in the centre and parallel to the axis of the tube. Then, with the help of an electric ventilator, air was sucked through the tube and the carriers of [radioactive] induction were precipitated on the T-rod, which was charged with a Zamboni column [i.e., a battery] to a potential of −200 volts. (Kohlrausch, 1906b, p. 1322)

Kohlrausch's new apparatus avoided the errors pointed out by Gerdien: by producing an artificial, controllable air current in the tube the accidental currents existing in the atmosphere were removed, and the application of a strong negative potential forced all radioactive decay products of the emanation, notably Ra A, Th A and Ac A (which have a positive charge) to deposit on the rod, independent, to a large extent, of the individual velocity of the carrier. Kohlrausch used his apparatus to determine not only the absolute charge collected by the T-rod, but also its decreases in the course of time (due to radioactive decay, especially $\alpha$-decay). From observations, carried out in July 1906 at Gleinstätten near Graz, he found in particular, 'that the emanation and its decay products provide the main reason for the permanent ion-content of the atmosphere' (Kohlrausch, 1906b, p. 1324). While Jakob Salpeter used Kohlrausch's method to determine the influence of the earth's electric field on the emanation distribution in the atmosphere (Salpeter, 1909, 1910), Victor Franz Hess raised serious objections (Hess, 1910a); he estimated that Kohlrausch's observed values had to be multiplied by a factor of 2.63 (see Hess, 1910b, p. 510). Kohlrausch then took into account Hess' criticism; after improving his method he made, in August and

September 1910, new observations of the induced radioactivity in the air, this time at Seeham on the Obertrummer Lake near Salzburg (Kohlrausch, 1910). Three years later his friend Erwin Schrödinger again made observations in Seeham, using Kohlrausch's method.

Schrödinger's task in 1913 was to determine how much Ra A, the first decay product of the so-called radium emanation, the atmosphere in Seeham contained, and how this amount changed in the course of days and weeks.[286] Schrödinger had available, as he described in his paper '*Radium-A-Gehalt der Atmosphäre in Seeham 1913*' ('Radium-A Content of the Atmosphere in Seeham'), 'nearly the same set-up, which had been devised in detail by Kohlrausch [Kohlrausch, 1910] and had been applied three years ago at the same place' (Schrödinger, 1913b, pp. 2023–2024). This method of Kohlrausch deviated from his original apparatus of 1906 in several aspects: first, in 1910 he used a tube of slightly different dimensions (70 cm in length and 28 cm in diameter), which was closed at one end by a ventilator device operated by hand (for sucking in air from outside); second, the rod in the centre (for collecting carriers of electricity) could now be brought to a much higher negative potential, i.e., −2000 to −3000 volts, at which it became more attractive to positively charged ions in the air; third, a second, shorter tube of 16 cm diameter could be inserted. Schrödinger replaced Kohlrausch's electrometer device by a more sensitive one, providing 'perhaps the utmost sensitivity for currents which one is able to achieve with movable apparatus in country stations' (Schrödinger, 1913b, p. 2025). In order to obtain increased intensity, he connected a symmetric supplementary circuit arrangement, involving two additional tubes, to the electrometer; thus he compensated for the effects of unwanted currents disturbing the measurement, i.e., of currents arising from penetrating atmospheric radiation in the electrometer device.[287]

With this equipment Schrödinger performed in August 1913 an extended series of experiments consisting of '196 single measurements, among them 105 absolute determinations and 91 additional parallel experiments to estimate the mobility of the carriers of Ra A' (Schrödinger, 1913b, p. 2023).[288] The quantity, which he

[286] 'Radium emanation' was called then the gaseous decay product of the element radium—now it is called 'radon'. Radon decays under the emission of $\alpha$-rays into Ra A, which in turn decays under $\alpha$-emission into Ra B, while Ra B transformed by emitting $\beta$-rays into Ra C (which again is an $\alpha$-active substance). The substances Ra A, Ra B and Ra C together were called then 'active precipitation' or occasionally 'induced radioactivity' (see, e.g., the book of Stefan Meyer and Egon von Schweider: *Radioaktivität*: Meyer and Schweidler, 1916).

[287] Schrödinger claimed that his electrometer was so well compensated and sensitive that he could interpret small, totally irregular, fluctuations of the electrometer string of the isolated system [i.e., of the system without the effects of induced radiation] . . . as real Schweidler fluctuations of the $\gamma$- or secondary radiation' (Schrödinger, 1913b, p. 2027).

[288] Each single measurement in Schrödinger's series required as a first step the determination of the electrometer sensitivity (which took six minutes and involved six distinct observations). Then one (of two available) rods was fixed in the Kohlrausch aspiration tube to collect radioactive precipitation during an aspiration time of five minutes. (During these five minutes the observer had to turn a crank about 43 times per minute to the rhythm of a metronome.) The measurement of the current created by the radioactivity charged rod in the electrometer device took another six minutes. Finally, Schrödinger concluded the experiment with a further measurement of the null-effect of the electrometer. (This was done by replacing the charged rod by a second uncharged one.)

derived from the measurement, was the so-called 'saturation current' $i_0$, that is, the current maintained by the radioactive precipitation. This current $i_0$ can be put proportional to the amount of radioactive substance $\mathscr{E}$, contained in a cubic centimetre of the aspirated air, or

$$\mathscr{E} = \frac{k \cdot i_0}{\Phi}, \tag{25}$$

where $\Phi$ is the volume of air sucked per second through the aspiration tube, and $k$ is a constant depending on the aspiration time and the nature of the radioactive substance (especially, the decay constants).[289] Schrödinger applied special care to obtain the correct value of $i_0$, which must be extrapolated from the current determined a few minutes after the end of the aspiration period.[290] He referred to a graphical solution of this problem; that is, he drew a plot of the values obtained at intervals of 1.5, 2.0, 2.5, . . . , 6 minutes after the end of the aspiration period and derived from it the current values at times 0 and $\infty$, whose difference then yielded $i_0$.[291] He also made, as stated above, 'parallel experiments' with the aspiration tubes of 8 and 14 cm radius, respectively, obtaining the saturation currents $\mathscr{E}_8$ and $\mathscr{E}_{14}$. Since in the wider tube not all carriers of radioactivity having small mobility would reach the rod and be deposited, the ratio $q = \mathscr{E}_8/\mathscr{E}_{14}$ provided 'a rough measure of the average mobility of the carriers [of radioactivity], such that high values of $q$ denote a low mobility, while values near unity denote a large mobility' (Schrödinger, 1913b, p. 2041).[292]

Schrödinger's observation of $\mathscr{E}_8$, $\mathscr{E}_{14}$ and $q$ exhibited essentially the same daily variation as Kohlrausch's results of 1910: he noticed that two periods were superposed, one with larger amplitude exhibiting a maximum before 6 a.m. and a sharp minimum between 8 and 10 p.m., the other a maximum in the afternoon and a minimum during the night. While he explained the second period as being caused by the daily variations of atmospheric pressure, he suggested that 'the [strong] minimum in the evening is in any case a consequence of condensation processes' (Schrödinger, 1913b, p. 2045). That condensation phenomena could lead to a reduction of the carriers of radioactivity in the atmosphere and therefore to a decrease of $\mathscr{E}$, seemed to be confirmed by the high $q$-values (i.e., low mobility of the carriers) observed at the time of the minimum. In one important aspect, however, Schrödinger's results deviated from Kohlrausch's: he obtained much lower values for the saturation and also for the Ra A-content of the Seeham atmosphere; for example, Schrödinger found for the daily average of $\mathscr{E}_8$ a value

[289] Fritz Kohlrausch had given earlier (Kohlrausch, 1912) a detailed theoretical derivation of the constank $k$, which Schrödinger took over.

[290] The charged rod had to be taken out of the aspiration tube and brought into one of the tubes connected with the electrometer. (See the above description of the electrometer device.)

[291] Schrödinger did not take single observations for the graphical plot; he rather averaged over 157 observations for a given instant of time.

[292] Schrödinger repeated the same experiment as Kohlrausch had performed three years earlier, except that he took three times as many observations.

of $2.68 \times 10^{-11}$ electrostatic units as compared to Kohlrausch's $20.2 \times 10^{-11}$ in 1910. He wrote: 'I have discussed this strange result frequently and thoroughly with my friend Kohlrausch. Both of us have checked the evaluation of our reduction factors most carefully, we have even compared single measurements directly, without arriving at any other conclusion than this, namely that the Ra A-content [of the atmosphere] as determined by the aspiration method was on average really much greater in August 1910 than in August 1913 at observation points 100 to 200 steps apart' (Schrödinger, 1913b, p. 2052). Hence the only imaginable explanation lay in the wet weather prevailing in the summer of 1913, which suppressed the production of emanation on the ground (which in turn supplied the radioactivity to the atmosphere above it).

Schrödinger's experimental work on the Ra A-content of the atmosphere constituted more or less an exercise of the type performed in an advanced laboratory course. It provided him with the opportunity to join the active research of his colleagues at Exner's institute and at the *Institut für Radiumforschung*. Thus he confirmed the earlier results of Victor Franz Hess (Hess, 1910a), namely 'that, even at the time when the Ra A-content reaches a maximum, only about a fifth of the total ionization [of the atmosphere] is supplied by the decay-products of radium' and 'also that the fluctuation of [electrical] conductivity is much greater than can be explained on the fluctuations of radium products alone' (Schrödinger, 1913b, p. 2048). These conclusions necessitated the exploration of the sources of ionization of the atmosphere other than the $\alpha$-radiation from Ra A and its decay products. A major role, as was already known, had to be attributed in this respect to the $\gamma$-radiation emitted from the radioactive substances deposited on the ground, because $\gamma$-radiation—which is different from $\alpha$- and $\beta$-radiation—had been found to have a range of several hundred metres in the atmosphere. The study of this so-called 'penetrating radiation' represented an issue of growing importance in the research on atmospheric electricity in the period around 1910.

## The Penetrating Radiation and a Paper of Schrödinger

As we have mentioned earlier, Elster and Geitel had noted the fact—also observed independently by Charles Thomson Rees Wilson in Cambridge, England—that air, even when kept *completely* separated from all radioactive substances in a closed vessel, was weakly ionized (Elster and Geitel, 1901). On 31 December 1902, at a meeting of the American Physical Society in Washington, D.C., Ernest Rutherford from Montreal reported an experiment by H.L. Cook and himself (Rutherford and Cook, 1903). The authors observed the amount of 'spontaneous' ionization of the air contained in a testing vessel of about 1 litre capacity, surrounded by metal screens of varying thickness; while 'screens of thickness of 2 mm of lead had very little effect on the rate of discharge' and 'a thickness of 5 cm of lead was found to cut down the rate of discharge by 30%,' they noticed a surprising fact, namely: 'A greater increase of thickness had no effect, although 5 tons of pig lead was placed around the apparatus' (Rutherford and Cook, 1903,

p. 183). The same reducing effect on the ionization of air as 5 cm of lead was exerted by screens of 5 cm of iron and 70 cm of water, but then the remaining ionization remained unchanged with whatever further shielding material had been used. On the removal of the screens the discharge values returned to those values which had been obtained originally without a screen. Hence Rutherford stated in Washington: 'These results show that about 30 per cent of the ionization inside a closed vessel is due to an external radiation of great penetrating power. The radiation appears to come equally from all directions and is probably due to the excited activity on the surface of the room in which the observations were made.' (Rutherford and Cook, 1903, p. 183).

The residual radiation thus observed showed a similarity to 'some rays of an extremely penetrating character from naturally radioactive bodies' (Rutherford and Cook, 1903, p. 183), i.e., $\gamma$-rays from radium and thorium, although none of these substances had been placed in the room where the experiment had been carried out. At the same American Physical Society meeting a second team from Canada, consisting of John Cunningham McLennan (1867–1935) and Eli Franklin Burton (1879–1948) of Toronto, also reported about their independent discovery of the same effect (McLennan and Burton, 1903a). These authors presented further experimental results several weeks later, at the meeting of the Royal Society of Canada: they especially noticed then that, if different shielding metals were brought around the air vessel, different final equilibrium states of ionization (i.e., of residual ionization) were achieved (McLennan and Burton, 1903b).[293] During the following years many physicists in Europe and abroad thoroughly studied the detailed effects of the shielding substances, others investigated the influence of the density of the air on its ionization by penetrating radiation. It was found that the strength of the penetrating radiation depended on where it was observed. For instance, as Meyer and Schweidler wrote: 'In general, a noticeable decrease occurs if the apparatus [consisting of the vessel and instruments registering the ionization current] is brought from inside a building into the open air; and a rather dramatic one, if one observes above the surface of a lake or below the water' (Meyer and Schweidler, 1915, pp. 443–474).

The results just mentioned seemed to imply that the source of the penetrating radiation had to be searched for on the ground. Consistent with such a conclusion Theodor Wulf, a teacher at the Ignatius College in Valkenburg, Holland, found his extended observations, which he performed in the summer of 1909 at various places in Germany, Holland and Belgium.[294] Thus he closed the publication of his results with the following summary:

[293] The observation of the dependence of the residual ionization apparently contradicted the first experiment by Rutherford and Cooke; McLennan and Cunningham refined the experimental method.

[294] Theodor Wulf, one of the pioneers of the experimental study of penetrating radiation, was born on 28 July 1868 in Hamm, Westphalia. From 1904 to 1905 he served as professor of physics at the Ignatius College. He died on 14 June 1946 in Hallenberg in the Sauerland. Wulf is also known for the construction of an electrometer named after him, an instrument with two threads reaching a very high sensitivity for the measurement of electric charges.

[i] Experiments have been reported which prove, that at a given location the observed penetrating radiation is caused by primary radioactive substances lying in the upper-most layers of the earth, extending down to about 1 m below the surface. [ii] If part of the radiation stems from the atmosphere, then it is still too small to be detected by the methods that have been used. [iii] The temporal fluctuations in the $\gamma$-radiation can be explained by shifts of the emanation-rich air masses under the surface of the earth into larger or smaller depths on account of fluctuations in atmospheric pressure. (Wulf, 1909, p. 1003)

Wulf's conclusions agreed perfectly with the earlier observations of McLennan on and at the Great Lakes in Canada (McLennan, 1908), as well as those made by Alfred Gockel and Wulf at high altitudes near Zermatt, Switzerland (Gockel and Wulf, 1908). In contrast William Walker Strong (1883–1955) of The Johns Hopkins University, Baltimore, Maryland, found a totally different result, i.e., he derived from his observations, 'that a large part of the ionization in closed vessels, which is created by external penetrating radiation, is subject to large fluctuations; and that this radiation probably consists of $\gamma$-rays which not only emerge from radioactive substances in the ground, but rather from those existing in the atmosphere' (Strong, 1908, p. 119). How does one decide between these conflicting results?

Obviously, the clue to the answer had to be sought in an examination of the change of intensity of the penetrating radiation with increasing altitude from the ground. If the radioactive substances contained in the surface layers of the earth were responsible, then one should in general expect to find a substantial weakening of the intensity with increasing relative altitude. Observations on towers or on natural elevations yielded a preliminary, though not yet a definite answer. They stimulated the physicists to go up higher, i.e., carry out measurements during balloon flights, which reached heights of several thousand metres above ground. We shall discuss the results of these developments in the next section; here we shall mention the important Austrian contributions to the early research on penetrating radiation.

In the early summer of 1906 Heinrich Mache and Travis Rimmer observed the variation of penetrating intensity in Vienna (Mache and Rimmer, 1906). The authors confirmed the existence of a daily period having two maxima in the morning and evening, respectively; they noticed increasing intensity after rain; and they found that under normal weather conditions the intensity variations corresponded to variations of the potential of the earth's electric field. Mache continued the investigations after October 1907 in Innsbruck and summarized the results in a paper, '*Messungen über die in der Atmosphäre vorhandene radioaktive Strahlung von hohem Durchdringungsvermögen*' ('Measurements on the Radioactive Radiation of High Penetrating Power Existing in the Atmosphere'), communicated to the meeting of 13 January 1910 of the Vienna Academy of Sciences (Mache, 1910).[295] The new data enabled Mache to detect, besides the daily period,

[295] The observations in Innsbruck extended over a year, being carried on, after Mache's return to Vienna, by his successor Friedrich von Lerch and his assistant Dr. Kofler. (See Mache, 1910, p.55.)

a yearly one also. The latter showed a clear maximum in the months of fall and a minimum in February; stronger winds—except the *Föhn* (a warm wind)—brought an increase in intensity, but no unique effect could be attributed to atmospheric pressure; snow fall and a closed snow cover decreased and rain increased the intensity. After a careful analysis of all known factors influencing penetrating radiation, Mache confirmed the intensity increase with an increase in the potential of the earth's field. He finally concluded:

> The penetrating radiation is composed mainly of two parts. The first part emerges from the radioactive substances contained in the upper layers of the earth and their decay products.... The second part is provided by the decay products which have been created from the emanations distributed in the atmosphere, i.e., primarily if they have been deposited from the atmosphere on the ground by condensation products and by the action of the [electric] field of the earth. (Mache, 1910, p. 73)

Mache's results were essentially confirmed by investigations, which Egon von Schweidler performed in the summers of 1910 and 1911 at the *Luftelektrische Station* of the Vienna Academy in Seeham and about which he delivered a report to the academy meeting on 4 July 1912 (Schweidler, 1912). Five months later a paper of his former student Erwin Schrödinger, entitled '*Über die Höhenverteilung der durchdringenden atmosphärischen Strahlung* (*Theorie*)' ('On the Altitude Distribution of Penetrating Atmospheric Radiation—Theory'), was communicated again to the academy; in it Schrödinger presented a detailed theoretical analysis of the problem of penetrating radiation (Schrödinger, 1912c).

In particular, Schrödinger attempted with his paper to throw light on the origin of penetrating radiation, a question 'not easily answered, since because of the penetrating power rather distant sources of radiation also may provide noticeable contributions to radiation intensity at a given space point' (Schrödinger, 1912c, p. 2391). Hitherto, he wrote, three possible explanations had been suggested for the origin of the observed penetrating radiation in the atmosphere, but the opinions of the experts disagreed vehemently about the respective weight of the contributions from: '1. The radioactive substances contained in the ground or precipitated onto the surface of the earth; 2. The radioactive substances suspended in the atmosphere; and 3. hypothetical extraterrestrial sources of radiation' (Schrödinger, 1912c, p. 2391). While Rutherford's former collaborator Arthur Steward Eve (1862-1948) and the American Louis Vessot King (1886-1956) had previously examined the first contribution (Eve, 1911; King, 1912), Schrödinger now analyzed the second contribution, arguing at the same time: 'The third source of radiation is ... completely hypothetical, and should be introduced, only then with some justification, if one should find the other two absolutely insufficient in explaining observations' (Schrödinger, 1912c, pp. 2391-2392). As the proper method to perform his task, Schrödinger proposed 'to establish a calculation, as exact as possible, of the action of the radiators suspended in the atmosphere', in order to arrive at conclusions on the 'altitude distribution of the radiating substances' and thus also on the 'altitude distribution of the [penetrating] radiation' (Schrödinger, 1912c, p. 2392). This theoretical altitude distribution could then be compared to the data obtained from balloon flights.

In order to get the altitude distribution of the radioactive substances in the atmosphere—about which very little was known empirically at that time—Schrödinger made an assumption: he claimed that the distribution in question was 'at any rate primarily determined by the up- and down-moving air currents [circulation], which cause a thorough mixing of the atmosphere in a comparatively short time' (Schrödinger, 1912c, p. 2392).[296] On the basis of this simple assumption he found immediately that the density of the radiating substances suspended in the air ($q$) decreased with increasing height ($z$) in exactly the same way as the density of the air ($\rho$). Hence $q$ obeyed the equation,

$$q = q_0 \exp(-\alpha z), \tag{26}$$

where $q_0$ denotes the source concentration on the ground and $\alpha$ the (known) constant in the density distribution of the atmosphere. With this distribution of the sources of penetrating radiation, Schrödinger proceeded to calculate the total intensity of the radiation, $i(z)$, at a space joint of given altitude $z$ by summing over the contributions from radiators suspended at different altitudes and taking into account the absorption of the emitted $\gamma$-rays (in the air between the radiator and the space point $z$). He thus obtained the result:

$$i(z) = \frac{2\pi q_0}{\mu_o}\left\{\phi\left(\frac{\mu_o}{\alpha}[1 - \exp(-\alpha x)]\right) + \phi\left(\frac{\mu_0}{\alpha}\exp(-\alpha z)\right)\right\}, \tag{27}$$

where $\phi(x)$ denotes the function

$$\phi(x) = 1 - \exp(-x) + x\int_x^\infty \frac{\exp(-x)}{x}\,dx. \tag{27a}$$

In the derivation of Eq. (27) the absorption coefficient $\mu$ ($= \mu(z)$) was assumed to be proportional to the density, or given by $\mu = \mu_0 \exp(-\alpha z)$ with $\mu_0$ the ground value.[297] Since the function $\phi(x)$, Eq. (27a), increases rapidly from zero for $x = 0$ to unity, approaching the latter value already for $x = 3$, the radiation density takes on identical values on the ground (for $z = 0$) and at very high altitudes (for $z = \infty$). In detail, $i(z)$ starts from the value $(2\pi q_0/\mu_0)\phi(\mu_0/\alpha)$ on the ground, then increases rapidly to reach 'at the altitude of half the height of the barometer (at about 5,500 m) a maximum, whose value is exactly twice its value on the

[296] Schrödinger also discussed the contribution of the usually considered effects of diffusion, radioactive decay and gravity to the altitude distribution of penetrating radiation; he concluded: 'These factors alone would—as a simple calculation shows—just allow the emanations and [radioactive] inductions, originating from ground pores, to ascend to an altitude of a few metres above ground' (Schrödinger, 1912c, p. 2392).

[297] The quantity $\mu_0/\alpha$ entering Eq. (27) assumed the value 36.

ground' (Schrödinger, 1912c, p. 2401), and then remains constant up to an altitude of 20 km when a slow decrease sets in until the ground value is reached again at infinity.

Schrödinger was perfectly aware of the fact that several corrections had to be applied to his result, Eq. (27), before one could compare it to the real situation in nature. On the one hand, the exchange of air by circulation—which was assumed to exist throughout to establish Eq. (26)—was known to prevail only in the lower parts of the atmosphere, i.e., in the troposphere which goes up to about 11,000 m altitude. Below 1,000 m altitude, on the other hand, the radiation intensity had to be increased by substantial amounts originating from the ground activity, i.e., by the contribution treated earlier by Eve and King. Schrödinger evaluated this contribution—he took King's recent result (King, 1912)—to yield the extra intensity $i'(z)$ of the penetrating radiation,

$$i'(z) = \frac{2\pi q'}{\mu}\left\{1 - \phi\left(\frac{\mu_0}{\alpha}[1 - \exp(-\alpha z)]\right)\right\}, \tag{28}$$

where $q'$ and $\mu'$ denotes the density of the radiating substances (radiators) and the absorption coefficient in the groundlayers of the earth.[298] Finally he added two expressions, Eqs. (27) and (28), yielding the result,

$$i(z) + i'(z) = \frac{2\pi q'}{\mu'} + \left(\frac{2\pi q_0}{\mu_o} - \frac{2\pi q'}{\mu'}\right)\phi\left(\frac{\mu_0}{\alpha}[1 - \exp(-\alpha z)]\right)$$
$$+ \frac{2\pi q_0}{\mu_0}\phi\left(\frac{\mu_0}{\alpha}[\exp(-\alpha z)]\right). \tag{29}$$

Equation (29) changed the result especially for altitudes up to 1,000 m above the ground. Depending on the relative magnitudes of the contributions from ground radioactivity (i.e., $q'/\mu'$) and from the radioactivity suspended in the air immediately above the ground (i.e., $q_0/\mu_o$), respectively, Schrödinger derived the following conclusions from his theory: (i) for equal contributions $q'/\mu' = q_0/\mu_0$) the radiation intensity should be fairly constant from $z = 0$ *to* $z = 1{,}000$ m; if the radiation from sources suspended in the atmosphere dominates ($q_0/\mu_0 \gg q'/\mu'$), one would observe an increase in radiation intensity by a factor of about 2 from $z = 0$ to $z = 1{,}000$ m—this is roughly the situation described by Eq. (27); and for a dominating ground distribution ($q'/\mu' \gg q_0/\mu_0$) the radiation intensity would be reduced in 1,000 m altitude to a fraction of the value at $z = 0$.

As a last step in his paper Schrödinger referred briefly to the available data on penetrating radiation, which yielded either a fairly constant value for the intensity or a small decrease for the first 1,000 m. Therefore he derived from a comparison of his theory with observation, 'that the ground radiation and the

[298] Schrödinger assumed that all ground layers contribute to ground radiation.

air radiation are of the same order of magnitude, but the former appears to be a little greater' (Schrödinger, 1912b, p. 2406). He refrained from a similarly conclusive statement about the higher altitude distribution, although some interesting data were available which, together with his theoretical analysis, led to an important discovery, that of '*Höhenstrahlung*' or cosmic radiation. We shall discuss this discovery in the next section.

## The Hasenöhrl School

In his early papers Schrödinger often treated theoretical topics. Later he occasionally referred to connections with Fritz Hasenöhrl, who had been his teacher in theoretical physics at the University of Vienna. Here we shall have a brief look at what happened in Viennese theoretical physics between 1911 and 1914, during which time Hasenöhrl developed a school of his own. Two aspects seem to be remarkable in this context. First, quantum theory finally entered Hasenöhrl's institute; and second, the modern theory of solids was taken up and propagated in Vienna.

As we have reported earlier, Arthur Haas was the first person in Vienna to make an active contribution to quantum theory with his treatment of the hydrogen atom. Although the physicists in Vienna received Haas' ideas with some reservation, the fairly positive judgment about it by the famous Dutch theoretician Hendrik Antoon Lorentz made some impression. Still the first Viennese to respond to Haas was Arthur Schidlof, then at the physical laboratory of the University of Geneva.[299] In the introduction to his paper '*Zur Aufklärung der universellen elektrodynamischen Bedeutung der Planckschen Strahlungskonstanten h*' ('On the Clearing-up of the Universal Electrodynamical Significance of Planck's Radiation Constant *h*'), which he sent to *Annalen der Physik* at the end of March 1911, he referred to Arthur Haas' interpretation of the quantum of action by the atomic model, commenting:

> One may indeed raise an objection against this attempt at an explanation that the hypotheses used [by Haas], and partly also the data chosen, appear to have been selected *ad hoc*. Still, Lorentz [1910b] has connected Haas' hypotheses with our usual understanding, and he has pointed out that certain facts, whose theoretical interpretation had been given by Einstein [(1909a)], seem to favour very much the conception of Haas. (Schidlof, 1911, p. 90)

[299] Arthur Schidlof was born in 1877 in Vienna, but during his *Gymnasium* time he went to Geneva, where he studied at the university, obtaining in 1903 a *Diplôme* in chemical engineering. He then continued the study of experimental physics and performed his thesis work on magnetic hysterisis under the direction of C.-E. Guye (1904/1905). He became a permanent resident of Geneva (obtaining his naturalization in 1907) and was employed at the physical laboratory of the University of Geneva as assistant, lecturer, extraordinary professor and finally (from 1930) full professor of theoretical physics. He died on 27 November 1934 in Geneva. (See his obituary by A. Mercier; Mercier, 1935.)

Schidlof's first scientific papers dealt with experimental topics of magnetism and with measurements of the electron's charge; later he changed almost completely to theory, treating properties of fluids, thermodynamics, quantum and relativity theories and their consequences, and finally the application of wave mechanics to nuclear physics.

In his paper Schidlof tried to remove the

> important objection against Haas' hypothesis [i.e., that] it does not provide any explanation of the universal character of the elementary quantum of action, but restricts itself to pointing to a very remarkable agreement. (Schidlof, 1911, pp. 90–91)

To accomplish his purpose, Schidlof first introduced the following assumptions: the electrons in Thomson's model of the atom can be separated into two groups; the inner ones being highly concentrated and lying close to the centre of the positive sphere, and the outer ones representing the valence corpuscles; only the valence electrons can absorb radiation, upon which they leave the positive sphere; radiation will be emitted if an electron penetrates into the sphere. Schidlof then calculated the frequency $\nu$ of the eigenoscillations of the sphere of radius $a$ and charge $Ne$ containing $N$ electrons of mass $m_e$ (and charge $-e$); and he divided the energy necessary to emit an electron ($\varepsilon = e^2/a$) by this $\nu$, putting the quotient equal to Planck's constant $h$, i.e.,

$$\frac{\varepsilon}{\nu} = 2\pi e \left(\frac{am_e}{N^{1/3}}\right)^{1/2} = h. \tag{30}$$

Since $Ne/a^3$ represented the density of the positive charge, which Schidlof assumed to be the *same for all atoms*, Eq. (30) indeed yielded a *universal* constant $h$. By using the data from the dispersion of long electromagnetic waves by atoms of different substances—which provided the unknown charge density—Schidlof confirmed the constancy of the right-hand side of Eq. (30) and found the value $h = 6.05 \times 10^{-27}$ erg sec, in very good agreement with Planck's value of $6.5 \times 10^{-27}$ (Planck, 1906, § 157).[299a]

While Schidlof replaced Haas' quantum hypothesis with other assumptions, but still used a model—J. J. Thomson's—for the atom, the Viennese physics professor Fritz Hasenöhrl attacked a different problem, namely that of obtaining the line spectrum from atoms without referring to a particular model. In his lecture '*Über die Grundlagen der mechanischen Theorie der Wärme*' ('On the Foundations of the Mechanical Theory of Heat'), presented on 25 September 1911 at the 83rd *Naturforscherversammlung* in Karlsruhe, he tried to generalize Max Planck's *Ansatz* for the frequency of an harmonic oscillator (resonator), namely,

$$\tau(E_{n+1} - E_n) = h, \tag{31}$$

[299a] It must be mentioned, in this context, that the frequency $\nu$ entering into Schidlof's Eq. (30) is *not* the frequency of any optical line. The author stated in particular (towards the end of his paper): 'We have based our calculation on the assumption that the atoms are able to act as Planck resonators. One should, however, consider the fact that these resonators are in general not atoms. The eigenoscillations of atoms lie far in the ultraviolet part of the spectrum. Already those resonators, which react on visible light, possess diameters considerably larger than the atoms. Thus, for example, the diameter of the optical resonator for the sodium line ($\nu = 5 \times 10^{14}$ sec$^{-1}$) is about three times as large as that of the mercury atom [which should be much larger than the sodium atom]' (Schidlof, 1911, p. 99).

where $\tau$ denotes the period ( $= 1/\nu$) and $E_{n+1}$, $E_n$ are a pair of neighbouring energy states) in such a way as to describe the observed discrete spectra of atoms and molecules (Hasenöhrl, 1911; especially, pp. 933–935). He suggested that the equation

$$\int_{E_n}^{E_{n+1}} \tau \, dE = h \tag{32}$$

for the period $\tau$ might provide a suitable basis. That is, if one inserted for the energy dependence of $\tau$ a certain function, then the Deslandres formula of band spectra followed; with another function one could obtain the Balmer formula for hydrogen. When Arnold Sommerfeld in the dicussion of Hasenöhrl's talk suspected that the Balmer formula might result from a situation corresponding to a pendulum with finite amplitude in mechanics, Hasenöhrl admitted: 'The functional dependence [of the period of oscillation on the energy] is in the case of the pendulum very similar to that which I assumed [in order to obtain the Balmer series]; it would also result in a "series", but not exactly the one of Balmer [for hydrogen]' (Hasenöhrl, 1911, p. 935).

In his talk Hasenöhrl avoided any reference to Arthur Haas. Arnold Sommerfeld, who spoke at the Karlsruhe meeting immediately before Hasenöhrl, at least hinted at the papers of Haas and Schidlof implicitly by mentioning 'a further connection between the quantum of action $h$ and the molecular size, as has already been suggested repeatedly in recent times' (Sommerfeld, 1911b, p. 1066).[300] But he immediately added, 'However, I do not want to go so far as to see in this connection the real origin of $h$. . . . Apart from the uncertainty of our knowledge about the molecular dimensions, it appears to me that such a point of view does not do justice to the universal significance of $h$. I rather prefer the opposite point of view, i.e., not to explain $h$ from molecular dimensions, but to consider the existence of the molecules as a function and consequence of the existence of an elementary quantum of action' (Sommerfeld, 1911b, p. 1066). Sommerfeld stressed the same attitude in his report to the first Solvay Conference on Physics in Brussels a few weeks later. Still his concern with Haas' ideas, and Hendrik Lorentz' repeated favourable opinion about them, left a good impression with Hasenöhrl, who not only saw to it that Arthur Haas finally obtained his *Habilitation* in Vienna, but also had his student Karl Herzfeld work on a model that accounted for the hydrogen spectrum.[301]

[300] We have reported the content of Sommerfeld's Karlsruhe talk, entitled '*Das Plancksche Wirkungsquantum and seine allgemeine Bedeutung für die Molekularphysik*' ('Planck's Quantum of Action and Its General Significance for Molecular Physics') in some detail in Volume 1, Section I.6.

[301] Hendrik Lorentz, president of the conference, addressed Haas' model explicitly in the discussion of Max Planck's report. He also replied to Sommerfeld's plea in favour of 'a general hypothesis about the quantity $h$ rather than the special atomic models' by stating: 'Mr. Sommerfeld does not question the existence of a connection between the constant $h$ and the atomic dimensions. One may express this fact in two ways: The constant $h$ is either determined by these dimensions, or the dimensions which one ascribes to the atoms depend on the quantity $h$. I do not see a big difference between them' (Lorentz, in Eucken, 1914, p. 103).

Karl Ferdinand Herzfeld was born on 24 February 1892 in Vienna. After attending the *Schottengymnasium* in his hometown from 1902 to 1910, he began to study mathematics and physics at the University of Zurich and finally in Göttingen (1913–1914) returning, however, to Vienna to receive his doctorate. In his third semester in Vienna he submitted a paper, entitled '*Über ein Atommodell, das die Balmer'sche Wasserstoffserie aussendet*' ('On an Atomic Model Emitting the Balmer Series of Hydrogen'), to the Imperial Academy of Sciences (Herzfeld, 1912a). Herzfeld's task was to construct a detailed model of the hydrogen atom, from which a function $\tau(E)$ followed in such a way that it yielded, via Hasenöhrl's Eq. (32), the well-known Balmer series. For that purpose he modified Thomson's atomic model by assuming that the positive charge density was not uniform but rather depended on the distance from the centre of the sphere, arriving indeed at the desired solution.[302] In Herzfeld's model, the electron can move on quite different orbits; a given line of the Balmer spectrum was then connected with two limiting radii, say $R_n$ and $R_{n+1}$, which determined the energies $E_n$ and $E_{n+1}$ entering Eq. (32). Although Herzfeld discussed detailed consequences from his model—he speculated, e.g., about the structure of a line series emitted by the electron when moving on a radius larger than that of the atom—he concluded with the warning: 'One cannot, of course, state that the carriers of the hydrogen series actually look like this, but only that the model described here would yield the hydrogen series' (Herzfeld, 1912a, pp. 600–601). That is, he wished to consider his model as a mathematical exercise to obtain the hydrogen series rather than as a genuine picture of what really goes on in the hydrogen atom.

The papers of Hasenöhrl and Herzfeld thus did not progress along the ideas pioneered by Arthur Haas. Nevertheless, they show how the concepts of quantum theory crept into physics in Vienna. The theoretician Hasenöhrl played a central role in this process, especially by propagating discussions of the topics treated at the first Solvay Conference, to which he was invited as the only Austrian physicist—besides, of course, Albert Einstein, then professor at the German University of Prague. Returning from Brussels, Hasenöhrl passed on the reports of the conference to his interested students. Karl Herzfeld then analyzed one of the problems that had been treated by James Hopwood Jeans in his report (Jeans, 1914). Jeans had argued, in particular, that the experimentally observed blackbody radiation law (which had been adequately represented by Max Planck's quantum-theoretical formula of 1900) did not correspond to the situation of thermodynamical equilibrium, which would rather follow from the classical Rayleigh–Jeans formula.[303] Herzfeld, however, showed in a paper on the '*Beiträge zur statistischen Theorie der Strahlung*' ('Contributions to the Statistical Theory of Radiation'), which was communicated to the Vienna Academy of Sciences at the meeting of 13 June 1912, that Jeans' claim led to contradictions with

[302] Then both the period of the electron performing a circular motion round the centre of the sphere—as in Haas' model—and its energy depended on the charge distribution; as a consequence a dependence of the period on the energy also resulted.

[303] We have discussed the arguments of Jeans in detail in Volume 1, Sections I.4 and I.6.

experimental facts (Herzfeld, 1912b, § 1).[304] Interesting as these results were for quantum theory, it was clear that quantum theory in Vienna moved away from the model considerations towards general statistical problems, and when Niels Bohr in Copenhagen took the next decisive step in this direction with his theory of the atomic structure of hydrogen and its line-spectrum on the basis of Ernest Rutherford's atomic model, physicists in Vienna were as surprised as those elsewhere in the scientific world.

The news about Bohr's theory arrived in Vienna during the 85th *Naturforscher-versammlung* in September 1913.[305] Several months later Karl Harzfeld studied the Zeeman effect, i.e., the splitting of the spectral lines in a magnetic field, in the 'two theories which attempt to use the quantum concept in order to explain the laws of the series spectra, the first by Hasenöhrl [Hasenöhrl, 1911], the second by Bohr [Bohr, 1913b]' (Herzfeld, 1914, p. 193). He found that under suitably simple assumptions in both theories a normal Zeeman triplet emerged from a single spectral line (when a constant magnetic field was applied); in the case of Bohr's theory an effect proportional to the square of the magnetic field strength also occurred, which gave rise to a shift of the centre of the triplet and to the widening of the line widths, 'which becomes considerable for lines with higher series number and which is a necessary consequence of the fact that the orbital radius of the electron is large in comparison with the radius of the atom' (Herzfeld, 1914, p. 198).

Herzfeld belonged to the younger generation of physicists educated by Fritz Hasenöhrl; Hans Thirring, on the other hand, senior only by four years, had been among his first students. Thirring was born in Vienna on 23 March 1888, the son of a school teacher; he studied mathematics and physics at the University of Vienna from 1907 to 1911, receiving his doctorate in 1911.[306] Since the fall of 1910 he served as assistant at the theoretical institute of Hasenöhrl, supplementing the latter's activities. That is, while Hasenöhrl attracted the students mainly by his teaching and instruction, his assistant Thirring assembled a circle of friends for recreation. As Ludwig Flamm reported later:

> Ladies and gentlemen met under his guidance for excursions into the countryside around Vienna, for rock-climbing and skiing. In this circle the writer of these lines also learned rock-climbing and skiing. This happened by short practical training

[304] In the later sections of his paper Herzfeld tried to develop, starting from the validity of Planck's law, a statistical theory of resonators (Section 2), and tried to use this theory in order to account for several of the known difficulties in the electron theory of metals (Section 3). He addressed a wide spectrum of phenomena, about which he would hear, less than a year later, the reports and discussions of the outstanding experts Max Planck, Peter Debye, Walther Nernst, Arnold Sommerfeld and Hendrik Lorentz, at the so-called 'Kinetic Gas-Congress' (*Kinetischer-Gas-Kongress*') in Göttingen (in the last week of April 1913).

[305] We have mentioned in Volume 1, Section I.4, the first reactions of the physicists to Bohr's theory of atomic structure. At the 85th *Naturforscherversammlung* in Vienna, for example, Albert Einstein heard about Bohr's results from George de Hevesy. He found the fact that in the theory of Niels Bohr 'the frequency of the light [emitted by the atom] does not depend on the frequency of the electron [rotating in the atom]' remarkable and called it 'an enormous achievement' (see Hevesy to Bohr, 23 September 1913, quoted in Volume 1, p. 201).

[306] For the biography of Hans Thirring see the article of Ludwig Flamm on Thirring's seventieth birthday (Flamm, 1958).

without any particular preparation. Thus the Englishman [Robert] Lawson, who came to Vienna to work at the *Radiuminstitut*, was also taken by Thirring without much ado on a skiing tour. With skis bought the previous day he joined the party on the Sonnwendstein on Sunday; and then the new participant had to go downhill to Steinhaus at the Semmering. The newcomer could slow down the ride only by falls, but even he arrived safely at the foot of the mountain. During the summer vacations there were extended climbing excursions to the Dolomites, which ended at Lake Garda. I still recall with pleasure the fine activities of Thirring. (Flamm, 1958, p. 2)

Sportive entertainment was not Thirring's only interest at Hasenöhrl's institute. He also actively contributed to problems in theoretical physics. Thus he submitted in July 1913 a paper to the *Physikalische Zeitschrift*, entitled '*Zur Theorie der Raumgitterschwingungen und der spezifischen Wärme fester Körper*' ('On the Theory of Spatial Lattice Vibrations and of the Specific Heats of Solids'), by which he opened a new chapter in kinetic and quantum theory in Vienna (Thirring, 1913).

Spatial lattices had been considered for several years in Göttingen. First Erwin Madelung used them for the purpose of calculating the frequencies of the eigenvibrations of solids from the observed elastic constants (Madelung, 1909; 1910a, b). These frequencies entered into the quantum formulae for the specific heats of solids (at constant volume $v$), such as the equation

$$c_v = \frac{3R\left(\frac{h\nu}{kT}\right)^2 \exp\left(\frac{h\nu}{kT}\right)}{\left[\exp\left(\frac{h\nu}{kT}\right) - 1\right]^2} \tag{33}$$

which Albert Einstein had suggested earlier for monatomic gases (Einstein, 1906g). (In Eq. (33) $R = Nk$ denoted the universal gas constant; $k$ and $h$ Boltzmann's and Planck's constants, respectively; and $T$ the temperature in degrees absolute.)[307] The experimental data obtained from 1910 by Walther Nernst and his collaborators had shown that only a few characteristic frequencies seemed to play a crucial role in quantum theory, although an enormous spectrum of eigenvibrations should exist in solid crystals (see, e.g., Nernst and Lindemann, 1911a). Then Max Born and Theodore von Kármán (1881–1963) studied the theory of spatial lattices in detail; computing in particular their eigenvalue spectrum by strict mathematical methods, they arrived at an equation for the specific heats of cubic lattices which was very similar to Einstein's, namely,

$$c_v = 3R\frac{3}{(2\pi)^3}\int_0^{2\pi} \frac{\left(\frac{h\nu_0}{kT}\right)^2 \sin^2\left(\frac{\omega}{2}\right) \exp\left(\frac{h\nu_0}{kT}\sin\frac{\omega}{2}\right)}{\left[\exp\left(\frac{h\nu_0}{kT}\sin\frac{\omega}{2}\right) - 1\right]} \omega^2\, d\omega, \tag{34}$$

[307] We have discussed the development of the thoery of specific heats of solids in some detail in Volume 1, Sections I.6 and I.7.

where $\nu_0$ represents a limiting frequency of the vibrational spectrum (Born and von Kármán, 1912).[308] At about the same time Peter Debye proposed an alternative approach to the theory of specific heats of solids, yielding instead of Eq. (34) the formula (Debye, 1912b, c),

$$c_v = \frac{9R}{x_m^3} \cdot \int_0^{x_m} \frac{x^4 \exp(x)\, dx}{[\exp(x) - 1]^2} \tag{35}$$

with

$$x_m = \frac{h\nu_m}{kT} = \frac{\Theta}{T} \cdot \tag{35a}$$

The main difference with the Born–von Kármán theory was that Debye considered the crystal not as an atomic lattice, but as a composite molecule or an elastic continuum, in which $N$ atoms (in a gram-atom of the substance) perform collective oscillations and the molecular constitution of the solid only entered into the calculation of the maximum frequency $\nu_m$.[309]

Both theories described the data fairly well; in the high-temperature limit the classical value ($c_v = 3R$) was reached; in the low-temperature region Debye's formula showed an increase of $c_v$ proportional to the third power of the ratio $T/\Theta$, which seemed to be in excellent agreement with the available observation. Debye's theory provided a simpler description by a universal function of $\Theta/T$, given by the integral on the right-hand side of Eq. (35).[310] A careful analysis of the full material available at that time, undertaken by Nernst's collaborator Arnold Eucken (1884–1950), pointed to 'deviations [between the observed data points and Debye's formula] of several percent, which obviously cannot be explained anymore by experimental errors'; these deviations, however, had to be expected from the point of view of the 'more general theory of Born, von Kármán and Thirring,' which provided no such universal function (Eucken, 1914, p. 385).

For a member of the former Boltzmann institute in Vienna, like Hans Thirring, it was of course obvious that the Born–von Kármán theory had to be preferred above the quasi-continuum approach of Debye. Still he found some point for improvement in the first paper of the former authors. They had obtained there

[308] In a couple of papers, published in January 1913, Born and von Kármán worked out further details and generalizations of their lattice theory (Born and von Kármán, 1913a, b).

[309] Debye assumed, in particular, that the sum of all frequency modes up to $\nu_m$ should be equal to the number of degrees of freedom in the crystal, i.e., $6N$.

[310] In his Wolfskehl lecture of April 1913 Debye extended his theory to include the anharmonicity of crystal oscillations; he found that the expression for $c_v$, Eq. (35), did not have to be altered, but the characteristic temperature $\Theta$ had to be replaced by $\Theta_1 = \Theta[1 - a(\Delta/V_0) + \ldots]$, where $\Delta/V_0$ denoted the relative volume increase of the solid under investigation and $a$ a constant proportional to the anharmonic term in the crystal oscillations (Debye, 1914).

a general formula for the energy of a cubic solid, namely,

$$E = \frac{RT}{(2\pi)^3} \sum_{k=1}^{3} \iiint_0^{2\pi} \frac{x_k}{\exp(x_k) - 1} \, d\phi \, d\psi \, d\chi, \tag{36}$$

where $x_k = h\nu_k/kT$ and $\nu_k$ denote the eigenfrequencies of the spectrum (see Eq. (50) in Born and von Kármán, 1912, p. 308); they then introduced an approximation, by which the three-fold integration reduced to a single integration, and thus also obtained expression (34) for the specific heats. Thirring discovered: 'This simplifying approximation involves, on the one hand, an assumption which cannot be fully justified; on the other hand, by it the essential feature of Eq. [(36)] is washed out, namely the dependence of the specific heats on the elastic constants of the crystal' (Thirring, 1913, p. 868). He therefore attempted to evaluate the rigorous expression (36) of Born and von Kármán without any doubtful approximation. By making use of the identity

$$\frac{x}{\exp x - 1} = 1 - \frac{x}{2} + \frac{B_2}{2!}x^2 - \frac{B_4}{4!}x^4 + \ldots, \tag{37}$$

where $B_i$ denotes Bernoulli numbers, and exchanging in the expression for $E$, Eq. (36), summations and integrals, Thirring eventually found, for specific heat, the expression,

$$c_v = \frac{R}{(2\pi)^3} \iiint_0^{2\pi} \left[3 - \frac{B_2}{2!}\left(\frac{h}{kt}\right)^2 s_1 + \frac{B_4}{4!}\left(\frac{h}{kT}\right)^4 s_2 - \ldots\right] \cdot d\phi \, d\psi \, d\chi, \tag{38}$$

with

$$s_\lambda = (\nu_1^2)^\lambda + (\nu_2^2)^\lambda + (\nu_3^2)^\lambda. \tag{38a}$$

The lattice theory expansion, Eq. (38), had an interesting and important practical property: for not too small temperatures, i.e. (with $s_1^{\max}$ the largest value of $s_1$),

$$T > \frac{h}{k}\sqrt{s_1^{\max}}/2\pi, \tag{39}$$

the right-hand side did converge absolutely, yielding

$$c_v = 3R\left[1 - \frac{B_2}{2!}\left(\frac{h}{kT}\right)^2 I_1 + \frac{B_4}{4!}\left(\frac{h}{kT}\right)^4 I_2 - \ldots\right], \tag{40}$$

$$I_k = \frac{1}{3(2\pi)^3} \iiint_0^{2\pi} s_k \, d\phi \, d\psi \, d\chi. \tag{40a}$$

In this case all coefficients $I_k$ grew approximately like the $k$th power of $I_1$, Eq. (40) coincided with Debye's result, Eq. (33)), with the characteristic temperature being given by $\Theta = (h/k)\sqrt{I_1}$. However, the agreement between lattice theory and continuum theory was not perfect even in this case because the $I_k$ depended on the elastic constants of the crystal.[311] When Thirring finally used the known constants for copper, rock-salt and sylvite and compared the theoretically derived—via Eq. (40)—and the observed specific heats of these materials in the temperature range between 50° and 80° absolute, he found an overall fair description.[312]

Thirring continued to improve on his evaluation of the Born–von Kármán lattice theory; he presented, e.g., a talk on this subject at the *Naturforscherversammlung* in Vienna on 25 September 1913[313]; in December of the same year he again submitted to the *Physikalische Zeitschrift* a detailed paper on the theory, which was subsequently published in two parts (Thirring, 1914a, b).[314] He then contributed essentially to the dynamical lattice theory of solids, also using his work to obtain his *Habilitation* at the University of Vienna in 1915.

## Schrödinger and the Atomic Structure of Solids

In his article dedicated to the sixtieth birthday of Erwin Schrödinger, Hans Thirring recalled that he frequently discussed theoretical topics with his colleague from Exner's institute, 'with Schrödinger being mostly the giving part' (Thirring, 1947, p. 106). This later remark also referred to the work on lattice dynamics; indeed, Thirring pointed out in his first paper on the theory that he had, 'following a suggestion of E. Schrödinger,' attempted 'to turn the rigorous formula [Eq. (36)] into a [practically] useful expression by expanding it into a series' (Thirring, 1913, p. 868). The two young Viennese physicists, in spite of belonging to different

[311] See especially Eqs. (18) of Thirring, 1913, p. 870.

[312] The agreement was not complete, but the author noted: 'One must consider, however, that for NaCl and KCl one would on first inspection not expect any agreement, because the theory applies only to monatomic solids; and further, that the elastic constants of copper used . . . can only be viewed as preliminary results' (Thirring 1913, p. 871). In addition, the elastic constants used in all cases were those observed at room temperature, while one might expect higher values at lower temperatures, which would improve the agreement between the theoretical and the experimental values for specific heats.

[313] Thirring presented his talk, entitled '*Zur Theorie der Raumgitterschwingungen und der spezifischen Wärme kristallinischer Körper*' ('On the Theory of Spatial Lattice Vibrations and of the Specific Heat of Crystalline Bodies') on 25 September 1913. (See *Verh. d. Deutsch. Phys. Ges.* (2) **15**, p. 923, 1913.)

[314] In these papers Thirring treated especially the theory of diatomic lattices, arriving at a generalization of Eq. (36)—the generalized equation contained six instead of three integrals (which Born and von Kármán had derived for monatomic solids). Again he could give an expansion of the type of Eq. (38) to describe the specific heats of diatomic solids—in that case the coefficients $I_k$, Eq. (40a), had to be altered slightly. A comparison with the data for rock-salt and potassium chloride revealed much better agreement than provided by all previous formulae.

In the last section of the second paper (Thirring, 1914b) the author even suggested an extension to triatomic crystals, e.g., $CaF_2$; again he noticed reasonable agreement between theory and experiment, as far as specific heats were concerned.

institutes, clearly took the opportunity to discuss problems of mutual interest, and the atomic or molecular structure of solids was such a topic, which excited the collaborators of Hasenöhrl and Exner alike.[315]

One may, of course, ask the question why Thirring and Schrödinger first approached the lattice theory of Born and von Kármán, which involved mathematically advanced methods of the eigenvalue theory for infinite systems. However, both being students of Hasenöhrl, who had treated in his lecture courses 'the higher theoretical schemes of mechanics as well as the problem of eigenvalues in continuum physics' (Schrödinger, 1935, p. 87), they did not meet any great difficulty in understanding the papers of Born and von Kármán. Soon they were able to contribute investigations of their own to that topic. While Thirring—the assistant at the theoretical institute—embarked on an improvement of the theoretical formula for the specific heat of solids, Schrödinger—the assistant at the experimental institute—chose to study a primarily experimental question, namely whether the dynamical theory of lattices did account for the observed interference pictures created by the penetration of X-rays through crystals.

The discovery of the X-ray interference patterns by Max von Laue (1879–1960), Walter Friedrich (1883–1968) and Paul Knipping (1883–1935) in the spring of 1912 had been a sensation in the scientific world, extending far beyond physics (Friedrich, Knipping and Laue, 1912).[316] The interpretation given to the observed effect, namely that it represented the diffraction of electromagnetic radiation of very short wavelengths from the spatial gratings formed by the atoms in the crystals, had solved two problems at the same time: it demonstrated beyond doubt the electromagnetic wave nature of Wilhelm Conrad Röntgen's X-rays; and it proved the atomic lattice structure of crystals. It stimulated many physicists in Germany and abroad to investigate the situation further, yielding soon new, interesting and far-reaching results.[317]

The observed interference patterns of Friedrich and Knipping exhibited several peculiar features which demanded a theoretical explanation.[318] Especially, the

[315] Schrödinger probably left the topic of specific heats of solids entirely to his friend Thirring in 1913, since the latter planned to use it for his *Habilitation* thesis.

[316] On 21 April 1912 Friedrich and Knipping, at Arnold Sommerfeld's theoretical institute in Munich, directed, on a proposal of Max von Laue, X-rays (from a tube having essentially a continuous spectrum, plus a few characteristic lines) on crystals and observed on a photographic film (placed behind the crystal) more or less symmetric patterns of X-ray spots.

[317] William Henry Bragg (1862–1942) and his son William Lawrence Bragg (1890–1971) in England developed a slightly different method for obtaining diffraction patterns by reflecting X-rays from crystals (W. L. Bragg, 1913a, b; Bragg and Bragg, 1913). We shall discuss the status of X-ray diffraction in 1913 in the following section.

[318] With respect to the general interpretation of the effect two different ideas were proposed; besides the original interference idea suggested by von Laue (in Friedrich, Knipping and Laue, 1912), William Lawrence Bragg suggested his reflection idea (W. L. Bragg, 1913a). Upon closer inspection, however, it could be demonstrated that the two explanations did not contradict each other (von Laue, 1913a; Wulff, 1913). Peter Paul Ewald (1888–1985), a younger colleague of von Laue at Sommerfeld's institute—who had been working on dispersion and double refraction for his doctoral thesis of 1912—established the connection further in a paper '*Zur Theorie der Interferenzen der Röntgenstrahlen*

intensity maxima on the film were not more or less circular spots—as had been expected—but rather consisted of extended strokes perpendicular to the direction, connecting the point of the crystal which was hit by the primary X-rays to the position of the maxima. Von Laue and his student Franz Tank (born 1890) investigated this problem in experiments at the University of Zurich and arrived at a reasonable explanation (Laue and Tank, 1913).[319] In the last section of their paper, which was submitted in March 1913 to *Annalen der Physik*, the authors hinted at another problem: 'For the interference theory of the phenomena occurring when crystals are penetrated by X-rays, the only really important difficulty consists in estimating the role that one should attribute to the influence exerted on it by the heat motion [of the atoms in the crystal]' (Laue and Tank, 1913, p. 1010). They proposed to check first a suggestion of Max Planck, who had expressed earlier 'the opinion that perhaps considerable parts of the space lattice perform nearly the same vibrations, as far as amplitude and phase are concerned' (Laue and Tank, 1913, p. 1010);[320] if this suggestion was correct, then the heat motion should reduce the size of the crystal domain acting uniformly; especially since the elongation of the individual spots in the Laue diagram depended on the size of the crystal domain contributing to them, for low temperatures the elongations had to be observed 'for those distances [of the crystal] from the anti-cathode, for which they cannot be seen anymore in the case of usual temperatures' (Laue and Tank, 1913, p. 1011). A later quantitative calculation of the situation by von Laue, however, did not seem to confirm Planck's ideas and the consequences following from them (Laue, 1913e).

In deriving the latter result, von Laue had referred to previous work of Peter Debye in Utrecht. In two papers, entitled '*Über den Einfluß der Wärmebewegung auf die Interferenzerscheinungen bei Röntgenstrahlen*' ('On the Influence of the Heat Motion on the Interference Phenomena with X-rays') and '*Über die Intensitätsverteilung in den mit Röntgenstrahlen erzeugten Interferenzbildern*' ('On the

---

*in Kristallen*' ('On the Theory of X-ray Interferences in Crystals'), published in the issue of 1 July 1913 of the *Physikalische Zeitschrift* (Ewald, 1913a); in the same paper he discussed the shape of the interference spots. Leonard Salomon Ornstein, on the other hand, claimed that Bragg's explanation was superior; in a paper submitted to the Amsterdam Academy and published again (in German) in the issue of 1 October 1913 of *Physikalische Zeitschrift* he derived from it the intensity of the interference patterns, obtaining a result different from von Laue's (Ornstein, 1913a). The following debate between von Laue and Ornstein, carried on in the issues of 1 November, 1 and 15 December 1913 did not lead to any agreement between the opponents but clarified the respective assumptions (von Laue, 1913b; Ornstein, 1913b; von Laue, 1913d). The comparison between the theoretically calculated intensities, on the one hand, and the experimentally observed ones, on the other, which had to decide between von Laue and Ornstein, was complicated by the peculiar features of geometrical and physical origin as we shall discuss.

[319] Von Laue's Zurich experiments were supported by the *Institut International de Physique Solvay*. When working with crystals of fluorspar, he and Tank found that the elongation effect disappeared if the distance between the anti-cathode and the crystal was increased; hence the authors explained this phenomenon by assuming that for small distances from the anti-cathode to the crystal the curvature of the wavefront had to be taken into account.

[320] Planck expressed this opinion in a discussion of the talk by Max von Laue on X-ray interferences, presented on 14 June 1912 before the German Physical Society in Berlin.

Intensity Distribution in the Interference Pictures Created with X-rays') and published in the August issues of *Verhandlungen der Deutschen Physikalischen Gesellschaft* (Debye, 1913a, b), the author treated the problem of detailed intensity distribution in the X-ray interference patterns.[321] This investigation had interested Debye not only in providing a contribution to the theory of Laue patterns, but also to the quantum-theoretical problem of zero-point energy. He argued in particular:

> One might think of a method which allows one to conclude, from the sharpness of the interference pictures when corrected for all other influences, the heat motion. For this purpose one assumes, of course, that the energy sources of the secondary [X-]radiation are tightly connected to the atoms and therefore participate in its heat motion, a point of view, which one might admit to be very probable *a priori.* If one follows, as we do, this assumption, then for example an experimental test of the existence of zero-point energy on the basis of this effect [of heat motion on the sharpness of the interference pictures] might have a good prospect of being successful. (Debye, 1913a, p. 678)[322]

In order to carry out his intentions, Debye simplified the three-dimensional problem considerably; especially, he assumed a linear lattice of points, which are placed at variable positions, $na + \zeta_n$, where $n$ denotes an integral number, $a$ the lattice constant and $\zeta_n$ the thermal motion (position change) of the scattering electron.[323] He then calculated the intensity of the scattered X-rays on the assumption that the thermal motions of the electrons in the $x$-direction are statistically independent of each other and $N$ lattice points contribute to the interference picture; he thus arrived at the following formula for the averaged intensity at the point $r$:

$$J = N\left[1 - \exp\left(-\frac{kT}{f}\omega^2\right)\right] + \frac{\exp\left(-\frac{kT}{f}\omega^2\right)\sin^2\left(\frac{\omega a}{2}\frac{x}{r}N\right)}{\sin^2\left(\frac{\omega a}{2}\frac{x}{r}\right)}, \qquad (41)$$

[321] This problem had already been addressed by von Laue in his first publication on X-ray diffraction patterns, in which he calculated the intensity distribution according to the electromagnetic theory of diffraction of short waves by an extended, but fixed spatial grating. He then remarked, after comparing theory and experiment and finding a reasonable agreement: 'In spite of this agreement [of theory and observation] it should not be concealed that our theory must, in any case, be improved considerably. The thermal motion of the molecules indeed displaces them at room temperature by a noticeable fraction of the lattice constant; this fact must certainly be taken into account. Hence it would be too early to derive conclusions from expression (6) about the sharpness of interference maxima' (Friedrich, Knipping and Laue, 1912, p. 309). By expression (6) he referred to his theoretical formula for the intensity of interference patterns.

[322] The problem of providing an experimental decision in favour or against zero-point energy, which Planck had postulated to exist on the basis of his second quantum hypothesis (Planck, 1911a), concerned at that time—i.e., in 1913—the most active quantum physicists including Albert Einstein, Otto Stern and Paul Ehrenfest. (See our discussion in Volume 1, Section I.7.) Therefore Debye was happy to join research at the forefront of quantum physics with his study of the intensity of X-ray interference patterns from crystals.

[323] Debye assumed essentially that the electrons in the atom performed an oscillatory motion, reflecting the temperature motions of the atoms in a lattice.

where $k$ denotes Boltzmann's constant, $f$ the force constant of the atomic lattice and $\omega = 2\pi/\lambda$ ($\lambda$ being the wavelength of the secondary X-rays). (See Debye, 1913a, p. 686, Eq. (16).)[324] For small temperatures the first term on the right-hand side may be neglected; the second provides the usual diffraction result for a more or less fixed grating, having maximum intensity (proportional to $N^2$) at positions where the quantity $a\omega(x/r)$ assumes integral multiples of $2\pi$. For large $T$, on the other hand, the second term approaches zero; hence, 'the interference pattern passes, as had to be expected from the beginning, into a brightening independent of the direction [i.e., of $\cos\theta = x/r$]' (Debye, 1913a, p. 687); however, the change of the interference pattern with increasing temperature 'does not happen by a flattening and widening of the maxima of intensity', as 'the temperature motion does not change the sharpness of the interference stripes' (Debye, 1913a, p. 687).[325]

In his second paper of the summer of 1913, Debye extended his previous treatment to the three-dimensional lattice situation, deriving a generalized Eq. (41) which implied qualitatively similar consequences. Especially, he obtained an expression for the ratio of the 'interference' radiation intensity (II), i.e., the intensity generalizing the second term on the right-hand side of Eq. (41), to the 'dispersed' radiation intensity (I), i.e., the intensity generalizing the first term, namely,

$$\frac{(\mathrm{II})}{(\mathrm{I})} = \frac{N}{\exp\left[\dfrac{2kT}{f}\omega^2(1-\cos\theta)\right]-1}, \tag{42}$$

where $\cos\theta (= x/r)$ denotes the cosine of the angle formed by the directions of incident and secondary radiation, respectively. Since the factor $(1-\cos\theta)$ grows from zero for forward ($\theta = 0$) to 2 for backward ($\theta = 2\pi$) diffraction, the interference patterns from three-dimensional lattices are restricted in general, due to Eq. (42), to a cone of small aperture (around the forward direction), except when the force constant assumes a large value (as in the case of diamond). Debye further suggested taking care of the influence of quantum theory, especially on the low-temperature behaviour as follows: either replace the product $kT$ in the experimental factor $(kT/f)\omega^2$ by one-third times the average energy for an oscillating atom, or perform correctly the temperature average in a quantum-theoretical calculation. He admitted at the same time that both treatments provided 'only a rough approximation to the real case,' because of the 'as it is known

[324] Debye took a co-ordinate system, whose $x$-axis coincided with the linear lattice and whose origin was given by the middle point in the lattice. Then $r$ denoted the distance of the observation point from the origin.

[325] Debye derived still another consequence of the temperature motion in his approach: since, for small wavelength $\lambda$ of the secondary radiation, the effect of the temperature factor $(kT/f)\omega^2$ ($= (kT/f)(2\pi/\lambda^2)$) quickly increased, he concluded that wavelengths smaller than $10^{-9}$ cm would not contribute to the observed intensity. On the other hand, a large value of the force constant suppressed the temperature motion; hence the observed fact that X-ray interference patterns occur for diamond even in the backward direction, could easily be explained. (See Debye, 1913b, p. 738.)

unjustified assumption about the mutual independence of atomic vibration' (Debye, 1913b, p. 750).

In a third, long paper on the same subject, submitted in October 1913 to the *Annalen der Physik* and published in the first issue of 1914, Debye removed the above-mentioned restriction: he assumed in particular collective oscillations of the crystal having frequencies which 'run through all values in the elastic spectrum of the body,' i.e., he applied 'a method which has proved to be already successful in the theory of specific heats' (Debye, 1914a, p. 50). After having carried out the detailed calculations in this case, he obtained an expression for intensity similar to the one given by Eq. (41), i.e.,

$$J = \frac{A}{r^2}[N(1 - \exp M) + L \exp M], \tag{43}$$

where $A$ is an essentially constant factor and $r$ and $N$ denote the distance between the centre of the crystal and the interference spot and the number of atoms in the crystal, respectively. The temperature dependence is contained in the quantity $M$, which is zero for $T = 0$ and approaches $-\infty$ for large $T$. Clearly, the heat motion of the atoms exerts an influence on both terms of the secondary radiation, i.e., the 'dispersed' radiation (described by the first term $(A/r^2)N(1 - \exp M)$) and the 'interference' radiation (described by the term $(A/r^2)L \exp M$). Equation (43) could be evaluated in the limiting cases of very low and very high temperatures, as well as for intermediate ones, and compared with experimental data.[326]

More than six weeks after Debye submitted the above-mentioned paper, Erwin Schrödinger sent his first note on the same topic, '*Über die Schärfe der mit Röntgenstrahlen erzeugten Interferenzbilder*' ('On the Sharpness of the Interference Pictures Created by X-rays') to *Physikalische Zeitschrift*; it was printed in issue No. 2 of the year 1914, dated 15 January 1914, and thus appeared a couple

[326] Debye thus also hoped to be able to decide the problem of the existence of zero-point energy. Since no data were yet available, and Debye could not carry out experiments himself in Utrecht—where he held the professorship at that time—he restricted himself to summarizing at the end of his paper the theoretical results as follows: '1. The thermal movement of the atoms has an essential influence on the interference phenomena with observed X-rays. 2. The sharpness of the interference maxima is not affected; their intensity, as well as its spatial distribution, however, is affected. 3. As a consequence of thermal movement, the interference intensity decreases exponentially (a) with increasing angle between the direction of incidence and the direction of observation; (b) with increasing temperature; (c) with decreasing wavelength. (4) The exponent of the exponential function just mentioned vanishes at $T = 0$ in the absence of zero-point energy; it remains finite and assumes a substantial value if zero-point energy exists. (5) The exponent is always inversely proportional to the square of the wavelength. (6) The interference intensity is always accompanied by a scattered intensity, which has its highest intensity value where the interference intensity has its lowest value and *vice versa.* (7) The course of the phenomena can be approximately predicted by computations, provided data on the behaviour of the specific heat as a function of temperature are available. (8) In this approximation a similarity law holds, as for the specific heat of molecular bodies, according to which also in this instance the temperature dependence is a function only of the ratio of the characteristic temperature $\Theta$ and the temperature of observation' (Debye, 1914a, pp. 91-92; English translation, p. 36).

of weeks before Debye's paper (Schrödinger, 1914a). Schrödinger based his treatment exclusively on the dynamical lattice model of Born and von Kármán, which seemed to him to represent 'the most suitable [model] in order to estimate the probable influence of the heat motion on the interference phenomena, obtained by the action of X-radiation on regular crystals' (Schrödinger, 1914a, p. 79). He claimed, in particular, its superiority over Debye's earlier treatment of the same problem, based on the assumption of independent heat oscillations of the individual atoms (Debye, 1913a, b); since in Debye's results, say Eq. (41), the monochromatic eigenfrequency (i.e., $\nu = \{(1/2\pi)(f/m)\}^{1/2}$, with $m$ the mass of the atom) entered crucially, he expected 'that the transition to the certainly more justified *polychromatic* model changes the result, and, if the change is *essential*, then it will *better* describe the empirical facts' (Schrödinger, 1914a, p. 79).[327]

In his note Schrödinger followed, with respect to computational steps, to a large extent the procedure which Debye had taken several months earlier. In particular, he began by treating 'for the moment a one-dimensional atomic lattice with one-dimensional heat motion, since the three-dimensional problem presents rather serious mathematical difficulties if one drops the assumption of mutually independent atomic vibrations' (Schrödinger, 1914a, p. 79). He also used two further assumptions which Debye had made earlier, namely that the electronic oscillations in atoms were fast in comparison with the atomic period, hence one could average the phases of atomic motion. The averaged X-ray intensity emitted by the $N$ atoms of a linear lattice was then given by a double sum over the atoms in an exponential—corresponding to the amplitude of the secondary radiation emitted by a single atom multiplied by a Boltzmann factor (denoting the probability of the atoms to assume derivations from their rest positions) and integrated over all possible derivations $\xi_1, \ldots, \xi_N$—i.e.,

$$J = \text{const.} \sum_m \sum_n \exp[i\omega a(n-m)\cos\theta] \cdot \int_{-\infty}^{+\infty} \cdots \int_{-\infty}^{+\infty} \exp\left[-\frac{f}{2kT}\sum_l (\xi_{l+1}-\xi_l)^2 + i\omega\cos\theta(\xi_n - \xi_m)\right] d\xi_1 \cdots d\xi_N, \tag{44}$$

where $\theta$ denotes the angle between the lattice line and the direction from the centre of the lattice to the observation point.[328]

To evaluate the integrals in Eq. (44), Schrödinger introduced—like Born and von Kármán—normal coordinates for the displacement coordinates $\xi_n$. Thus he

[327] As we have indicated above, Debye's 1914 paper, which extended his earlier treatment to include a full spectrum of atomic vibrations (as in Debye's theory of specific heats), was not yet known to Schrödinger. But even if it were he would prefer—as Thirring did in his papers on the specific heat problem—the Born-von Kármán model over Debye's.

[328] The other symbols have the same meaning as in Debye's Eq. (41). Especially $a$ is the lattice constant.

finally found

$$J = \sum_m \sum_n \exp\left[ i\omega a(n-m)\cos\theta - \frac{kT}{2f}\omega^2 \cos^2\theta \sum \frac{(\alpha_{nl} - \alpha_{ml})^2}{p_l} \right], \tag{45}$$

where $\alpha_{nl}$ denote the coefficients of the transformation matrix and $p_l$ the momentum variables connected with the eigenfrequencies $\nu_l$ of the linear lattice by the equation ($m$ being the mass of the lattice atoms)[329]

$$p_l = \frac{m}{f}\nu_l^2. \tag{46}$$

Schrödinger computed the sum $\sum_l (\alpha_{ml} - \alpha_{ml})^2/p_l$ in the limit of an infinite point lattice as $|n-m|$. Finally, he carried out the remaining double sum over indices $n$ and $m$, arriving at the intensity formula

$$J = \frac{\frac{1}{2}[\exp(\mathscr{E}) - \exp(-\mathscr{E})]N}{\frac{1}{2}\left[(\exp(\mathscr{E}/2) - \exp(-\mathscr{E}/2))^2 + 2\sin^2\frac{\alpha}{2}\right]^{-2}}$$

$$- \left[\tfrac{1}{2}(\exp(\mathscr{E}/2) - \exp(-\mathscr{E}/2))^2 + 2\sin^2\frac{\alpha}{2}\right]^{-2}$$

$$\cdot \left\{ \left[\tfrac{1}{2}(\exp(\mathscr{E}/2) - \exp(-(\mathscr{E}/2))^2 - (\exp(\mathscr{E}) - \exp(-\mathscr{E}))\sin^2\frac{\alpha}{2}\right] \right.$$

$$\cdot \left[ (1 - \exp(-N\mathscr{E}) + 2\exp(-N\mathscr{E})\sin^2(N\alpha/2)) \right]$$

$$\left. + \tfrac{1}{2}\exp(-N\mathscr{E})(\exp(\mathscr{E}) - \exp(-\mathscr{E}))\sin\alpha \sin(N\alpha) \right\}, \tag{47}$$

with

$$\alpha = \omega a \cos\theta \tag{47a}$$

[329] By the transformation of $\xi_n$ to normal co-ordinates $a_n$, the energy for the lattice mass point number $n$, i.e.,

$$\frac{f}{2}\sum_n (\xi_{n+1} - \xi_n)^2 + \frac{m}{2}\sum_n \dot{\xi}_n{}^2,$$

assumed the normal form

$$\frac{f}{2}\sum_n p_n a_n{}^2 + \frac{m}{2}\sum \dot{a}_n{}^2.$$

(See Born and von Kármán, 1912, pp. 305–307, and Schrödinger, 1914a, p. 81, Eqs. (11) and (12).)

and

$$\mathscr{E} = \frac{kT}{f}\omega^2\frac{\cos^2\theta}{2}. \tag{47b}$$

Evidently Schrödinger's Eq. (47) replaced Debye's much simpler Eq. (41). For $\mathscr{E} = 0$—i.e., for zero temperature—only the second term survived and yielded, like Eq. (41), the usual grating formula of the optical theory. For finite $\mathscr{E}$, even one that still satisfied the relation

$$\exp(-N\mathscr{E}) \ll 1, \tag{48}$$

the first term in Eq. (47) became dominant; and Schrödinger demonstrated that in the vicinity of the intensity maxima—i.e., for $\alpha = 2n\pi + \mathscr{E}z$, with integral $n$ and $\mathscr{E} \ll 1$—the intensity was given by the formula

$$J = 2\frac{N}{\mathscr{E}}\frac{1}{1+z^2}. \tag{49}$$

This equation implied the following consequences:

> 1. The interference pictures [patterns] become, with rising temperature, increasingly wide and smeared out in an approximately symmetrical manner. Their width, defined as above [i.e., given by the condition that the intensity of the interference picture dropped at its rim to 1% of the central maximum intensity], is proportional to the absolute temperature.
> 2. The central intensity of an interference picture is, according to [Eq. (49)], inversely proportional to $\mathscr{E}$, i.e., is inversely proportional to the absolute temperature.
> 3. The total radiation intensity contributing to an interference picture [i.e., to the interference spots associated with a given order] does not depend on the temperature. (Schrödinger, 1914a, p. 85)

Finally, for large values of $\mathscr{E}$, the right-hand side of Eq. (47) became equal to $N$—as in Debye's formula (41)—that is: Schrödinger also found 'a uniform brightening of considerably lower intensity than the original one in the maxima [namely $N^2$] which were *unperturbed* [by the heat motion]' (Schrödinger, 1914a, p. 85).

Schrödinger did not go into any detail in comparing his results with experimental data, as he intended to leave the check 'until the three-dimensional problem has been treated' (Schrödinger, 1914a, p. 86). Nevertheless he claimed: 'There can, however, be hardly any doubt that the laws established so far will be also valid for the three-dimensional case, though only for high enough temperatures, since we have calculated the probability of a given configuration according to the laws of statistical mechanics without taking into account the quantum theory' (Schrödinger, 1914a, p. 86). But Schrödinger's conclusion excited the opposition of a colleague, Johann Kern, who argued: 'This suggestion must be considered the more surprising, because a simple rough estimate shows that—in the one-dimensional case—the ratio of $J_{max}/J_{min}$, given by Schrödinger's formula, for a [Laue] spot (at room temperature, $\lambda \simeq 10^{-9}$ cm, $f \simeq 10^5$ dyn/cm—i.e., say for rock-salt—and at an angle of 30° between the directions of observation and the incident ray) would be $\frac{1}{2}$, which would correspond to a nearly complete washing-out and is certainly far from the actual situation' (Kern, 1914b, p. 337).[330] Even if the extension of Schrödinger's approach to the three-dimensional lattice did remove this difficulty, Kern continued his criticism, nature would contradict another consequence of Eq. (47), namely the fact that it implied a temperature-independent total intensity of the diffraction spot.

The crucial mistake in Schrödinger's theory lay, in Kern's opinion, in the wrong evaluation of the second term on the right-hand side of Eq. (45). In evaluating the expression $\sum_l (\alpha_{nl} - \alpha_{ml})^2/p_l$, Schrödinger assumed that the boundary condition at both ends of the linear lattice played no role; Kern now claimed that the opposite was true, noting: if the end points of the linear lattice consisting of $N$ atoms were not kept fixed, Schrödinger's value $|m - n|$ followed; however, if they were, one found rather a different value $|m - n| - (m - n)^2/(N - 1)$. Hence he concluded:

> The result derived by Schrödinger is *strictly* valid *for an infinite one-dimensional crystal irradiated only on a finite piece.* The larger this irradiated fraction of the crystal, the more effective becomes the influence of the boundary conditions.... However, the ambiguity of the solution shows *that the one-dimensional problem might hardly be suitable in this case in providing an adequate description of the influence of the heat motion on the X-ray interference picture.* (Kern, 1914b, p. 342)

Kern's critique appeared in the issue of 1 April of *Physikalische Zeitschrift.* On 4 April 1914 Schrödinger signed and submitted a paper '*Zur Theorie des Debyeeffekts*' ('On the Theory of the Debye Effect') to the same journal, where he answered his opponent (Schrödinger, 1914b). Although he agreed with the conclusion that the sum in question depended on the boundary conditions chosen, since the 'normed eigenfunctions' $\alpha_{nl}$ did so, he emphasized: 'The physical conclusion of the *final result* for $J$ nevertheless remains *untouched* by this, provided the number of atoms is sufficiently large; and $J$ can be evaluated with desirable precision by a simple approximation method, which does not depend on boundary conditions' (Schrödinger, 1914b, p. 498).[331] Schrödinger also con-

[330] Kern was at the University of Leyden, where meanwhile Hendrik Lorentz had become very interested in the theory of X-ray diffraction. In a paper, submitted in December 1913, Kern entered into the discussion of the intensity distribution of X-ray interference pictures, based on a formula presented by Lorentz in his Leyden lectures (Kern, 1914a, especially p. 137).

[331] Schrödinger proved this assertion by analyzing in detail the consequences of the additional term, $(m - n)^2/(N - 1)$, in the exponential of the expression for the intensity. He thus showed, by a careful *epsilontic* method, that the contribution from this term could be made arbitrarily small.

sidered different boundary conditions, but argued that they would not change the intensity formula, Eq. (47), either.[332]

Encouraged by this success, Schrödinger hoped that Eq. (47) might retain the same structure if extended to the three-dimensional lattice. He therefore went on to derive further conclusions from his formula. Thus he noticed first that in the $l$-sum contributing to the intensity, Eq. (45), small differences, $|m - n|$, played a decisive role, 'a physically extremely sympathetic' result, since it implied that the 'interference phenomenon is caused essentially by the co-operation of relatively close atoms' (Schrödinger, 1914b, p. 500). Second, he found that he had substantiated the ideas of Planck, von Laue and Tank—referred to earlier—on the origin of X-ray interference phenomena through the synchronous vibrations of large parts of the space lattice, and especially the suggestion of von Laue and Tank that the elongation of the interference spots could be reduced by the temperature effect.[333] Third, 'The fact that only pairs of atoms, which are not too far distant from each other, contribute noticeably to the interference intensity, also removes the suspicion, arising from natural instinct, that one might, by applying the method of normal vibrations—where the sine and cosine waves, in which the wave motion can be thought of as decomposed, are considered *as coherent in the entire lattice*—do violence to the physical problem' (Schrödinger, 1914b, p. 501).

After clarifying the above points of principle, Schrödinger entered into a detailed discussion of the practical consequences of his Eq. (49), describing the intensity of interference spots in the vicinity of maxima. Thus he confirmed again that the total intensity of a given spot did not depend on the temperature, declaring subsequently: 'One should not, however, conclude from this result that the photographically determined intensity of the spot must be independent of temperature' (Schrödinger, 1914b, p. 502). The actual situation was this: for any given exposure time (of the photographic plate) a minimum intensity ($J_0$) is required to create a noticeable darkening;[334] hence the actual extension, $2\alpha_0'$, of a photographically registered Laue spot was not infinite, but followed from the equation

$$2\alpha_0' = 2\mathscr{E}\sqrt{\frac{J_{\max}}{J_0} - 1}, \tag{50}$$

[332] A natural boundary condition was perhaps the assumption that the lattice end-points were bombarded by gas molecules; but since both extreme cases of fixed and completely free end-points yielded the same intensity, all intermediate situations should do so *a fortiori*.

[333] According to Debye's theory of the temperature effect this suggestion could not be confirmed (see Laue, 1913e). Schrödinger, however, now claimed: 'Debye's theory appears to me, due to certain things neglected in it so far, not to be in a position to describe such fine details of the interference picture' (Schrödinger, 1914b, p. 501). In this context he pointed to Debye's assumption of independent vibrations of neighbouring atoms, which resulted in '*the same* attenuation factor for all pairs of atoms' and therefore in a 'missing influence of the thermal motion on the "elongation"' (Schrödinger, 1914b, p. 501, footnote 2).

[334] $J_0$ depended on the exposure time and on the properties of the photographic plate.

with

$$J_{\max} = \frac{2N}{\mathscr{E}}. \tag{50a}$$

Since $\mathscr{E}$ depended—via Eq. (47b)—on the temperature, it followed immediately that: 'The photographic width of the interference picture must, for a constant time of exposure, first increase and later decrease again with increasing temperature, until it disappears at a definite temperature—which is dependent on the exposure time [via the dependence on $J_0$]' (Schrödinger, 1914b, p. 502).

For the photographically observed brightness Schrödinger was also able to derive conclusions, namely (i) that the complete fading away of the interference spots predicted in the earlier paper (Schrödinger, 1914a, p. 85) could not be observed in reality, and (ii) that for a constant time of exposure the photographic intensity of a given Laue spot decreased with increasing temperature.[335] Schrödinger was perfectly aware of the fact that experience would not confirm all the details of his calculation, because the exact form of the Laue spots

> must necessarily still depend on so many other factors—such as the finite distance between anti-cathode and crystal, between crystal and photographic plate; the finite extension of the focal point [of the X-ray emitting anti-cathode]; the inhomogeneity of the incoming [X-ray] wave; structural defects, the roughest of which present themselves in a "striation"—that the influence of temperature might possibly be totally masked. (Schrödinger, 1914b, p. 503)

Nevertheless, he proudly referred to the recent experiments of von Laue and J. Stephan van der Lingen, who had observed the disappearance of the interference spots for a rock-salt crystal at a temperature of 320°C (Laue and van der Lingen, 1914), as confirming his theory.

## Lattice Dynamics and a Proof of the Atomic Structure of Solids

Schrödinger's two contributions dealing with the influence of temperature on the Laue spots revealed several characteristics of the young scientist. On the one hand, the assistant of Franz Exner fully displayed his particular interest in and concern with the intimate relationship between the observed phenomena and their theoretical description, presenting himself as a good connoisseur of the experimental situation. On the other hand, he derived theoretical results from first principles and basic assumptions in order to establish the connection between data and physical interpretation as directly as possible. Unlike Peter Debye, however, he did not show any particular eagerness to deal with subtle features of quantum

[335] Schrödinger assumed, for that purpose, that the blackening of the photographic plate was proportional to the average value—taken over the interference spots—of the intensity of the secondary radiation.

theory, such as the existence or non-existence of zero-point energy. For Schrödinger the main physical problem of the day was still whether one could establish beyond any doubt the atomistic structure of matter and refute decisively the continuum interpretation of matter underlying the so-called phenomenological theories of the properties of matter, which were supported in Austria and elsewhere by the partisans of Ernst Mach.

The final solution of this problem posed several tasks in physical theory, of which Schrödinger was very much aware. Thus he opened a paper, entitled '*Zur Dynamik elastisch gekoppelter Punktsysteme*' ('On the Dynamics of Elastically Coupled Point Systems') and submitted in early March 1914 to *Annalen der Physik*, with the words:

> It has been stated very often, and belongs so-to-say to the creed of the atomist, that all partial differential equations of mathematical physics, which connect the spatial and temporal variations of any physical variables (temperature, deformation, field strength, etc.) are incorrect in the strict mathematical sense. Because the mathematical symbol of the differential quotient prescribes the limiting process to *arbitrarily* small spatial variations, although we are convinced that we have to stop, when performing such "physical" differential quotients, at "physically infinitely small spaces," i.e., at those, which still contain very many molecules. (Schrödinger, 1914c, p. 916)[336]

If one went on, however, to deal with smaller spaces than the 'physically infinitely small spaces' addressed above, then a new situation would arise, described 'not even distantly by those simple laws, which can be expressed by partial differential equations' (Schrödinger, 1914c, p. 916).

In Schrödinger's opinion, the atomistic view demanded as a first task that 'all those differential equations, which have been obtained preliminarily by treating a continuous medium by differential equations in the strict sense, must now also be derived in the above indicated sense as difference equations on the basis of a model built of molecules' (Schrödinger, 1914c, pp. 916–917). While this task had already been solved in the past 'in most of the cases'—in fact, historically the derivation of the properties of solids from the atomistic point of view usually preceded the one based on the continuum concept—the other task, 'whose solution at first enables one to prove the exclusive validity [of the atomistic view] over the phenomenological theories' (Schrödinger, 1914c, p. 917), was not. It consisted of 'finding and predicting *such* conditions, under which the differential equations based on the continuum concept actually lead to observable incorrect results' (Schrödinger, 1914c, p. 917).

Schrödinger claimed that so far 'the only successes in this direction lie in the field of kinetic gas theory' (Schrödinger, 1914c, p. 917), and he mentioned explicitly the temperature jump in the heat conduction and the finite viscosity

[336] In substantiating the need of the atomists, Schrödinger referred to statements of Ludwig Boltzmann (1897a, e).

of rarefied gases as examples of quantities which could be derived from the atomistic theory, but not from any phenomenological theory (based on the continuum hypothesis) of heat conduction or inner friction. The discussion of other phenomena on the basis of the atomistic view required dealing with the dynamics of a mechanical system of an enormous number of degrees of freedom, a more or less impossible task. As a contribution in handling this principal difficulty, Schrödinger now proposed a new integration method for ' a system of mass points, which provides in the limit indicated above the partial differential equation of the vibrating string, the one-dimensional wave equation' (Schrödinger, 1914c, p. 918). 'It appears to me,' he emphasized, 'that the relations between the laws of motion of the point system and the propagation of waves in a one-dimensional elastic medium become more explicit in the solution which will be given below' (Schrödinger, 1914c, p. 918).

Evidently, Schrödinger selected the one-dimensional point lattice not only because of his familiarity with the problem, but also because of the physical importance that it had assumed so far in the theory of specific heats of solids and in the debate on the influence of temperature on the Laue interference spots. Schrödinger now wished to abandon the method of eigenvibrations, which had been used previously with great success by Born, von Kármán, Thirring and himself, because of the following shortcomings:

> The method exhibits similarities to Fourier's method when applied to the string. It implies the disadvantage that, as in the former, an arbitrarily given initial state must first be Fourier analyzed, before one is able to consider the further motion, and that the motion itself is given by a superposition of a continuous sequence of normal modes, hence the resulting vibration can hardly be recognized. (Schrödinger, 1914c, p. 920)

Schrödinger believed that a method had to be developed, where the essential features of the solution might be represented more explicitly. For that purpose he went back to the equations of motion of a point lattice, in which he assumed that only the nearest neighbours on each side of a given mass point (index $n$, mass $m$) exert forces on it, the latter being proportional to the difference between the deviations from their equilibrium positions of the respective mass points, i.e., $f(\xi_{n+1} - \xi_n)$ and $-f(\xi_n - \xi_{n-1})$. By introducing the new variables $x_n$, instead of the dynamical variables, $\xi_n$ and $\dot{\xi}_n$, of the system, with $x_{2n} = \sqrt{m}\dot{\xi}_n$ and $x_{2n+1} = \sqrt{f}(\xi_n - \xi_{n+1})$, the equations of motion could be rewritten as

$$\frac{dx_n}{dt} = -\frac{\nu}{2}(x_{n+1} - x_{n-1}), \qquad n \text{ integer.} \tag{51}$$

Schrödinger now recognized Eq. (51) as, 'apart from the factor $\nu = [2\sqrt{f/m}]$, one of the two fundamental functional relations between three Bessel functions, which lie in a sequence having the parameter difference of unity' (Schrödinger,

1914c, p. 921). Consequently he wrote the solution of Eq. (51) in the form

$$x_n = \sum_{k=-\infty}^{+\infty} x_k^0 J_{n-k}(\nu t), \qquad n = -\infty, \ldots, +\infty, \tag{52}$$

where $x_n^0$ denotes the value of $x_n$ at the time of origin.

To illustrate the nature of the above solution, Schrödinger discussed the special case of a linear lattice, with the initial condition that all mass points are in their respective equilibrium positions, except the middle point (with index 'zero') of the lattice is displayed by the distance $\varepsilon$. The solution evidently became

$$\xi_l = \varepsilon J_{2l}(\nu t), \tag{53}$$

which implied, on account of the properties of the even-number Bessel functions, the following motion: a given mass point is perturbed, then returns to its rest position and oscillates beyond it, and soon performs an harmonic vibration with an amplitude decreasing in course of time like $1/\sqrt{t}$. However, Schrödinger also noticed:

> *With all that the maximum of the intensity of motion remains at the place of* [*the initial*] *perturbation*, while it decreases on both sides gradually to zero. The entire process seems *to bear rather a distant resemblance to the propagation of a temperature perturbation than to the propagation of a sound wave.* (Schrödinger, 1914c, p. 925)

By further analyzing the dependence of the general solution, Eq. (52), of the motion of the lattice points on the initial conditions imposed on them, Schrödinger derived the result that a wave-like motion did not occur unless 'both systems of quantities, $\xi_n^0$ and $\dot{\xi}_n^0$, displayed sufficient resemblance with the value distributions of two continuous functions of $x$, which show noticeable changes only over distances that are large in comparison with the distance $a$ of the lattice points' (Schrödinger, 1914c, p. 930). Thus he concluded that in real physical situations those quantities of a crystal lattice, which depend on the average values of the lattice point displacements $\xi$, might be described by wave-like solutions; others, especially the propagation of a temperature homogeneity in a solid, certainly would not do. He finished his article by saying:

> All conclusions, which one might be tempted to derive from the analogy of such point lattices—whose simplest representation we have discussed here—with the continous elastic media on the propagation of *temperature perturbations* in such systems, need, it seems to me, to be taken with great caution. (Schrödinger, 1914c, p. 932)

With the above results Schrödinger had certainly reached his goal of presenting further evidence for the superiority of the atomistic structure of solids over the continuum assumption, which certainly did not yield the correct behaviour for temperature perturbations. Besides, he was in a position to criticize some conclusions derived by his old competitor Debye and by Max Born. Both had suggested previously that lattice defects would crucially influence if not create some properties of solids; Born, especially, had just explained the phenomenon of heat conduction on the basis of such impurities (Born, 1914). Debye, on the other hand, claimed in his Göttingen Wolfskehl lecture (of April 1913) that only the deviations of the interatomic forces from Hooke's law provided a finite value for heat conductivity (Debye, 1914, pp. 43-46). In the light of his last investigations on the propagation of a temperature inhomogeneity in a linear point lattice, their reasoning appeared to Schrödinger as being rather unlikely to hold in nature. He hoped that the hitherto successful integration method in those cases, involving the use of Bessel functions, might also work to explain other interesting properties of solids. To prepare further applications of his methods he treated in a second paper, '*Zur Dynamik der elastischen Punktreihe*' ('On the Dynamics of an Elastic Point Sequence'), which he submitted in June 1914 to the Vienna Academy of Sciences, the case of a linear point lattice, on each of whose mass points acted a time-dependent force (Schrödinger, 1914d). Again he found a general solution, namely,

$$x_n(t) = \sum_{k=-\infty}^{+\infty} \left[ x_{n-k}^0 J_k(\nu t) + \int_0^{\nu t} J_k(z) \psi_{n-k}(\nu t - z)\, dz \right], \tag{54}$$

where the functions $\psi_{n-k}(\nu t)$ were, up to a constant factor, identical with the applied time-dependent forces.

Schrödinger specialized the general solution (54) to certain cases. First, he discussed a linear lattice, in which just one point is excited by a periodic force: for an excitation frequency $\nu'$ smaller than the eigenfrequency $\nu$ of the system he found a propagating wave proceeding in both directions, right and left, of the originally perturbed point; and for $\nu' > \nu$ he obtained a standing vibration, with neighbouring points moving with opposite phase and strongly falling amplitudes with increasing distance from the centre of perturbation. Second, he treated the cases of a finite point lattice and of a point lattice which is infinite only in one direction. All these results yielded, as far as previous solutions had been derived in special cases, basic agreement with these. For Schrödinger they seemed to open up a large field for further fruitful studies in the kinetic theory of solids for years to come.

## I.4 From Brilliant Pre-War to Dark War Days

The years immediately preceding World War I have often been called one of the brightest periods in European culture and civilization. Enormous creative powers seemed to be active on the Old Continent: they could be seen in the establishment of a bourgeois lifestyle reflecting a more widely distributed wealth than had ever existed before; in the liberal governments prevailing in most countries; in the people supporting nearly unlimited freedom in arts and sciences. Thus it was a time of great movements in the arts and literature; from impressionism to symbolism, fauvism, expressionism, cubism and abstractionism in painting; from decorative to functional architecture; from tonal to atonal music. It was a time of great experiments in literature and in theatre and of the first successes of a new art form, the cinema. Light filled the cities day and night; electric street cars, the first underground railways and automobiles enabled people to get to work and to pleasure faster. In one word: it was a great and *belle* epoch, in which many movements like liberalism and socialism united people far beyond the borders of the old countries. From this open atmosphere and quick communication, science, in particular, profited enormously: news and discoveries were quickly exchanged; international honours—like the Nobel prizes—or fellowships in national academies were bestowed on eminent people of all nations; scientists felt as members of one family working optimistically for the benefit and brighter future of mankind.

Among the pre-war centres Vienna played its appropriate role in culture and science. We may select as examples of the latter two events intimately connected with the Austrian capital: the discovery of cosmic radiation and the 85th *Versammlung deutscher Naturforscher und Ärzte* in September 1913. The young Schrödinger was very much a child of this wonderful and stimulating time; his career at the University of Vienna proceeded steadily, and his scientific work offered promising aspects for the future. The outbreak of the great war, though not completely unexpected and not carefully avoided by the politicians and the lobbies behind them, changed everything: it interrupted work and lives; it took away economic welfare from wide classes of citizens: and it changed the outlook of people so deeply that hope was increasingly lost that the halcyon pre-war days might never return. In the war the Austrian Empire was completely and irreversibly destroyed; and the opportunities for the young scientist Schrödinger faded like the memory of the once glorious days in Vienna.

## The Discovery of Cosmic Radiation (*Höhenstrahlung*)

As we have mentioned earlier—in Section I.3—Franz Exner had emphasized the necessity of balloon observations at an early stage of research in atmospheric electricity. We have further pointed—also in Section I.3—to the importance of investigating the distribution of penetrating radiation at greater altitudes above ground, which again implied at that time—around 1910—the use of balloons. The first balloon flight undertaken for this purpose was arranged in Switzerland: Albert Gockel, professor of physics at the University of Fribourg, had available during an international balloon week in December 1909 the balloon *Gotthard* of the Swiss Aeroclub.[337] On 11 December the *Gotthard* ascended with three passengers. Gockel, in his report on '*Luftelektrische Beobachtungen bei einer Ballonfahrt*' ('Observations of Atmospheric Electricity During a Balloon Ride'), which he submitted in February 1910 to *Physikalische Zeitschrift* (Gockel, 1910, p. 280), said: 'The purely meteorological observations were performed by Dr. de Quervain, who was assisted in piloting the balloon by Lieutenant Müller, the observations on atmospheric electricity were made by the author.' The balloon ride took four hours from Zurich to the Jura west of Bienne and reached a maximum height of 4,500 m above sea level. Gockel measured, in particular, the amount of positive and negative ions with an aspiration apparatus and the intensity of the penetrating radiation. He described the apparatus for the latter measurement as follows:

> The instrument used for this purpose consisted of an electrometer *à la* Wulf, on which a cylindrical ionization chamber made of 2-mm copper foil having 22 cm diameter and 18.5 cm height—i.e., a volume of 7.03 litres—was screwed on. The pin serving as dispersion body [i.e., for measuring the ionization current of the enclosed air] was 70 mm high and had a diameter of 9 mm. The capacity of the total apparatus was 6.25 cm. All screws were sealed with leather and grease. (Gockel, 1910, p. 281)

Unfortunately, he did not have available a second apparatus to control his measurements; thus the insulating qualities especially of the instrument could not be checked. As a result of the measurements he claimed, 'that although in the free atmosphere a diminution of the penetrating radiation [with increasing altitude] occurs, this diminution does not reach the amount which one would expect if the main part originates on the ground' (Gockel, 1910, p. 282). Two further balloon rides, the first carried out on 15 October 1910 from Zurich to Olten, and the second on 2 April 1911 from Berne to Lake Bienne, enabled Gockel to supply the necessary checks of his first observations; they confirmed all the previous conclusions. These conclusions, he argued, were in essential agreement with those derived by Mache in observations on the ground, 'namely that a non-negligible part of the penetrating radiation is independent of the direct action of the radioactive substances contained in the uppermost layers of the earth' (Gockel, 1911, p. 597). However, it already contradicted the result of Karl Bergwitz, who had found 'on a balloon ride [Bergwitz, 1910] that at 1,300 m altitude the ionization in a sealed closed vessel was reduced to 24 percent of its value obtained on the ground' (Gockel, 1911, p. 597).[338] The person to clear up

[337] Albert Wilhelm Friedrich Eduard Gockel was born in Stockach, Baden, on 27 November 1860. He studied physics at the universities of Freiburg/Breisgau and Würzburg, at the *Technische Hochschule* in Karlsruhe and the University of Heidelberg, obtaining his doctorate from the latter in 1885. After working as a secondary-school teacher until 1895, he became assistant to Joseph von Kowalski at the University of Fribourg. He joined the faculty of this university in 1901, was promoted to extraordinary professor in 1903 and full professor in 1909. Gockel contributed to various topics on the borderline between physics and meteorology, on natural radioactivity of the earth, storm electricity and atmospheric interferences of radio waves. He died on 4 March 1927 in Fribourg.

[338] Karl Friedrich August Bergwitz was born on 7 November 1875 in Wolfenbüttel. He became *Privatdozent* at the *Technische Hochschule* in Braunschweig in 1909 and extraordinary professor at the same institution in 1915. He died in Braunschweig on 14 November 1958.

the situation by a long series of balloon flights would be Victor Franz Hess of Vienna.

Victor (or Viktor) Franz Hess was born on 24 June 1883 in *Schloß Waldstein* near Deutsch-Freistritz, Styria.[339] He attended the *Gymnasium* in Graz (1893-1901), after which he studied physics (especially with Leopold von Pfaundler) at the University of Graz. Receiving his doctorate '*sub auspiciis Imperatoris*' (meaning 'under the auspicies of the Emperor', i.e., with highest distinction) on 16 June 1906, he planned to go to Berlin in order to perform research in optics with Paul Drude. On the latter's death he joined instead Franz Exner's *Zweites Physikalisches Institut* in Vienna, turning his attention to the subjects of radioactivity and atmospheric electricity.[340] He got his *Habilitation* in 1910 with a paper, entitled '*Absolutbestimmungen des Gehaltes der Atmosphäre an Radiuminduktion*' ('Absolute Determinations of the Content of Radium Inductions in the Atmosphere,' Hess, 1910a). He then joined Stefan Meyer as first assistant in the *Institut für Radiumforschung* later the same year. Hess now embarked upon systematic and quantitative investigations of the radioactive substances contained in the air, refining and intensifying the methods of previous and contemporary authors. He devoted special interest to the phenomenon of penetrating radiation. In a note, published in the 15 September 1910 issue of the *Physikalische Zeitschrift*, Theodor Wulf provided the results of four days of observation, during the spring of the year, on the ionization of the atmosphere around the Eiffel Tower in Paris; he concluded that, in comparison with the values on the ground (i.e., below the observation point on the tower), the intensity of radiation 'decreases at nearly 300 m [altitude], not even to half [of its ground value],' while with the assumption that radiation emerges from the ground there would remain at the top of the tower 'just a few percent of the ground radiation' (Wulf, 1910, p. 812). Wulf's observations were of great value, because he could take many data at a fixed location—he observed for four days between 11 a.m. and 5 p.m.; his apparatus exhibited none of the difficulties which the balloon riders had occasionally experienced, hence they [Wulf's observations] were rightly considered as the most reliable source of information on the altitude effect in the penetrating radiation existing so far.

Hess therefore started his programme by comparing Wulf's results with the theoretical description, which Artur Eve had presented in his paper 'On the Ionization of the Atmosphere due to the Radioactive Matter,' published in the recent January issue of *Philosophical Magazine* (Eve, 1911). If one assumed a

[339] Biographical information about Hess can be found in the 'biography' given in *Nobel Lectures: Physics 1922-1941*, Amsterdam, New York: Nobel Foundation (Elsevier, 1965) pp. 363-364; and in articles by Rudolf Steinmaurer, 1953, 1958, 1964, 1965.

[340] Hess' first publication was on a modification of a formula for the index of refraction for a mixture of two fluids (Hess, 1906), which also served as his doctoral dissertation. Although he wrote another paper on the same topic in Vienna (Hess, 1908), he soon contributed to the solution of problems of radioactivity—first with a work on the properties of a decay product of uranium, UX (Hess, 1907).

uniform distribution of Ra C on the surface and in the uppermost layers of the earth, 'an elevation to 100 m should reduce the effect [i.e., the intensity of the penetrating radiation] to 36 percent of the ground value,' concluded Hess and added: 'This is such a serious discrepancy [with Wulf's results] that its resolution appears to be of the highest importance for the radioactive theory of atmospheric electricity' (Hess, 1911a, p. 1207). Since in the evaluation of Eve's formula the absorption coefficient of the $\gamma$-radiation in air entered crucially, which had so far not been determined directly but had just been extrapolated from data for solid and fluid substances with the help of the density law (i.e., the constancy of the quotient of the absorption coefficient over the density of the absorbing substance), Hess—who had worked earlier on problems of refraction—desired to close this gap by 'direct measurements of the absorption of $\gamma$-rays in air' (*loc. cit.*, p. 1208). He carried out the experiment in a meadow in the vicinity of the *Institut für Radiumforschung*, using probes of about 1 g $RaCl_2$ and distances up to 90 m and obtained an absorption coefficient of $0.477 \times 10^{-4}$, in fair agreement with the previously extrapolated value. Hence the contradiction between Eve's theory, in which the penetrating radiation was emitted only from radioactive substances on the ground, and Wulf's observation remained. Since the existing intensity data at higher altitudes were still poor and uncertain, Hess concluded: 'A clarification can only be expected from further measurements of the penetrating radiation in balloon ascents' (Hess, 1911a, p. 1212).

Hess continued his programme quickly by his own balloon observations. 'On 28 August [1911] at 8 a.m. the balloon "*Radetzky*" of the Austrian Aeroclub ([having a volume of] 1,100 $m^3$) with *Oberleutnant* S. Heller as pilot and me as sole passenger was lifted by the crew of the *Militäraeronautische Anstalt* under the command of *Hauptmann* W. Hoffory' (Hess, 1911b, p. 1578). It took Hess up to a height of 1,070 m above ground and allowed measurements during the four hours of flight. A second ride in another balloon ('*Austria*') during the night of 12 to 13 October 1911 took him to an altitude of 360 m above ground. During both balloon rides he observed that the intensity of the penetrating radiation remained practically constant, independent of height. He concluded the report on the two rides by suggesting two improvements: first, 'it will be necessary to perform parallel measurements with a thick-walled and a very thin-walled apparatus in the balloon' in order to enable one 'to study separately the behaviour of $\beta$- and $\gamma$-radiation'; second, one had to extend the observations to 'very great altitudes up to 7,000 m' (Hess, 1911b, p. 1584). From April to August 1912 he had the opportunity of undertaking seven balloon flights up to 5,200 m above ground, about which he reported in a detailed paper communicated to the Vienna academy meeting of 17 October 1912 (Hess, 1912). During all these rides he used three apparatuses, two with thick walls and one with thin walls (which also allowed him to register the effect of $\beta$-rays). As a result of his measurements Hess stated:

> [i] Immediately above ground the total radiation decreases a little . . . . [ii] At altitudes of 1,000 to 2,000 m there occurs again a noticeable growth of penetrating

radiation. [iii] The increase reaches, at altitudes of 3,000 to 4,000 m, already 50 percent of the total [$\gamma$-]radiation observed on the ground. [iv] At 4,000 to 5,200 m the radiation is stronger by [producing] 15 to 18 [more] ions [that is, more than 100 percent] than on the ground. (Hess, 1912, pp. 2026–2027)

These results seemed to him 'most easy to explain by the assumption that radiation of very high penetrating power enters the atmosphere from above and creates, even in the lowest layers [of the atmosphere], a part of the ionization observed in closed vessels' (Hess, 1912, p. 2025).[341] Hess further found time variations of radiation intensity, but excluded 'the sun as the direct source of this hypothetical penetrating radiation,' due to the fact that he observed on his balloon flights 'neither during the night nor during a solar eclipse any weakening of the radiation' (Hess, 1912, p. 2029).

Hess' observations of 1912, which yielded an increase in intensity of the penetrating radiation with increasing height, could hardly be accounted for by the hitherto suggested assumptions that it was caused by the radioactive substances contained in the ground and the air above the ground. Irregular changes of the high altitude component, independent of any meteorological conditions, also pointed to an extraterrestrial origin. In a paper '*Über den Ursprung der durchdringenden Strahlung*' ('On the Origin of the Penetrating Radiation'), which Hess submitted in June 1913 to the Vienna Academy of Sciences, he carefully analyzed all possible sources of error of his measurements; especially, he again gauged his apparatus (for observing the ionization of air) and he determined as accurately as possible the radiation emerging from its walls (Hess, 1913a). The results of this analysis supported his previous conclusion, namely, 'that a large part of the total penetrating radiation does not emerge from the known radioactive substances of the earth and of the atmosphere' (Hess, 1913a, p. 1077).[342] Finally, a new flight in the balloon '*Astarté*' on 1 June 1913, provided to him free of charge by the Austrian Aeroclub and its proprieter (E. C. Sigmundt of Trieste), which went up to 4,050 m above sea level, confirmed the previous observation of August 1912: namely 'that at altitudes of about 4,000 m there occurs a very fast increase in penetrating radiation' (Hess, 1913b, p. 1484). Hess therefore exclaimed, 'the theory developed recently by E. Schrödinger [Schrödinger, 1912c] for the altitude distribution of penetrating radiation fails *vis-à-vis* the increase of radiation with height' (Hess, 1913b, p. 1486).

This result may be taken as a definite statement of the existence of a hitherto unexplained type of penetrating radiation. Hess published a summary of his first

[341] The opinion that an extraterrestrial source might contribute essentially to penetrating radiation had been put forward by the English physicist Owen Willans Richardson (1879–1959) several years before (Richardson, 1906a, b), but it had been discarded because one expected it to be essentially absorbed until it became effective (Kurz, 1909).

[342] In order to explain the balloon observations at altitudes between 1,000 and 2,000 m, the Ra C content of the atmosphere had to be about twenty times larger at those altitudes than on the ground, an obviously impossible assumption; even the values of the intensity of $\gamma$-radiation on the ground seemed to be too big to be due to the radium content of the uppermost layers of the earth's crust (Hess, 1913a, p. 1077).

paper of 1913 (Hess, 1913a) in the *Physikalische Zeitschrift* (Hess, 1913c), which reached a wider public than the Vienna *Sitzungsberichte* and carried (since its first issues in 1899) many of the pioneering papers in the field of atmospheric electricity. The same journal also would publish the reports presented at the 85th *Naturforscherversammlung* held in September 1913 in Vienna. At this splendid scientific assembly Hess' discovery would be confirmed.

## The 85th *Versammlung Deutscher Naturforscher und Ärzte*

Pre-war Vienna housed many great scholars, even pioneers, in medicine, physiology, biology and meteorology, who partly created influential schools spreading throughout the world. Chemistry, for example, was represented at the highest level by Carl Ritter Auer von Welsbach, as much a scientist as an industrialist.

Auer von Welsbach was born on 1 September 1858 in Vienna.[343] From 1880 to 1882 he studied with Robert Bunsen in Heidelberg, where he became interested in the main field of his later research, the chemistry of rare earths. Continuing his studies in Vienna at the chemical institute of Adolf Lieben, he developed in particular the method of fractionated crystallization, succeeding by this method in 1885 in separating the substance didymia into the chemical elements neodymium and praesodymium.[344] The detailed knowledge of the chemistry of rare earths enabled Auer von Welsbach to create an industry producing these substances for use in gas lamps. His plant, situated at Atzgersdorf near Vienna became world famous. In 1892 he invented the so-called Auer light (*Auerglühstrumpf*) which revolutionized the techniques of illumination as much as his later (1902) introduction of the first metallic-filament lamp, the osmium filament lamp. Auer von Welsbach's recognition of cerium–iron as fire-steel inaugurated a new industry.[345]

The Atzgersdorf plant added not only to the fame of Austrian industry, but to Austrian science as well. Many investigations of the fundamental properties of chemical elements performed in Viennese laboratories, and at other places, made use of its products, e.g., the magnetic and spectroscopic measurements at Franz Exner's institute to which we have already referred.[346] We have also

[343] His father, Aloys Ritter Auer von Welsbach was then director of the *Hof- & Staatsdruckerei.* (See the biography of Auer von Welsbach by D'Ans, 1928.)

[344] Twenty years later, in 1905, he again reported the separation of a substance, namely ytterbia into two elementary ones, aldebaranium and cassiopeium. A little later, Georges Urbain also described the separation of ytterbia, denoting the chemical elements ytterbium and lutetium. These names became standard for the chemical elements No. 70 and 71.

[345] Auer von Welsbach received many honours, both in science and in public life, during which his industrial enterprises went through good and bad times including bankruptcy. He died on 4 August 1929 at *Schloß Welsbach* in Carynthia.

[346] Spectroscopic measurements with substances produced in Atzgersdorf were also carried out in Vienna by Joseph Maria Eder and Eduard Valenta (and their collaborators) at the photochemical laboratory of the *K. K. Graphische Lehr- und Versuchsanstalt* (e.g., Eder and Valenta, 1910).

previously drawn attention to the fact that the research on radioactivity, supported especially by the Vienna Academy of Sciences, received vital help from Atzgersdorf. It was Auer von Welsbach's collaborator Ludwig Haitinger, who actively participated in the preparation of radium substances which provided the basis for pioneering work at the *Institut für Radiumforschung.*

Besides Auer von Welsbach pre-war Vienna provided work for two younger chemists, who would later acquire great reputations through their fundamental contributions to science, namely George Charles de Hevesy (Georg von Hevesy) and Friedrich Adolf Paneth, the former stemming from Budapest (born on 1 August 1885); the latter from Vienna (born on 31 August 1887). Paneth had studied at the Universities of Munich, Glasgow and Vienna and received his doctorate in Vienna in 1910; he then joined the *Institut für Radiumforschung* in 1912.[347] A year later he obtained a very able collaborator in de Hevesy who came as a research associate from Ernest Rutherford's laboratory in Manchester. Paneth made the first application of the radioactive indicator method by determining the solubility of lead sulfide (with lead being replaced by its radioactive isotope Ra D) in water.[348] The method of radioactive indicators or tracers in chemistry and biology was subsequently developed mainly by George de Hevesy who would win with it the 1943 Nobel prize in chemistry.[349]

It is thus fair to say that in Vienna chemistry was as worthily represented as the other sciences at a time when Vienna saw one of the most illustrious and brilliant science congresses in Europe, the 85th *Versammlung Deutscher Naturforscher und Ärzte,* which took place from 21 to 28 September 1913. A contemporary report described it in the following words:

> The high expectations, which one connects with the beautiful imperial city of Vienna and which consequently found adequate expression in the extraordinarily high number—far above 7,000—of participants, were fully satisfied; the highlights of the festive events were a reception at the Imperial Court and a banquet given by the City of Vienna in the town hall. (*Physikalische Zeitschrift* **14**, p. 1074)

[347] See the article of de Hevesy (1957).

[348] In 1917 Paneth made another great discovery, namely polonium-hydrogen, and then followed the discovery of the hydrides of other metals. He left the *Radiuminstitut* in 1918 to assume a chair in chemistry at the Technical University of Prague. The following year he moved to Hamburg and in 1922 to the University of Berlin. In 1929 he became director of the chemical institute at the University of Königsberg, a position which he left in 1933. He then went to England (1933–1938: visitor at the Imperial College of Science and Technology, London; 1938: reader in atomic chemistry, University of London; 1939–1953: professor of chemistry at the University of Durham). From 1953 to his death on 17 September 1958 (in Vienna) he directed the *Max-Planck-Institut für Chemie* in Mainz. Besides his discovery of the metallic hydrides and his pioneering work in the radioactive tracer method, Paneth also became known through his improvements of the techniques of isolating and measuring minute amounts of helium; he applied these techniques, for example, to determine the age and origin of meteorites. He received considerable honours, including a fellowship of the Royal Society of London, and prizes, e.g., the 1916 *Lieben* prize of the Vienna Academy of Sciences, and the 1956 *Liebig* medal of the German Chemical Society.

[349] For the biography of George de Hevesy, see Volume 1, p. 184, footnote 274.

As usual there were general sessions and special sessions of the different sections of the great German association. The contemporary report stated, in particular:

> The physical section was well attended, in spite of the fact that many of the better-known German names were missing. The main interest concentrated on the address given by Einstein in the joint session [of the physical section] with the mathematical and astronomical sections, which stimulated a very lively discussion, as well as the joint session with the chemical and mineralogical sections, in which Laue, Friedrich, Wagner, Tammann, Zsigmondy and Stark reported. (*Phys. Zs.* **14**, pp. 1073-1074)

Unlike the previous Vienna *Naturforscherversammlung* in 1894, the Austrian Empire made a substantial contribution to the success of the physics meeting; earlier the physicists had been ashamed to show their poor institute, but now in the fall of 1913 they could present their new, modern physics institute situated next to the *Institut für Radiumforschung*. The contemporary report mentioned:

> The sessions were held in the lecture hall of the *Physikalisches Institut*, whose rooms and equipment were most willingly supplied by *Hofrat* [Ernst] Lecher; he and *Hofrat* [Victor] von Lang [i.e., Lecher's predecessor], who chaired the first sessions of the physics section, deserve the particular thanks of the physics section for organizing the sessions. In particular, they and the other Viennese physicists also deserve thanks for the arrangement of the social meetings; all participants will especially remember the party held on Tuesday evening in the rooms of the *Physikalisches Institut*, where members of the section were hosted most amiably by the Viennese physicists. The equipment of the newly built institute excited great interest—as well as that of the *Institut für Radiumforschung*—which has been in operation for three years and is richly equipped. (*Phys. Zs.* **14**, p. 1072)

Like other scientists and physicians, the physicists in Vienna also enjoyed a full programme. The physics section held six sessions, the first on the afternoon of Monday, 22 September 1913, and the last on the afternoon of Thursday, 25 September. The programme of the first session included reports by Robert Wichard Pohl from Berlin on the photoelectric electron emission, of James Franck and Gustav Hertz—also from Berlin—on the connection between ionization by impact and electron affinity, and of the Viennese Karl Herzfeld on free electrons in metals. While Einstein's lecture on gravitation dominated the second session (on Tuesday morning, 23 September 1913), the third session was devoted to topics that were particularly dear to the Viennese physicists, i.e., radioactivity and penetrating radiation in the atmosphere. Hans Geiger (1882-1945) of Berlin, the former long-time collaborator of Ernest Rutherford in Manchester, opened the session by presenting a practical experimental method for counting $\alpha$- and $\beta$-particles emitted from radioactive substances (Geiger, 1913a).[350] A little later in

[350] In his talk Geiger essentially described the apparatus which was later called the Geiger counter and played a great role in nuclear physics and beyond.

the session, Victor Franz Hess spoke on an improvement in Wulf's apparatus for determining the *absolute* value of the $\gamma$-ray intensity emitted from radium (Hess, 1913d). This improvement would also serve the research on penetrating radiation in the atmosphere, on which Theodor Wulf presented the first talk, entitled '*Einige Ergebnisse von Simultanmessungen der in der Atmosphäre vorhandenen Strahlung hoher Durchdringungsfähigkeit*' ('Some Results from Simultaneous Measurements of the Radiation of High Penetration Power Existing in the Atmosphere,' Benndorf, Dorno, Hess, von Schweidler and Wulf, 1913). He compared, in particular, observations made on the ground in Graz (by Benndorf and Veith), in Davos (by Dorno), in Vienna (by Hess and Kofler), in Innsbruck (by Kruse) and in Valkenburg (by Wulf), and arrived at the following conclusions:

> 1. That the suggestion of an extraterrestrial source in order to explain the fluctuations of the $\gamma$-radiation is not substantiated by these observations. If there is any such radiation [of extraterrestrial origin] existing at higher altitudes, it is not noticeable in the atmospheric layers close to the ground.
> 2. The fluctuations of the daily average [of penetrating radiation intensity on the ground] seem to be generated by local influences. (Wulf, 1913, p. 1143)

Hess, the expert on high-altitude penetrating radiation, did not report on this topic at the Vienna meeting.[351] However, he was excellently replaced by Werner Kolhörster of Halle, who spoke on '*Messungen der durchdringenden Strahlung im Freiballon in größeren Höhen*' ('Balloon Measurements of the Penetrating Radiation at Higher Altitudes,' Kolhörster, 1913b).[352] He had earlier in the year constructed an apparatus, similar to Wulf's, which was particularly suited for balloon flights, because it could be sealed airtight, hence the local air pressure would not affect the measurements (Kolhörster, 1913a). He then used this apparatus in three balloon rides, all starting from Bitterfeld (about 25 km northeast of Halle), and going to Schüttenhoven in Bohemia, Neutomischl in Posen and close to Halle, reaching altitudes of 3,600 m, 4,000 m and over 6,000 m, respectively.

[351] Hess had spoken on the subject at the two previous *Naturforscherversammlungen* in Karlsruhe (Hess, 1911c) and Münster (Hess, 1912); at the latter he especially summarized the results of his balloon measurements of the summer of 1912.

[352] Werner Heinrich Gustav Kolhörster was born on 28 December 1887 in Schwiebus and studied physics at the University of Halle under E. Dorn. He then joined the *Aerophysikalische Forschungsfonds* in Halle, devoting himself to balloon observations on penetrating radiation, both at great altitudes and under water. In 1914 he continued his studies in physics at Hans Geiger's laboratory in the *Physikalisch-Technische Reichsanstalt*, Berlin. The plan that Kolhörster might take over the direction of the Physical-Meteorological Observatory in Davos, Switzerland, was prevented by the outbreak of World War I. During war service he performed measurements of atmospheric electricity at the Bosporus in Turkey (1916–1918), and after the war he became a school teacher for several years. In 1922 he joined the *Physikalisch-Technische Reichsanstalt.* Eight years later the Prussian Academy provided financial assistance to establish a laboratory in Potsdam, in which Kolhörster would carry out research on cosmic radiation. In 1935 he was finally appointed ordinary professor of geophysics and director of the *Institut für Höhenstrahlungsforschung* in Berlin-Dahlem. He died on 5 August 1946 in Munich from the consequences of an automobile accident. (See Kolhörster's obituary by Siegfried Flügge: Flügge, 1948.)

He presented the results obtained during these rides in his Vienna talk, which he summarized in the following statements:

> Although the measurements of the penetrating radiation, in particular during balloon flights, are still plagued by various errors—I would estimate the error of my measurement to be 10 to 15 percent—the earlier observations up to about 4,500 m height should be considered as being substantiated; further, the strong increase of radiation up to the achieved altitude of 6,300 m has been measured for the first time with a more or less incontestable apparatus. The assumption that the origin of this penetrating radiation might not be sought in the known radioactive substances of the earth or the atmosphere, therefore, gains a considerably higher probability. (Kolhörster, 1913b, p. 1155)[353]

These results were supported at the Vienna meeting by Alfred Gockel, who mentioned in the discussion of Kolhörster's talk his own observations on Swiss glaciers also yielding high $\gamma$-ray intensity. When Karl Bergwitz suggested that temperature differences might have influenced the ionization measurements at high altitudes—especially a higher temperature due to increased radiation from the sun which would stimulate a larger amount of penetrating radiation in the atmosphere—Kolhörster replied: 'The appartus became so cold [at higher altitudes] that I was not permitted to touch it with my bare hands without perturbing the measurements. According to their [i.e., the measurements'] results my conclusions must be valid *a fortiori*' (Kolhörster, 1913b, p. 1156).

Thus Kolhörster's observations confirmed and extended the conclusions obtained by Hess the previous year, and one may actually call the 85th *Naturforscherversammlung* the birthday of *Höhenstrahlung* or cosmic radiation. Both Hess[354] and Kolhörster[355] would continue their pioneering investigations of

[353] The measurements of the first balloon flight included those above regions where the ground contained rich deposits of radioactive materials (especially between Annaberg and Sachorz in Bohemia). The other two flights, however, did not pass any such regions. Kolhörster found that the higher altitude intensity of penetrating radiation was not strongly influenced by great radioactivity on the ground.

[354] World War I interrupted Hess' plan to conduct a systematic programme of measurements of penetrating radiation with international collaboration. Instead he turned his attention to specific problems of radioactivity and its application to medicine. In 1920 he left the Vienna *Institut für Radiumforschung* and accepted a call to an extraordinary professorship of physics at the University of Graz. A little later he joined, on a two-year leave of absence from Graz, the U.S. Radium Corporation in Orange County, New York. In 1925 he was promoted ordinary professor in Graz; in 1931 he moved to a chair at the University of Innsbruck. Having returned to Graz in 1937, he had to leave Austria the following year. Again he took a position in the United States, becoming a professor at Fordham University in New York. Hess, who shared the 1936 Nobel prize in physics with Carl David Anderson—for his discovery of cosmic radiation—died on 17 September 1964 in Mount Vernon, New York.

[355] In the 1920s Kolhörster introduced an essential improvement to the apparatus for cosmic ray research—for which he was honoured with the *Leibniz* medal of the Prussian Academy. In 1928 he developed with Walther Bothe the coincidence method for cosmic radiation; this method allowed one to build 'cosmic ray telescopes' to observe directional dependence and opened a new era in research of the phenomena. Kolhörster contributed basically to the discovery and registration of big cosmic ray showers which played a decisive role in the discussions of physicists during the 1930s and 1940s.

penetrating radiation of extraterrestrial origin, which would also become, two decades later, one of the most fruitful topics of fundamental physical research.

The main physical issue dealt with in the fourth session of the physics section, held together with the sections of chemistry, electrochemistry, mineralogy and geology on the morning of 24 September 1913, was as new and perhaps even more exciting than the phenomena of cosmic radiation. With the Berlin physicist Heinrich Rubens in the chair, the following speakers reported: Max von Laue from Zurich and Walter Friedrich from Munich on '*Röntgenstrahlinterferenzen*' ('X-ray Interferences,' Laue, 1913c, Friedrich, 1913), and Erich Wagner—also from Munich—on an '*Experimenteller Beitrag zur Interferenz der Röntgenstrahlen*' ('Experimental Contribution to the Interference of X-rays,' Wagner, 1913).[356] These talks, of which Karl Scheel gave a detailed summary in the *Naturwissenschaften* (issue of 5 December 1913), constituted the first conference presentation on the Continent of the exciting discovery made one-and-a-half years earlier by Walter Friedrich and Paul Knipping at the instigation of Max von Laue (Friedrich, Knipping and Laue, 1912).[357] This discovery had stimulated the British physicist William Henry Bragg (1862-1942) and his son William Lawrence Bragg (1890-1971) several months later to develop a slightly different method for obtaining diffraction patterns by reflecting X-rays—with a continuous spectrum—from crystals (W. L. Bragg, 1913a, b; Bragg and Bragg, 1913).[358]

Max von Laue, in his Vienna talk, now presented the theoretical principles of the different methods and discussed the successes achieved thus far in disentangling complicated crystal structures, such as quartz and pyrite. In spite of these successes he admitted a limitation of the methods by concluding:

> With crystals we are in a similar situation as if we had to investigate an optical grating without a microscope, using only the spectra it produces.... But the knowledge of the positions and intensity of spectra does not suffice for the construction. The phases, with which they vibrate relative to one another, enter essentially. In order to obtain a 'microscopic picture' of crystal structure one should at least also determine the phase differences between the different interference spots of a photogramme, and this task might certainly be rather difficult. (Laue, 1913, p. 1079)

Walter Friedrich focused, after Laue, on the experimental setup for obtaining crystal diffraction patterns; he discussed several details such as the influence of

[356] The programme of the fourth session mentioned further talks by the Göttingen professors Gustav Tammann '*Über die Theorie des Polymorphismus*' ('On the Theory of Polymorphism') and Richard Zsigmondy '*Über Gelstrukturen*' ('On the Structure of Gels') and by the Aachen experimentalist Johannes Stark '*Über die elektrische und die damit verbundene optische Änderung der Atome*' ('On the Electrical Change of Atoms and the Optical Change Connected with It').

[357] See especially Scheel, 1913, pp. 1205-1206. Von Laue's talk had a slightly different title there, i.e., '*Über Interferenz von Röntgenstrahlen in Kristallen*' ('On X-ray Interferences in Crystals').

[358] William Henry Bragg spoke on the same subject at the Birmingham meeting of the British Association for the Advancement of Science, 10-17 September 1913. A little later X-ray interference phenomena and the analysis of crystal structure was the central topic at the second Solvay Conference in Brussels, 27-31 October 1913. Among the speakers who contributed to the subject were Max von Laue, William Henry Bragg and Marcel Brillouin of Paris.

the characteristic X-ray lines on the diffraction pattern, the effect of temperature on the observed intensity, and the possibility of an absolute determination of X-ray wavelengths. Erich Wagner (1876-1928) finally reported investigations, made together with his student Richard Glocker, in order to test William Lawrence Bragg's hypothesis. The results did provide 'a numerical proof for the correctness of the interference theory in the form of the reflection interpretation of Bragg, which must be applied here' (Scheel, 1913, p. 1206). While the fourth session documented the formation of the new, interdisciplinary field of the structure of crystals uniting physical, chemical and mineralogical methods, in the fifth and sixth sessions of the physics section alone—held on the afternoons of 24 and 25 September 1913—specific topics of recent physical research were treated. Thus, for example, the renowned British guest Charles Glover Barkla of London talked in the fifth session on 'Characteristic Röntgen Rays'; in the sixth session Max Born presented a '*Theorie des Gesetzes von Eötvös, eine Anwendung der Quantenlehre*' ('Theory of Eötvös' Law, an Application of Quantum Theory'), in which he tried to explain the recently observed derivations from a law that Lorand von Eötvös had given for the temperature dependence of the surface tension of fluids; in the same session Hans Thirring of Vienna also gave a report of his results '*Zur Theorie der Raumgitterschwingungen und der spezifischen Wärme kristallinischer Körper*' ('On the Theory of Spatial Lattice Vibrations and of the Specific Heat of Crystalline Bodies').[359]

Altogether the sessions of the physics section at the 85th *Naturforscherversammlung* represented the high level of physics in German-speaking countries. Let us now turn to that very talk, on which—as the contemporary rapporteur mentioned—the main interest of the physicists was concentrated.

## Einstein's Talk at Vienna and Its Consequences

In the second physics session on Tuesday, 23 September 1913, starting at 9 a.m. Albert Einstein spoke first on the topic '*Zum gegenwärtigen Stande des Gravitationsproblems*' ('On the Present Status of the Problem of Gravitation,' Einstein, 1913b). The Viennese public had not yet had the opportunity to hear this already very famous scientist, who had discovered special relativity and made fundamental contributions to molecular and quantum theory. He now was attempting to construct a new relativity theory of gravitation, a theory on which he had begun to work actively more than two years earlier while in another city of the Austrian Empire, the Bohemian capital of Prague.

Back in 1907 Einstein had written a review article, entitled '*Das Relativitätsprinzip und die aus demselben gezogenen Folgerungen*' ('The Principle of Relativity and Its Consequences') for the *Jahrbuch der Radioaktivität und Elektronik* (Einstein 1907d, 1908) in which he proposed certain possibilities of generalizing special relativity to a more general theory on the basis of two assumptions: first, the exact proportionality between the inertial and the gravitational mass; second,

[359] For the contents of the talks see the report of Scheel, 1913, pp. 1206-1208.

the validity of the principle of relativity for systems accelerated relative to each other. Then, however, he had discontinued working on relativity theory, becoming deeply involved in the problems of quantum and radiation theory for about four years. The situation, however, changed and he accepted the call to the chair for theoretical physics at the German University of Prague.[360] The tradition of this place, where Ernst Mach had developed his critique of Newton's mechanics and composed his book *Die Mechanik in ihrer Entwicklung* (Mach, 1883), co-operated with the fact that Einstein's programme of quantum and radiation theory did not progress to turn the new professor's attention back to relativity.[361] A further stimulus was provided by the interest of his colleague, the mathematician Georg Alexander Pick (1859–1929), who had helped to bring Einstein to Prague and now became his closest friend; in particular, Einstein often sought his advice on mathematical questions.[362] The Prague atmosphere proved to be very stimulating, and in June 1911 Einstein submitted a paper '*Über den Einfluß der Schwerkraft auf die Ausbreitung des Lichtes*' ('On the Influence of Gravitation on the Propagation of Light,' Einstein, 1911) to *Annalen der Physik*, in which he showed, on the basis of the equivalence of relatively accelerated frames of reference, that the velocity of light *in vacuo* was not a constant, as hitherto assumed, but depended on the gravitational potential, such that 'a light ray passing close to the sun would therefore suffer a bending by the amount of $4 \times 10^{-6} = 0.83$ seconds of arc' (Einstein, 1911, p. 908). Max Abraham (1875–1922) then developed a theoretical scheme, different from Einstein's, in which the light velocity also depended on the gravitational potential (Abraham, 1912). Einstein subsequently criticized this work because of the fact 'that Abraham's system of equations cannot be brought into accord with the equivalence hypothesis, and his concepts of space and time cannot be upheld from the purely formal mathematical point of view' (Einstein, 1912, p. 355).[363] Later in 1912 Gunnar Nordström of Helsingfors, stimulated by Einstein's and Abraham's work, derived a theory of gravitation, in which the constancy of the velocity of light could be upheld independently of the gravitational potential, thus avoiding one of the main points of the Einstein–Abraham disagreement (Nordström, 1912). By that time Einstein was already back in Zurich, in a chair for theoretical physics at the *Eidgenössische Technische Hochschule* (E.T.H.), having left Prague on 25 July 1912.[364] In Zurich he developed with his friend and colleague at the E.T.H., Marcel Grossmann (1878–1936),

[360] In Prague Einstein succeeded Ferdinand Leppich, professor of mathematical physics. The chair was changed to one of theoretical physics, and Einstein obtained his own institute consisting of a few rooms. He started the position in Prague in April 1911. (For the detailed circumstances of the Prague professorship and Einstein's life in the Bohemian capital, see Illy, 1979.)

[361] It might be recalled at this point that Ernst Mach himself showed at that time some interest in Einstein's work on relativity, as well as in the modifications of Newton's theory of gravitation.

[362] See Illy, 1979, p. 78, especially footnote 19.

[363] The discussion between Einstein and Abraham went on during the year 1912 and even later, with both writing further papers on the topic but not reaching agreement. For this and other developments in general relativity theory we refer to the article by Mehra, 1973a.

[364] See Illy, 1979, p. 84.

professor of mathematics, the mathematical formulation of a theory of generalized relativity and gravitation, using the methods of invariant theory and the absolute differential calculus of Elwin Bruno Christoffel, Curbastro Gregorio Ricci and Tullio Levi-Civita, and others (Einstein and Grossman, 1913). It was on this work that Einstein reported at the Vienna *Naturforscherversammlung*.

Einstein began by pointing to the necessity of modifying Isaac Newton's theory of gravitation in the light of relativity theory, i.e., if one wanted to exclude signals or physical actions transmitted with velocities faster than $c$, the velocity of light *in vacuo*. In searching for such modifications, Einstein continued, one had to apply several restrictions or 'obvious physical hypotheses on the gravitational field' (Einstein, 1913b, p. 1250). Einstein listed four of those: (1) energy and momentum conservation; (2) the equivalence of inert and heavy masses of closed systems; (3) the validity of special relativity; (4) the independence of the physical laws from the absolute value of the gravitational potential. He admitted: 'I am completely aware of the fact that the postulates (2)-(4) resemble more a physical confession of faith than a safe foundation; but I may nevertheless say that they represent the most natural [consistent] hypotheses with our present state of knowledge' (Einstein, 1913b, p. 1251). On the basis of postulate (3)—which implied the validity of special relativity in regions of constant gravitational potential—he excluded Abraham's theory. Nordström's theory, on the other hand, satisfied this postulate in the narrowest sense, i.e., its equations were covariant even under Lorentz transformations; it further satisfied all the other postulates that Einstein listed. Nevertheless, after a detailed presentation Einstein argued: 'The fact remains, unsatisfactory as it is, that although the inertia of bodies is *influenced* in this theory by other bodies, it does not appear to be *caused* by them; the reason is that in this theory the inertia of a body becomes larger, the farther we remove other bodies from it' (Einstein, 1913b, p. 1254).

Einstein's verdict against Nordström's otherwise fully satisfactory theory of gravitation—in which the gravitational field was represented by a scalar potential and the light rays remained straight in an arbitrary potential—rested on his firm belief in Mach's idea that inertia depends on the mutual interaction of matter.[365] Besides establishing agreement with the thoughts of the Viennese Ernst Mach, Albert Einstein brought into the foundation of his new theory of gravitation the result of another scientist of the Austrian Empire, that of Lorand von Eötvös of Budapest.[366] Eötvös had, since the late 1880s, performed experiments with sensi-

[365] This idea is usually referred to as Mach's principle, which Mach formulated in his *Mechanik* as: 'For me only relative motions exist.... When a body rotates relative to fixed stars, centrifugal forces are produced, when it rotates relatively to some different body not relative to fixed stars, no centrifugal forces are produced. I have no objection to calling the first "rotation" as long it be remembered that nothing is meant except relative motion with respect to fixed stars' (Mach, 1883, Chapter I; English translation, 1892). For a discussion of Mach's principle we refer to Dicke, 1964.

[366] Lorand (Roland) von Eötvös was born on 27 July 1848 in Budapest, the son of the writer and statesman Baron József von Eötvös (1813-1871). In 1872 he became Professor of Physics at the University of Budapest, and served as president of the Hungarian Academy of Sciences from 1889 and as minister for education (in Hungary) from 1894 to 1895. He died on 8 April 1919 in Budapest. Besides his work on mass equivalence, Eötvös contributed to the physics of fluids and of gases (critical point).

tive torsion balances proving the exact equivalence of inert and heavy mass. For Einstein this result demonstrated that one could not decide about the absolute acceleration of a physical system. 'One sees,' he said, 'that in this respect the Eötvös experiment plays a similar role as the Michelson experiment for the question of the physical evidence of a *uniform* motion' (Einstein, 1913b, p. 1255). The theory, which he built on these assumptions and ideas, made use of the absolute differential calculus—one which Friedrich Kottler, a student of Friedrich Hasenöhrl, had first applied in a paper on the formulation of Maxwell's equations of electrodynamics *in vacuo*, submitted in July 1912 to the Vienna academy (Kottler, 1912).[367]

Einstein now reported the most recent results which he and Marcel Grossmann had obtained (Einstein and Grossmann, 1913). These implied, in particular, the deflection of light when passing near the sun. Hence he suggested finally: 'Now, whether the first approach [of Nordström] or the second approach [of Einstein and Grossmann] corresponds essentially to the situation in nature, must be decided by taking photos of the stars showing up next to the sun during a solar eclipse. Let us hope that the eclipse of the year 1914 will allow us to obtain already the important decision' (Einstein, 1913b, p. 1262).

The first speaker in the discussion of Einstein's talk was Gustav Mie (1868–1957), who pointed out that he had developed a theory of gravitation within the context of a general theory of matter (Mie, 1912a, b, c). Einstein then remarked: 'I have not mentioned the theory of Mie, because in it the equivalence of the heavy and inert mass is not strictly realized' (Einstein, 1913b, p. 1263). Mie replied that this was not the case in the theory proposed by Einstein and Grossmann; he also criticized Einstein's generalized relativity postulate (3), especially its formulation derived from Mach, 'according to which it might not even be possible to prove accelerations absolutely' (Mie in Einstein, 1913b, p. 1264). He again argued that this principle was not satisfied in the Einstein–Grossmann theory either; but this did not matter, for Mie believed that 'it did not have physical significance' ('*keinen physikalischen Sinn hat*,' Mie in Einstein, 1913a, p. 1264). Finally the discussion (in which, among others, Max Born and Eduard Riecke of Göttingen and Friedrich Hasenöhrl and Gustav Jäger of Vienna participated) concentrated on possible experimental tests. Einstein again emphasized the effect of bending light rays in the gravitational field of the sun, which seemed to be observable by astronomers. (As Einstein said: 'According to the view of the astronomers, of whom I inquired, the observation of such a bending lies within the range of present possibilities,' Einstein, 1913b, p. 1264.) Mie, however, drew attention to still another fact: 'According to Einstein's theory the period of vibration of atoms must be at a space point of high gravitational potential different from that at a point of gravitational potential zero. The lines

[367] Kottler was born on 10 December 1886 in Vienna. He studied physics at the University of Vienna under Franz Exner and Friedrich Hasenöhrl. In 1916 he was appointed *Privatdozent* and in 1923 extraordinary professor for theoretical physics at the University of Vienna. After his dismissal in 1938, he went to the United States, joining the Eastman Kodak Research Laboratory in Rochester, New York. In 1956 he returned to the University of Vienna as honorary professor.

of a series spectrum on a star of large mass must therefore be shifted against the lines observed on the earth. This is not so in my theory' (Mie in Einstein, 1913b, p. 1266). Einstein agreed that both in Nordström's and his own theory of gravitation such line-shifts had to occur, but concluded: 'Unfortunately also other changes may generate line-shifts, and it is therefore very difficult to check whether such a shift originates exactly from this reason [i.e., from gravitation]' (Einstein, 1913a, p. 1266).[368]

Einstein's lecture at the *Naturforscherversammlung* left a lasting impression on the Viennese physicists.[369] It especially encouraged Friedrich Kottler to continue his own investigations on relativity theory, which he had begun earlier with the formulation of electrodynamics in the Minkowski space (Kottler, 1912). In a series of papers he now tried to establish 'a relativity principle for accelerated motions' (Kottler, 1914a, b; 1916). For this purpose he developed in the first two papers a mathematical tool for orthogonal transformations in Minkowski space; it allowed him to relate motions with constant curvature, i.e., motions under the influence of a four-force with constant absolute magnitude—such as free fall (on earth) or the uniform rotation around a fixed axis in (three-dimensional) space. Kottler obtained in his theory, as Einstein did in his previous extensions of relativity theory, a bending of the light rays in a gravitational potential (Kottler, 1914b, § 4). Military service in World War I interrupted his further efforts; however, in a paper submitted in July 1916 he compared his approach to the one that Einstein had meanwhile taken leading to the establishment of the General Relativity Theory (Kottler, 1916). Kottler pointed out in particular that he had always attempted to satisfy Einstein's original equivalence hypothesis, namely that 'a real field of force and an apparent field of acceleration are completely equivalent in physics' (Kottler, 1916, p. 957); on the other hand, he claimed that 'Einstein has since then abandoned the hypothesis of equivalence' (*loc. cit.*, p. 955), because he introduced in his General Relativity Theory (Einstein, 1916c) the gravitational field as an independent dynamical concept. Kottler preferred to consider motion in a gravitational field as being force-free; this implied that Galilei's law of inertia had to be changed, and gravitation had to be interpreted as a pure phenomenon of inertia. As a consequence, in his theory the three-dimensional co-ordinate space kept its Euclidean structure and the gravitational field remained isotropic, all different from Einstein's theory. Einstein, in replying

[368] Nearly two years earlier, when still at Prague, Einstein discussed (with the astronomer Erwin Finlay-Freundlich of Berlin-Babelsberg) the possibility of testing the deflection of light rays in the gravitational field of the sun by star observation during an eclipse. Freundlich then wanted to use the solar eclipse in South Russia in the late summer of 1914 for that purpose. However, the outbreak of World War I prevented the work of the expedition, and confirmation of Einstein's prediction had to be postponed until the British Solar Eclipse Expedition succeeded in 1919.

[369] It certainly persuaded them of the superiority of the relativity approach to gravitation theory over other, more conventional ones, such as was suggested by Gustav Jaumann of Brünn (Jaumann, 1912). Jaumann, back in 1910, had been a competitor of Einstein for the chair in Prague. (See Illy, 1979, p. 76.)

to Kottler's criticism, admitted the competence and serious concern of his younger colleague (see Einstein, 1916g).[370] Still he argued that his equivalence principle was different from Kottler's; it stated rather that the same laws of nature were valid in all systems of reference related to each other by uniform acceleration. This specific equivalence principle was satisfied in the General Relativity Theory, because: 'The postulate of general covariance of the equations implies the principle of equivalence as a particular special case' (Einstein, 1916g, p. 641).

Einstein's theory would also make its way in Vienna superseding all other approaches there. Viennese physicists, like Hans Thirring or the young Wolfgang Pauli, would contribute to the scheme by studying detailed questions. Erwin Schrödinger also became—as we shall discuss in the following section—deeply involved in Einstein's theory, studying it in detail and trying to solve interesting problems. We may go even further and say that the Vienna *Naturforscherversammlung* stimulated a lot of his future work. This refers as much to his investigations of X-ray interference patterns—which we have already analyzed in the last section—as to studies in relativity—undertaken between 1916 and 1918—and even physiology of vision—undertaken as late as 1924.[371]

## Schrödinger as *Privatdozent*

On 9 January 1914 Erwin Schrödinger joined the illustrious circle of academic teachers at the University of Vienna by obtaining his *Habilitation* in physics. He announced for the following summer semester a special course on a topic connected with his own current research, i.e., on '*Interferenzerscheinung der Röntgenstrahlen*' ('Interference Phenomenon of X-rays').[372] Thus he demonstrated right away his willingness to teach and propagate the most recent results in physics.

At the time when Schrödinger received his *Habilitation* it was by no means clear that he would develop to become a representative of pure theoretical physics. His scientific work covered a wide range of topics, extending from experimental investigations (such as those on atmospheric electricity and radioactivity) to theoretical descriptions of experimental phenomena (such as those on penetrating radiation and X-ray interference patterns) and to genuinely theoretical problems (such as the dynamics of point lattices). It appears that the young Schrödinger

[370] Einstein began the note '*Über Friedrich Kottlers Abhandlung* . . . .' ('On Friedrich Kottler's Paper, . . . ') by saying: 'Among the papers dealing critically with General Relativity Theory, those of Kottler are especially remarkable, because this colleague has really dived into the spirit of the theory' (Einstein, 1916g, p. 639).

[371] Thus the Munich zoologist Karl Ritter von Hess spoke in a general session of the *Naturforscherversammlung* on '*Entwicklung von Lichtsinn und Farbensinn im Tierreich*' ('Development of Light and Colour Vision of Animals'). In 1924 Schrödinger came across the work of von Hess again, and it confirmed his views on the philogenetic evolution of colour vision. (See our discussion of Schrödinger's paper, 1924e, in Section II.4 of the next chapter.)

[372] See *Physikalische Zeitschrift* **15**, p. 486.

had not yet decided on his particular path, or rather preferred to become a non-specialized scientist like his teacher Franz Exner, who had treated both experimental and theoretical topics. In any case, Schrödinger's teaching duties were not restricted to either field, and he announced topics for his courses that reflected this non-specialization.[373] His research programme in 1914 embraced not only theoretical work, which we discussed in the previous section, but also purely experimental studies. The programmes demanding greater experimental skill, he performed together with his friend Fritz Kohlrausch; others he did completely on his own.

At the meeting of 2 July 1914 the Vienna Academy of Sciences received a communication, signed by Kohlrausch and Schrödinger, entitled '*Über die weiche* ($\beta$) *Sekundärstrahlung von* $\gamma$*-Strahlen*' ('On the Soft $\beta$-Secondary Radiation from $\gamma$-Rays,' Kohlrausch and Schrödinger, 1914). In those days the secondary $\beta$-radiation emerging by the impact of $\gamma$-rays on matter was normally taken as a measure of the original $\gamma$-ray intensity. But in order to obtain a reliable relation between primary and secondary rays, the authors proposed to investigate further the specific properties of the excited $\beta$-radiation, especially since previous observations had revealed several unclear, even contradictory results. They announced at the close of their introduction: 'In the following we shall report the results of our measurements on soft secondary radiation, which is created by the $\gamma$-rays of Ra C.' They added: 'The experiments, which have been carried out with the support of the Imperial Academy of Sciences at Vienna's *Radiuminstitut* are, we believe, performed under better defined conditions than have been achieved up to now; and they allow a quantitative description' (Kohlrausch and Schrödinger, 1914, p. 1322).

Kohlrausch and Schrödinger suspected that 'a considerable part of the often very contradictory results of electrical determination of the $\gamma$-ray [intensities] is caused by the fact that the experimental apparatus used is, due to insufficient and incomplete covering of the $\gamma$-radiation source, not protected enough against the influence of secondary radiation originating in the entire surrounding [of the source] and varying with it [i.e., the surrounding]' (Kohlrausch and Schrödinger, 1914, p. 1322). Hence they put their $\gamma$-ray source (Ra C) in a spherical vessel filled with liquid mercury, leaving only a small tube through the centre for the emission of a directed $\gamma$-ray. This directed $\gamma$-ray struck an ionization chamber connected to a sensitive electrometer measuring the ionization current.[374] A metal plate of variable thickness was placed in the ionization chamber, such that the

[373] For example, he planned to lecture in the winter semester of 1914/1915 on '*Ausgewählte Kapitel aus der statistischen Mechanik und Quantentheorie*' ('Selected Topics of Statistical Mechanics and Quantum Theory'; see *Physikalische Zeitschrift* **15**, p. 863), a topic which was more on the theoretical side.

[374] The authors suggested the use of different ionization chambers and electrometers—see Kohlrausch and Schrödinger, 1914, pp. 1324–1329—but presented in the paper only the results obtained with one specific choice.

production of soft secondary ($\beta$-) radiation could be observed with the incident $\gamma$-ray coming in at different angles. The authors then determined the ionization currents arising from the metallic plates in the centre of the chamber after correcting for the $\beta$-radiation coming from different sources (e.g., from the natural ionization of the air, from the walls of the ionization chamber, etc.). They registered results for different materials, ranging from aluminium to lead, and they observed the strong dependence of the ionization—which is a measure of the created soft $\beta$-radiation—on the type of material, its thickness and the angle of the incident $\gamma$-ray with the plate.

These results then were interpreted theoretically by involving three assumptions: both the primary $\gamma$-radiation and the secondary $\beta$-radiation are absorbed according to an exponential law with different mass-absorption coefficients; and thirdly, the secondary radiation produced for the unit volume of the plate (struck by the primary $\gamma$-rays) is a function only of the plate's type and thickness, and the angle of incidence. The data followed the theoretical description satisfactorily yielding absorption coefficients in fair agreement with those derived from other experiments. Especially, the authors found that:

> The secondary $\beta$-radiation excited in a given material [of the plate] may be characterized by three coefficients, say: mass-radiation coefficient, asymmetry coefficient, and absorption coefficient. From the measurements with seven metals we conclude that the mass-radiation is approximately constant for low atomic weights up to zinc; then it grows to achieve more than twice that value for lead. The asymmetry coefficient [i.e., the parameter describing the angular dependence, especially the forward-backward difference of the mass radiation] decreases fast [from positive values for metals of low atomic weight] with increasing atomic weights, disappears for tin and becomes negative for lead. The absorption coefficient of the secondary radiation does not seem to depend on the nature of the radiating substance but only on the absorber and—according to earlier observations—on the hardness of the primary ray. (Kohlrausch and Schrödinger, 1914, pp. 1366–1367)

The study of the properties of soft $\beta$-radiation created by $\gamma$-rays impinging on given materials (metals) provided a fine example of the careful work performed at the Vienna *Radiuminstitut.* We may assume—in agreement with Schrödinger's confession, 'I learned ... in close collaboration with my friend K. W. F. Kohlrausch what experimenting is, without, however, learning it by myself' (Schrödinger, 1935, p. 87)—that Kohlrausch was the leading person in devising the details of the apparatus, while Schrödinger carried out the details of the theoretical evaluation, which indeed was very elaborate and made use of results previously obtained by him (Schrödinger, 1912c). In spite of this division of work between the two of them, we must not assume that Schrödinger's interests centred at that time purely on theory. Quite the contrary, from Schrödinger's publications and the testimony of his colleague Hans Thirring we know of further involvement in experiments on the penetrating power of $\beta$-rays and on needle radiation.[375]

Schrödinger undertook the latter experiment in 1919 and it constituted to some extent the last public indication of his personal involvement in experiments, before he shifted completely to theory. In 1914, however, he had not yet decided. Thus he started, for example, in July of that year an experimental investigation, with the goal of studying a particular method of determining the capillary constant of fluids. The idea of this method, which had been employed nearly twenty years earlier by the Viennese physicist Gustav Jäger, is the following: a bubble of air is slowly pushed through a capillary tube, which dips vertically into a fluid and maintains the maximum additional (capillary) pressure to perform the task of pushing a bubble of a given dimension. Schrödinger noticed that the existing formula describing the situation—and yielding a determination of the capillary constant—needed some improvement. By carefully taking into account all physical processes involved when the bubble of gas leaves the capillary tube, he arrived at a new formula. This relation contained—as compared to the previous one—a term taking into account second-order corrections in the radius of the capillary tube. He checked his new term by some experiments with extremely wide capillary tubes and registered quite satisfactory quantitative agreement. He submitted the paper, a '*Notiz über den Kapillardruck in Gasblasen*' ('Note on the Capillary Pressure in Gas Bubbles'), in late October 1914 to *Annalen der Physik* (Schrödinger, 1915a). The place from where he submitted—he mentioned Raibl in Carynthia—was where he performed military duties after the outbreak of World War I.[376]

The days before World War I were certainly among Schrödinger's happiest and more relaxed periods in his life. He lived in a community of able and self-confident scientists, among theoreticians and experimentalists guided by Exner and Hasenöhrl. He collaborated successfully with his closest friend Kohlrausch, who also entered his private life at an important juncture by making Schrödinger acquainted with Annemarie Bertel of Salzburg. It was in Seeham in the summer of 1913, when Schrödinger measured the Radium-A content of the atmosphere, that he first met his future wife.[377] Six years later, on the occasion of his engagement, he presented a reprint of the published paper to his bride-to-be with a handwritten dedication saying: 'Addendum on 1 October 1919: As I just became aware, at that time [in the summer of 1913] there must have been several other things contained in the atmosphere at Seeham besides Radium-A, B and C, although my electrometer did not indicate any trace of it. It owes its discovery

[375] See Thirring's remarks on p. 110.

[376] Schrödinger stated at the end of his paper: 'I intend to continue the experiments, which had to be interrupted at the end of July [1914], later on with the idea of obtaining ample material to test the formula' (Schrödinger, 1915a, p. 418). However, he never returned to these experimental tests.

[377] We have discussed this work and the publication arising from it (Schrödinger, 1913b) in the previous section.

alone to Miss Bertel of Salzburg, who drew the author's attention to it.'[378] Mrs. Schrödinger recalled their first meeting with the words: 'I was very impressed by him because, first of all, he was very good-looking. He had a remarkable face.... I was already impressed because Kohlrausch had told me about him' (AHQP Interview with Mrs. Schrödinger, 5 April 1963, p. 2). All the signs in the summer of 1914 seemed to point to a bright future, a *revoir* of Schrödinger with Miss Bertel in Seeham, but then the situation changed as war broke out.

## World War I and the Physicists

Since the Austro-Prussian War in 1866 and the Franco-Prussian (German) War in 1870/1871 Europe had enjoyed peace, except for several smaller martial actions in her southeastern corner.[379] During this period, lasting several decades, enormous technological, scientific and cultural progress had provided Europe and the entire Western hemisphere with a considerable improvement in human living conditions, despite the fact that simultaneously social and political unrest had built up in many countries. The entire period could not be considered as a politically smooth one though; e.g., there remained strong hostility between France and the new German Empire, expressing itself basically in the French desire to avenge the defeat and recover her losses.[380] The Franco-German differences, the increasing competition of old and new colonial powers, and also difficulties in South and Southeastern Europe, required the full statesmanship of politicians like the German Chancellor Otto von Bismarck; various treaties and alliances between European countries had been established, finally separating Europe into two camps.[381] On the one hand, there was the so-called '*Entente*

[378] Schrödinger connected the memories of his happy pre-war days with an optimistic view of his future marriage, by announcing humorously: 'A joint publication of the above-mentioned discoverer with this author will soon be presented elsewhere' (quoted in Hermann, 1963, p. 177). Erwin Schrödinger and Annemarie Bertel were married in Vienna on 6 April 1920.

[379] We shall mention these in a little more detail later.

[380] In many aspects this time of peace could not be compared to the previous similarly-peaceful period following the Napoleonic wars (i.e. ranging from 1815 to 1848), in which the great powers (Austria, Britain, France, Prussia, and Russia) agreed to watch that no disturbance occurred in the restored stability of European countries and governments.

[381] After the war with France, Bismarck succeeded in achieving good relations with the two other continental powers, Austria and Russia. While the increasing German–Austrian co-operation had settled all resentments remaining from 1866 (i.e., the Austro-Prussian War), the German–Russian '*Reassurance Treaty*' ('*Rückversicherungsvertrag*') prevented the possibility of an alliance between France and Russia. But politics, as well as the defensive *Triple Alliance* which Germany and Austria had entered into with Italy, had not been trouble-free. There remained unresolved, for example, the problems between the Austrian Empire and Italy concerning the Italian speaking Austrian provinces (i.e., Trentino and Friaul, including the city of Trieste); in particular Austrian and Russian interests conflicted heavily in the Balkan region. With the retirement of Bismarck in 1890 the situation changed drastically, and the subsequent discontinuation of the German–Russian *Reassurance Treaty* resulted in France approaching Russia. On the other hand, the German Empire threw itself into the cauldron of colonial conflicts, which had previously existed only between Great Britain, France, Japan and Russia.

*Cordiale*,' a secret treaty between Great Britain and France, arranged in 1904.[382] Then Russia, after her defeat in the war with Japan (1904–1905), turned again to active politics in Europe, especially in the Balkan region, and approached France for support against Austria in the following years. On the other hand, there stood the old Austro-German-Italian '*Triple Alliance*,' in which German-Austrian ties had been strengthened over the decades, while the bonds with the smaller partner, Italy, had weakened. Still the increasingly unstable political situation prevailing since the turn of the twentieth century, with governments on all sides becoming less resistant to deciding political problems by war actions, did not automatically terminate the extended period of peace in Europe. It was rather an accident, an action carried out by a few individuals at an extremely remote place in the distant, southeastern corner of the continent, which set ablaze the torch of a universal war, such as had not shaken Europe and the world since the times of Napoleon.

On 28 June 1914 Archduke Franz Ferdinand, nephew and designated heir of Emperor Franz Joseph, and his wife were assassinated in Sarajewo, the centre of the Austro-Hungarian administration in the province of Bosnia. The assassin, Gabriel Princip, belonged to a group of Bosnian students of Serbian nationality, who protested against the Austro-Hungarian repression in this region by terrorist actions and who received some support from Belgrade, the Serbian capital. In order to restore the honour of the Habsburg house, the government of the Austro-Hungarian Empire issued on 23 July 1914 an ultimatium to Serbia; it demanded the dissolution of the political associations working for an enlarged Serbia, i.e., one including Bosnia and other parts of the Austro-Hungarian countries, and the participation of Austrian officials in the investigations (connected with the Sarajewo murder) against Serbian citizens.[383] Since the answer from Belgrade did not satisfy the imperial government in Vienna and the military leaders there urged several retaliations, the *Dual Monarchy* declared war on Serbia on 28 July 1914. Then Russia, the foremost ally and protector of Serbia, answered with a partial mobilization of her armed forces on the southern front of the country bordering Austria. Germany, after having stood firmly with her

[382] The post-Bismarck colonial politics of the German Empire—whose economic expansion had called for the opening up of new resources and markets for industrial goods—connected with a considerable build-up of her fleet, led to an alienation with Great Britain. As a consequence, France and Great Britain got together and settled their former differences, concerning colonial politics, with a diplomatic agreement achieved on 8 April 1904 and supporting the Moroccan politics of France. Later on, both countries enlarged the accord, also agreeing to discuss mutual action in case one country was militarily attacked.

[383] In order to understand the background of the Sarajewo murder, we shall briefly sketch the outline of the history of, what later became, Yugoslavia. (For more details, see, for example, the article on Serbia in *Encyclopaedia Britannica*, Volume *20*, Chicago-London-Toronto-Geneva: Encyclopaedia Britannica, Inc., 1962, pp. 341–351.)

After the decline of the Roman Empire in the fifth century B.C., and after the subsequent transition of the Germanic tribes, Slavonic tribes arrived in the Balkans: i.e., the Bulgarians in the northeast regions and the Serbs and Croats in the western and northern regions. The latter two peoples were closely related ethnically and by language; however, they became separated by religious faith and history. Thus the Croats, settling essentially in the west and on the Dalmatian coast, became Roman Catholics, while the Serbs adopted the Greek Orthodox church.

Habsburg ally in the critical days following the assassination of Archduke Franz Ferdinand, now declared war against Russia and France. On 2 August 1914 Germany opened the war actions by sending an ultimatum to the Belgian government requesting free passage for her troops on the way to France; this passage then occurred, with Belgian resistance, on 4 August, upon which Great Britain—which had guaranteed the sanctity of the Belgian borders—declared war on the German Empire. The machinery of marching into a great European war had been started, with the two alliances—the *Triple Alliance* and the *Entente Cordiale* plus Russia—opposing each other. Italy stayed out in the beginning, leaving the two powers, Austria-Hungary and Germany, henceforth called the Central Powers, alone.

In view of the rapid, explosive development of this situation, one may question the motivation of civilized European peoples, or at least their responsible governments, in breaking into a war so easily. Had they not *all* benefited enormously from the previous decades of peace? Was it not due to this period that one owed such enormous progress in nearly every aspect of life: the unprecedented advance in culture, science and technology, achieved by the joint efforts of all European nations, the British, the French, the Germans, the Italians and the Russians? Indeed, the achievements were great and conspicuous, such that they made European civilization foremost in the world: for instance, electrical power stations had been established, replacing human labour and less reliable water power and making man independent of natural conditions by illuminating cities and working places, thus enabling him to expand his creative hours around the clock; the railway system and new modes of transportation, such as the automobile and

---

The Serbs, after having been under Bulgarian and Byzantine dominance, formed a kingdom in the eleventh century, embracing (later) not only central Serbia but also Bosnia, Montenegro, Albania and parts of Macedonia. In the Battle of Kosovo (1389) the Turks annexed the Serbian kingdom and incorporated it, seventy years later, into the Ottoman Empire. Parts of Croatia and Dalmatia, which had been under Venetian and Hungarian influence, were also seized by the Turks, following the defeat of the Hungarians in the Battle of Mohács (1526). During the following centuries the Austrian Habsburgs, who had kept some regions of Hungary, defended the southeastern borders of Christian Europe against the Turkish invasions; however, from 1683 they started a strong move eastward, taking back European countries from the Ottoman Empire, especially Hungary, Croatia and Transylvania. Serbia and Bosnia, as well as the southeastern parts of the Balkans remained, more or less, firmly under Turkish rule, especially after Austria made peace with the Ottoman Empire in the middle of the eighteenth century. The Serbs therefore turned to Russia for help against the Turkish yoke. In the late eighteenth century they started a war of independence, which also conflicted with Austrian and Hungarian interests—Hungary had meanwhile absorbed most of the Croatian and Transylvanian regions. After several upheavals a Serbian state around Belgrade was created in the early nineteenth century, supported by Russia. Serbia was defeated by the Turks in 1813, but was recovered two years later by a Serbian army led by Miloš Obrenović; in 1830 Serbia obtained autonomy from Istanbul, although still remaining formally part of the Ottoman Empire. In 1878 the independence of Serbia was proclaimed at the Berlin Congress, while Bosnia—which was desired by Serbia as well—was given as a mandate to the Austro-Hungarian Empire. This decision aggravated the difficulties between Serbia and the *Dual Monarchy*. In 1908 the latter finally annexed Bosnia and Herzegovina. Serbia created, with Bulgaria and Greece, the so-called Balkan League, which defeated Turkey in the First Balkan War (October 1912–May 1913). On the partition of Macedonia, the members of the Balkan League quarrelled in the Second Balkan War (June–July 1913); as a result, Bulgaria left the alliance and turned to Turkey and the *Triple Alliance* of Austria, Germany and Italy.

the aeroplane, had been developed; chemical fertilizers had been introduced in order to increase crops all over the world; new, more productive and hardy forms of domesticated plants had been bred; several diseases that had plagued mankind had been conquered by recognizing their origin in microscopic bacteria and discovering appropriate medicines against them. All these and other achievements had been reached by international co-operation, a largely free exchange of knowledge and methods. Powerful movements, which had spread out far beyond the European borders, in the fine arts, music and literature provided new forms of human expression. Should this common Western civilization—as it extended from Europe to America, China and Japan—now be completely forgotten for the sake of a barbaric war, the outcome of which was at best uncertain, and whose cruelty would be unparalleled due to the abuse of the most recent technologies?

What did peoples, or their governments, expect to gain from the war? It seemed perhaps most evident from the point of view of the Austrians, who wanted to re-establish the honour of their emperor and defend their multinational state. One may also understand the reasons of the Russian Empire, which wished to establish on her southwestern flank a ring of Slavonic satellites and attain her old dream of free access to the Mediterranean Sea by controlling the water routes held so far by the Turkish Empire. France, on the other hand, had never given up the idea of getting back the land lost to the German Empire in 1871 and reducing and surpassing again the latter's power. Each of the German and British Empires finally sought to settle the colonial and economic competition in its own favour. Japan, which finally joined the *Entente*, expected to gain the German colony of Tsing-Tao in China in order to gain a firmer foothold on the Asian Continent. One must admit that most motivations and goals, even the more ambitious ones, were shared by governments and peoples alike. Hence most of the humble soldiers, called to military service, marched to the battlefields with pride and enthusiasm, desiring nothing more than to win the victory for their native land. Even in the multinational Austro-Hungarian Empire, with its many unresolved conflicts, hardly any important opposition arose against the governmental politics throughout most of the difficult years of the war.

In Austria and Germany there existed—as it did in France—the same system of general conscription; i.e., basically all men able to perform military service could be drafted. This implied that a large proportion of the male population of these countries, between eighteen and sixty years of age, was called to bear arms, with younger and previously untrained men applying as *Kriegsfreiwillige* (volunteers of the war). Scientists and artists, previously a supranational class of people, were no exception to this rule. The list of Austrian and German physicists, published under the heading '*Übersicht über die Kriegsbeteiligung der Deutschen Physiker*' ('Overview on the War Participation of German Physicists') in the *Physikalische Zeitschrift* in 1915 and 1916 is long and distinguished; it contains over 150 names, nearly all of them university teachers and many of them volunteers. Practically none of the known younger representatives is missing, and some of the senior professors and *Geheimräte* (Privy Councillors), like Theodor Des Coudres of Leipzig, Ernst Grimsehl of Hamburg, and Jonathan Zenneck of Munich, also served their country.[384] Most of them served the common patriotic purpose by participating in fights on the battlefields; only a few stayed away from the front, in service or in instructing positions. Special plans to employ physicists in technical or scientific war projects did not exist on the Austrian and German side, at least not at the beginning of the war. The so-called *Artillerie-Prüfungs-Kommission* (Artillery Testing Commission) in Berlin employed physicists and related scientists only later: that is, from 1916, Max Born, Alfred Landé, Erwin Madelung and some other physicists worked under the guidance of Rudolf Ladenburg in order to investigate sound-ranging methods.[385] A situation similar to that of Germany also existed in France.[386] Only in Great Britain, where no general military draft system existed, especially not in peacetime, did scientists soon become involved in particular positions connected with warfare; one may refer in this context notably to the Anti-Aircraft Experimental Section of the Munitions Inventions Department, under the direction of the physiologist Archibald Vivian Hill, which employed the physicists Ralph Fowler and Eward Arthur Milne.[387]

The first military actions of World War I were undertaken by the German army in the west: powerful troops marched, as had been planned many years earlier by the former Chief of the German General Staff, Count Schlieffen, through Belgium into France in order to avoid the line of French fortresses, from Lorraine to the Belgian border, protecting the interior of France against Germany. The Austro-Hungarian army, on the other hand, first became engaged in conquering Serbia and in trying to keep the Russian army busy, because German headquarters had initially left only a few troops to defend the eastern border. However, the Central Powers did not achieve their goals, either in France or in Serbia. Indeed, the Serbian army even drove back the invading troops and themselves entered Hungarian territory in the south. Strong Russian troops penetrated into East Prussia, where they were, however, defeated in late August 1914; but in Galicia the Russians scored great successes at the same time, driving the Austrians back to Cracow, so that the Austrians had to request German help in order to start a counter-offensive in 1915. With the Bulgarians joining the Central Powers in September 1915, Serbia would be conquered; then Turkey, having joined the war against Russia and Britain in October 1914, succeeded in stopping actions—such

[384] See *Phys. Zs* **16**, pp. 142–145, 215–216, 268, 312, 330, 367, 387, 428, 488 (1915); *ibid.* **17**, p. 28 (1916). In later issues of the *Physikalische Zeitschrift* no such lists were published; however, this does not imply that no further physicists then entered war service.

[385] For a more detailed description of the *Artillerie-Prüfungs-Kommission* and the work of the physicists there, see Volume 1, Section III.2.

[386] In telling the story of Louis de Broglie, we have mentioned that both he and his brother Maurice and their colleague Léon Brillouin became involved in the development of wireless telegraphy techniques. (See Volume 1, Section V.4, p. 582.)

[387] See the Fowler story in Volume 4, Section II.2.

as the famous British landing at Gallipoli—on the Bosporus. A heavy blow to the Austrian-German side was that Italy broke the *Triple Alliance* on 24 May 1915 by declaring war on Austria and attacking her southern borders on the Isonzo River.

## Artillery Officer Erwin Schrödinger

Like so many of his colleagues Schrödinger was immediately drafted at the beginning of the war. In an autobiographical sketch he recalled simply: 'Then came the war, in which I participated as an artillery officer at the southeastern front, without getting wounded and without illness and with little distinction' (Schrödinger, 1935, p. 87). He was stationed in several places, first in Raibl—at that time in southern Carynthia, now in Italy—and later in the Hungarian Danube port of Komárom.[388] Later places included Prosecco and the region of Trieste.[389] The most eventful part of Schrödinger's war service is detailed in his diary: in the '*Tagebuch* 1915' he described military actions in which he participated from 26 July to 27 September 1915 and which brought him, as an artillery officer, from Komárom on the Danube to Görz (or Goricia, as it is now called) on the eastern Austro-Italian front.[390]

Schrödinger began his entries in the diary with the following words: 'Monday, 26 July 1915, 10 a.m.: departure from Komárom. We are supposed to reinstall a 12 cm L40 marine-battery. Nobody knows exactly where it should go.' A secret order brought Schrödinger and his subordinates by rail, via the town of Bruck-on-the-Mur in Styria, Marburg (now Maribor in Slovenia) and Laibach (now Ljubljana in Slovenia), into the Görz region; on 28 July they reached Oveia Draga, a town there. Schrödinger met his first difficulties, as he wrote in his diary: 'The men have not eaten all day; nothing has been prepared, although the railway station officer of Ogeina has ordered food. There is no food supply station at all and I have no idea where he ordered the food from. Weigl has preserves bought, and a good-natured regimental physician allows us to stay at the ambulance station and gives us wine.' The following day they marched to St. Peter, where officers were awaiting them, and who told them that the battery in question had already been installed and was firing. The soldiers were housed on a farm and Schrödinger, like the other officers, was in the castle of a local count, which was very comfortable, except that all the windowpanes had been broken

[388] See the places of submission of the two papers, the first (Schrödinger, 1915a) being dated 27 October 1914 in Raibl, the second (Schrödinger, 1915b) being dated July 1915 in Komárom.

[389] In Prosecco he concerned himself with Einstein's theory of gravitation (see Hermann, 1963, p. 176); while Annemarie Schrödinger remembered later: 'I visited him once [during wartime] at his post near Trieste' (AHQP Interview with Mrs. A. Schrödinger, 5 April 1963). (This was possibly in the vicinity of Görz—see below.)

[390] Komárom, the starting point of Schrödinger's war tour, was an important Hungarian port on the Danube, half way between Pressburg (Bratislawa) and the bend in the Danube. The major part of the city—i.e., the part north of the Danube—is now called Komárno and lies in Czechoslovakia; today's Komárom is still a notable part of the heavily industrialized northwestern region of Hungary.

due to the occasional action of a 30.5 cm heavy howitzer in the castle's garden. Towards noon Schrödinger went to Görz by car, and later arrived at his battery, where he was supposed to serve as one of three officers conducting the firing. Hence he had to be present for a part of each day at the battery.

In the beginning, Schrödinger's operations were not successful. Thus he reported: '2 August: We fire badly. I am scolded [by superior officers], upon which I check all the elements [of the gun operation] myself, creating a delay in firing. In the afternoon things go better. I do not understand why we are not or scarcely have been fired upon, since we use *strongly*-smoking gun-powder. Aeroplanes are searching for us.' The same situation continued during the following days; however, on 6 August they came, for the first time, under the fire from Italian batteries.

What Schrödinger described in his war diary was part of the battles that continued from the end of May 1915. After a first advance by the Italians a fairly stationary front had crystallized on the Isonzo River between Italian and Austrian troops. In order to improve their positions the Italian army had fought in two battles, the first Isonzo battle from 23 June to 7 July, the second from 17 July onwards, without gaining any advantage. Towards the end of the second Isonzo battle Schrödinger and his men arrived at the front, but meanwhile the fighting had faded into sporadic actions.[391] Thus he witnessed more or less a trench war, with insignificant moves forwards and backwards; the Austrian batteries fired, the Italian batteries answered, and *vice-versa*, each occasionally scoring on the other side. On 7 August 1915 Schrödinger's battery ran into trouble when the previously fixed foundations began to crumble. He wrote in his diary: 'Stop firing, back into the grenade shelter .... At least two batteries fire at us and are shooting well. In front of and behind the connecting passage there are grenade craters of 2-3 m; cement and dust drifts into the shelter. The men are very brave, we are however well covered. If a bomb does not hit the centre of the [shelter's] roof, not much can happen.' Between 8 and 20 August Schrödinger's battery was moved to a new position, which was only 100 m from the old one. A basement was made of 38 cubic metres of concrete, and another safe shelter was built to protect them against Italian fire. He noted: 'It seems completely out of the question that our gun will do as much harm to the enemy as will compensate for the efforts we have put in.' Basically no heavy fighting occurred at that time, and Schrödinger therefore expected that the Italian troops were preparing a new offensive.

Quiet as the fighting was, so also was Schrödinger's life. He wrote in his diary: 'Every second afternoon I spend in Görz, in the cafes and strolling about, today (20 August) I was vaccinated. By the way, enteritis seems to be attacking me. In

[391] In October 1915 the Italians launched a new attack, which resulted in great losses on their side and was broken off on 5 December without achieving success. Only in the summer of the following year, on 8 August 1916, would Italian troops march into Görz which a year later, in October 1917, would fall back into Austrian hands.

a little spot on the other [Italian] side (Broglia) cholera is said to have broken out.' Turning again to war actions, he remarked: 'Losses (in the artillery) are reported like unlucky accidents. Normally nothing happens. Then all of a sudden a grenade finds a shelter or an observation station, and 4 or 5 people are torn to shreds. One should not have bad luck.'

News came in from other fronts of the war. The Austrians celebrated the successes of the 1915 battles against the Russians at Warsaw and other places. But in the circle of officers, assembled at the castle of St. Peter, Schrödinger also heard details of the previous (1914) debacles in Galicia and notably in Serbia. Thus he noted in his diary: 'Especially in Serbia, everybody [Austrian] seems to have lost his head; rushed, planless, chasing retreat. Some men here report that they have been concerned for *days* with destroying supplies of all kinds . . . . The Belgrade bridge has been blown up with some troops still on it, out of pure mindlessness . . . No attempt was made to keep the city.'

While he listened to the war stories, Schrödinger's life at the Italian front did not change any. He wrote on 19 September 1915: 'External events zero. Sometimes it was completely quiet here; now there are infantry attacks on the plateau. Extremely boring. I concern myself, if I do not have anything else to do, with the psychology of the basic acts of conscience: reminiscence, association, concept of time.' He dreamt repeatedly of a short episode with a young lady back in September 1912; it had not been a success for him. He consoled himself now by the fact that he had behaved correctly in the situation and recalled the essential maxims of his attitude: 'Calm, calm, calm. Silence, and do not get duped. Silence like metal, like concealed stone.'

The final pages in Schrödinger's war diary are filled with accusations against the war. He wrote: 'It is really terrible and I am homesick for [scientific] work. If this lasts for much longer, I shall peter out bodily and mentally . . . . Every reasonable idea is swallowed by another; what use is it all, if the war is not finally terminated?' Although he considered the previous two months—about which he kept the diary—as 'still the best' ('*noch die besten*'), he was tired of the war, which had made his life consist only of sleeping, eating and playing, in short, a lazy and inactive life. He wrote: 'What is the reason, why is it only me who has nothing to do? Is it my fault? Is it the standards which I apply; or is it due to the particular nature of my expectations for activity, especially in the surroundings in which I am living now?' In spite of this inactivity, he claimed: 'I am young and not at all desperate.' However, he clearly felt that the question to ask was not '*when* will the war be over, but *will* the war be over at all?' In other words, fifteen months of war had already killed any thought about the end of the war, because 'people get used to war conditions' and they begin to ask: 'How shall we stand peace?' Schrödinger even feared that an irreversible transformation had already taken place in the population: they had become used to suffering due to the war much more easily than one would have expected earlier. Schrödinger concluded: 'We have war. The word sounds like a joke . . . . Because it sounds as if this were an exceptional state.'

At this point the entries in the diary stopped. We do not know what further military actions Schrödinger, the artillery officer, experienced. We just know that in 1917 he was back in Vienna and completed the rest of his war service at home.

## Heavy Losses

The initial offensive moves of World War I caused great losses among the patriotic soldiers of all the nations involved. The Austrian and German side especially were soon grieving the death of several outstanding men from scientific and cultural life. For example, the German painter August Macke (born 1877), a frequent visitor to Paris (where he had many friends), fell on 26 September 1914 in the Champagne; his colleague and friend Franz Marc (born 1880) died two years later in the murderous Battle of Verdun (on 4 March 1916).[392] Among the poets the German Ernst Stadler (born 1883) was slain by a grenade at Zandvoorde on 30 October 1914; his countryman August Stramm (born 1874) fell in an assault near Horodec (in Russia) on 1 September 1915, after having participated in over seventy battles and skirmishes.[393] The Austrian poet Georg Trakl (born 1887), on the other hand, served as a pharmaceutical orderly in the Lemberg battles; after one major fight (the Battle of Grodek), he had to treat during the night nearly ninety badly injured people without proper medication and he could not help them, upon which he tried to shoot himself; brought to a hospital in Cracow, he died on the night of 3/4 November 1914 of a drug overdose.[394]

Death also reaped a murderous harvest among the physicists, of whom most were decorated with the highest war medals, e.g., the '*Eiserne Kreuz*' ('Iron Cross') in Germany. On 4 September 1914 Maximilian Reinganum (born 1876), professor of physics at the University of Freiburg and one of the most talented German

[392] Franz Marc wrote a moving obituary of August Macke (on 25 October 1914). He was deeply touched by his friend's death. His obituary represents a great human document from the early war period, and may equally apply to many of the outstanding personalities of European civilization who were killed in military action. (See Marc, 1948, pp. 167–168.)

[393] Stramm described in several short expressionistic poems, with titles like '*Kriegsgrab*' ('Tomb of the War'), '*Patrouille*' ('Patrol'), or '*Sturmangriff*' ('Assault') his experiences of the war. See for example, '*Sturmangriff*' in '*Lyrik des expressionistischen Jahrzehnts*,' Wiesbaden: Limes-Verlag, 1955, p. 191.

[394] In two poems, '*Im Osten*' ('In the East') and '*Grodek*' Trakl described the horror of the battlefields. In the last verse of the former he described the effects of the war as:

'*Dornige Wildnis umgürtet die Stadt.*
*Von blutenden Stufen jagt der Mond*
*die erschrockenen Frauen.*
*Wilde Wölfe brachen durchs Tor.*'
(In *Lyrik des expressionistischen Jahrzehnts, loc cit.*, p. 133.)
'A thorny wilderness envelops the city
From bloody stains the moon haunts
The terrified women
Wild wolves have broken through the door.'

It appeared to the poet that a new wilderness coming from the East had begun to erase Western civilization.

experimentalists, was pierced in the heart by a shell splinter near Le Ménil in the Vosges; Georg Lutze (born 1887), a pioneer of atmospheric electricity research and assistant at the wireless telegraphy station in Brussels-Laeken, was killed in an assault near Châlons-sur-Marne on 8 September; in the first fight in which he participated, at Avocourt, Friedrich Bidlingmaier (born 1875), observer at the earth-magnetic station of the Munich Observatory and *Privatdozent* for geophysics at the University of Munich, received a mortal wound, of which he died on 26 October; four days later, on 30 October, the senior professor Ernst Grimsehl (born 1861) of Hamburg was killed while leading his company into battle at Langemark and Wydendreft; F. L. Kohlrausch (born 1879), professor of radiology at the *Bergakademie* in Freiberg, Saxony, also died on 30 October in an assault on Le Bassée, northern France; Wilhelm Michl (born 1886), the physics assistant at the Vienna *K. & K. Tierärztliche Hochschule* and collaborator at the *Institut für Radiumforschung*, did not recover from his wounds suffered in a battle near Chyrów in Galicia and died on 16 November in hospital at Troppau; Werner Planck (born 1889), substitute assistant at Woldemar Voigt's institute in Göttingen, fell on 19 November in a heavy skirmish near Lutomirsk; Gustav Rümelin (born 1882), *Dozent* for physical metallurgy at the *Technische Hochschule*, Aachen, was killed in the trenches at Zillebeke, southeast of Ypern on 16 December; Max Behacker (born 1885), the assistant at Anton Lampa's institute at the University of Prague, who had continued his studies after obtaining his doctoral degree with Max Planck in Berlin (1913) and with Albert Einstein in Zurich (winter 1913/1914), was shot in the stomach while Austrian troops advanced in the summer of 1914 after taking back Lemberg from the Russians (he was buried at the beginning of August 1915 at Kamionka-Strumilova); Mathias Cantor, extraordinary professor of theoretical physics at the University of Wurzburg, died on 23 May 1916 of a serious wound which he obtained in the Dolomites; on 25 October 1916, Friedrich Neumeier (born 1889), doctoral candidate at the University of Königsberg, fell in the Somme battles near Le Transloy.[395]

The physicists just mentioned had already started on the ladders of their professional careers; by sacrificing their lives they added some military fame to their scientific reputations. On the other hand, the number of talented students, who also died in the war, was by far the larger one. These students passed away without having become recognized by the scientific community, with perhaps just one exception. In the issue of the German scientific journal *Naturwissenschaften*, dated 12 April 1918, Max Born, then professor at the University of Berlin—who himself was prevented from performing military service at the front and worked instead at the *Artillerie-Prüfungs-Kommission* (Artillery Testing Commission)—wrote an obituary of a Herbert Herkner (born 1894), whom he knew as a mathematics student at the University of Göttingen and who had been killed

[395] For the details of the Austro-German losses in World War I see the obituaries, published under the title '*Dem Andenken der im Kriege gefallenen deutschen Physiker*' ('In Memory of the German Physicists Who Fell in the War'), in *Physikalische Zeitschrift* **16** (1915), **17** (1916) and **18** (1917).

in the Battle of Cambrai on 22 November 1917.[396] Born claimed that the case of this student constituted 'an event of secular rarity' ('*ein Ereignis von säkularer Seltenheit*'), because he was 'a blessed genius' ('*ein begnadeter Genius*') who distinguished himself 'so clearly above the average, that the teacher to whom the education of his mental abilities was entrusted possesses the happy conscience of supporting one of the really great men on the first steps of the ladder to fame' (Born, 1918a, p. 179). He deplored 'the [passing of the] unique personality of this noble man, with whom a portion of the spiritual future of Germany has been destroyed' (Born, *loc. cit.*, p. 179).

While Herkner had only entered upon his studies and had not yet left a finished product of his genius, the war took away another four, already internationally acknowledged, geniuses in their best years. Although not on the battlefield, Otto Sackur (born 1880), department leader of Fritz Haber's *Kaiser-Wilhelm-Institut für physikalische und Elektrochemie* and perhaps the most talented one among the younger German physico-chemists, died on 17 December 1914, during an explosion in an experiment carried out for military purposes 'in the service of the community' ('*im Dienst der Allgemeinheit,*' Herz, 1915, p. 115).[397] On 11 May the astrophysicist Karl Schwarzschild (born 1873) died, at home in Potsdam, 'of a malicious illness caught at the war front' (Runge, 1916, p. 545). For more than twenty years he had made first-rate contributions to his field; and just recently he had presented two fundamental papers on two different, very deep physical topics—one on Einstein's recent theory of gravitation (Schwarzschild, 1916a) and the other on quantum theory (Schwarzschild, 1916c). 'One could still expect great contributions from him,' wrote his friend and former Göttingen colleague Carl Runge in his obituary, 'because he united the mastery of mathematical analysis with a deep understanding of physical phenomena, and with experimental and technical skill' (Runge, 1916, p. 545).[398]

In spite of the nationalistic tendencies which had been stirred up by the war—and which led to such actions as the manifesto '*An die Kulturwelt*' ('To

[396] Born had tried earlier to get Herkner back from active service and place him in the *Artillerie-Prüfungs-Kommission.* He reported in his *Recollections*: 'My "staff" [at the *Kommission*] . . . consisted of students of mathematics and physics whom I claimed from other army units. The first was my pupil [Alfred] Landé . . . . I began to regard this chance of saving gifted young men from being wasted at the front as a major task and pursued it with vigour. In several cases I succeeded, in others I had grievous disappointments. The worst was the case of Herbert Herkner, who had studied mathematics during my last year in Göttingen and showed the greatest talent to appear for many years. He was now an infantryman in the fighting lines. I had a long struggle in getting my application through for his transfer to our department. At last I succeeded. The order reached Herkner a day before a great battle, where he was killed (Battle of Cambrai, 22 November 1917). I was terribly upset by this news, accusing myself of not having pressed my application for his release more strongly. I was convinced that mathematics had lost in Herkner a genius of the first order, and I expressed this view in an obituary which I offered Arnold Berliner for his *Naturwissenschaften.* He accepted it . . . . It is perhaps the only case where a young student has been honoured by an article like those usually devoted to great scholars' (Born, 1978, p. 171).

[397] We have presented a biographical note on Sackur in Volume 1, p. 145, Footnote 198.

[398] We have described Schwarzschild's life and work in Volume 1, especially in Section II. 4, pp. 225-227.

the Civilized World'), signed by ninety-three German scientists[399]—an Englishman, who fell against Germany, received an honourable obituary in a German scientific journal. Kasimir Fajans, *Privatdozent* for physical chemistry at the *Technische Hochschule*, Karlsruhe, wrote in the *Naturwissenschaften* issue of 27 July 1916—at a time when the bloody battles of Verdun and the Somme were terrifying the nations on both sides—the following: 'On 10 August 1915 Henry Moseley fell, twenty-seven years of age, in the Dardanelles. The death of this young physicist, who was among the ablest of his generation, will raise the deepest regret in wide scientific circles . . . . The death of this extremely talented man, from whom one could still expect many important discoveries, constitutes a heavy loss to science' (Fajans, 1916, pp. 381-382).[400]

Even as the death of the young experimentalist Moseley (born 1887) affected the scientific life of his native land of England, Austrian science was probably more affected by the death of Friedrich Hasenöhrl, thirteen years his senior. Hasenöhrl had, immediately following the outbreak of the war, voluntarily put himself on military duty and had been placed on several missions in Przemysl and later in Cracow. As Stefan Meyer reported in his obituary:

> Since the inactivity in the fortress headquarters of Cracow did not satisfy him, he tried being an aeronautical observer in a plane. Then, in May of that year, came the ignominious treason of our former ally Italy, and now he could no longer bear to be in the north. He joined the infantry to help defend "his" Tyrolian mountains, of which he had climbed so many and whose specific nature he knew like that of good friends, and where he had found recreation in summer year after year. (Meyer, 1915, p. 429)

Hasenöhrl held the command of one *Marschbatallion* of the 14th Infantry Regiment at Monte Piano, where he was wounded on 20 July 1915; he recovered in a Salzburg hospital. Meyer wrote: 'Hardly and poorly recovered, he was ordered to go to the region of Vielgereuth; there, in a heroic and successful fight on 7 October [1915], he was killed in action by a grenade while leading his company' (Meyer, 1915, p. 429). He was buried as a simple *Oberleutnant*, Meyer reported, but added: 'A deep recognition of what had been lost with him, then became clear. The deep and lasting impression of this irreplaceable loss felt by everybody, may best be seen by the fact that His Majesty the Emperor of Austria telegraphed his heart-felt condolences to the widow—which was unusual in the case of a

[399] The manifesto, which was issued early in October 1914, besides disclaiming Germany's war guilt, justified the violation of the Belgian borders and stated that the existence of German culture rested on German militarism. Among those who signed were microbiologist Paul Ehrlich and physicists Max Planck and Wilhelm Conrad Röntgen. Among those few, who refused to sign, was the mathematician David Hilbert. Albert Einstein and Georg Nicolai, professor of physiology at the University of Berlin, tried to issue a counter-manifesto, entitled '*Manifest an die Europäer*' ('Manifesto to Europeans'), in which they stressed the blessings of the previously existing European co-operation; they obtained few signatures. (See Clark, 1972, pp. 228-230.)

[400] We have presented an account of the life and work of Henry Gwyn Jeffreys Moseley in Volume 1, Section II.3, pp. 194-196.

*Landsturmoberleutnant*—and awarded him posthumously the medal *Eiserne Krone* with *Kriegsdekoration*' (Meyer, 1915, p. 430). Austria had indeed lost 'one of her best physicists in the full power of his creativity' ('*einen seiner besten Physiker in der Vollkraft seines Wirkens,*' Meyer, 1915, p. 429); when eighteen years later Erwin Schrödinger received the Nobel prize for physics, he would state in the autobiographical sketch submitted to the Nobel Foundation: 'At that time [in 1915] Hasenöhrl was killed in battle, and I feel that otherwise today his name would now stand instead of mine' (Schrödinger, 1935, p. 87).

## The Later War Years

The year 1916 constituted a period of World War I in which it was completely uncertain who would finally win the war. A German offensive was started in France, leading to the extended Battle of Verdun (since February), the forces of the *Entente* answered with an offensive on the Somme, while strong Russian armies under General Brussilov attacked the southeastern front in June—stimulating the Rumanian government to join the war on the side of the *Entente*. These actions and counteractions resulted in a complete stalemate; however, Russian military power became essentially exhausted. While Rumania was conquered by the combined Austro-Hungarian, German and Bulgarian efforts, the *Entente*—notably the British supported by Arab troops—gained territorial successes in Mesopotamia from the Turks. The same stalemate might have continued, possibly for years to come, had not the United States of America declared war on Germany on 6 April 1917 and, with fresh manpower and enormous material involvement, strengthened the worn-out *Entente* and decided the outcome of the entire war. Although the Russians, after their October Revolution of 1917, finally made peace with the Central Powers (in March 1918 at Brest-Litovsk), and although the German armies renewed their offensive on the western front, since the summer of 1918 the complete loss of the war became increasingly inevitable for Germany and Austria. In quick succession their allies surrendered—first Bulgaria in September 1918, then Turkey in October—and finally Austria (on 3 November) and then Germany herself collapsed on 11 November 1918.

During this second period of the war the artillery officer Schrödinger occupied a much less dangerous post than before. Although still on military duty, he spent his time essentially at home. We do not know exactly when he returned to Vienna, but it was certainly before the summer of 1917.[401] In his curriculum vitae (of July 1938) he simply remarked that he 'later [in the war] was active as a teacher

[401] Schrödinger used to state in his scientific publications the exact place of submission; thus he signed papers, submitted in late 1914 and in the spring of 1915, from Raibl and Komárom, respectively (Schrödinger, 1915a, p. 418; 1915b, p. 295). The next of Schrödinger's publications does not indicate a submission date; it appeared, however, in the issues of 24 and 31 August 1917 of the German journal *Naturwissenschaften* (Schrödinger, 1917a). In it, and in later publications to which he contributed during World War I, Schrödinger gave his Vienna institute address as the place from where he had submitted the paper. Hence we may assume that he was stationed in the later war period in Vienna, which is in agreement with his involvement in teaching meteorology. (See below.)

of meteorology.'[402] Schrödinger's recollection is confirmed by the existence of a notebook, entitled '*Vorträge über Meteorologie an der Fl. O. Sch. in W$^{r.}$ Neustadt*'—which should be translated as 'Lectures on Meteorology at the *Fl*[*ak*]-*O*[*ffiziers*]-*Schule* in Wiener Neustadt,' a town lying about 30 km south of Vienna.[403]

In these lectures he attempted to acquaint anti-aircraft officer-candidates 'in a few hours with the basic concepts and facts of meteorology' (*Vorträge*, p. 1). These basic concepts included discussions on the constitution of the atmosphere; radiation from the sun; temperature distribution in the atmosphere and its daily and annual changes; gas pressure; high and low pressure zones; circulation of the atmosphere, especially the winds above oceans, continents and mountains; weather divides; thunderstorms; cloud structure; and a detailed discussion and explanation of weather charts.[404]

Schrödinger's meteorological lectures to the students attending the military school in Wiener Neustadt did not cover much material beyond that presented in any introductory course on meteorology at the University of Vienna or elsewhere, just as he had taken in his own student years with Julius Hann. However, the preparation and delivery of this course led him to continue his former scientific activity, the absence of which he deplored so much in his diary while stationed at the Isonzo front in September 1915. A paper, '*Zur Akustik der Atmosphäre*' ('On the Acoustics of the Atmosphere'), which was received by the *Physikalische Zeitschrift* on 31 July 1917, bears the first fruits of his renewed involvement in physical problems, although the essential idea was conceived before Schrödinger returned to Vienna (Schrödinger, 1917b).[405]

The particular problem treated in this paper must have been of great interest to the active artillery officer Erwin Schrödinger, because it had some relevance to his military assignment, i.e., the problem of recognizing the position of explosions. Back in 1910 G. von dem Borne, of the *Kgl. Erdwarte* Krietern, near Breslau, had tried to provide an explanation for a well-known phenomenon: loud explosions (like those caused by volcanic eruptions or heavy bombardments)

[402] The *curriculum vitae* (*Lebenslauf*) has been filed on AHQP Microfilm No. 39, Section 1.

[403] Wiener Neustadt housed several military schools and barracks.

Schrödinger's meteorology notes have been filed on AHQP Microfilm No. 39, Section 2.

[404] At the end of his lecture notes on meteorology, Schrödinger added sets of questions dealing with the entire material presented, as well as with aspects on the organization of weather stations.

[405] In a postscript to this paper, Schrödinger remarked: 'The essay was composed on the [battle] field' ('*Der Aufsatz ist im Feld entstanden*,' Schrödinger, 1917c, 567). Since the paper owes its origin, as we shall discuss in more detail below, to a paper of another author which appeared in an August 1916 issue of the *Physikalische Zeitschrift*, we can safely conclude that Schrödinger was at that time still performing military service on the battlefield. On the other hand, the author also cited literature published earlier than 1917; and from the variety of situations we derive that he did not complete the article on the battlefield, but at a place where he had available a good scientific library, notably at home in Vienna.

Actually, the date of submission of this paper provides the earliest documented evidence that Schrödinger was back in Vienna. However, the paper published in the August 1917 issue of *Naturwissenschaften*, referred to in Footnote 401, must have been submitted by the author considerably earlier also from Vienna.

can be heard very well at a certain large distance—called the 'zone of anomalous audibility' and which extends approximately to a distance of 114 to 300 km—while being far less noticeable at smaller distances ('zone of silence'); he argued in particular that this phenomenon was caused by a reflection of sound waves from the layers of the upper atmosphere about 100 km above ground (Borne, 1910). While Borne's explanation had been supported by others—among them the Viennese meteorologist Alfred Wegener—Wilhelm Schmidt of the *K. & K. Zentralanstalt für Meteorologie und Geodynamik* claimed, six years later, in his paper '*Zur Fortpflanzang des Schalles in der freien Atmosphäre*' (On the Propagation of Sound in the Free Atmosphere') that it was 'physically totally useless, because it fails completely as far as intensity relations are concerned' (Schmidt, 1916, p. 334).[406] His claim rested on the following argument: if an acoustic wave did reach the upper layers of the atmosphere, in which density is very low, then the energy transported by this wave must also assume a low value; the thus weakened wave was further damped on the way back to the earth's surface, hence it 'could under no circumstances be loud enough to be audible compared with sound propagating horizontally [i.e., in the lower layers of the atmosphere having high density]' (Schmidt, 1916, p. 334). Schrödinger attacked this argument; he argued that Schmidt's result, namely that 'the final intensity [of the sound wave reflected from the upper atmosphere] occurred ... multiplied with the square of the ratio

$$\frac{\text{density of the air at the culminating point}}{\text{density of the air above the earth's surface}}\text{'}$$

(Schrödinger, 1917b, p. 445) violated the principle of energy conservation, hence it could not be correct.[407] As a consequence he stated 'that sound propagation, even in an isothermal atmosphere, is *not* governed by the [usual] wave equation ... but rather by slightly different equations, which should be derivable by applying general hydrodynamical principles to the present situation' (Schrödinger, 1917b, p. 446).

In order to prove his statement Schrödinger first studied a special case, namely the propagation of a vertical plane acoustical wave in an isothermal atmosphere, in which the density $\rho$ obeyed the barometric formula, i.e.,

$$\rho(x) = \rho_0 \exp(-\alpha x), \tag{55}$$

with

$$\alpha = \frac{g\rho_0}{p_0}, \tag{55a}$$

[406] As we have mentioned, Schmidt was also a student of Franz Exner at about the same time as Schrödinger. (See Section I.2.)

[407] In the introduction to his paper Schrödinger searched for a possible mechanism to explain the energy loss of the acoustical wave predicted by Schmidt. He concluded, however—from an investigation of the special case of an isothermal atmosphere, for which Schmidt's result had to be correct as well—that no such mechanism was available.

provided no disturbance was present. In Eqs. (55) and (55a), $x$ denotes the vertical distance from the ground; $\rho_0$ and $p_0$ give density and pressure, respectively, on the earth's surface, and $g$ describes the acceleration caused by the gravitational field of the earth. On introducing Lagrangean variables into the problem, Schrödinger showed that any small disturbance of the atmosphere, as caused by the acoustic wave under consideration, had to satisfy the following equation of motion

$$\frac{\partial^2 \eta}{\partial t^2} - v^2 \frac{\partial^2 \eta}{\partial x^2} + g\gamma \frac{\partial \eta}{\partial x} = 0, \tag{56}$$

where $v$ ( $= \sqrt{\gamma p_0/\rho_0}$) denotes the velocity of sound in the isothermal atmosphere, and $\gamma$ denotes the ratio of the specific heats of air (i.e., of the specific heat at constant pressure over the specific heat at constant volume). Equation (56), in which $\eta$ describes the position variable of the acoustical wave as a function of space and time, deviated from the usual wave equation by the last term proportional to the earth's acceleration $g$. Its solution, namely,

$$\eta = A \exp\left(\frac{\alpha x}{2}\right) \cdot \cos\left\{ \nu \left[ t + \frac{x}{c(\nu)} \right] \right\}, \tag{57}$$

where

$$c(\nu) = v \left[ 1 - \frac{\alpha^2 v^2}{4\nu^2} \right]^{-1/2}, \tag{57a}$$

exhibits—for not too high frequencies $\nu$—a negligible dispersion of the acoustical wave; and also the property that 'the amplitude increases or decreases inversely proportional to the ratio of the square roots of the densities' (Schrödinger, 1917b, p. 447).[408]

Schrödinger generalized his considerations of the propagation of acoustic waves also to include the case of propagation in three-dimensional space. Even this extension had no effect on the previous result, namely, 'that the decrease of the density [of air] in the upward direction is exactly compensated, as far as [sound] energy transport is concerned, by the increase of the amplitude [of the sound wave], hence the propagation of energy proceeds in the same way as in a homogeneous medium [i.e., in air of constant ground density]' (Schrödinger,

[408] The fact can be recognized directly be comparing the exponential factors for $\eta$—i.e., $\exp(\alpha x/2)$—and $\rho(x)$ in Eq. (55)—i.e., $\exp(-\alpha x)$.

In a postscript to his paper, dated November 1917 and published in the 1 December issue of the *Physikalische Zeitschrift*, Schrödinger pointed out that the characteristic amplitude factor, $\exp(\alpha x/2)$, had earlier been derived by Siméon Denis Poisson (in 1807), and later by Lord Rayleigh and Horace Lamb. He added,however, that his own calculation appeared to him not to be 'superfluous' ('*überflüssig*,' Schrödinger, 1917c, p. 567), because Poisson's treatment rested on several wrong physical assumptions, and Lord Rayleigh and Lamb had studied only the one-dimensional case, while he—Schrödinger—had also extended the result to apply to the three-dimensional propagation of an acoustical wave. (See below.)

1917b, p. 450).[409] Hence Schmidt's objection to von dem Borne's theory of anomalous audibility seemed to be removed. Schrödinger, however, did not stop at this point, but investigated further possible damping mechanisms of sound waves, namely, those due to heat conduction and internal friction, which played a role especially in the upper atmosphere consisting of rarefied gases. By applying the known results of kinetic gas theory he found the following situation: with decreasing gas density the damping effects caused by heat conduction and internal friction led to a more drastic attenuation of the amplitude of acoustic waves, the smaller the wavelengths became. Hence he concluded ultimately:

> According to these results the possibility that sound waves return from the border of the hydrogen layer [i.e., the layer where they are reflected], still with finite intensity, appears to me to be rather small if we cannot place the border of this layer at an even lower altitude than presently assumed (i.e., at a height of 70–100 km above the ground) . . . . In any case, a harmonic analysis of one and the same explosion wave in the inner and outer zone of audibility should be, in my opinion, of particular interest. Because of the pronounced selective character of the absorption [of the sound wave], which changes rapidly with altitude, one might arrive—by studying the intensity ratios of the wavelengths that are still observed in both regions—at a direct and even rather accurate determination of the altitude of the layer at which the return [reflection] occurs. (Schrödinger, 1917b, p. 453)[410]

The paper on the acoustics of the atmosphere constituted only one of several studies which Schrödinger published during the second half of World War I. Indeed he worked at that time on quite a few topics, and became scientifically as productive as in the pre-war period. In spite of this fact and his comparatively comfortable and safe position, Schrödinger desired nothing more than the end of the war. Thus he wrote:

> *Da sind zwei Armeen ausmarschiert seit 1914.*
> *Die eine kämpft noch.*
> *Die von der anderen haben Frieden gemacht unter der Erde.*
> *Wähle! Zu welcher willst Du?*
> (Two armies marched out in 1914.
> One is still fighting.
> Those belonging to the other have made peace underground.
> Choose! Which one do you want to join?)[411]

[409] Schrödinger further proclaimed that changes in temperature and chemical composition, such as occurred in the upper atmosphere—it was assumed that sound reflection occurred in the transition layer between the lower atmosphere (mainly composed of nitrogen and oxygen) and the upper atmosphere (consisting mainly of hydrogen)—did not alter his conclusions.

[410] While the first part of Schrödinger's treatment of acoustic waves in the atmosphere was later recognized as providing 'the valid formulae for the situation in question' (Wegener, 1928, p. 193), doubts arose with respect to the applicability of his conclusions concerning damping by temperature conduction and internal friction. It was argued that Schrödinger had not paid attention to the fact that light gases prevailed in the upper atmosphere, and had not taken into account the existence of ionization (see Wegener, *loc. cit.*). Both characteristics tended to produce a far lower damping effect than Schrödinger had suggested. Hence the actual situation in the atmosphere co-operated to substantiate von dem Borne's suggestion on the origin of the phenomenon of anomalous audibility.

[411] Note found in the *Nachlaß*, filed on AHQP Microfilm No. 39, Section 7.

## I.5 Schrödinger's Scientific Work in World War I

In the years between the outbreak of war in the summer of 1914 and its end in late 1918, Schrödinger completed—if one also includes a paper communicated to a meeting of the Vienna Academy of Sciences on 16 January 1919 (Schrödinger, 1919a)—ten scientfic papers. Taking into account the fact that only two of the papers were submitted in the early phases of the war (i.e., up to July 1915, before Schrödinger was called to the Italian front), and then an interruption occurred until the summer of 1917—although the paper on the acoustics of the atmosphere, which we have already discussed, had been devised in the summer of 1916—these publications constitute a respectable record of the intensive scientific productivity of an author, who simultaneously performed military service during the entire period. Certainly, some of Schrödinger's papers during the war represented smaller contributions such as the first one, the '*Notiz über den Kapillardruck in Gasblasen*' (Schrödinger, 1915), or the two short contributions on General Relativity Theory published in early 1918 (Schrödinger, 1918a, b).[412] There were also two further papers, both on the problem of specific heats (Schrödinger, 1917a, 1919b), in which the author mainly reviewed the work of others. Still, one should not forget the fact that Schrödinger belonged to those scientists who published only a part of their results, while dealing in private notes with a variety of further questions and problems. This is confirmed by the existence of several notebooks and memoranda, which he wrote during World War I and later kept in his possession. In his *Nachlaß* especially, three notebooks can be found, dealing with questions of, or related to, General Relativity Theory; a further three notebooks discussing the problems of statistical mechanics; and several shorter notes on particular theoretical topics.

### An Analysis of Smoluchowski's Work on Fluctuations

Foremost among the problems which Schrödinger treated, in his unpublished notebooks and notes, is that of fluctuations. It may be that he inherited his interest in those phenomena from his teacher Egon von Schweidler. In any case, Schrödinger, in late 1914, composed the text of a notebook, entitled '*Schwankungsopaleszenz*' ('Fluctuation Opalescence'), in which he concerned himself with the most important papers on the optical consequences of fluctuations, especially those of Lord Rayleigh, Smoluchowski and Einstein.[413] Schrödinger emphasized the intimate connection between the optical phenomena and the deviation of the

[412] We have reported on the content of the paper on capillary pressure (Schrödinger, 1915a) in the previous section.

[413] The notebook is undated and has been filed on AHQP Microfilm No. 39, Section 4. However, Schrödinger quoted in it a paper of Leonard Ornstein and Frits Zernike in the *Proceedings of the Royal Academy of Sciences*, Amsterdam, which appeared in the second half of 1914. Therefore, one might assume that Schrödinger composed his notes on fluctuation opalescence in late 1914, probably on a furlough from military service (perhaps around Christmas 1914).

system (exhibiting the optical phenomena) from the dynamical equilibrium.[414] The demonstration of the existence of such deviations in the pioneering work of Albert Einstein and Smoluchowski on the Brownian motion and similar fluctuation phenomena had decided, as Smoluchowski pointed out in his Wolfskehl lecture of 1913, 'the conflict between the thermodynamical-energetic and the atomistic-kinetic worldview' (Smoluchowski, 1914a, p. 89). Schrödinger, in acquainting himself systematically with these topics during the war years, felt especially attracted by the work of Smoluchowski, who combined theoretical and experimental skill in approaching physical problems.[415] However, he did not come into any closer contact with Smoluchowski, as the latter died on 5 September 1917, a victim of a dysentery epidemic. 'Everyone who followed his brilliant scientific activity, regarded him as the true scientific heir of the Boltzmann spirit in the description of nature,' wrote Arnold Sommerfeld in his obituary of Smoluchowski, adding: 'His name will be connected forever with the most recent rise of atomistics. He has been torn away from the wealth of his successful work; nobody will be able to replace his ingenious nature' (Sommerfeld, 1917b, p. 533).

Marian von Smoluchowski was born on 28 May 1872 in Vorder-Brühl, near Vienna; he grew up and studied in Vienna. Josef Stefan and Franz Exner had been his teachers; he obtained his doctorate in 1895. Ludwig Boltzmann had influenced him through his writings. While he did not have a close personal contact with Boltzmann, Friedrich Hasenöhrl had become his dear friend.[416] As Sommerfeld mentioned: 'Not only did scientific interests bring the two together, but also an equal love and taste for music and sincere fellowship in mountain hiking tours and in skiing' (Sommerfeld, 1917b, p. 533).

Hasenöhrl's high esteem for Smoluchowski passed to his students, including Erwin Schrödinger. After Smoluchowski's death Schrödinger filled two notebooks with '*Besprechung der letzten Arbeiten Smoluchowskis*' ('Review of the Last Papers

[414] Schrödinger carried through the fluctuation calculations for fluids with the help of a method, which showed some similarity to that used by Peter Debye for obtaining the specific heats of solids with which he was very familiar. He arrived at the well-known result of Lord Rayleigh implying a scattering of light by small particles proportional to the inverse fourth power of the wavelength. He was also interested in the case of the scattering of acoustical waves and found a slightly different situation; he concluded especially: 'The main difference, as compared to the case of electric waves, lies in the fact that we are allowed to assume there [i.e., in the electric case] that the "fluctuations" [of the molecules] occur quite independently and are not modified in any way by the transmission of the wave. Here [i.e., in the acoustical case] this is not so; here it is quite *essential* that *associated with* the transition of the [acoustical] wave a heat current and a momentum current arise. Hence we shall find *here* dispersion [of the sound wave], but not *there* [of the optical wave].'

[415] In the Schrödinger *Nachlaß* another manuscript of fifteen pages has been kept, entitled '*Brownsche Schwankungen des Drehwinkels*' ('Brownian Fluctuations of the Angle of Rotation') and filed on AHQP Microfilm No. 39, Section 4. This manuscript on the special fluctuation phenomenon was probably written not much later than the notebook, quoted above, on opalescence. Schrödinger developed in it a solution to the Einstein–Smoluchowski equations, which describe the Brownian rotation of small particles; this solution involved the use of Legendre polynomials.

[416] Smoluchowski left Vienna soon after Boltzmann returned to his hometown (in 1894) and continued his studies in Paris (with Gabriel Lippmann), Glasgow (with Lord Kelvin) and Berlin (with Emil Warburg). He obtained his *Habilitation* in Vienna (in 1898), but left in 1900 to accept a professorship in Lemberg (now Lwov in Russia) and finally moved to Cracow in 1913.

of Smoluchowski'), analyzing not only the work of the deceased but also showing in detail its relation to the work of Ludwig Boltzmann.[417] Thus he stated: 'If one wished to characterize the field of activity of Smoluchowski, in which lay his greatest achievements, with a short, suitable phrase, then it probably would be this: "Proof that a certain type of conception of nature, which is assumed as having been proved to provide a correct picture of nature in the *first* approximation, can be shown to provide a correct picture in the second approximation also."' Of course, Schrödinger associated the conception of nature in question with the name of Boltzmann, defining it by three features: (i) the state of any finite piece of matter can be described *exactly* by a *finite*, but large number of variables ('microstate'). (ii) In reality, nature does not admit dissipative processes, although this fact seems to contradict *every* experience. (iii) The inconsistency implied in the above statement [(ii)] can be resolved by the observation that the small number of observed variables (the 'macrovariables' defining the 'macrostate') are only averages over the many variables describing the microstate, hence each macrostate is composed of a huge number of microstates. Schrödinger continued: 'Among these imaginable microstates there must exist, due to logical necessity, also such which lead to a subsequent change of the macrovariables that differs totally from what has been predicted by experience and classical thermodynamics. However—as Boltzmann states—if one takes into account *the totality of all* possible microstates corresponding to a macrostate, *then a fraction of it approaching unity possesses the property that leads to the development, which is found by experience and expected on account of the laws of thermodynamics*' (*Notebook, No. 1*).[418]

Schrödinger went on, in his characterization of the Boltzmann view of nature, by mentioning the reversibility and recurrence arguments, which had been raised by Joseph Loschmidt and Henri Poincaré, respectively, as hindrances blocking the kinetic interpretation of the second law of thermodynamics, and their interpretation by the statistical hypothesis concerning the macrostate of matter. He continued: 'Now, these two objections have not been refuted at all until today. *But our point of view with respect to them has been changed fundamentally.* It is the main merit of Smoluchowski to have laboured with unshakable energy, with immutable conviction and with a fabulous working power to obtain this shift in the point of view acknowledged.' The main aspect in this change was the following:

[417] The two notebooks, together containing fifty-one pages, have been filed on AHQP Microfilm No. 39, Section 5. We do not know whether Schrödinger intended to publish his notes on Smoluchowski, or at least a part of them; or whether he used them in lectures.

[418] Schrödinger argued that the one-sided thermodynamical processes observed in nature could be easily explained. He wrote: 'Every housewife knows this when she prepares a chocolate cake. For this purpose she must "mix", e.g., $\frac{1}{4}$ kg powdered sugar and $\frac{1}{4}$ kg cacao. This can easily be achieved by stirring.' However, the opposite process of de-mixing does not occur, 'because what we call the "mixed" state, is not *one given* state, but, if we consider the total number of permutations, then the *majority* of them belong to the class of the "mixed" states—and a state, which is just in a given way de-mixed, or even the totality of all such states, ... corresponds to such a negligible fraction of the totality of all permutations that it is practically never established' (*Notebook No. 1*).

while originally the partisans of Ludwig Boltzmann and Josiah Willard Gibbs' statistical theory had attempted to prove the thermodynamical behaviour of systems, one later found that the deviations from the thermodynamical behaviour are not so utterly improbable; they can be calculated and shown to be in quantitative agreement with experimental observation. Schrödinger noted: 'And this fact gives us, in spite of the large conceptual defects which undoubtedly are still connected with the deductions of statistical mechanics, the conviction that it [i.e., statistical mechanics] is nevertheless correct.... Now it is to the field of observable deviations from the thermodynamical state or development, to which Smoluchowski turned his special attention during the early years of this century. With this understanding I meant [my above statement]: proof that Boltzmann's conception of [the nature of matter] provides a real picture of nature even in *second* approximation, that is: including the so-called fluctuation phenomena' (*Notebook No. 1*).

Since contributing his paper '*Über Unregelmäßigkeitein in der Verteilung von Gasmolekülen und deren Einfluß auf die Entropie und Zustandsgleichung*' ('On Irregularities in the Distribution of Gas Molecules and Their Influence on Entropy and the Equation of State') to the *Boltzmann-Festschrift* (Smoluchowski, 1904), Smoluchowski had indeed been a pioneer in the theoretical and experimental investigations of fluctuation phenomena. He had explained the phenomena of Brownian motion and critical opalescence. In his last papers—the ones which Schrödinger wanted to discuss and analyze in his two notebooks—his goal had shifted to the following: 'Besides the desire to establish experimental proofs [for fluctuation phenomena and their theoretical description], the desire becomes prominent [in the last papers] to obtain a clear and definite picture, a perception of the "transition region" between the completely reversible fluctuation region and the completely irreversible region of "normal" processes' (Schrödinger, *Notebook No. 1*).

Smoluchowski had first entered into the debate on the validity of the law of thermodynamics in a lecture, entitled '*Experimentell nachweisbare, der üblichen Thermodynamik widersprechende Molekularphänomene*' ('Experimentally Verifiable Molecular Phenomena Contradicting Usual Thermodynamics'), presented on 18 September 1912 at the *Naturforscherversammlung* in Münster, Westphalia (Smoluchowski, 1912). There he had, after a discussion of the known fluctuation phenomena and their theoretical description in terms of statistical mechanics (Brownian motion, opalescence, density and electricity fluctuations), turned to a careful analysis of the second law of thermodynamics on the basis of reversible microscopic processes and arrived at specific conclusions. He stated: 'The molecular fluctuation phenomena do not provide today any reason to overthrow completely the second heat theorem, like so many other dogmas of physics. They only force us to a deviating formulation, if we demand the universal validity of the laws of thermodynamics. Perhaps an apparently quite insignificant restriction of the phrasing might be sufficient, by saying: "There cannot exist any automatic mechanism, by which *continuously* useful work might be created at the expense of heat of lowest temperature"' (Smoluchowski, 1912, p. 1079). Seven months later, at the Göttingen Wolfskehl week in April 1913 he again addressed the '*Gültigkeitsgrenzen des zweiten Hauptsatzes der Wärmetheorie*' ('Limits of Validity of the Second Law of Thermodynamics,' Smoluchowski, 1914a). He departed from a theoretical analysis of the density fluctuation data, obtained by the Swedish chemist Theodor Svedberg (1884–1971) by observing a colloidal gold solution with an ultramicroscope (Svedberg, 1912). He noticed especially that the larger was the period of observation the larger the deviations from the thermodynamical 'normal' state—i.e., the average density of the suspended particles—could be expected. Consequently the situation in the colloidal solution resembled that of Brownian motion, which macroscopically could be described by the diffusion equation, in which only microscopically reversible processes had to be involved. Smoluchowski thus stated the conclusion that, 'irreversibility is only a subjective concept of the observer, whose applicability does not depend on the type of the natural process, but rather on the position of the initial point and on the duration of the observation,' because 'such processes will appear to us as being irreversible, whose initial point lies far beyond the average fluctuation range and which are observed only within a period of time that is short compared to the time of recurrence' (Smoluchowski, 1914a, pp. 112–113).

Smoluchowski also continued his studies of density fluctuations in colloidal solutions during the following years, focusing his attention on the understanding and description of the time dependence of the group-formation of suspended particles. (See, e.g., Smoluchowski, 1914b, 1915b.) He summarized the main results of these studies in '*Drei Vorträge über Diffusion, Brownsche Molekularbewegung und Koagulation von Kolloidteilchen*' ('Three Lectures on Diffusion, Brownian Motion and Coagulation of Colloidal Particles'), given at the invitation of the *Wolfskehl-Stiftung* in Göttingen from 20 to 22 June 1916 (Smoluchowski, 1916b).[419] The essentially new topic presented in the 1916 lectures constituted his new theory of coagulation, which he had developed, 'on the stimulation of Professor R[ichard] Zsigmondy, who had communicated by letter the results of his beautiful experimental investigations on coagulation of gold solutions and the theoretical problems involved [in the explanation]' (Smoluchowski, 1916b, p. 593).[420] He based his approach on a suggestion of Zsigmondy, namely to describe the mechanism of coagulation by associating with the particles a 'sphere of action $R$, such that the Brownian motion of the particles occurs undisturbed as long as the distance of their centres is larger than $R$; however, two particles must immediately stick together, as soon as the distance of their centres comes down to $R$' (Smoluchowski, 1916b, p. 593). By extending the theory of Brownian

[419] In the previous publication, submitted in the fall of 1915, Smoluchowski had treated in detail the Brownian motion of particles which are under the influence of external forces (Smoluchowski, 1916a).

[420] Smoluchowski had, in his Münster lecture, already mentioned the difficulties of describing the fluctuation phenomena in colloidal solutions having a rather small density of suspended particles (Smoluchowski, 1912, p. 1073).

motion to include this simplified interaction *Ansatz*, Smoluchowski obtained several results which agreed perfectly with Zsigmondy's observations. Especially, he found that the velocity of coagulation increases proportionally to the diffusion constant of the solvent, $D$, and the time development of the coagulation is described basically with the help of the characteristic period $T$,

$$T = (4\pi D R \nu_0)^{-1}, \tag{58}$$

denoting the average time in which exactly one particle gets stuck to another given one ($\nu_0$ is the average number of particles per unit volume).

Smoluchowski's work exhibited three features which attracted his younger colleague Schrödinger.[421] One was the clarity in discussing the theoretical foundation and the physical interpretation; an example might be seen in Smoluchowski's analysis and improvement of the second heat theorem, which we have quoted above, and to which Schrödinger devoted several pages in his first notebook. Schrödinger derived from Smoluchowski's formulation a general requirement that had to be satisfied, in his opinion, by all the laws of nature. He wrote in particular: 'Whether we *aim* at a process in nature, or we *do not aim* at it, is irrelevant for the process in question. No law of nature, including the theorem of the impossibility of the *perpetuum mobile* of the second kind, should be formulated in such a manner as to refer to the intentions or desires of the observer.'

The second important feature of Smoluchowski's work for the Boltzmann partisan Schrödinger constituted the further triumph of the kinetic theory of matter by the derivation of a consistent description of the time development of the coagulation process. The central role that time played in Smoluchowski's approach in describing the fluctuation phenomena, left a particularly deep impression on Schrödinger, in whose philosophical thinking the time concept had begun to assume a central place.[422] On the other hand, for his investigations of physical problems, a third aspect dominating Smoluchowski's work served as a genuine guide: the very close connection of theoretical reasoning and experimental verification. Like Smoluchowski, Schrödinger not only sought to derive subtle formulae, but also studied, immediately afterwards, whether the results really did describe the fine details of experimental data. The papers, which Schrödinger published on fluctuation phenomena between 1915 and 1919, provide excellent examples of this fact. Schrödinger earned with these papers considerable merit as an heir to the great Viennese tradition in kinetic theory, established by

[421] It might be of some interest to mention at this point that Schrödinger did not include, in his discussion of Smoluchowski's last works, the very last paper, entitled '*Über den Begriff und den Ursprung der Wahrscheinlichkeitsgesetze in der Physik*' ('On the Concept and the Origin of the Probability Laws in Physics'), which appeared posthumously in issue No. 17 of 26 April 1918 of *Naturwissenschaften* honouring Max Planck's sixtieth birthday (Smoluchowski, 1918). This fact would agree with dating Schrödinger's notebooks in late 1917 or early 1918, i.e., soon after the death of Smoluchowski.

[422] The earliest written document, showing Schrödinger's fundamental and philosophical interest in the concept of time, are the notes which he composed in the summer of 1915 while staying at the Austro-Italian front. (They are attached to the war diary, which we have discussed previously. Later we shall return to the role of philosophical aspects in Schrödinger's thinking.)

Joseph Loschmidt and Ludwig Boltzmann and continued so brilliantly by Marian von Smoluchowski.

## On the Evaluation of Statistical Data

The concern with the kinetic theory of matter and a deeper understanding of statistical phenomena enabled Schrödinger to treat several problems, which emerged from experimental discussions, and to submit papers on them. The first of these, bearing the title '*Zur Theorie der Fall- und Steigversuche an Teilchen mit Brownscher Bewegung*' ('On the Theory of the Fall- and Rise-Experiments with Particles Exhibiting Brownian Motion,' Schrödinger, 1915b), owed its origin to a question which had troubled the physics community of Vienna for years. The question was: To what extent did Brownian motion affect the experiments of Felix Ehrenhaft and others in determining the elementary charge, by measuring the velocity of falling or rising small droplets in a homogeneous electric field? Several authors, especially Harvey Fletcher of Brigham Young University in Provo, Utah—a former student of Robert Andrews Millikan's—developed theoretical descriptions of the experiment correcting for the Brownian motion, notably for extremely small droplets (Fletcher, 1911). Finally, D. Konstantinowsky used Fletcher's theory in his doctoral dissertation, completed at the *I. Physikalisches Institut* of the University of Vienna; he thus confirmed by his own measurements Ehrenhaft's earlier result that subelectron charges existed in nature (Konstantinowsky, 1914, 1915a). Schrödinger, on learning about the latter's work, discovered a serious error in the basic theoretical formula applied in the evaluation of the experiment; he wished not only to improve the 'presently fashionable method of calculation' ('*gegenwärtig übliche Rechenverfahren*'), but also to establish the still missing 'theoretical measure for accuracy' (Schrödinger, 1915b, p. 289).

The crucial point when dealing experimentally with the problem of Brownian motion, Schrödinger argued, was the fact that the observed fluctuations did not obey the known law of error distribution, because the actual set of observations did not consist of all spatial displacements of a given particle, $x_1, \ldots, x_n$, during a given time interval $t$, but rather of time intervals, $t_1, \ldots, t_n$, during which the particle progressed a given distance $l$.[423] The theoretical description of the experimental procedure, therefore, implied first of all the calculation of the probability $p(t)\,dt$ associated with the event 'that the particle just crossed, for the first time, a mark put at the distance $x = l$ during the time interval between $t$ and $t + dt$' (Schrödinger, 1915b, p. 290). Konstantinowsky had used for this probability the expression

$$p(t)dt = \tfrac{1}{2}\left(\frac{\alpha}{\pi}\right)^{1/2} \exp\left\{-\frac{\alpha(l - vt)^2}{t}\right\}\cdot(l\cdot t^{-3/2} + v\cdot t^{-1/2})\,dt, \qquad (59)$$

[423] The following discussion deals with the one-dimensional case, since we are only interested in the motion in the direction of the fall or rise of the droplet.

where $v$ denotes the velocity of the charged particle due to the electric field, and $\alpha$ a constant (whose inverse is equal to twice the square of the value of the displacement per second of the particle). Schrödinger now claimed this result was wrong for two reasons: first, it described not the desired quantity but rather the difference of the probabilities for the particle to go beyond the mark $x = l$ in the positive and negative directions respectively; second, it embraced those cases in which the particles reached the mark $x = l$ repeatedly. He therefore derived a new formula for $p(t)\,dt$, denoting the probability for the first passage in the time interval between $t$ and $t + dt$; he obtained

$$p(t)\,dt = l\left(\frac{\alpha}{\pi}\right)^{1/2} \exp\left\{-\frac{\alpha(l - vt)^2}{t}\right\} t^{-3/2}\,dt, \tag{60}$$

instead of the one given by Eq. (59).[424]

Equation (60) determined the desired probability for any given time $t_i$. Since the (observed) events connected with any of the observed time values can be considered as being independent of each other, the probability that the time values, $t_1, \ldots, t_n$, occur will be the product $\prod_i p(t_i)\,dt_i$. The experimental problem then consists of the following two tasks: given the data $t_1, \ldots, t_n$, what is (i) the most probable value, $v_w$, for the velocity $v$ of the microscopic particle under the action of the electric field, and what is (ii) the most probable $\alpha_w$ and, connected with it, the most probable average square of the particle displacement $\Lambda_w^2$ $(=1/2\alpha_w)$? Evidently these values had to be found by differentiation of the expression $\prod_i p(t_i)\,dt_i$ with respect to $v$ and $\alpha$, respectively, and putting the results equal to zero. This procedure yielded[425]

$$v_w = \frac{l}{\overline{t_i}}, \quad \text{and} \quad \Lambda_w^2 = l(\overline{v_i} - v_w), \tag{61}$$

with

$$\overline{t_i} = \frac{1}{n}\sum t_i, \quad \text{and} \quad \overline{v_i} = \frac{1}{n}\sum \frac{l}{t_i}. \tag{61a}$$

[424] For the derivation of Eq. (60), Schrödinger used a trick; he assumed that it was not the microscopic particles that moved—besides exhibiting Brownian motion—with uniform velocity $v$, but they rather moved with velocity zero; instead, it was the mark that moved with the inverse velocity $-v$ starting from the position $l$ at time 0, hence it assumed at time $t$ the position $x = l - xt$. The density distributions of the particles on the left and on the right of this mark then satisfied the diffusion equation with appropriate boundary conditions. Schrödinger then calculated the number of particles which had not yet crossed the mark (the left particles). He transformed this result back into the original system of reference, and finally differentiated the expression thus obtained to arrive at Eq. (60).

[425] Schrödinger also derived further results from Eq. (60), i.e.: 1. The average time for the first passage of the mark $l$ is, for very many trials, very close indeed to $\bar{t} = l/v$, as was to be expected. (From Fletcher's Eq. (59) one obtains $\bar{t} = l/v + 1/4\alpha v^2$.) 2. The most probable time of passage is $l/v - 3/4\alpha v^2$, which assumes a finite value (i.e., $2\alpha l^2/3$) even for zero velocity. 3. The average square time deviation $\sqrt{\overline{\Delta t^2}}$ is proportional to $\sqrt{l}$ (as has been verified experimentally). 4. The probability of a return, i.e., that a particle which has gone a distance $\varepsilon$ beyond the mark $l$ returns to the mark, is given by the expression $\exp\{-2\varepsilon v/\Lambda^2\}$. (Schrödinger proposed to test the last result experimentally.)

The second important problem, which Schrödinger set out to solve, was the calculation of the error in the determination of $v$ and $\Lambda^2$. He derived two relations

$$\frac{\sqrt{\overline{(v_w - v)^2}}}{v} = \frac{\Lambda}{\sqrt{lvn}}\sqrt{1 + \frac{3\Lambda^2}{lvn}} \approx \frac{\sqrt{\left(\frac{1}{n^2}\sum\frac{1}{t_i}\sum t_i\right) - 1}}{\sqrt{n}} \tag{62a}$$

and

$$\frac{\sqrt{\overline{(\Lambda_w^2 - \Lambda^2)^2}}}{\Lambda^2} = \sqrt{\frac{2n-1}{n}}. \tag{62b}$$

These formulae could be compared with the previous error estimates based on the 'criterion of smallest oscillations' ('*Kriterium der kleinsten Oszillationen,*' Schrödinger, 1915b, p. 295). The latter stated, say, in the case of a determination of $\Lambda^2$ that: one calculates $\Lambda_w^2$ from 20, 30, 40, . . . , 70 time measurements, and then assumes that the most probable error of the last $(\Lambda_w^2)^{70}$ is approximately given by the difference, $|(\Lambda_w^2)^{70} - (\Lambda_w^2)^{60}|$. Schrödinger now claimed, on account of his results contained in Eqs. (62a) and (62b): 'This assertion is, however, completely wrong. The two series of experiments, *one of which is completely contained in the second*, even *constitutes the main part of the experiments of the second series*, are not at all independent of each other; hence they "probably" will provide far less deviating results [than two really independent ones]' (Schrödinger, 1915b, p. 294). In an example of 67 determinations of falling and rising times he showed that the error of $\Lambda^2$, as calculated with the help of Eq. (62b), was about 17 per cent, while the criterion of smallest oscillations, using the difference $(\Lambda_w^2)^{67} - (\Lambda_w^2)^{60}$, indicated an error of only 4 per cent. Hence he emphasized: 'As the measure of accuracy obtained one must take the formulae [(62a)] and [(62b)] . . . , not the "criterion of smallest oscillations"' (Schrödinger, 1915b, p. 295).[426]

Schrödinger's results were well received by his colleagues interested in that field. For example, D. Konstantinowsky, who had previously applied Fletcher's theory to analyze his own rise and fall experiment, got to know of Schrödinger's work prior to publication and immediately accepted the new treatment.[427] In a note entitled, '*Zur Theorie der Berechnung der Steig- und Fallversuche an Teilchen mit Brownscher Bewegung: Ein Beitrag zur Frage der Existenz des Elementarquantums der Elektrizität im Anschlusse an die Abhandlung von E. Schrödinger*' ('On the Theory of the Evaluation of the Rise and Fall Experiments with Particles Exhibiting Brownian Motion: A Contribution to the Problem of the Existence of the Elementary Quantum of Electricity in Connection with the Paper of E. Schrödinger'), which he submitted in September 1915 (and which was published

[426] The latter criterion had been used earlier by Ehrenhaft, Konstantinowsky and others.

[427] Schrödinger, in an addendum to the proofs of his paper, remarked that 'Mr. Konstantinowsky drew my attention in a letter to a fatal mistake' (Schrödinger, 1915b, p. 295). The 'fatal' mistake was that Schrödinger had ascribed Eq. (59) to Konstantinowsky, while the latter had just taken it from Fletcher's paper of 1911.

in the 15 October issue of *Physikalische Zeitschrift*), he mentioned earlier doubts on the validity of Fletcher's theory and pointed to his own previous plan 'to derive a complete theory of the time distribution of the experiment in question, i.e., to improve Fletcher's corrections' (Konstantinowsky, 1915b, p. 370); however, the outbreak of war and his subsequent military duty had prevented this. Now he frankly admitted: 'Because of the elegant calculations of Schrödinger, who has succeeded in solving the problem mentioned and since published the results, I am now relieved of this task' (Konstantinowsky, 1915b, p. 370). In his own note Konstantinowsky suggested a simple derivation of Schrödinger's results by applying Einstein's theory of Brownian motion to the fall and rise experiment.[428] The main advance of Schrödinger's investigation, however, consisted of the fact that 'one now succeeds ... in estimating the average errors of the results obtained' (Konstantinowsky, 1915b, p. 372). For the particular problem in which Konstantinowsky was interested—i.e., the determination of the electric charge carried by microscopic particles—the new evaluation formulae (namely Eq. (61), together with the error estimates Eqs. (62)) confirmed Ehrenhaft's and his own previous result: 'Thus follow essentially the same *noticeable deviations to the lower side of the value* [of the electric charge] *of the electron down* [*to values of*] *the order of magnitude of* $10^{-11}$ *electrostatic units*' (Konstantinowsky, 1915b, p. 372).

Even more than Konstantinowsky's acceptance of his results, Schrödinger was satisfied by another fact. A '*Notiz über die Berechnung der Brownschen Molekularbewegung bei der Ehrenhaft-Millikanschen Versuchsanordnung*' ('Note on the Calculation of Brownian Motion in the Ehrenhaft-Millikan Experiment') appeared in the 15 September issue of *Physikalische Zeitschrift*; the author of this note, Marian von Smoluchowski of Cracow, attempted 'to clarify a certain point of the ... mathematical theory, developed especially in greater detail by Fletcher' (Smoluchowski, 1915a, p. 318). As *the* point to clarify he quoted the formula, Eq. (59), by which Fletcher had tried to describe the observed probability $p(t)\,dt$. Smoluchowski then calculated what in his opinion was the correct probability expression and obtained Eq. (60). Finally, he discussed several practical consequences of the correct formula. Both the result and its derivation agreed very strikingly with Schrödinger's result; thus Smoluchowski added a *Nachtrag* (addendum) to his note, published about a month later, in which he wrote: 'Later I discovered in issue No. 16 of this journal the article by E. Schrödinger ([on] p. 289) ..., in which the author derived results agreeing essentially with my own. Had I known of Schrödinger's calculations obtained a few weeks earlier, then my note would not have been written; nevertheless, the differences in the point of departure and in the method of presentation might provide some interesting aspects. Perhaps one could take such strange coincidences as a sign of the necessity of scientific development' (Smoluchowski, 1915c, p. 375). Schrödinger certainly felt a little differently; he must have been very pleased by the observation that

[428] For this purpose Konstantinowsky referred to a paper of Maurice de Broglie in *Physikalische Zeitschrift* (M. de Broglie, 1910).

Smoluchowski, the great authority on fluctuations, acknowledged and respected his calculations.

Nearly three years later, in March 1918, artillery officer Schrödinger submitted a second paper on the evaluation of statistical data—like the first one—to *Physikalische Zeitschrift*: a '*Notiz über die Ordnung von Zufallsreihen*' ('Note on the Ordering of Chance Sequences,' Schrödinger 1918d). In it he discussed the correct use of a statistical relation that Smoluchowski had obtained in the theoretical description of emulsions. This relation could be written as

$$\overline{n^* - n'} = (n^* - \nu)\cdot P, \tag{63}$$

where $\nu$ denotes the average particle number in the emulsion; $n^*$ denotes a given number of particles existing in an optically-bound subvolume and $n'$ the number of particles observed subsequently after a given interval of observation time $\tau$; and $P$ represents the probability ($\leq 1$) that an individual particle, which was in the subvolume at the beginning of the time interval, has left it at the end. (The average value, denoted by the bar on the left-hand side of Eq. (63), has to be taken for a large number of experiments with fixed initial $n^*$ and all resulting $n'$—see Smoluchowski, 1914b; 1916b, especially Eq. (13) on p. 562.)

Miss Agnes Podjed had applied Eq. (63), in a paper published in the 1 February 1918 issue of *Physikalische Zeitschrift*, for the purpose of interpreting statistical data on the relative number of female births in the Prussian Rhineland, taken over a period of 396 consecutive months during the years 1880-1912 (Podjed, 1918). She noticed that these data satisfied the equation rather well if $P$ assumed the value 1, and concluded further that the situation under investigation constituted 'a good illustration for Boltzmann's concept of irreversibility' (Podjed, 1918, p. 42), i.e., it could represent the so-called $H$-curve.

However, Schrödinger did not agree with such conclusions. He claimed that

> notwithstanding the doubtless existing *analogies* between two number sequences [i.e., the sequence following from the observation of the diffusion effect—as observed by Svedberg and others—and the birth rate sequence considered by Miss Podjed], a very radical *difference* does not seem to have been emphasized quite as sharply as was necessary with respect to its importance for the full understanding of Svedberg's experiments achieved by Smoluchowski. (Schrödinger, 1918d, p. 218)

Smoluchowski's relation, Schrödinger emphasized, made sense only for $P < 1$; if applied to a sequence of data with $P = 1$—as attempted by Miss Podjed—which corresponded to a sequence of *completely disordered* numbers, then it could hardly be interpreted as a model for irreversibility.[429] He argued:

[429] With Miss Podjed, Schrödinger assumed that the sequence of considered birth rates required that $P = 1$. The fact that a better fit of the data had been obtained by Miss Podjed by putting $P = 0.9$ did not remove the objection to her conclusion. Rather, Schrödinger claimed that the evaluation of $P$ rested on a slightly erroneous use of statistical data—i.e., 'that the manner of forming averages of all $n^*$-sets also associates with the less numerous and therefore less reliable ones the same weight, thus a larger error arises than would correspond to the amount of data used' (Schrödinger, 1918c, p. 219). Improving on the evaluation method for $P$ he obtained the value $P = 0.986$ which was rather close to unity.

Such a sequence is 'insensible' against stirring. In it, after an extremely improbable, say abnormally large, number nearly always follows a more probable number, closer to the average value $\nu$; this is caused *only* by the fact that the numbers near the average value occur quite frequently; they occur *more* frequently, the closer they lie to the average value. An *influence* of $n^*$ on the following number $n'$ *does not exist* [in the case of $P = 1$]. (Schrödinger, 1918d, p. 219)[430]

## The Ultimate Proof of Schweidler's Fluctuations

While Schrödinger so far had discussed only single equations and their applications to statistical data, he supplied, in two papers forwarded to the Vienna Academy of Sciences, a complete theory of one of the most fundamental experiments in early twentieth-century physics. We have mentioned earlier—in Section I.2—that Egon von Schweidler had suggested, in 1905 at the Liège Congress of Radiology, the essentially statistical nature of the radioactive-decay process (Schweidler, 1905), and that soon afterwards Fritz Kohlrausch and Hans Geiger had proved this suggestion to be correct by demonstrating experimentally the existence of fluctuations (Kohlrausch, 1906a; Geiger, 1908). We have also reported that later experiments of von Schweidler had revealed similar fluctuations occurring in $\gamma$-rays (Schweidler, 1910a, b), and that Norman Campbell had some criticism against the experimental methods in general (Campbell, 1909a, b).[431] Kohlrausch and von Schweidler had then avoided the defects of the earlier charge-discharge methods (criticized by Campbell) by applying the unifilar electrometer of Elster and Geitel; they examined quantitatively the fluctuations of the $\alpha$-particles emerging from very weak uranium and polonium sources (Kohlrausch and von Schweidler, 1912). The physicists in Vienna had also analyzed different fluctuation phenomena, such as the range fluctuations in air of $\alpha$-rays emitted from a given source.[432] During the war Victor Hess and his collaborator Robert Lawson continued the studies on the fluctuations of $\gamma$-rays by looking at their ionization effects (Hess and Lawson, 1916). Still, the entire question of the existence and non-existence of fluctuations had not been completely decided.

[430] The same situation—i.e., a complete disordering (with $P = 1$)—occurred, in Schrödinger's opinion, when one studied the number of $\alpha$-particles emitted during equal time intervals by a radioactive probe of constant activity in a given space angle. On the other hand, he suggested that in some population statistics partial ordering might exist; as a consequence they would indeed provide an illustration of Boltzmann's $H$-theorem. In still other data, such as that connected with disease statistics, a situation might arise corresponding to the case of non-ideal gases (i.e., including Van der Waals' forces) in statistical mechanics.

[431] Campbell had also proposed using a different experimental method for the investigation of possible $\gamma$-ray fluctuations, namely that of converting $\gamma$-rays into electrons via the photoelectric effect (Campbell, 1909b).

[432] Ludwig Flamm, e.g., had concluded in a paper, entitled '*Theoretische Untersuchungen über Ursache und Größe der Reichweiteschwankungen bei einzelnen α-Strahlen eines homogenen Bündels*' ('Theoretical Investigations about the Cause and Magnitude of the Range Fluctuations for Single $\alpha$-Rays of a Homogeneous Bundle'), that the observed fluctuations might be essentially explained by assuming reasonable fluctuations of the density of electrons and nuclei in the air through which the $\alpha$-particles are passing (Flamm, 1914).

Schrödinger, through his own earlier work, his acquaintance with the Viennese pioneers in radioactive research (Stefan Meyer and Egon von Schweidler), and the close ties between his own *II. Physikalisches Institut* with the neighbouring *Institut für Radiumforschung*, was well aware of these results, as well as of the principal importance of the experiments and their weak points—emphasized so strongly by Campbell.[433] When he decided to deal with the physical, mathematical and apparatus problems in these experiments, he provided the following motivation:

> With the subsequent considerations we hunt a two-fold goal: on the one hand, to present a nice, sufficiently complicated example of the general statistical theory, which can be realized comparatively easily, even say for the purpose of demonstrating it before a larger audience; and on the other hand, to start developing a rational, secure theory for this experimental set-up, which will be appropriate in deciding these fundamentally important problems—a theory, the lack of which has been disturbing up to now. (Schrödinger, 1918c, p. 242)

The appropriate set-up discussed by Schrödinger in his paper '*Über ein in der experimentellen Radiumforschung auftretendes Problem der statistischen Dynamik*' ('On a Problem of Statistical Dynamics Occurring in Experimental Radium Research')—presented to the Vienna Academy of Sciences at the meeting of 14 March 1918 (Schrödinger, 1918c)—was, of course, the one devised by his friend, i.e., 'the *electrometric compensation method for measuring the fluctuations of radioactive decay* (*Schweidler's fluctuations*), invented [in 1906] by K. W. F. Kohlrausch and used ever since by numerous researchers with various—mostly inessential—modifications' (Schrödinger, 1918c, p. 241).

The main features of the method, which Kohlrausch would summarize in a later review article, were as follows: 'The principle of the measurement is that one compensates the average value of the ionization current, created by a strong $\alpha$-radiating substance [A], either with the help of a high resistivity shunt (Bronson resistance), or via the action of a second equally strong probe [B], which leads, in an identically constructed ionization chamber (due to an electric field applied [to the chamber] in the opposite direction), to a saturation current having on average the same magnitude but opposite sign; both currents are put on a measuring device (a highly sensitive electrometer). If both probes A and B are exactly equal, then, provided there were no statistical fluctuations in the probes, the de-earthed electrometer would indicate no deflection. If, sometimes in A and sometimes in B, more atoms decay per second, then the balance is disturbed; and then fluctuations arise whose magnitude and properties can be observed and compared with the effect computed from Schweidler's hypothesis' (Kohlrausch, 1926, pp. 201-202).

Now, due to Schweidler's hypothesis for a radioactive probe, emitting $Z$ $\alpha$-particles per unit time, the deflections of the electrometer indicator exhibited

[433] He would refer to the literature on fluctuation experiments in some detail in his paper (e.g., footnote 8 on p. 241, and footnote 1 on p. 262 of the paper: Schrödinger, 1918c).

the following fluctuation in time $t$

$$\overline{s^2} = a\sqrt{2Zt}, \tag{64}$$

where $a$ denotes the deflection created by one (uncompensated) $\alpha$-particle.[434] Campbell argued, however, that such a simple relation would never be found in reality, for two reasons: first, the electrometer indicator possesses a mechanical inertia, hence it does not respond instantly to the fluctuation; and, second, the insulation of the apparatus is not perfect, resulting in the fact that a surplus $\alpha$-particles per unit time, the deflections of the electrometer indicator exhibited or second ionization chamber, i.e., the one to which a positive voltage is applied and the other to which a negative voltage is applied (Campbell, 1909a). In attempting a 'rational, secure theory' of Kohlrausch's method, Schrödinger therefore had to take into account the above-mentioned situation. Simultaneously, he wished to make use of 'the general statistical theory,' i.e., the theory of fluctuations, with which he had become thoroughly acquainted during the previous years.

In the theoretical description of fluctuation phenomena a generalized diffusion equation played an important role. Albert Einstein had presented the simplest version in his second paper on Brownian motion (Einstein, 1906b, §3, especially the fourth equation on p. 378). The Dutch physicist Adriaan Daniel Fokker—when he was with Einstein in Zurich—had generalized Einstein's fluctuation equation and applied it to compute the average energy of rotating electric dipoles in a radiation field (Fokker, 1914, p. 812).[435] More than two years later Max Planck became interested in what he called the 'Einstein-Fokker theorem' ('*Einstein-Fokkerscher Satz*,' Planck, 1917b, p. 325), in connection with his theory of the rotational spectra of molecules (Planck, 1917a, c); he had especially extended Fokker's formulation to apply to systems having several degrees of freedom, and he provided a proof of the equation based on elementary principles of statistical mechanics (Planck, 1917b).[436] Planck's equation determines the time development of the probability $W(q_1, \ldots, q_m, t)$ of a system described by the $m$ parameters (variables), $q_1, \ldots, q_m$, which might undergo changes in the time interval $\tau$ due to the action of external forces (resulting in changes $q_i \rightarrow q_i + r'_i$, $i = 1, \ldots, m$) and of the irregular molecular motion ($q_i \rightarrow q_i + r_i$, with averages $\overline{r_i}$ and $\overline{r_i^2}$). It

[434] Equation (64) describes the result of von Schweidler's hypothesis, as reported earlier (in Section I.2).

[435] A. D. Fokker was born on 18 August 1887 in Buitenzorg, Java. He studied with Hendrik Lorentz in Leyden, obtaining his doctoral degree in 1913. He then went to Zurich to collaborate with Einstein on General Relativity Theory. In 1928 he succeeded his teacher Lorentz as Curator of the *Fysisch Kabinet* of *Teylers-Stichting*; the same year he became adjunct professor at the University of Leyden (from which position he retired in 1955).

[436] Fokker had announced already in December 1913, 'a detailed presentation of the derivation of this equation ... soon to appear in the "*Archives Néerlandaises*"' (Fokker, 1914, p. 813); however, the presentation did not appear in print (see Planck, 1917b, p. 324).

can be written (see Planck, 1917b, Eqs. (16) and (34)) as[437]

$$\tau \frac{\partial W}{\partial t} = -\sum_{i=1}^{m} \frac{\partial}{\partial q_i}[(r_i' + \overline{r_i}) \cdot W] + \tfrac{1}{2} \sum_{i=1}^{m} \sum_{k=1}^{m} \frac{\partial^2}{\partial q_i \partial q_k}[W \cdot \overline{r_i r_k}]. \tag{65}$$

In 1918 Schrödinger further generalized what he called the 'Fokker–Planck theorem' ('*Satz von Fokker–Planck*,' Schrödinger, 1918c, p. 242) to include 'internal' actions in the system, which change the variables $q_i$ by the amounts $r_i''$ during the time interval $\tau$; the new generalization did not alter Eq. (65), except that one had to replace (in the first term on the right-hand side) $[(r_i' + \overline{r_i}) \cdot W]$ by $[(r_i'' + r_i' + \overline{r_i}) \cdot W]$. He further sketched the derivation of his equation, which 'allowed for a somewhat greater freedom in its application' (Schrödinger, 1918c, p. 240). This freedom he needed, of course, to deal with the experimental apparatus of Kohlrausch.

The crucial part in the apparatus was a highly sensitive quadrant electrometer used in a quadrant connection, i.e., with a high potential applied to the needle, while one quadrant pair was connected directly to earth and the other pair was connected to earth via a high resistance $\omega'$.

Schrödinger described this system by the three variables $q_1 = s$, $q_2 = v$ and $q_3 = \phi$, where $s$ and $v$ denote the deflection and velocity of the indicator, respectively, and $\phi$ the potential of the quadrant. Then he first discussed the application of the extended Eq. (65) in the special case of no external changes (i.e., $r_1' = r_2' = r_3' = 0$), and with only irregular changes of the potential $\phi$ (i.e., $r_3'' = 0$). The latter should be created by the fluctuations of two $\alpha$-radiating polonium sources of equal average strength, emitting $(n\tau + \nu_+)$ and $(n\tau + \nu_-)$ $\alpha$-particles in the time interval $\tau$, respectively, which in turn create saturation currents (of saturation degree $\eta(|V|)$) in two ionization chambers—operated on high voltage of opposite sign, $+V_0$ and $-V_0$—that are connected with the quadrant electrometer. The irregular change of $\phi$ is obtained to a very good approximation as

$$r_3 = \frac{E}{C}[-2n\tau\phi\eta' + (\nu_+ - \nu_-)], \tag{66}$$

where $E$ denotes the charge transferred to the quadrant, $C$ its capacity, and $\eta'$ the differential quotient $(d\eta(V)/d|V|)_{|V|=V_0}$, yielding the averages

$$\overline{r_3} = -\frac{2nE\eta'}{C}\phi\tau, \tag{67a}$$

[437] Special cases of Eq. (65) had already been used earlier, in particular, by Schrödinger and Smoluchowski in their papers describing the influence of Brownian motion in the Ehrenhaft–Millikan experiment (Schrödinger, 1915b; Smoluchowski, 1915a). Later Reinhold Fürth (born 1893) in Prague applied the methods of Smoluchowski and Schrödinger in his detailed investigations of Brownian motion (Fürth, 1917).

and

$$\overline{r_3^2} = \frac{2nE^2}{C^2}\tau. \tag{67b}$$

The Fokker–Planck equation for the above problem could be written as

$$\frac{\partial W}{\partial t} = -v\frac{\partial W}{\partial s} + \frac{\partial}{\partial v}[(ks + \mu v - K\phi)\cdot W]$$

$$+\frac{\partial}{\partial \phi}\left[\frac{1}{C}\left(\frac{1}{\omega'} + 2nE\eta'\right)\phi W\right] + \frac{\partial^2}{\partial \phi^2}\left[\frac{nE^2}{C^2}W\right], \tag{68}$$

with $W = W(s, v, \phi, t)$ describing the probability of meeting the system at time $t$ in the phase space region $(s, ds;\ v, dv;\ \phi, d\phi)$. The quantities $k$, $\mu$ and $K$ denote the different force constants of the electrometer.[438]

Schrödinger stated: 'The just-derived Fokker differential equation constitutes the exact foundation for solving *all* probability problems connected with Kohlrausch's compensation method,' adding that 'any of these is equivalent to a definite *boundary value problem* of Eq. [(68)]' (Schrödinger, 1918c, p. 248). As examples of the application and solution of this equation he turned to two special cases. In the first, the stationary case (with $\overline{r_3} = 0$), he was able to confirm a previous result of Campbell—'the only fact *hitherto* known about our experimental set-up,' as he admitted (Schrödinger, 1918c, p. 248)—namely, the equation for the average square $\overline{s^2}$, i.e.,

$$\overline{s^2} = \frac{nE^2\omega}{C}\frac{K^2}{k^2}\frac{k(\mu + \lambda_3)}{\mu(\lambda_3^2 + \mu\lambda_3 + k)}, \tag{69}$$

with $1/\omega = 1/\omega' + 2nE\eta'$ and $\lambda_3 = 1/C\omega$. (See Campbell, 1909b, p. 316, Eq. (11).) The experimental investigations of the second example, namely the stationary case in which the electrometer operated aperiodically, played the major role. In that case, Schrödinger found for $\overline{s^2}$ the result

$$\sqrt{\overline{(s - s')^2}} = \frac{K}{k}\sqrt{\frac{D}{\lambda_3}[1 - \exp(-2\lambda_3 t)]}, \tag{70}$$

where $s' = s_0 \exp(-\lambda_3 t)$ describes the variable centre of the indicator position $(D = nE^2/C^2)$.

Schrödinger postponed 'the most complete exhaustion of the just-obtained point of view for the purpose of experimental physics, as well as a full discussion of other detailed questions which were important for exact, quantitative

[438] The time interval $\tau$ has dropped out of both sides of Eq. (68).

observation [of Schweidler's fluctuations]' (Schrödinger, 1918c, p. 242) to a later paper. This paper developed into a long memoir, bearing the title '*Wahrscheinlichkeitstheoretische Studien, betreffend Schweidlersche Schwankungen, besonders die Theorie der Meßanordnung*' ('Probability Theoretical Investigations Concerning Schweidler's Fluctuations, Especially the Theory of the Experimental Method'), and was finally presented to the Vienna Academy of Sciences at the meeting of 16 January 1919 (Schrödinger, 1919a). Schrödinger organized his results in two parts, discussing in part I 'three detailed questions dealing mainly with the removal of intellectual difficulties met by the author when studying literature on the topic' (Schrödinger, 1919a, p. 177), and continued in part II with the general theory of Kohlrausch's compensation method.

The first detailed question was an 'intellectual' problem, namely, whether the decay irregularities of the equal radioactive probes can really be observed in an experiment. Since it seemed to be impossible to have two radioactive sources with exactly the same strength, Schrödinger proposed to avoid the difficulty by taking two approximately equal sources and observing carefully the deflection of the electrometer indicator corresponding to the regular decay (without fluctuations) due to the exponential law; having obtained the deflection he called it the 'neutral' point $\sigma$. Evidently $\sigma$ shifted with time, hence he concluded that Schweidler's fluctuations could not be measured 'unless the average decay time of the probes was *not* very large in comparison with the time needed by the electric charge [$E$], placed on the insulated system [the electrometer], to decrease to its $e$th part' (Schrödinger, 1919a, p. 177).

In the second detailed question he treated the problem of whether, in forming the averages of $n$ square fluctuations, one had to divide the sum of the latter by $n$ or $n-1$. Campbell had argued earlier in favour of $n$ (Campbell, 1909a, p. 132), but Schrödinger now showed that it must be $n-1$ (Schrödinger, 1919a, Section 4). He admitted, by pointing to the small error in the case of large $n$:

> This question does not play any practical role .... Yet, if the most probable value is also still associated with an error and is thus only the most probable value: and yet, the title [most probable value] should be given only to that value which is *strictly* defined. It is against the conscience of a natural philosopher (scientist) to admit, due to lack of consideration, a *systematic* error, however small it may be. I am especially called upon to clarify this matter, because I have in another, though strictly analogous, case supported the (as will be shown) incorrect division by $n$ based on a deduction, which was intended to be particularly rigorous but in reality was cheating. (Schrödinger, 1919a, p. 189)[439]

This short, but nevertheless unreserved, admission and the appeal for a clean intellectual conscience—down to the smallest detail—characterized the entire work of the scientist Schrödinger.

In the third detailed question of part I the author investigated the question of whether the observed integral fluctuation effect could be influenced by other 'individual' fluctuation effects; they might have their origin in the fact that only

[439] Schrödinger referred to his earlier treatment of Brownian motion (Schrödinger, 1915b, p. 292 ff).

the $\alpha$-rays emitted from the radioactive source in a given space angle can be counted, or that $\alpha$-rays stem from different depths of the radioactive probe. He found that both effects do indeed add to the total observed effect.

In part II of his article Schrödinger presented a systematic survey of all the methods used to observe radioactive fluctuations. He distinguished '*three essentially different statistical methods*, depending on whether one observes: (i) the *stationary* probability of the indicator deflections; (ii) the fluctuation of the *indicator path* performed during a given *time*; or (iii) the fluctuation of the *time interval* necessary for the indicator to pass through a given distance' (Schrödinger, 1919a, p. 178). So far, he continued, only a proper theoretical description of the first method [by Campbell] existed, but 'only methods (ii) and (iii) provide immediately absolute results' (Schrödinger, 1919a, p. 178). Schrödinger, therefore, developed in detail the theories of those methods. As a consequence it followed, e.g., that Hans Geiger's method of 1908 could be—in contrast to Campbell's criticism (Campbell, 1909a)—justified, and that Fritz Kohlrausch's first measurements needed only small corrections. To obtain the very complicated theory of method (iii) in section 9, he made use of his own results achieved in his paper on Brownian motion (Schrödinger, 1915b).[440]

One might wonder whether Schrödinger's two papers on the measurement of radioactive fluctuations represented more than just a brilliant theoretical exercise—performed with great skill and tenacity—or just an *a posteriori* justification of the already obtained experimental results. None of this is true, however. Schrödinger's work must be seen rather as an integral and vital part of active research in Vienna. Thus he mentioned at the end of his first publication:

> At this institute [i.e., the *II. Physikalisches Institut*] measurements of Schweidler's fluctuations are being carried out with the help of an Elster-Geitel unifilar electrometer; the description given above will be used to analyze their statistics and will—I believe—be experimentally confirmed. This confirmation has already been partly achieved, and I should emphasize that the full information in the hitherto obtained experimental results, kindly given to me by my friends and colleagues, has helped me most effectively in working out the theory. (Schrödinger, 1918c, p. 262)

The experiments to which he referred were carried out under the supervision of his friend Fritz Kohlrausch. Miss Elisabeth Bormann, a student at Exner's institute, analyzed the experimental details of Kohlrausch's compensation method for her thesis.[441] Miss Bormann used what Schrödinger called method (ii). She

[440] In Sections 10 and 11 of his paper Schrödinger derived the most general theory of method (ii), even taking into account the inertia of the indicator. Before that the only person to consider such a correction had been Edgar Meyer's Tübingen student A. Ernst, who had used method (i) (see Ernst, 1916).

[441] See Schrödinger, 1919a, p. 206. Miss Bormann's paper, entitled '*Zur experimentellen Methodik der Zerfallserscheinungen*' ('On the Experimental Methodology of Decay Phenomena'), was communicated on 12 December 1918 to the Vienna Academy (Bormann, 1918). She referred, in the introduction, in detail to the theoretical description of fluctuation experiments, referring especially to the 'much more general point of view' of Schrödinger's theory, and stated: 'It is the task of the present work to test this theory experimentally' (Bormann, 1918, p. 2353).

investigated especially various points which had been criticized earlier; for example, she showed that the insulation leakages indeed gave rise to a different behaviour of the fluctuations than was expressed by Eq. (64).[442] Kohlrausch emphasized later: 'Bormann's measurements culminate in demonstrating that the number $Z$ of the $\alpha$-particles emitted per second was, on one hand, calculated from the observed fluctuations to be 143,000; while it was determined, on the other hand, by direct observation to be 141,000' (Kohlrausch, 1926, p. 210). Kohlrausch found this perfect agreement between measurement and the theory of his friend most convincing, hence he concluded: 'The work of Miss Bormann has, based on this theory [of Schrödinger], demonstrated agreement between experiment and [theoretical] expectation and has thus provided—as it appears to the reviewer—the first *unobjectionable* proof for the correctness of Schweidler's hypothesis on the statistical nature of the radioactive decay of atoms' (Kohlrausch, 1926, p. 211).

## Schrödinger and Quantum Statistical Theory in Vienna

Quantum theory in general had *not* been, as we have emphasized earlier in Section I.3, a major field of interest for the Viennese physicists, despite several original studies devoted to some aspects of it, e.g., those of Arthur Haas or Karl Herzfeld. Even the eagerness of the leading theoretician of Vienna after 1906, Friedrich Hasenöhrl, must be called comparatively lukewarm, so far as his active involvement was concerned. The papers of his student Hans Thirring, on the other hand, only dealt with the well-established application of the quantum of action in explaining the specific heats of solids. A characteristic sign of the relatively reserved reception of Planck's theory, and its many generalizations by Einstein and others, may be concluded from the fact that Thirring, in his first paper on the specific heats, had inserted into the Born–von Kármán formula (for comparison with experiment) not the value $\beta = h/k$ derived from the blackbody law, but rather the value obtained from molecular measurements. 'It appeared adequate', he argued, 'to take Perrin's value here, where we are dealing with a purely molecular theory' (Thirring, 1913, p. 871).[443]

Like Thirring, Schrödinger also became involved in quantum theory via his interest in the problem of the specific heat of solids, a problem that indeed had the right of domicile in Vienna. After all, the Viennese physicist Boltzmann had formulated, about forty years earlier, the equipartition theorem as a fundamental

[442] The average square fluctuation $\overline{s^2}$, rather than increasing linearly with time, reached a saturation value in agreement with Schrödinger's theory. (See Schrödinger, 1919a, pp. 209–210.)

[443] Thirring noticed that inserting Planck's value would have resulted in less agreement between theory and experiment. We should add, though, that in his second paper Thirring had shown a better appreciation for the value of quantum theory, and had written in the same context: 'For the number of molecules in a gram-atom the value recently given by Millikan, i.e., $6.06 \times 10^{23}$, was used, although the value of Perrin, i.e., $6.82 \times 10^{23}$, would yield better results [in agreement with experiment]. Millikan's value seems to me, however, to be the more accurate one; *in addition it also coincides better with the value following from radiation theory*' (Thirring, 1914b, p. 182, our italics).

consequence of the kinetic theory of matter, giving each degree of freedom of an atom or of a molecule the energy $\frac{1}{2}kT$ at temperature $T$ ($k$ denotes Boltzmann's constant); hence, e.g., a gram-atom of a monatomic solid should possess the specific heat of $3R$ ($R$ denotes the universal gas constant), i.e., exactly the value prescribed by the rule of Pierre Louis Dulong and Alexis Therèse Petit (1819). Later, while in Munich, Boltzmann also proposed various clever and artistic arguments for those cases, in which the equipartition theorem seemed to fail, especially for gases having complex molecules (Boltzmann, 1892c). Einstein's application of Planck's quantum hypothesis—in order to explain the observed deviations of the specific heats of certain solids from the Dulong-Petit rule, notably at low temperatures—opened a general method of getting around the difficulties of the equipartition theorem (Einstein, 1906g). The extensions of Einstein's approach, suggested by Walther Nernst and Frederick Alexander Lindemann (1911a, b), by Peter Debye (1912b, c) and, finally, the systematic elaboration of the lattice theory by Max Born and Theodore von Kármán (1912; 1913a, b) constituted the essential steps forward towards a consistent kinetic theory of the specific heats of solids. Schrödinger described these developments in two review articles: the first, the more elementary one, entitled '*Die Ergebnisse der neueren Forschung über Atom- und Molekularwärmen*' ('The Results of the More Recent Research on Atomic and Molecular Heats'), was addressed to a wider public and appeared in the *Naturwissenschaften* issues of 24 and 31 August 1917 (Schrödinger, 1917a); the second, a rather detailed and technical review was received by *Physikalische Zeitschrift* on 5 July 1918 and published more than a year later in five successive issues (of 15 September, 1 and 15 October, 1 and 15 November 1919) of that journal (Schrödinger, 1919b).

Schrödinger had announced, as mentioned earlier, for the winter semester of 1914/1915 at the University of Vienna, a lecture course on selected topics of statistical mechanics and quantum theory, in which the problem of specific heats might have assumed an important, if not central role. One should conclude, therefore, that the essay in *Naturwissenschaften* represented a short, popularized exposition of what the young *Privatdozent* had intended to say in his lectures. In any case, he had had the material ready for some time when, in the spring or summer of 1917, he submitted the paper. The paper only included references to results published up to the summer of 1914, i.e., essentially up to the theories of Max Planck, Walther Nernst, Paul Ehrenfest, Albert Einstein and Otto Stern, which aimed at describing Arnold Eucken's measurements of the specific heat of hydrogen at low temperatures.[444]

The second, much more detailed and technical review in *Physikalische Zeitschrift* may be regarded as a genuine handbook article, with ample reference to the most recent publications up to the summer of 1918.[445] The virtue of

[444] We have discussed these items in detail in Volume 1, Section I.7.

[445] Evidently Schrödinger began, after submitting his *Naturwissenschaften* essay in the summer of 1917, a systematic study of the most recent literature, as well as a search for older papers, books and dissertations which were available in the Vienna library, dealing with both experimental and theoretical aspects of the specific heat problem.

Schrödinger's writing did not lie in the originality of the material presented, as he did not add any ideas or results stemming from his own research, but rather in the clear exposition and the thorough and critical analysis of the fundamental ideas of Boltzmann and of classical physics, of their specific failure and of their decisive improvement by the quantum-theoretical assumptions.[446] Schrödinger not only described (in Section 8) the essentials of the specific heats apparatus and measurements (notably those at low temperatures) carried out in Nernst's Berlin institute or in Heike Kamerlingh Onnes' Leyden laboratory, but also sought and established throughout the paper a subtle comparison between the available empirical data and the theoretical predictions. He exhibited enormous skill in selecting such representations (by figures and tables), in which the deviations of experiment from theory became emphasized rather than suppressed. (See, e.g., his presentation of the failure of the Nernst–Lindemann formula for the specific heat of copper in Table III, p. 454; or his critical analysis of the validity of Debye's $T^3$-law for solids at very low temperatures in Section 18.)

With respect to the clarity of the exposition of theoretical results and the careful, critical discussion applied to them, Schrödinger's review of the specific heats problem of 1918 showed some similarity to a much more famous review, i.e., that of Wolfgang Pauli on relativity theory. True, Schrödinger lacked the open, unperturbable sureness of his younger fellow countryman, replacing it with the considerable judgment of an experienced observer of experimental and theoretical progress. Yet, when one takes into account the specific differences in the character of the two Viennese physicists and their different backgrounds and upbringing, one must admit that both reviews share more than the same standard of quality.

One of the main, or perhaps the main, consequences of his concern with the specific heat problem was that it forced Schrödinger to dive into various fundamental aspects of quantum theory. Thus he discussed the considerations leading to the blackboy law (in Section 4), the subtle details of the differences between Planck's first and second quantum hypothesis (in Section 5), and their connection with Nernst's heat theorem (in Section 6). He also treated the possibility of avoiding or bypassing the quantum hypothesis altogether with the help of specific assumptions that remained within classical theory, such as the agglomeration hypothesis (in Section 3) and its final refutation by experiments (in Section 23). The latter result confirmed Schrödinger's acceptance of the quantum theory as a necessary description of natural phenomena, which had already occurred earlier in connection with his recognition of the lattice theory of Born and von Kármán as the proper kinetic description of solids. Schrödinger included in his review a thorough discussion of the defects of Debye's continuum approach to the frequency spectrum of solids, with detailed references to the data on elasticity and residual rays (Sections 13–18), before he went on to demonstrate how the

[446] Even the probably most unusual part of the review, the discussion of the specific heats at very high temperatures, especially the ferromagnetic solids in Section 26, summarized only the experimental and theoretical results known from other authors.

space lattice theory removed existing discrepancies in the experimental results (Sections 19–22). Thus the unification of Planck's quantum theory with the kinetic picture of solids served the higher cause of Ludwig Boltzmann's ideas.[447]

One should add that in Vienna quantum theory had meanwhile won a new active supporter in Ludwig Flamm, a fellow student of Schrödinger's and a member of Hasenöhrl's *Seminar*, who began to publish on topics related to quantum theory in 1917.[448] In September of that year he analyzed the values of the characteristic properties of the electron following from recent experimental data (Flamm, 1917). Besides Robert Andrews Millikan's newest precision determination of the electron's charge of 1917, he considered Friedrich Paschen's measurements of Rydberg's number from the spectrum of ionized helium (Paschen, 1916), which involved the quantum theory of atomic structure as developed by Niels Bohr (1913b) and extended by Arnold Sommerfeld (1915b, c). Several months later, on 8 January 1918, he also submitted a review paper, '*Zum gegenwärtigen Stand der Quantentheorie*' ('On the Present Status of Quantum Theory'), to *Physikalische Zeitschrift*, where it appeared about eight weeks later (Flamm, 1918a). He discussed there several, mostly recent, studies of quantum-theoretical topics, having in mind the goal 'of applying the more precise conceptions, obtained from the establishment of the theory of series spectra, back to

[447] Schrödinger indeed considered quantum theory as *the* most suitable method for removing the difficulties between Boltzmann's description of specific heats and experimental data. The point was that quantum theory did soften the hard discontinuity assumptions, which were involved in the sudden disappearance of certain degrees of freedom at certain temperatures. This situation applied especially in the case of molecular gases. After discussing the respective approaches, i.e., the classical and the quantum-theoretical, Schrödinger concluded: 'In this case also quantum theory provides a softening of the previously existing intellectual severity. *According to Boltzmann one has preliminarily*: the atoms must be considered as *material* points, possessing *only three degrees of freedom*. According to *quantum theory*, it is sufficient to consider them as having only a *sufficiently small moment of inertia*, so-to-speak, as *material points in sufficient approximation*' (Schrödinger, 1919b, p. 524).

[448] Ludwig Flamm was born in Vienna on 29 January 1885, the first of four children of the watchmaker Josef Flamm. He entered the University of Vienna in 1905; but shortly afterwards his father died and Ludwig had to support his studies by tutoring. After obtaining his doctorate in 1909, he was offered a position at the *Technische Hochschule* of Vienna by Gustav Jäger (1910). He collaborated with Heinrich Mache on experimental problems of radioactivity. In the spring of 1916 he became *Privatdozent* for physics at the University of Vienna; in the summer of the same year he changed to the *Technische Hochschule* again, advancing to an extraordinary (in 1919) and a full professorship (in 1922). He retired in 1956 and died on 4 December 1964 in Vienna.

Although Flamm suffered from the consequences of an eye infection since his *Gymnasium* years—hence he was prevented from participating in military service during World War I—he was a sporting man; he joined Hasenöhrl, together with other members of the *Seminar*, on his alpine tours. Shortly after his graduation he worked as an assistant at the Austrian *Gradmessungsbüro*, being involved in the length measurements of the distance Pola-Tieste (in 1909). In July 1920 he married Elsa, the youngest daughter of Ludwig Boltzmann, upon which Albert Einstein remarked: 'Here one sees clearly, which of the Vienna physicists of the rising generation deals most amiably with the works of Boltzmann' ('*Da sieht man jetzt deutlich, welcher von den Wiener Nachwuchsphysikern sich mit den Werken Boltzmanns am liebevollesten beschäftigt*,' quoted by Thirring, 1966, p. 4). He occupied honourable positions at the *Technische Hochschule* of Vienna (1929/1931, dean; 1950/1951, rector) and was a corresponding member (from 1928) and a full member (from 1940) of the Austrian Academy of Sciences. (See the article on his seventieth birthday and the obituary, both by Flamm's friend and colleague Hans Thirring: Thirring, 1955; 1966.)

the theory of specific heats and cavity radiation, and thereby treating the entire field of quantum theory from a unified point of view' (Flamm, 1918a, p. 117).

The more precise conceptions emerging from the theory of spectra consisted essentially of the hypothesis that any system of $n$ constituents may assume (countably many) quantized, stationary states, each of which is associated with a given probability (see Einstein, 1916d). Given this information, one could write down the partition function (or *Zustandssumme*) and derive from it the properties characterizing the system (e.g., entropy, energy, specific heat, at a given temperature, or the occupation numbers for all states in the thermodynamical equilibrium). As examples Flamm presented treatments of the harmonic oscillator, of Debye's elastic solid and of the rotation of a diatomic molecule.[449]

The same emphasis on the relationship between quantum-theoretical and statistical problems was also exhibited by Flamm's second paper on quantum theory, which was properly entitled '*Bemerkungen zu den statistischen Grundlagen der Quantentheorie*' ('Remarks on the Statistical Foundations of Quantum Theory') and received by *Physikalische Zeitschrift* on 5 February 1918 (Flamm, 1918b). In this paper Flamm simplified several steps leading to a quantum-theoretical expression for the entropy, thus completing a derivation of 'the fundamental quantum-theoretical formulae for the calculation of energy, entropy, and specific heat of a substance at constant volume, on the basis of a minimum number of special assumptions in the most general way possible' (Flamm, 1918b, p. 168).

Flamm's approach to quantum-theoretical problems exhibited features common with those of Hasenöhrl, Thirring and Schrödinger. Like them he was primarily interested in extending statistical mechanics to include aspects of quantum theory, in short, he continued and enlarged the heritage of the great Ludwig Boltzmann.

## Problems and Ideas Connected with General Relativity

Einstein's lecture of 1913 at the *Naturforscherversammlung* on gravitation theory left a great impression in Vienna. The physicists in Vienna continued to follow Einstein's further work, also when he left Zurich for Berlin in April 1914 to accept a special research position with the Prussian Academy of Sciences. In Berlin, Einstein completed the new Theory of General Relativity in November 1915. Less than four months after the long and final memoir elaborating the theory, entitled '*Die Grundlage der allgemeinen Relativitätstheorie*' ('The Foundation of General Relativity Theory,' Einstein, 1916c) had appeared in print, in September 1916 Ludwig Flamm submitted his paper '*Beiträge zur Einsteinschen*

[449] In the latter case he proved that the amount of inertia calculated for the hydrogen molecule agreed with that found in Niels Bohr's dumb-bell model. The value thus obtained disagreed with Paul Ehrenfest's (1913b) by a factor of 4, because Flamm used a different quantization rule. (See Flamm, 1918a, p. 126, Eq. (37).)

*Gravitationstheorie*' ('Contributions to Einstein's Theory of Gravitation') to *Physikalische Zeitschrift* (Flamm, 1916).[450]

Flamm dealt with two questions: first, he discussed the exact spherically symmetric solutions of the equations of gravitation—that is, the field inside an incompressible fluid and the (external) field of a mass point—given by Karl Schwarzschild (1916a, b), with the intention of rendering 'really illustrative the strange properties of the gravitational field,' and to let 'the physical prerequisites (assumptions) of General Relativity Theory become perhaps still better evident' (Flamm, 1916, p. 448); second, he provided a rigorous numerical calculation of the physical constants which determine the gravitational field of the sun. Besides illustrating the solution of Schwarzschild in a pedagogical geometrical form, Flamm derived an important physical consequence: in an arbitrary gravitational field the ratio of, say, the wavelength of the red cadmium spectral line (serving as a chronometer), over the lattice constant of rock-salt (serving as a rod for length measurements), is an absolute constant of nature (Flamm, 1916, p. 451). As a consequence of this conclusion one could describe, even in General Relativity Theory, the motion of a mass point by the variational equation

$$\delta \int_{P_1}^{P_2} ds = 0, \tag{71}$$

where $ds$ denotes the four-dimensional linear element and the integral extends from the space-time point $P_1$ to $P_2$. Flamm noted:

> The influence of the gravitational field on the motion of a mass point acts, therefore, in the same way as the changes exhibited by rods and clocks; the behaviour of the phenomena relative to each other has remained the same as before. (Flamm, 1916, p. 452)

Flamm's clear analysis of the characteristic aspects of General Relativity Theory attracted Schrödinger, who had become interested in the subject at about the same time as his colleague.[451] Back in Vienna he discussed the matter frequently with Flamm.[452] He soon discovered problems to work on for himself; in late 1917 he was able to submit two notes containing original results to *Physikalische Zeitschrift.* In the first of these, entitled '*Die Energiekomponenten des Gravitationfeldes*' ('The Energy Components of the Gravitational Field'), and received by the journal on 22 November 1917, Schrödinger analyzed—like Flamm earlier—Schwarzschild's solution of Einstein's equation for a massive sphere

[450] Actually Friedrich Kottler, another Viennese physicist and former contributor to the topic of generalizing relativity, might have been more qualified to be the first to respond to Einstein's new theory of gravitation. But at that time (1916) he was in military service.

[451] Schrödinger studied Einstein's new theory in 1916 while he was at Prosecco in military service. (See Hermann, 1975, p. 218.)

[452] See the footnote in Schrödinger's second contribution to General Relativity (Schrödinger, 1918b, footnote 1 in the left column of p. 22), in which the author referred to 'repeated oral discussions' ('*wiederholte mündliche Diskussionen*') with Ludwig Flamm.

(Schrödinger, 1918a). He calculated explicitly the sixteen so-called energy components $t_{\sigma}^{\alpha}$ of the gravitational field, observing the covariance of that field with respect to linear transformations; and he found a special co-ordinate system, in which 'the $t_{\sigma}^{\alpha}$ . . . *vanished everywhere* (outside the gravitating sphere) *identically in all components*' (Schrödinger, 1918a, p. 6). This result appeared strange to him, as it might imply one of two consequences: either the impossibility of formulating the conservation of energy and momentum, or the existence of genuine gravitational fields—i.e., those which cannot be transformed away—having vanishing energy components.

Schrödinger's note immediately received the attention of the highest authority in the field—Albert Einstein (who had arrived, independently of Schrödinger, at similar conclusions: Einstein, 1918a) and who replied, in a note, also published in *Physikalische Zeitschrift*, that he agreed with the result of the above calculation, but did not after all find it surprising (Einstein, 1918b). He argued:

> As far as Schrödinger's objection is concerned, it owes its convincing power to the analogy with electrodynamics; in the latter both the tensions and the energy density of each field must be different from zero. However, I do not see any reason for the assumption that this must be the same in the case of the gravitational field. There may certainly exist fields of gravitation possessing neither (finite) tensions nor energy density. (Einstein, 1918b, pp. 115-116)

In a later issue of *Physikalische Zeitschrift*, the Viennese physicist Hans Bauer (1891-1953) also showed that the inverse of Schrödinger's result was correct, namely 'that the quantities $t_{\sigma}^{\alpha}$, if expressed in a suitable co-ordinate system, *do not vanish* in the case *of absence of a gravitational field*' (Bauer, 1918a, p. 163). The final clarification of the situation was achieved by Einstein and Felix Klein; in two papers presented to the Berlin and Göttingen academies, respectively, they demonstrated conclusively that well-defined energy and momentum conservation laws could be formulated in General Relativity Theory (Einstein, 1918d; F. Klein, 1918). But, as Wolfgang Pauli summarized later:

> One must therefore not associate any physical intepretation with the values of the $t_{\sigma}^{\alpha}$; i.e., one does not succeed in localizing energy and momentum of the gravitational field in a generally covariant and physical manner. (Pauli, 1921b, p. 742)

Schrödinger soon followed his first relativity note by a second one, '*Über ein Lösungssystem der allgemeine kovarianten Gravitationsgleichungen*' ('On a System of Solutions of the Generally Covariant Equations of Gravitation'), which was received by *Physikalische Zeitschrift* on 30 November 1917 and published in the second issue of January 1918 (Schrödinger, 1918b). In it he treated a spatially finite, spherical universe, but without a cosmological constant, such as Einstein had recently introduced (Einstein, 1917c). Schrödinger calculated, in particular, the energy tensor $T_{\mu}^{\nu}$ of matter, obtaining a diagonal representation, with elements given by the relation

$$T_1^1 = T_2^2 = T_3^3 = \tfrac{1}{4}T_4^4 = \frac{1}{\gamma R}, \tag{72}$$

where $\gamma$ and $R$ denote Einstein's constant of gravitation and the radius of the universe, respectively. This solution could be interpreted as representing a compressible fluid at rest, having constant density and constant inner tension. Schrödinger was not bothered by the fact that the gravitational mass density $(=(T_4^4)^2 - (T_1^1)^2 - (T_2^2)^2 - (T_3^3)^2)$ became zero; this fact had to be expected, he argued, 'from a theory, in which the concept of mass is also *relative*, i.e., only determined by the *interrelations* between the bodies' (Schrödinger, 1918b, p. 22). In order to establish real masses, he claimed, the assumption of constant density must be given up, as:

> It appears to me to be entirely in the spirit of the postulate of relative mass if the interaction function, which we call inert or heavy mass, is created by or occurs only through deviations from that distribution that is uniform and constant in time. (Schrödinger, 1918b, p. 22)

Again Einstein took the time to reply personally to Schrödinger's suggestion. He pointed out in a note, published in the issue of 15 March of the *Physikalische Zeitschrift*, that he had already considered Schrödinger's solution, but dismissed it, because it implied a negative mass density—Schrödinger had to take the negative roots in Eq. (72)—whose space-time dependence could not be derived without further hypothetical assumptions (Einstein, 1918c). Hence he concluded simply:

> The path taken by Schrödinger does not seem to me to be an accessible one, since it leads too deeply in to the thicket of hypotheses. (Einstein, 1918c, p. 166)

In spite of Einstein's rejection of his particular suggestions in General Relativity Theory, Schrödinger did not retire from dealing with that theory. In doing so, he concentrated on more general features and the fundamental conceptual aspects of General Relativity, leaving the discussion of special problems to his colleague and friend Hans Thirring.[453] He took great interest in the principal mathematical method used in Einstein's theory, namely, the tensor calculus. Schrödinger thus developed a systematic approach to mechanics within the framework of General

[453] Hans Thirring first treated successfully the effect of rotating masses in General Relativity Theory (Thirring, 1918a; 1921). In a second paper, written jointly with his student Josef Lense, he calculated the influence of the eigenrotation of a central body, such as the sun, on the motion of planets and moons; they found what was later called the Thirring-Lense effect (Lense and Thirring, 1918). In a third paper Thirring finally drew attention to the formal analogy between the Maxwell-Lorentz equations of electrodynamics and Einstein's equations of gravitation in first approximation (Thirring, 1918b).

Relativity in three (unpublished) notebooks, entitled '*Tensoranalytische Mechanik* I, II, III' ('Tensor Analytical Mechanics I, II, III').[454]

In *Notebook I*, he started from the fundamental metric tensor form $\mathcal{T}$, given by

$$2\mathcal{T}\, dt^2 = g_{ik}\, dq_i\, dq_k = ds^2, \tag{73}$$

where $dt$ and $dq$ denote the time and four-dimensional co-ordinate differentials, $ds$ the line element, and $g_{ik}$ the fundamental metric tensor. He then derived the result that the equations of motion of a mass point, having energy $E$ and acted upon by a potential $V$, followed from the variational principle

$$\delta \int_{P_1}^{P_2} \sqrt{2(E-V)}\, ds = 0, \tag{74}$$

thus showing that the known mechanical principles, such as the principle of the straightest path or the principle of minimum constraint, were also satisfied in General Relativity.[455] In *Notebook II*, Schrödinger discussed the items of curvature (according to Bernhard Riemann), Augustin Louis Cauchy's differential calculus, space curvature, the problem of volume in curved spaces, the concept of time, and finally the motion of the planets in Einstein's theory of gravitation. *Notebook III* contained several, partly sketchy calculations, involving the use of and the transition to special co-ordinate systems, e.g., cyclic co-ordinates. Schrödinger further displayed the analogies between mechanical and optical description—especially the relation between Christiaan Huygens' principle and the partial differential equation of William Rowan Hamilton—and he treated Josiah Willard Gibbs' thermodynamics of heterogeneous systems as an example of the application of tensor calculus.

Schrödinger certainly wrote the notes mainly for the purpose of acquainting himself with the mathematical methods used in General Relativity Theory, and then applying it to a wide range of topics in theoretical physics. In turn, he then exploited the results of such applications to draw his own conclusions on the nature of the new theoretical description and on its physical interpretation. He occasionally added, after discussing a specific derivation, a more general remark; for example, he concluded that the quantum conditions depend on the integrated form of the equations of motion. In these remarks he sometimes wrote down

[454] The notebooks have been filed on AHQP Microfilm No. 39, Section 3. The date, 1914, suggested for the composition of the notes, cannot be upheld. For example, we find in the first notebook a reference to a paper by Karl Schwarzschild published in 1916 (Schwarzschild, 1916c); in the second notebook, on the other hand, Hermann Weyl's book *Raum-Zeit-Materie* (Weyl, 1918c) is quoted and which appeared in the middle of 1918—the preface is dated Easter 1918. Altogether, we conclude that Schrödinger did not start writing the notes on tensor analytical mechanics before the middle of 1917 (when he became re-established in Vienna)—probably even later.

[455] We have mentioned above that Ludwig Flamm, in his first contribution to General Relativity Theory, had also made use of a variational principle, i.e., Eq. (71).

particular points of view revealing his own concepts or philosophical thoughts. As he said, after dealing with the (two-dimensional) Kepler problem in atomic physics (i.e., in the Bohr–Sommerfeld theory): 'What we call "co-ordinates of the electron," are certain "generalized co-ordinates" of the ether's structure. The ether obtains "structure" by the field (analogue: para- and ferromagnetism)' (*Notebook I*). Upon discussing the cyclic co-ordinates in General Relativity Theory, he described his 'new view of gravitation' ('*neue Auffassung der Gravitation*') with the words: 'An important consequence of my view is this: the energy contained in a massive body is the kinetic energy of the hidden cyclic "gravitational motion." It is this latter motion, which in the case of planets orbiting around the sun, or in the case of a tossed stone, absorbs or provides, respectively, the "sizable" kinetic energy that can be seen to disappear and reappear, exactly as the kinetic energy of the molecules of a gas enclosed in a cylinder which creates the driving force on the piston or receives it.' A little later he went on to say: 'I imagine that the cyclic gravitational motion is an electromagnetic one, similar to that of Maxwell's ether mechanism' (*Notebook III*).

The ideas of Heinrich Hertz played an important role in Schrödinger's presentation of tensor analytical mechanics, ideas which Hertz had developed in his *Principien der Mechanik* (Hertz, 1894), especially his analysis of the fundamental concepts.[456] Hertzian mechanics represented, as the Viennese physicist Franz Paulus stated in 1916, 'a certain conclusion of theoretical mechanics, which attempts to eliminate from theoretical mechanics the concept of force as that fundamental, "not further explainable," concept, which is associated with unavoidable uncertainties due to its metaphysical connotation' (Paulus, 1916, p. 835). General Relativity now seemed to dissolve in some way the perhaps most fundamental force in nature, gravitation; hence one could imagine the existence of a deeper relation between Einstein's recent and Hertz' old ideas. With this in mind, Schrödinger composed in 1918 a manuscript, entitled '*Hertz'sche Mechanik und Einstein'sche Gravitationstheorie*' ('Hertzian Mechanics and Einsteinian Gravitation Theory').[457]

Schrödinger opened the introduction of Section 1 of his manuscript by the programmatic statements: 'If one should state, to which fundamental act of consciousness the present epoch of electrodynamics and mechanics (including the theory of gravitation) owes its characteristics, then one must, from a *physical* point of view (the philosopher or the mathematician might perhaps have another,

[456] Since its publication Hertz' book had left a deep impression in the Vienna of Ernst Mach and Ludwig Boltzmann. Boltzmann had just at that time been involved with writing his own *Vorlesungen über Principe der Mechanik* (Boltzmann, 1897f). A decade later Boltzmann's student Paul Ehrenfest treated, for his doctoral thesis, the formulation of continua mechanics in the framework of Hertz' mechanics. (For details we refer to Martin J. Klein's Ehrenfest biography: M. J. Klein, 1970, Chapter 4, and to Arthur I. Miller's article on Boltzmann's mechanics, presented at the International Boltzmann Meeting 1981 in Vienna: Miller, 1982.)

[457] The manuscript is filed on AHQP Microfilm No. 39, Section 3, with the remark, 'Undated, but perhaps circa 1915.' We do not agree, however, with this dating, as one finds in Section 3 a reference to Hermann Weyl's *Raum-Zeit-Materie*, a book which appeared in the spring of 1918.

even mutually different, opinion) answer: this important act of consciousness lies in the identification of the three concepts: *energy*, *inertia* and *gravitation*.'

Schrödinger gave, in the first section of the manuscript, basically an account of the historical development from Hertz' book to the completion of General Relativity. Hasenöhrl had shown that localized energy in space can be associated with inertia. Now General Relativity implied, with its identification of inert and gravitational masses, a further reduction of the number of fundamental concepts in mechanics. In order to understand the deeper nature of gravitational action, Schrödinger proposed to apply the Hertzian mechanics, working without any real forces. He claimed: 'With these obvious points of contact between Hertzian mechanics and Einstein's theory of gravitation, I have difficulty in not assigning importance to *the* fact, that *in both theories the "forces" are dressed in exactly the same mathematical terms, namely that of the Riemann–Christoffel three-index-symbols of a quadratic form of the differentials of the position co-ordinates*.' The formal reason justifying Schrödinger's claim was the following: in both mechanical theories the equations of motion were solved by paths, which were geodetic lines in a general Riemannian continuum. Schrödinger further added the speculative hope that Hertz' mechanics might be used to describe the thermodynamical, ponderomotive actions of a given heated body on another following from kinetic theory. He concluded: 'The stimulus for me to enter into such considerations, lies in the fact that gravitation and electromagnetic actions may be taken to be, in principle, of the same type as the thermodynamical actions.'

He did not go into any further details or elaboration of this last remark however, but instead studied, in Section 2 of his manuscript, the principle of least action, Eq. (74), with the quantity $2(E - V)$ being replaced by $ds^2/dt^2$, where $ds^2 = \sum_{i,k} g_{ik}\, dq_i\, dq_k$. On the one hand, the time taken by the passage from points $P_1$ to $P_2$ was given by what he called the 'chronometer integral,' i.e.,

$$t = \int_{P_1}^{P_2} \frac{ds}{\sqrt{2(E - V)}}. \tag{75}$$

On the other hand, one could eliminate time altogether from the solution of the action principle, Eq. (74), and construct the latter as a field of extremal orthogonal trajectories: Schrödinger showed that this solution indeed satisfied all the conditions necessary to represent a solution of the Hamilton–Jacobi differential equation. In Section 3, he finally investigated another point by demonstrating that the time $t$ in the action principle for a system of $n$ degrees of freedom might be replaced by a further $(n + 1)$th co-ordinate, the so-called 'Liouville co-ordinate' $q_0$, an idea which he owed—as he indicated in his notebooks on tensor analytical dynamics—to his friend Ludwig Flamm.

## Physics and Philosophy

The notes on the more conceptual aspects of General Relativity, although remaining hidden in unpublished manuscripts, helped Schrödinger to formulate his own ideas. In the war years he often pondered about the concept of time, which may not appear strange for an artillery officer who was kept for months in one place on the stationary Austro-Italian front. Indeed, he designed, again in 1915, an incomplete essay on elementary and philosophical aspects of the concept of time.[458] In the manuscript, Schrödinger discussed the problem of properly defining the concepts of sensation ('*Sinnesempfindung*') and recollection ('*Erinnerung*'). On referring to the example of the sensation and recollection of the colour 'red,' he concluded that 'the sensation "red" embraces the recollection of "red" totally,' and also that the sensation must possess an additional property which he called 'palpability' ('*Sinnfälligkeit*') and 'which itself is *not* the subject of recollection' ('*selbst ist nicht Gegenstand einer Erinnerung*'). By 'palpability' he denoted a certain quality of reality, and for him it constituted a straightforward task in proving that recollections do not necessarily share the quality of reliability, as they easily can be distorted, for example, by the effects in physiological optics. After such considerations, he asked: 'What have we gained for the concept of time?' His reply: 'I said earlier that the essential, so-to-speak "time-generating", characteristic of our consciousness is the *repeated occurrence of the same element*. . . . In this context one should understand occurrence not just as *sensual* occurrence, but *occurrence of recollection* as well.'

The difficulty in that conclusion consisted, of course, in how to define the repeated occurrence of an element, be it a sensation or a recollection. The definition necessarily had to imply the ability to notice the 're-occurrence' of the element in question, and that presupposed the ability to recognize the non-occurrence of the event. Schrödinger solved the problem by claiming the connection of the same and different elements, connections, which he argued were of the type established by mathematical theorems. Since the latter do not imply the concept of time, he argued in turn, 'that the *repetition* of elements may create the concept of time, and it is solely the *connection* of elements that suggests their repetition; and also that the root of the concept of time lies in the connection of elements (as something immediately given), not *vice-versa*.'[459]

His philosophical and psychological thoughts, written in 1915 on the concept of time, not only connected well with the physical-technical considerations within the framework of Hertzian mechanics and General Relativity three years later, they also provided evidence for a continuing and increasing interest in the fundamental philosophical problems of physics. A short note, dated 10 September

[458] These notes were included in the '*Tagebuch 1915*', along with the war events (which we have reported earlier—in Section I.4).

[459] These concepts are not the main topic of the manuscript; they rather serve as auxiliary tools to introduce time on a more elementary level. We must conclude that the first few pages of the manuscript have been lost or, at any rate, not filed on the microfilm.

1918, dealing with causality, represents another sign of the same interest.[460] Schrödinger started from a quotation of Hermann von Helmholtz, who claimed that the law of causality had to be acknowledged 'as a law of our thoughts, preceding any experience' ('*als ein aller Erfahrung vorausgehender Gesetz unseres Denkens*'). Schrödinger wrote:

> I deny that completely. [He continued:] As far as the *a priori* is concerned [we must say that] in this sense Kant is always again misunderstood. It [i.e., the *a priori*] is certainly not meant with the meaning of time, but of logical sequence. If Kant states that space and time are *a priori* forms of conception, this clearly does not imply that we are already in the possession of some mystical, inborn knowledge of space and time, clearly before any experience, but [only] means the fact that we, being in *the possession of these concepts or rather conceptions*, may and must state them as being the properties of all and of each thinkable experience; that then, *if* we have them [i.e., the concepts of space and time] and *only* them, can we predict with certainty the organization or necessity to organize every future experience in this sense. (*Manuscript* on *Kausalität*)

Schrödinger claimed that Henri Poincaré had subscribed to the same interpretation of *a priori* concepts. He continued:

> Just the same surely applies to the 'principle' of causality, in complete agreement with [David] Hume's classic analysis. [Essentially what Hume had done was that] he succeeded in analyzing the principle of causality so deeply that he was able to recognize that it was not necessarily valid. (*Manuscript* on *Kausalität*)[461]

Schrödinger attempted with his note to point out that the principle of causality might in fact 'turn out to be vulnerable in the course of development' ('*in Laufe der Entwicklung auch wieder verletzbar ist*'), a possibility which seemed to be realized in the laws of radioactive decay. The discussion of the principle of causality was connected with the discussion on kinetic and atomic theory, which had concerned Schrödinger for years.[462] On the other hand, it led him to deal in greater detail with philosophical schemes, such as those proposed by David Hume (1711–1776) or Immanuel Kant (1724–1804), or again the more recent philosophy of science, notably that of Ernst Mach and Richard Avenarius.[463] The idea of recurrence, which he used in connection with his foundation of the concept of time, reminded him of one of the guiding thoughts in Indian philosophy. In 1918, Schrödinger wrote a set of notes on the aspects and concepts

[460] The manuscript has been filed on AHQP Microfilm No. 39, Section 7.

[461] At this point he added: 'I have tried the same analysis in the case of time without arriving at a really satisfactory result.' Evidently, he referred to his philosophical-psychological considerations of 1915 which have been discussed above.

[462] In Schrödinger's *Nachlaß* a notebook can be found, entitled '*Über Atomistik, mechanische Naturerklärung und Phänomenologie*' ('On Atomistics, Mechanical Interpretation of Nature and Phenomenology'). It is undated, but may have been written before 1920. We shall discuss its content in Chapter II. (Filed on AHQP Microfilm No. 39, Section 8.)

[463] Schrödinger wrote a short note, '*Ad R. Avenarius Kr. d.r. E.* [ = *Kritik der reinen Erfahrung*, 1888, 1890] *und D. m. W.* [*Der menschliche Weltbegriff*, 1891]. In this note he tried to criticize a specific statement of Avenarius on human experience.

of this philosophy, such as '*nirvana*' and '*karma*,' the existence of the 'ego' and the 'world,' and their interpretation and relation to European thought and philosophy.[464]

The concern with philosophy represented, in the war years—or, at least, in the second half of this period—a major, all-embracing occupation of the physicist Schrödinger. Many years later he recalled:

> In 1918, when I was thirty-one, I had good reason to expect a chair of theoretical physics at Czernowitz (in succession to Geitler). I was prepared to do a good job lecturing on theoretical physics, with, as my supreme model, the magnificent lectures given by my beloved teacher Fritz Hasenöhrl, who had been killed in the war; but for the rest [of the time], to devote myself to philosophy, being deeply imbued at the time with the writings of [Baruch] Spinoza, [Arthur] Schopenhauer, [Ernst] Mach, Richard Semon and Richard Avenarius. My guardian angel intervened: Czernowitz soon no longer belonged to Austria. So nothing came of it. I had to stick to theoretical physics, and, to my astonishment, something occasionally emerged from it. (Schrödinger, 1961; English translation, p. viii)[465]

## Atomic and Quantum Theory

As it turned out, in 1913, by the time of the *Naturforscherversammlung* in Vienna, enormous progress in atomic theory had taken place, i.e., Niels Bohr's theoretical explanation of the hydrogen spectral lines. George de Hevesy, who had witnessed the development of the new theory in Manchester before joining the Vienna *Institut für Radiumforschung*, wrote to Bohr from the conference: 'This afternoon I spoke with Einstein ... then I asked him about his view on your theory. He

[464] Among other *Notes and Memoranda* in the Schrödinger *Nachlaß* eight notebooks are kept on '*Indische Philosophie und Religion*' ('Indian Philosophy and Religion'). They have been filed on Microfilm No. 39, Section 7, with the remark: 'The first appears to be from the *Gymnasium*. The last is dated 1925.' We find among them a list of '*eingegangene Literatur*' ('literature received'), which includes references to literature published up to the year 1910. Notebooks then follow, dated 2 July 1918 (the entries in this notebook started on 2 July 1918 and continued until 28 August 1918), 31 August 1918 (the last on 9 September 1918), and 11 September 1918 (entitled '*Samkhya-Philosophie*'), an issue whose entries date from 17 August 1919 to 27 September 1919. Finally, there is a longer manuscript dated 1925. One should mention that Schrödinger wrote at the beginning of these notebooks: '*Ältere Notizen, später zusammengefaßt in dem "Unterhaltungen" von 1925*' ('Older Notes, Later Summarized in the "Conversations" of 1925'). By analyzing the details of the microfilmed notebooks one realizes that the notes of 1918 and 1919 have been organized systematically later on. We shall discuss the entire topic of Schrödinger's concern with Indian philosophy in a later chapter (III), especially in the context of the notes of 1925.

[465] Among the more philosophical activities of Schrödinger we may count the composition of a longer essay with the title '*Über Atomistik, mechanische Naturerklärung and Phänomenologie*' ('On Atomistics, Mechanical Interpretation of Nature and Phenomenology'). (It is contained in a notebook that has been filed on AHQP Microfilm No. 39, Section 8.) The author provides in his notes a careful definition of what a phenomenological description of nature is—elaborating in particular the essential points of view of Ernst Mach and his followers; and he contrasted this view with the atomistic view held by Boltzmann and his followers, such as Max Planck and others—a view which was not only built on (classical) mechanics, but on electromagnetism as well, and implied as a characteristic basis the existence of atoms and the attempt to provide a unified explanation of *all* physical phenomena.

The notebook is not dated, but from the reference to quantum theory of specific heats we conclude that it was written after 1912—which was when this topic entered Schrödinger's scientific mind.

told me, it is a very interesting one, an important one if it is right and so on.... I told him then that it is established now with certainty that the Pickering–Fowler spectrum belongs to helium. When he heard this he was extremely astonished and told me: "Then the frequency of light does not depend at all on the frequency of the electron.... This is an *enormous achievement.* The theory of Bohr must then be wright. [*sic*].'' (Hevesy to Bohr, 23 September 1913).

Apparently the positive opinion of Einstein was not shared by the physicists in Vienna, as we cannot find any response to Bohr's theory of atomic structure from Vienna. It can be argued that the war came too fast, and soon most of the theoretical physicists, notably Friedrich Hasenöhrl and his student Karl Herzfeld—who had earlier been involved in the theory of atomic spectra—were called to military service at the front. A year later Hasenöhrl was killed, while Herzfeld remained in military service for the rest of the war. When Erwin Schrödinger returned to Vienna and restarted scientific work in 1917, Bohr's theory had been developed and generalized, especially in Germany by the Munich theoretician Arnold Sommerfeld (Boltzmann's successor) and the Berlin astronomer Karl Schwarzschild. In the light of new successes in explaining X-ray spectra, the hydrogen fine structure and the Stark effect, the theory could no longer be called doubtful. Still, the response from Vienna failed to materialize for years to come.[466]

Schrödinger systematically studied, after his return to Vienna, the recent publications on quantum theory, notably those of Max Planck, Albert Einstein and Karl Schwarzschild. However, he had not yet become an active contributor to the field of atomic and molecular structure. He began slowly to think about his own way of handling certain fundamental questions arising from it. A few remarks in Schrödinger's unpublished manuscripts and notebooks, composed at that time, throw light on his thinking. He stated, for example, when discussing the principle of the straightest path: 'In the case of quantum conditions, one may doubt whether they are not essentially related with the integral final result. Whether an orbit is allowed to remain stationary can possibly be finally decided, only after the system has been subdued to all opportunities of "deviations." In any case, the most elementary of all quantum conditions may be expressed in a differential formulation; moreover, the same applies, according to Schwarzschild, to all quantum conditions if expressed in "angular variables." The latter, however, seem to follow only from the integration of the equations of motion.'[467]

[466] After Hasenöhrl's death the Vienna faculty tried to get Sommerfeld for the theoretical chair; however, after lengthy negotiations, Sommerfeld declined. Herzfeld did not return to Vienna after the war, but went to Munich to become assistant to the chemist Kasimir Fajans (1887–1975). There he would actively participate in the rich scientific life at Sommerfeld's institute and become acquainted with many of the pioneers of the future atomic theory, like Gregor Wentzel, Wolfgang Pauli, Werner Heisenberg, Otto Laporte and Fritz London, before he went on (in 1926) to the United States (1926-1936, professor of physics at the Johns Hopkins University, Baltimore, Maryland; 1936–1968, professor at the Catholic University of America in Washington, D.C.) He retired in 1968 and died in June 1978.

[467] The quote is from the first notebook on tensor analytical mechanics, which we have discussed earlier. (See also Footnote 454.)

Altogether Schrödinger did not yet exhibit any willingness to enter into the investigation of atomic structure, though he showed an interest in questions of principle, such as the very nature of radiation. According to Einstein's pioneering work of 1909, the study of the fluctuations of radiation—we recall that Schrödinger's thesis adviser Egon von Schweidler had tried to measure the fluctuation of high-frequency $\gamma$-rays (Schweidler, 1910a, b)—should reveal the true nature of light and decide the validity of the light-quantum hypothesis. Schrödinger knew about Einstein's hypothesis, because he had worked through the latter's papers on statistical mechanics in preparing his own research on fluctuation phenomena. The recent analysis of the emission and absorption process of light, which Einstein presented in several papers of 1916 (Einstein, 1916d, e; 1917a), again applied fluctuation arguments to support the light-quantum hypothesis. Einstein's arguments attracted the attention of Schrödinger, who began to investigate the problem of the nature of light himself.

In the summer of 1919 Schrödinger stated: 'A series of theoretical considerations support the opinion that we should think of the emission of light as perhaps a process composed of *directed* elementary processes, i.e., in such a way that for each emission process radiation goes only into a small, eventually very small, angle of space' (Schrödinger, 1920a, p. 69). If the light-quantum hypothesis gave the correct picture of the nature of light, then two rays of light emerging from a given point of a light source should not interfere when emitted under a large angle. Schrödinger pondered about an experiment to test this conclusion by using a light source, whose linear dimensions were not large compared to the wavelength of the emitted light. True, the previous studies of Fresnel interference seemed to imply that this experiment would yield coherence and interference of large-angle bundles of light; however, Schrödinger argued that Fresnel interference did not really decide the nature of light, but rather the validity of the so-called Huygens-Kirchhoff principle.[468] The trick of optically diminishing the size of the light source (i.e., by forming smaller images with combination of lenses) would not help solve the problem in question. Schrödinger therefore decided to perform the experiment with a real source of small dimensions, namely with thin electrically annealed wires of Wollaston platinum of 2–4 $\mu$ diameter; the rays emitted from these wires had angles of up to 60°, which he produced with the help of narrow slits cut by razor blades in a tin foil that were then brought close to the radiating wire. He reported the experiment in a paper, '*Über die Kohärenz in weitgeöffneten Bündeln*' ('On the Coherence in Wide-Angle Bundles'), which was received by *Annalen der Physik* in August 1919 and published in early 1920 (Schrödinger, 1920a). The experiment again revealed the coherence of different

[468]The principle of Christiaan Huygens (1629–1695) allows the construction of propagating wavefronts by assuming that, at a time $t$, from all points of a given wavefront elementary wavelets arise, forming the wavefront at a future instant of time. Gustav Robert Kirchhoff (1824–1887) provided this principle with an adequate mathematical formulation (Kirchhoff's formula).

Schrödinger wrote detailed notes on various formulations and applications of Kirchhoff's principle which are kept in his *Nachlaß* and have been filed on AHQP Microfilm No. 39, Section 6.

rays, in spite of their wide-angle separation. Did the result contradict the light-quantum hypothesis for the emission process of light? No, said Schrödinger, commenting:

> Only later did it become clear to me that in this way nothing had been proved for or against the directedness of the elementary emission process; what has again been proved is just the validity of the Huygens–Kirchhoff principle for a cylindrical surface, which lies entirely in the air space and encloses the filament at a narrow distance. (Schrödinger, 1920a, p. 71)

This experiment was the last which Schrödinger performed in Vienna and also—apart from a simple investigation in physiological optics, about which we shall report later—the last experiment of the physicist Schrödinger ever. After staying about six months longer at the *II. Physikalisches Institut*, he left Vienna in 1920 to accept the position of an assistant to Max Wien at the University of Jena. The Austrian capital—after the complete defeat of the Austro-Hungarian Empire, and its dissolution into a series of independent countries—no longer offered a suitable position. Schrödinger hoped to promote his future scientific career in the likewise-defeated German Empire. Thus this chapter of Schrödinger's early life in Vienna ended; he would not return to his home university and his hometown (apart from short visits) for another thirty-six years.

# Chapter II
## Waves and Quanta: Preludes to Wave Mechanics

### Introduction

The preliminary steps leading to the creation of wave mechanics have been the subject of quite a few historical investigations. They occupy extensive sections in books dealing with the entire development of quantum and atomic theory, e.g., those of Max Jammer (1966), Friedrich Hund (1967) or Edward M. MacKinnon (1982). The origin of wave mechanics, as a topic in its own right, has also attracted the scholarship of numerous historians of science, who have devoted detailed articles to it, and of which we may mention the papers of Johannes Gerber (1969) and Helge Kragh (1979, 1982). Considerable emphasis has been laid on high-lighting the two personalities who have contributed the most crucial ideas to the theory, namely Louis de Broglie and Erwin Schrödinger.[1]

The comparatively small number of scientists involved in the creation of wave mechanics—besides de Broglie and Schrödinger, essentially Albert Einstein must be mentioned as an originator—seems to simplify the task of the historian of that subject, because he may concentrate on the efforts of these few persons. In addition it appears, on first inspection, that the history of wave mechanics can be understood by using less technical methods than those necessary in the case of the Göttingen-Cambridge version of quantum mechanics. Nearly all tricky references to and comparisons with empirical data are lacking, and no special mathematical tools, different from those used in common classical physical theories, are needed such as those that prevent in matrix mechanics the '*anschauliche*' ('intuitive') interpretation of the phenomena described. However, these apparent simplifications should not be overestimated. Neither can the original ideas of wave mechanics, their motivation and formulation, be separated from the rest of atomic and quantum physics that was treated at the same time; nor must Einstein and Schrödinger be considered as scientists into whose deep and real thoughts one can penetrate easily. As far as the '*Anschaulichkeit*' ('visualizability' or 'intuitiveness') of the wave mechanical concepts and description is concerned, one may recall the following historical fact: most physicists of the early and middle 1920s had serious difficulties in following or accepting them.

[1] From the existing extensive literature on de Broglie and Schrödinger, we refer for the former to the papers of Fritz Kubli (1970), Heinrich A. Medicus (1974) and Edward MacKinnon (1976); and for Schrödinger to the book of William T. Scott (1967) and to the papers of V. V. Raman and Paul Forman (1969), Paul A. Hanle (1975, 1977a, b; 1979), Linda Wessels (1975, 1979, 1980), Edward MacKinnon (1980) and Helge Kragh (1979, 1982, 1984).

In this chapter we shall try to outline the principal, preliminary steps towards the rise of wave mechanics. We shall not go back (as others have done, e.g., Jammer and Gerber, cited above), to earlier centuries when the present concepts of particles and waves were still being formed, but rather start *after* they became well defined at the turn of the twentieth century. At that time both the dynamics of corpuscles and the electromagnetic wave theory had been established.[2] New phenomena, like X-rays and radioactive radiation were then discovered, which presented difficulties when one tried to describe them within the framework of the standard theories. In particular, the difficulties which were manifest in the discussion of X-rays and their puzzling behaviour, observed in the first twenty-five years since their discovery, will be discussed in Section II.1.

The efforts towards a better understanding of the nature of radiation, which had occupied leading physicists at the beginning of the century, diminished somewhat in the decade between 1915 and 1925, when physicists shifted their interest to a detailed investigation of atomic and molecular structure based upon the pioneering ideas of Niels Bohr. Thus, especially after World War I, powerful schools of quantum and atomic theory developed in Copenhagen, Munich and Göttingen. The activity and productivity at these places was so great that Berlin lost its former leading role in quantum physics; other places, like Breslau and Stuttgart in Germany or Cambridge in England, served as subsidiaries of the dominant triangle Copenhagen–Göttingen–Munich, while still other places, like Paris, were considered as being off the track. In Section II.2 we shall discuss the role of the side-stages in quantum theory during the early 1920s, of which several, notably Leyden, Paris and Zurich, assumed crucial importance in the history of wave mechanics.

The origin of the concept of matter waves, the first big step toward wave mechanics, has already been told, together with the story of its discoverer Louis de Broglie, in Volume 1 (Section V.4). Here we shall concentrate mainly on the scientific development of Erwin Schrödinger in the 1920s. In Section II.3 his arrival in Zurich will be reported, together with the main ideas which he brought to the new environment. Unlike Louis de Broglie, quantum theory and the analysis of atomic structure did not lie at the centre of Schrödinger's scientific work. Schrödinger only gradually entered into the modern conceptions and even occasionally contributed to a specific question, as we shall discuss in Section II.4. Statistical mechanics, however, was a far more important and continuous field of research for Schrödinger. He had been interested in this subject since his student days in Vienna, and now in Zurich he published a series of papers on it (from 1922 to 1926). These investigations on statistical mechanics must be viewed as constituting the second crucial step, besides de Broglie's matter waves, in the genesis of wave mechanics; they will be treated in Section II.5.

[2] Let us include, for simplicity, here also the special relativistic extensions of the classical particle and wave theory, which occurred soon afterwards.

# II.1 The Dual Nature of Light

## The Origin of the Classical Concepts of Particles and Waves

The present concepts of particles and waves may be traced back to the last quarter of the seventeenth century, namely to the *Philosophiae Naturalis Principia Mathematica* of Isaac Newton (1642–1727) and the *Traité de la lumière* of Christiaan Huygens (Newton, 1687; Huygens, 1690). With the reference to the *Principia* we do not wish to imply that Newton was the first to discuss the motion of corpuscles; this problem had already been studied before, e.g., by Pierre Gassendi (1592–1655) and Robert Boyle (1627–1691).[3] But Newton provided the theory of particle dynamics which would be developed in the following centuries to describe the behaviour of corpuscles or material particles under the influence of arbitrary forces. Similarly, we refrain from claiming that Huygens created in his *Traité* the wave theory of light.[4] What he did there was basically to treat light, in analogy to sound, as the motion of a dilute form of matter, the ether; i.e., the fast moving particles of a luminescent body impinge upon the particles of ether, creating a shock which proceeds forward spherically from the centre of impact. While Huygens explained the phenomenon of refraction by assuming what was later called Huygens' principle, his treatment lacked the essential features of standard wave theory. These features included, for instance, a constant wavelength and the polarization of the propagating waves; all these were brought in during the following 120 years. Nevertheless, Huygens explained with his primitive idea the puzzle presented in those days by the double refraction of Iceland spar. Yet, this brilliant success met with the subtle criticism of Newton. The latter held a variety of views on the nature and propagation of light, preferring, especially in his younger days, the assumption that light consisted of tiny moving particles.

Newton's corpuscular view of light became the ruling doctrine in the eighteenth century, notably through the work of James Bradley (1693–1762) in England, but likewise supported on the Continent—e.g., by Johann (John) Bernoulli (1710–1791), Pierre Louis Moreau de Maupertuis (1698–1759) and most of their

[3] For details of the early history see the book of E. J. Dijksterhuis (1956), especially Part Four.

[4] Christian Huygens (or Huyghens) was born on 14 April 1629 in the Hague, in the Netherlands, the son of Contantijn Huygens, secretary to several Princes of Orange. He studied law and mathematics, the latter with Frans van Schooten at the University of Leyden and the Academy of Breda. He then travelled widely on the Continent and in England (e.g., accompanying Henry, Duke of Nassau, on his mission to Denmark in 1649), obtaining the title of a doctor of laws in Anjou (1655), and then he lived in Paris from 1661 to 1681; meanwhile his scientific reputation was growing (in 1663 he was elected fellow of the Royal Society of London, and in 1665 he was elected a member of the Paris Academy of Sciences). After the *Edict of Nantes* (tolerating the Protestants in France) had been revoked by King Louis XIV in 1681, Huygens returned to the Hague and lived there until his death on 8 June 1695. He worked on mathematical, physical and astronomical problems; he invented the pendulum clock in 1656 and constructed, together with his brother Constantijn, long telescopes with which he discovered, e.g., the first moon of the planet Saturn in 1656. (For details of his life and work, see Lommel, 1913.)

contemporaries—with only a little opposition, such as from Benjamin Franklin (1706–1790) and Leonhard Euler (1707–1783). The pioneering studies of Thomas Young (1773–1829), Augustin Jean Fresnel (1788–1827) and Etienne Louis Malus (1775–1812), at the turn of and in the first two decades of the nineteenth century, settled the case in favour of the wave theory of light. During the following decades the experimental researches of Michael Faraday (1791–1867) on electricity and magnetism provided the foundation of the theory of electrodynamics, which was then established—starting in 1860—by James Clerk Maxwell (1831–1879). In this theory, light was explained as representing electromagnetic waves (Maxwell, 1862a), an idea which was so brilliantly confirmed by Heinrich Hertz (1857–1894) through his discovery of long electromagnetic radiation (Hertz, 1888a, b).[5]

Towards the end of the nineteenth century the slightest doubt about the electromagnetic wave nature of light seemed to have been removed. The behaviour of light *in vacuo* had to be described either by a second-order partial differential equation of hyperbolic type in space $(x, y, z)$ and time $t$ for the vector $\mathbf{E}$ ( $= E_x, E_y, E_z$) of the electric field, i.e.,

$$\frac{\partial^2 \mathbf{E}}{\partial x^2} + \frac{\partial^2 \mathbf{E}}{\partial y^2} + \frac{\partial^2 \mathbf{E}}{\partial z^2} - \frac{1}{c^2}\frac{\partial^2 \mathbf{E}}{\partial t^2} = 0, \tag{1}$$

with $c$ denoting the velocity of light *in vacuo*, or by an equation of the same structure for the magnetic field vector $\mathbf{H} = (H_x, H_y, H_z)$, whose direction was perpendicular to that of $\mathbf{E}$. Similarly, no serious physicist denied at that time that small material particles obeyed the laws of Newtonian mechanics, especially the equations of motion in which Joseph Louis Lagrange (1736–1813) and William Rowan Hamilton (1805–1865) had generalized the ideas laid down in the *Principia*[6]. That is, the state of a particle with mass m is described by its position vector $\mathbf{x} = (x, y, z)$ and its momentum vector $\mathbf{p}$ $(= m(d\mathbf{x}/dt) = (p_x, p_y, p_z))$, whose dependence on time follows from the Hamilton equations, namely,

$$\frac{d\mathbf{x}}{dt} = \frac{\partial H}{\partial \mathbf{p}} = \left(\frac{\partial H}{\partial p_x}, \frac{\partial H}{\partial p_y}, \frac{\partial H}{\partial p_x}\right), \tag{2a}$$

and

$$\frac{d\mathbf{p}}{dt} = -\frac{\partial H}{\partial \mathbf{x}} = -\left(\frac{\partial H}{\partial x}, \frac{\partial H}{\partial y}, \frac{\partial H}{\partial z}\right), \tag{2b}$$

where $H$ denotes the Hamiltonian function, the value of which can be identified with the sum of the kinetic and the potential energies of the particle. The two sets of equations, i.e., Eqs. (1) and Eqs. (2a) and (2b), share a mathematical property: they are partial differential equations. However, they describe very different physical objects. The solutions of Eq. (1) represent periodic fields in

[5] The detailed historical development of the concepts of light may be found in the book of Edmund Whittaker (1951).

[6] We do not necessarily claim, at this point, that this implied the existence of smallest particles of matter or atoms. Indeed, the atomic constitution of matter was not accepted by several influential physicists, such as Ernst Mach and Wilhelm Ostwald. (See our discussion in the previous chapter.)

space and time, consisting of electromagnetic waves which propagate with the velocity $c$, each having a polarization determined by the direction of its electric field vector. The particles, whose behaviour is defined by Eqs. (2a) and (2b), on the other hand, follow trajectories whose curvature depends on the action of external forces $\mathbf{F}$ (which enter the equation via the potential energy $U$, such that $\mathbf{F} = \partial U / \partial \mathbf{x}$.

While it became evident that light was the typical representative of electromagnetic waves, the search for the most elementary objects representing corpuscles went on. They were first discovered in electric discharges, a field of experimental research, which developed from the time of Faraday, especially during the second half of the nineteenth century. Thus Julius Plücker (1801–1868) found that the glass tube near the cathode of a gaseous discharge emitted phosphorescent light (Plücker, 1858), and his student Johann Wilhelm Hittorf (1824–1914) showed that this phosphorescence was created by rays emerging from the cathode (Hittorf, 1869). A couple of years later Cromwell Fleetwood Varley (1828–1883) suggested that the rays in question consisted of ejected material particles (Varley, 1871). After another twenty-five years the debate on the real nature of the so-called cathode rays was settled by the experiments of Philipp Lenard (1862–1947), who first obtained cathode rays outside the gaseous discharge (Lenard, 1893), and then by Joseph John Thomson (1856–1940), who showed that the rays consisted of negatively charged particles of small size compared with the dimensions of ordinary atoms and molecules (Thomson, 1897a). At about the same time the so-called canal rays, which had first been observed by Eugen Goldstein behind the cathode of a gaseous discharge (Goldstein, 1886), were found to be positively charged particles having the mass of ions in electrolysis (Wien, 1897).[7] Both the

[7]The discovery of canal rays and their subsequent development were reported by Ernst Gehrcke on the occasion of the seventieth birthday of Eugen Goldstein (Gehrcke, 1920a), and by Eduard Rüchardt on the occasion of the centenary of their discovery (Rüchardt, 1936).

Eugen Goldstein was born on 5 September 1850 in Gleiwitz. He started to study medicine at the University of Breslau in 1869, but then went to Berlin, attending lectures in physics and chemistry (of August Wilhelm Hofmann) at the university. In 1871 he joined the physics laboratory founded by Hermann von Helmholtz; he obtained his doctorate (in 1879) with a thesis in which he showed the deflection of cathode rays in an electric field (Goldstein, 1880). In 1888 he obtained the position of physicist at the Berlin Observatory; from 1898 he worked in a laboratory of his own in Berlin-Schöneberg. He died on 25 December 1930 in Berlin.

Goldstein investigated many topics of experimental physics; his first paper was on the line- and band-spectra of molecules; in his second paper he dealt with the deflection of cathode rays, a topic which he discussed in detail in his doctoral thesis. The studies of cathode rays and gaseous discharges also led to his greatest discovery, that of canal rays in 1886. Besides investigating their properties, Goldstein also worked on colour phenomena created by the impact of cathode rays on solids and organic substances, and on spectroscopic questions. He studied, e.g., what he called the '*Grundspektren*' of alkali atoms, which were later interpreted as the lines of the alkali ions, and he was the first to discover the so-called 'second' helium spectrum, which may be associated with a weakly bound helium molecule ($He_2$). He further discussed problems of cosmic physics suggesting, as early as 1881, that the cathode rays emitted from the sun may create phenomena on earth such as polar lights, etc.

Goldstein received several honours: the Paris Academy of Sciences gave him a prize in 1903, and he was awarded the Hughes medal of the Royal Society of London in 1908. The German Physical Society elected him *Ehrenmitglied* in 1919. (For other biographical details see the article of Otto Reichenheim in the Goldstein issue of *Naturwissenschaften*: Reichenheim, 1920.)

cathode rays—or electrons, as they were later called by George Johnstone Stoney (1826–1911)—and the canal rays or positive ions obeyed the dynamical laws given by Eqs. (2a) and (2b); i.e., the motion of these corpuscular objects is influenced appropriately by electric and magnetic fields—indeed, it was through the deflection of the corpuscles in these fields that physicists gained crucial information about their nature, especially the ratio of electric charge to mass.

## X- and $\gamma$-Rays

The last decade of the nineteenth century clarified the situation with respect to the known types of corpuscles and radiation; and simultaneously new problems arose. In November 1895 Wilhelm Conrad Röntgen (1845–1923) discovered a new kind of radiation which he called X-rays (Röntgen, 1895). Soon afterwards Antoine Henri Becquerel (1852–1908), stimulated by Röntgen's discovery, found peculiar rays emitted by uranium salts (Becquerel, 1896a). The subsequent investigations of Friedrich Giesel (1852–1927), Ernest Rutherford (1871–1937), Henri Becquerel and Pierre Curie (1859–1906) revealed two components in the rays from uranium and other radioactive substances, which were given the names of $\alpha$- and $\beta$-rays. The latter could easily be deflected by magnetic or electric fields and were recognized as fast cathode rays or electrons (Becquerel, 1900). The identification of $\alpha$-rays with positive ions took a little longer. Not until the fall of 1902 did Rutherford succeed in showing that they were influenced by rather strong magnetic flelds and exhibited a similarity to canal rays (Rutherford, 1903); in 1908 Rutherford, together with Hans Geiger (1882–1945), again proved that $\alpha$-rays consist of doubly-ionized helium atoms (Rutherford and Geiger, 1908). Finally a third, very penetrating component was isolated in radium probes by Paul Ulrich Villard, the so-called $\gamma$-rays (Villard, 1900).[8] Even in the strongest magnetic fields $\gamma$-rays could not be deflected (Paschen, 1904b, c), and they were found by Artur Steward Eve (1862–1948) to show very similar properties to hard X-rays (Eve, 1904).

Many investigators, especially Rutherford and his associates (as, e.g., Eve) interpreted $\gamma$-rays as being closely related to X-rays, representing electromagnetic pulses of the shortest wavelength. A very different opinion, however, was put forward by Friedrich Paschen (1865–1947) in a series of papers published in 1904 (Paschen, 1904a, b, c). In the first of these papers, which he submitted in March 1904, Paschen drew attention to the fact that $\gamma$-rays ionized various gases quite differently from X-rays—the above-mentioned experiments of Eve demonstrating the similarity with *very hard* X-rays were not published until November of that year—and added: 'If one concludes from the absence of any electric or magnetic deflection [of the rays] that $\gamma$-rays possess no electric charge, then this conclusion is still not justified, since for the deflection by the electric and magnetic fields at

[8] Paul Villard was born on 28 September 1860 in Lyon. He was educated at the *École Normale* and became professor at the *Conservatoire des Arts et Métiers*. Villard, from 1908 a member of the Paris Academy of Sciences, died in Bayonne on 13 January 1934.

our disposal not only the charge is crucial, but also the apparent inertia of the quanta [of which the rays are supposed to consist]' (Paschen, 1904a, pp. 164–165).

Paschen now performed his own absorption experiments by covering a radium source with lead and found that the radiation, which was commonly defined as $\gamma$-rays [i.e., the rays penetrating through 1 cm of lead], carried a negative charge like $\beta$-rays. Hence, he concluded:

> It is therefore probable that the $\gamma$-rays are cathode rays of still larger velocity than the $\beta$-rays studied by [Walther] Kaufmann [1901, 1903] and for which the absorption coefficient [in lead] is still fifty times larger [than for the $\gamma$-rays from Paschen's radium probe]. These [i.e., Kaufmann's $\beta$-rays] possess, according to Kaufmann, different velocities between 2.13 and $2.83 \times 10^{10}$ cm/sec. The highest velocity represented in the sample approaches the velocity of light [*in vacuo*] to within 5.7 percent. Hence the negative electricity [i.e., the electrons] in the $\gamma$-rays has to move still faster. The small, constant limiting value of the absorption coefficient [observed by Paschen for $\gamma$-rays] should correspond to a high constant limiting value of the velocity. A fifteenth part of the cathode rays of arrangement I [in Paschen's experiments] possess this velocity. The main interest in these rays seems to lie in the following fact: the case of an electric charge moving with the velocity of light [*in vacuo*], which has been discussed theoretically several times recently, may be realized in $\gamma$-rays. (Paschen, 1904a, p. 171)

Paschen further investigated what he called '$\gamma$-electrons' in the following months. In August 1904 he submitted a paper to *Physikalische Zeitschrift*, which contained a different proof for his very fast $\beta$-ray hypothesis. In particular, he measured the energy of these rays with a calorimeter, finding that 'the energy of a $\gamma$-electron is ... larger than 3,200 times that of the fastest $\beta$-electron in the measurement of Kaufmann' (Paschen, 1904c, p. 567). Less than six months later, however, he again sent a short note to *Physikalische Zeitschrift*, in which he had to withdraw the result of the energy determination. There he stated:

> This winter I have repeated the experiments with a probe of 35 mg [radium bromide, as against 40 mg in the previous summer] under significantly better conditions without being able to obtain an unobjectionable result. Rather the large ice-calorimeter used showed itself to be subject to many perturbations, which are difficult to estimate and which can hardly be removed even by the previously applied differential method .... The earlier experiments must have been affected by larger perturbations than the present ones. (Paschen, 1905, p. 97)

The same result as Paschen's, namely that $\gamma$-rays created less heat in the calorimeter than if they were ultra-high energy electrons, was obtained by Rutherford and his associate Howard Turner Barnes (1873–1950) and published soon afterwards (Rutherford and Barnes, 1905).

### Bragg's Corpuscular-Pair Theory of $\gamma$-Rays

One aspect of this unsuccessful side-story may be of some historical interest, for Paschen had referred to it when he talked about the 'apparent inertia of the quanta [i.e., of the electrons with high speed].' In those days of 1904 the theory of fast-moving electrons attracted many physicists, among them Max Abraham (1875–1922) and Hendrik Lorentz; in their competing theories of rigid and deformable electrons, respectively, both of them predicted increases of inertial mass with growing velocity, although the dependence on velocity was different. Lorentz found, for a charged mass $m_0$ moving with velocity $v$, the formula

$$m_v = m_0 \frac{1}{\sqrt{1 - v^2/c^2}}. \tag{3}$$

This result was confirmed in the fall of 1905 by Albert Einstein's relativity theory (Einstein, 1905e). This theory required a suitable generalization of the Hamiltonian equations, which was first accomplished by Max Planck (1907). Experiments on the deflection of $\beta$-electrons only became accurate enough by 1909; then they started to confirm Eq. (3) as being the correct expression (Bucherer, 1909).

While the new relativistic mechanics (replacing Newton's) did not alter the decision against the identification of $\gamma$-rays with high energy electrons, William Henry Bragg (1862–1942) from Adelaide, Australia, developed—beginning in 1907—an alternative corpuscular theory of $\gamma$-rays. Unlike Paschen, Bragg did not doubt, to begin with, the close relationship between $\gamma$-rays and X-rays, but rather suggested 'a discussion of the possibility that $\gamma$- and X-rays may be of a material nature' (Bragg, 1907, p. 429). Bragg was impressed by the striking similarities of the properties and effects shared by *all* types of radioactive rays, which he had concluded from the existing literature as well as from his own experiments. Thus he found, for example, that 'the ionization which we measure in the ionization chamber is almost wholly due to the emission of slow-speed electrons [$\delta$-rays] from the atoms of the gas contained in the chamber, or of the chamber walls; and this is true for all forms of [radioactive] radiation [falling into the ionization chamber]' (Bragg, 1907, p. 434). He interpreted this fact by assuming that the radioactive substances emit, besides positive ($\alpha$) and negative ($\beta$) particles also 'neutral particles, such as, for example, a pair consisting of one $\alpha$ or positive particle and one $\beta$ or negative particle' (Bragg, 1907, p. 440). The penetrating power of such a pair would be very great compared to those of the constituent $\alpha$- and $\beta$-particles, and it seemed plausible to Bragg that the hardness or penetrability of the composite ray depended on the velocity of the pair. 'Assuming, then,' he concluded, 'that the neutral pair has great penetrating, but weak ionizing powers, is uninfluenced by magnetic or electric fields, and shows no refraction, it does so far conform to the properties of the $\gamma$-ray' (Bragg, 1907, p. 441).

Bragg's suggestion of the corpuscular-pair nature of $\gamma$-rays, which he explicitly extended to X-rays (Bragg, 1907, p. 429), may appear to be a very bold one. But, at the time he presented it, it was not completely at variance with existing experience. True, people had suggested since the discovery of X-rays that these rays represented some kind of electromagnetic pulses (Stokes, 1896; Wiechert,

1896; Thomson, 1898a); Wilhelm Conrad Röntgen, however, in his third communication on X-rays, had suggested a corpuscular theory (Röntgen, 1897). Then Hermanus Haga (1852–1936) and C. H. Wind's experiments, according to which X-rays when scattered by a thin slit showed some kind of diffraction fringes, shifted opinion in favour of the wave theory (Haga and Wind, 1899). Arnold Sommerfeld had analyzed the data of Haga and Wind theoretically and derived wavelengths of X-rays of the order of $10^{-8}$ cm (Sommerfeld, 1900). In the following years Haga and Wind had, upon some criticism of their methods and results, improved their apparatus (Haga and Wind, 1903); and yet, doubts still remained as to whether the diffraction experiments really proved the wave nature of X-rays (see, e.g., Walter and Pohl, 1909). Charles Glover Barkla (1874–1944) of Liverpool hoped to provide an independent proof of the electromagnetic wave or ether pulse theory of X-rays, when he demonstrated, in experiments on the scattering of X-rays in air and by light solids, that the rays possessed polarization (Barkla, 1905, 1906).[9] But, even the polarization effect did not impress such stubborn opponents of the wave theory as Bragg, who claimed in 1907 on his neutral-pair hypothesis: 'It is conceivable that it might show a one-sided or polarization effect, for if it were ejected from a rotating atom it would itself possess an axis of rotation' (Bragg, 1907, p. 441).

The debate between Bragg and Barkla continued during the following years.[10] Bragg and his collaborator J. P. V. Madsen found, for instance, that the $\beta$-particles excited by $\gamma$-rays exhibited a very noticeable lack of symmetry around the plane normal to the exciting $\gamma$-ray, which would not follow from the existing ether pulse theory (Bragg and Madsen, 1908); together with J. A. Glasson he obtained a similar result for X-rays (Bragg and Glasson, 1909). On the other hand, Barkla and Charles A. Sadler promoted an important argument in favour of the electromagnetic wave nature of X-rays; they discovered the existence of secondary homogeneous X-rays characteristic of the atomic weight of the substance exposed (Barkla and Sadler, 1908b). The characteristic X-rays could be easily explained by the analogy to the characteristic optical spectrum of chemical substances, while it presented difficulties for the corpuscular theory of X-rays. The complexity of these phenomena and the experimental subtleties involved in their investigation, resulted however, in the fact that both opponents, Barkla and Bragg, remained unshaken in their respective convictions about the nature of X-rays for many years. We might mention, in this context, that Bragg was also not impressed by the result of an older experiment, which Erich Marx (1874–1956) of Leipzig had expounded at the 77th *Naturforscherversammlung* in Meran in September

[9]In a previous experiment Barkla had observed the fact that the intensity of the secondary (cathode) radiation, stimulated by X-rays falling on metallic surfaces, depended on the orientation of the X-ray tube, an effect which pointed to a certain polarization of the primary (X-) rays (Barkla, 1904).

[10]Immediately after the appearance of Bragg's paper in October 1907, Barkla criticized the neutral-pair hypothesis of X-rays as being at variance with his own observation on X-ray scattering (Barkla, 1907); upon which Bragg replied that such a conclusion rested on an incorrect evaluation of the neutral-pair theory by Barkla (Bragg, 1908a).

1905, according to which X-rays should move with the same velocity as light or Hertzian waves (Marx, 1905). Bragg did not hesitate to call the experiment 'most ingenious' (Bragg, 1907, p. 446); still he concluded after an analysis of the details of Marx's experimental set-up: 'It is clear that the experiment of Marx is quite consistent with the hypothesis that X-rays are complex, and consist in part of aether pulses travelling with the velocity of light, and producing $\delta$-rays [i.e., comparatively slow electrons], and in part of material particles, or pairs, travelling at a speed as yet undetermined, and exciting high-speed cathode rays [or electrons]' (Bragg, 1907, p. 448).[11]

The Bragg–Barkla dispute was resolved in the years between 1909 and 1913. First Arnold Sommerfeld, who was an advocate of the ether pulse or electromagnetic wave theory of X-rays, developed a theory of X-ray production by the deceleration of electrons (at the anti-cathode of an X-ray tube) on the basis of pure electromagnetic theory; especially, he succeeded in explaining the spectrum and angular distribution of the intensity of the created X-rays in a satisfactory manner (Sommerfeld, 1909b). Sommerfeld's theory was taken over several years later by Ernest Rutherford who used it to account for the creation of $\gamma$-rays in radioactive substances (Rutherford, 1914c, 1915). He also explained the observed spectra of $\beta$- and $\gamma$-rays and the connection between both types of radiation with this theory; in particular, the $\beta$-rays should exhibit a continuous velocity spectrum if $\beta$-particles emitted from the atomic nucleus lose energy when passing through the electron cloud surrounding the nucleus, and this energy is turned over the $\gamma$-rays.

The second, even more important step towards the resolution of the Barkla–Bragg dispute came in April 1912 when Walter Friedrich and Paul Knipping, acting on the proposal of Max von Laue, discovered the diffraction patterns of X-rays on passing through crystals; these patterns were interpreted by von Laue as being due to the scattering of electromagnetic radiation by the three-dimensional grating of the atoms constituting a crystal, with a wavelength of the order of magnitude of the lattice constant (Friedrich, Knipping and Laue, 1912). Again Ernest Rutherford was the first to demonstrate, together with his student Edward Neville da Costa Andrade (born 1887), that a similar diffraction pattern would arise in the case of $\gamma$-rays of radium B associated with wavelengths of $0.799$ to $1.365 \times 10^{-8}$ cm (Rutherford and Andrade, 1914a), and for the penetrating $\gamma$-rays of radium B and radium C associated with wavelengths of $0.71$ to $4.28 \times 10^{-9}$ cm (Rutherford and Andrade, 1914b).[12]

[11] In his experiment of 1905 Marx compared the velocity of electromagnetic pulses with the velocity of X-rays in the following way. He observed the intensity of cathode rays created by the impact of X-rays on a saucer-shaped electrode, collected by a Faraday cylinder that was connected with an electrometer; if the electrode, from whose surface the cathode rays emerged, received an electric pulse travelling the same distance as the X-rays, then Marx obtained a maximum of the cathode ray intensity.

[12] Rutherford and Andrade, for their demonstration, used the X-ray diffraction method of William Henry Bragg and William Lawrence Bragg (1913).

Bragg, although he was aware of the fact that the discovery of X-ray diffraction considerably weakened the views he had held on the nature of $\gamma$- and X-rays, nevertheless emphasized in November 1912: 'The properties of X-rays point clearly to a quasi-corpuscular theory, and certain properties of light can be similarly interpreted. The problem then becomes, it seems to me, not to decide between two theories of X-rays, but to find . . . one theory which possesses the capacities of both' (Bragg, 1912, pp. 360-361).

## Heat Radiation and Its Description (1777–1900)[13]

The establishment of X- and $\gamma$-rays as electromagnetic radiation signified a major step forward in the recognition of the nature of radiation, because it extended the spectrum to wavelengths that were several orders of magnitude shorter than the ones known up to that time. On the other hand, the ultrashort wavelength (or extremely high frequency) radiation displayed properties, e.g., the ionization of gases, which had not been fully understood in terms of the electromagnetic wave theory. It was these properties which stimulated William Henry Bragg to claim that X- and $\gamma$-rays might be composed of material particles.

During the first decade of the twentieth century a question arose about the nature of another kind of radiation which, in contrast to the radiations mentioned above, could hardly be called a new. This was the phenomenon of 'heat radiation' or 'radiant heat,' as it was originally called by the Swedish chemist Carl Wilhelm Scheele (1742–1786) in 1777.[13] Johann Heinrich Lambert (1728–1777) in Berlin and Marc Auguste Pictet (1752–1825) in Geneva continued the study of Scheele's heat radiation in the eighteenth century; thus, for example, Pictet showed that heat radiation propagates with very high velocity, at least with that of sound and maybe with that of light (Pictet, 1790). His fellow citizen Pierre Prevost (1751–1839) formed, on the basis of Pictet's and his own experiments, a theory of heat radiation: according to this each warm body would emit heat rays that propogate straight into space until they hit another body, which either absorbs or reflects them or lets them pass (Prevost, 1809). In the beginning of the nineteenth century Friedrich Wilhelm (Sir William) Herschel (1738–1822) investigated the difference between light and heat radiation, and thus discovered the infrared radiation of the sun as being responsible for thermal effects (Herschel, 1800a, b). He arrived

[13] The history of the understanding of heat radiation goes back to ancient times. The connection between the radiating, light-emitting sun and the occurrence of heat effects on earth had been evident for a long time. On the other hand, it has been attributed to the Greek scientist Archimedes (287–212 B.C.) that he made use of burning reflectors to destroy hostile ships in the war for his hometown Syracuse against the Romans. In the late seventeenth century the Saxonian scientist and philosopher Ehrenfried Walter von Tschirnhaus (1651–1708) made particularly large burning-reflectors (of about 1 m in diameter) in order to produce locally high temperatures; he thus invented European porcelain. Scheele, who still adhered to some kind of material or 'phlogiston' hypothesis of heat, proposed the following explanation for the close relationship between heat and light: heat should be a chemical compound of oxygen with phlogiston, while light should have the same composition with just a higher amount of phlogiston (Scheele, 1777).

at the conclusion that both heat radiation and visible light possess the same nature, since they exhibit the phenomena of refraction and reflection alike, a conclusion, which was put on firmer grounds about three decades later by the Italian physicists Leopoldo Nobili (1784–1835) and Macedonio Melloni (1798–1854), and which was finally settled by Karl Hermann Knoblauch (born 1820) in Berlin in the mid-1840s.[14] Gustav Kirchhoff (1824–1887) then presented his fundamental theoretical investigation on the absorption and emission of radiation by arbitrary bodies. Among other results, he demonstrated that the ratio of the emissivity to absorptivity for a given body should only be a function of $\lambda$, the wavelength of the radiation considered, and of the body's temperature $T$ (Kirchhoff, 1859b).

The experimental determination of this universal function, $\Phi(\lambda, T)$, which could also be identified as the intensity distribution of the radiation of an absolute black body, had been undertaken in the following decades, with Friedrich Paschen (1865–1947) finally obtaining the formula (with $c_1$ and $c_2$ constants)

$$\Phi(\lambda, T) = c_1 \lambda^{-\kappa} \exp\left(-\frac{c_2}{\lambda T}\right), \tag{4}$$

where $\kappa$ assumed the value 5.5 (Paschen, 1896). Almost simultaneously Wilhelm Wien (1864–1928) derived the same formula on theoretical grounds, predicting an exponent $\kappa = 5$ (Wien, 1896). Wien's formula, i.e., Eq. (4) with $\kappa = 5$, became established during the following years by the work of several independent investigators until, in October 1899, Otto Lummer (1860–1925) and Ernst Pringsheim (1859–1917) noticed systematic deviations for very long wavelengths (Lummer and Pringsheim, 1896b). Heinrich Rubens (1865–1922) and Ferdinand Kurlbaum (1857–1927), by means of independent measurements extending to even longer wavelengths, fully confirmed the existence of such deviations and yielded accurate data within a year (Rubens and Kurlbaum, 1900), upon which Max Planck subsequently established his blackbody radiation law (Planck, 1900c), i.e.,[15]

$$\Phi(\lambda, T) = c_1 \lambda^{-5} \frac{1}{\exp\left(\frac{c_2}{\lambda T}\right) - 1}. \tag{5}$$

Less than two months later Planck completed a theoretical derivation of Eq. (5), based on electromagnetic radiation theory and on statistical considerations. In this derivation, Planck involved the new concept of finite energy-quanta, $h\nu$, that was associated with the emission and absorption of radiation of a given frequency $\nu (= c/\lambda$, $c$ being the velocity of light *in vacuo*) of radiation (Planck, 1900f).[16]

[14] For a discussion of the early historical development of the physics of heat radiation we refer to Ferdinand Rosenberger's *Geschichte der Physik*, Volume 3 (1890), and to Ernst Mach's *Principien der Wärmelehre* (Mach, 1896).

[15] A detailed discussion of the developments during the second half of the nineteenth century can be found in Volume 1, Section I.1.

[16] For details we refer to our discussion in Volume 1, Section I.2.

Max Planck's formula, Eq. (5), must be considered, in the light of the development of the physics of heat radiation, as the solution crowning the endeavours of several centuries and involving many brilliant investigators. The main point was that the complete identification of heat radiation and light finally seemed to have been proved. Yet, on this very point in Planck's treatment serious opposition arose. Wilhelm Wien, recognized as the greatest authority in the field of heat radiation, argued in October 1900:

> I must emphasize that I still stick, in contrast to Mr. Planck (see Planck, 1900a, p. 725], to my previously stated opinion [see Wien, 1893b, p. 633] that short and long electromagnetic waves—as far as their relations to heat radiation are concerned—differ more than just quantitatively from each other. With respect to the process of absorption it is generally assumed that longer waves can be described by a single vector, or, what amounts to the same, that matter can be regarded in that case as being continuous; however, for shorter wavelengths the influence of the molecular constitution of bodies comes into play. Exactly the same must be true for emission processes. Therefore, I believe it to be improbable from the very beginning to assume that a radiation law, which rests on the molecular hypothesis, should also be valid for very long waves. The agreement [of the radiation law, Eq. (4)] with the empirical data in the case of short wavelengths obviously shows that the assumptions made [in the derivation of that law] are approximately satisfied for wavelengths that are not too long. (Wien, 1900b, pp. 537–538)[17]

Wien's argument in favour of the existence of two types of heat radiation seemed to repeat a development that had occurred in the nineteenth century. Then many physicists had seriously discussed whether there existed two different ethers, one propagating light and the other propagating heat radiation; this view, to which for instance Melloni adhered, had been challenged by André Marie Ampère who believed that one ether served both purposes. Planck joined in the unitary view of Ampère, while Wien preferred the view of Melloni. The success of the new radiation law, Eq. (5), which brilliantly stood through all further experimental tests, seemed to prove Planck right. But, surprisingly, the idea of Wien, which appeared to have been overcome in 1900, also received a fresh revival a few years later, leading then to important progress in the understanding of the nature of radiation.

## The Hypothesis of Light-Quanta (1905–1909)

Albert Einstein was responsible for the revival of interest, and most of the following investigations concerning the nature of radiation. He first dealt with the problem of heat radiation in a paper submitted in March 1904; there he

[17] Planck had (from 1897) begun to concern himself with the theoretical foundation of the blackbody radiation law, and finally succeeded in deriving Wien's law on the basis of the electromagnetic theory and thermodynamics (Planck, 1899, 1900a). After generalizing Wien's law into Eq. (5) in October 1900, he looked for and obtained a theoretical derivation by again using a purely electromagnetic description of heat radiation.

established a relation between two known laws, i.e., between the law of Josef Stefan and Ludwig Boltzmann (stating the proportionality of the energy of the blackbody radiation and the fourth power of the absolute temperature) and the so-called displacement law of Wien (claiming that the product of absolute temperature and the wavelength, $\lambda_m$, associated with the maximum intensity, was a constant and independent of temperature); he did this by considering energy fluctuations of blackbody radiation following from kinetic theory (Einstein, 1904). Nearly a year later he sent another paper to *Annalen der Physik*, in which he treated Wien's radiation law, a law that he knew provided a satisfactory description of a large part of the energy distribution in blackbody radiation (Einstein, 1905b). First, if he assumed the usual electromagnetic description of heat radiation, then the theoretical frequency distribution

$$\rho_\nu(T) = \frac{8\pi\nu^2}{c^3}\frac{R}{N_0}T, \tag{6}$$

(where $R$ and $N_0$ denote the universal gas constant and the Avogadro or Loschmidt number, respectively) followed. The expression on the right-hand side of Eq. (6), however, neither agreed with what followed from Planck's law, Eq. (5), namely,

$$\rho_\nu(T) = \frac{8\pi h\nu^3}{c^3}\cdot\frac{1}{\exp\left(\dfrac{h\nu N_0}{RT}\right)-1}, \tag{7}$$

nor with what followed from Wien's law, Eq. (4), namely,

$$\rho_\nu(T) = \alpha\nu^3\exp\left(-\frac{\beta\nu}{T}\right), \tag{8}$$

with $\alpha$ and $\beta$ being, up to factors of $c$, the constants $c_1$ and $c_2$.

Still, Einstein noticed the fact that Planck's distribution law, Eq. (7), passed over, for large values of $T/\nu$, into his Eq. (6). Hence for large wavelengths and large radiation densities the usual electromagnetic theory indeed described the situation adequately, just as Wien had conjectured in 1900. The question remained, however, as to what happened for small values of $T/\nu$, representing short wavelengths and low density of the blackbody radiation. Empirically this density obeyed Wien's law, but this did not solve the problem. Here perhaps Wien's old hint might help i.e., the molecular structure might change the usual electromagnetic radiation picture.

In order to tackle the problem, Einstein used the fluctuation concept, by means of which he had succeeded a year earlier in illuminating the relationship between the Stefan–Boltzmann law and Wien's displacement law. Especially, he analyzed the volume fluctuation of radiation obeying Wien's law, Eq. (8), and compared it with the volume fluctuations of an ideal gas according to the kinetic theory. The result was very surprising, as he found: 'Monochromatic radiation of low density (within the range of Wien's radiation formula) behaves, in a thermodynamic sense, as if it consisted of mutually independent radiation quanta of magnitude $R\beta\nu/N_0$' (Einstein, 1905b, p. 143).[18] In other words, the short wavelength radiation described by Wien's law seemed to consist of light-quanta—their magnitude $R\beta\nu/N_0$ being equal to $h\nu$, with Planck's constant $h$ being properly inserted—which behaved like the molecules of a dilute gas in kinetic gas theory.

After obtaining this strange result, Einstein immediately pointed to several known phenomena, whose explanation seemed to imply a similar molecular structure of light, especially the photoelectric effect. Indeed, even earlier than Einstein, Joseph John Thomson had interpreted the fact that X-rays, when passing through gases, ionize only a few molecules, as being incompatible with the usual view of a continuous, uniform energy distribution existing at the front of an electromagnetic wave; he claimed rather that the fronts of X-rays consisted of localized spots of high intensity 'analogous to a swarm of cathode rays' (Thomson, 1904c, p. 63). Afterwards Johannes Stark presented further examples of the atomistic consititution of radiation (e.g., Stark, 1909b, c).[19] How did such an assumption, in spite of its obvious success, fit into the usual view of the constitution of radiation being an electromagnetic wave phenomenon? Especially, how could both views, the new light-quantum hypothesis and the classical wave picture, fit together with Planck's law of blackbody radiation?

Einstein answered the last question in a paper entitled, '*Zum gegenwärtigen Stand des Strahlungsproblems*' ('On the Present Status of the Radiation Problem') and submitted to *Physikalische Zeitschrift* in January 1909 (Einstein, 1909a). He had already suggested in 1906 that Planck's theory somehow contained his own concept of light-quanta (Einstein, 1906c); now he analyzed the derivation of the radiation law with the most critical care. First, he confirmed his result of 1905 that blackbody radiation, when treated according to the electromagnetic theory, led inevitably to the density distribution, Eq. (6)—meanwhile the same formula had been established by Lord Rayleigh (1842–1919), James Jeans (1877–1946) and Hendrik Antoon Lorentz—which did not agree with the empirical facts.[20] Hence Planck's derivation of the radiation law, Eq. (5), which also rested on the electromagnetic theory of heat radiation, could not be correct. Einstein even traced what appeared to him to be the error in Planck's treatment, namely his use of Boltzmann's statistical expression for the entropy. If Planck had used it

[18] We have discussed the details of Einstein's analysis in Volume 1, Section I.3.

[19] For details of Stark's contributions to the light-quantum, see our discussion in Volume 1, Section I.5.

[20] The problem of obtaining the radiation formula from classical theories—i.e., Maxwell's electrodynamics and the Boltzmann-Gibbs' statistical mechanics—has been discussed in Volume 1, Section I.4.

properly, contended Einstein, then he would have arrived at Eq. (6) instead of Eq. (7). 'As much as every physicist must be happy about the fact that Mr. Planck avoided this demand in such a successful manner,' he emphasized, 'it should also not be forgotten that Planck's radiation formula is not consistent with the theoretical foundation from which Mr. Planck started' (Einstein, 1909a, pp. 187–188).

Since, however, Planck's law described the data perfectly, one again had to ask the question as to what picture it implied of the structure of heat radiation. Einstein solved the problem by applying the method which had already helped him in 1904 and 1905. He now considered the energy fluctuations following from Eq. (7) and obtained the result

$$\overline{\Delta^2} = \overline{(E - \bar{E})^2} = h\nu\bar{E} + \frac{c^3}{8\pi\nu^2\, d\nu}\frac{\bar{E}^2}{V}, \tag{9}$$

where $E$ denotes the radiation energy in the volume $V$, and $\bar{E}$ is its average value. After observing that the second term on the right-hand side followed from the electromagnetic theory, he noted: 'However, we would have obtained this second term for $\overline{\Delta^2}$, alone if we had started from Jeans' formula. The first term of the above expression for $\overline{\Delta^2}$ . . . is therefore not compatible with the present [radiation] theory' (Einstein, 1909a, p. 189). The fluctuation formula, Eq. (9), rather indicated the existence of two separate phenomena, and Einstein stated further that 'due to Planck's radiation formula the effects of both fluctuation phenomena behave like fluctuations originating from mutually independent causes' (Einstein, 1909a, p. 190); the first term in particular had to be interpreted as a fluctuation arising in the case that 'radiation consisted of quanta of the size indicated [i.e., $h\nu$]' (*ibid.*, p. 191).

Einstein again presented these results, especially his dual electromagnetic wave-light-quantum picture of radiation—he did not, in aggreement with previous experience, confine this view to heat radiation alone—at the 81st Assembly of German Scientists and Physicians in Salzburg in September 1909. Max Planck, who was present and even chaired the session in which Einstein spoke, raised—in the discussion following Einstein's presentation—doubts as to whether the light-quantum hypothesis was really necessary; he rather proposed 'to shift the entire difficulty of the quantum theory [i.e., the difficulties involved in laying the foundation of the radiation law and in explaining such phenomena as the photoelectric effect] to the question of interaction between matter and radiant energy' (Planck, in Einstein, 1909b, pp. 825–826). On the other hand, Planck claimed that 'the processes in pure vacuum could then, for the present, still be explained with the help of Maxwell's equations' (Planck, in Einstein, 1909b, p. 826).

## Proof of the Light-Quantum (1916–1922)

Planck repeated the same critical statements on several later occasions (e.g., Planck, 1910a). For Planck, who had experienced in his younger years the triumph of Maxwell's electrodynamics through the discovery of the electromagnetic waves by Heinrich Hertz and who had himself achieved major successes by applying the theory (e.g., to the explanation of dispersion phenomena), any doubt in the wave nature of radiation appeared to shake the very foundations of contemporary physics. Planck had the greatest respect for the physical intuition and accomplishments of Einstein; still, when he tried to secure for Einstein a suitable position in Berlin, he wrote: 'That he [Einstein] may sometimes have missed the target in his speculations, as for example in his theory of light-quanta, cannot really be held against him' (Planck, in the draft of a letter of recommendation for Einstein to the Prussian Ministry of Education; quoted in Clark, 1971, p. 215).

Such opposition, which was shared by pratically all physicists interested in the quantum theory—even the former supporter of the light-quantum, Johannes Stark, who withdrew from further development after 1913—did not convert Einstein. He remained silent on the question of light-quanta for several years. Then, in the summer of 1916 he proposed a new derivation of Planck's radiation law, based on a consideration of heat radiation interacting with atoms that were described by Bohr's theory (Einstein, 1916d). In a subsequent investigation he analyzed the exchange of momentum between radiation and atoms; he concluded that 'we arrive at a consistent theory only when we consider those elementary processes [between radiation and atoms] as completely directed events' (Einstein, 1916e, p. 49). More explicitly, he stated in a letter to his friend Michele Besso: 'Thereby it follows that in each elementary energy transfer from radiation to matter, the momentum $h\nu/c$ is transferred to the molecule. Hence we conclude that every such elementary process is a completely directed event. With that, the light-quanta must be considered as good as being substantiated' (Einstein to Besso, 6 September 1916).

Einstein tried for several years to support the result of this analysis by a more direct demonstration of the directed light-quantum without getting any further.[21] A real breakthrough, however, occurred in an entirely different investigation. The American experimental physicist and expert in X-ray physics, Arthur Holly Compton (1892–1962), then working at Washington University in St. Louis, Missouri, studied in late 1922 the scattering of X-rays from atoms of light elements. In a talk, entitled 'A Quantum Theory of the Scattering of X-rays of Light Elements' and presented at the Chicago meeting of the American Physical Society, 1–2 December 1922, he revealed results showing an increase of the wavelength of scattered radiation, whose magnitude depended on the scattering angle (Compton, 1923a). Both the energy loss of the X-rays in scattering and the observed angular distribution of the scattered rays as a function of their wavelength, could well be described by what Compton called 'quantum theory,' i.e., the hypothesis of X-ray quanta having definite energy ($h\nu$) and directed momentum (of magnitude $h\nu/c$), just as Einstein had demanded. He also stated clearly in his published paper, submitted soon after the meeting: 'Such a change of wavelength is directly counter to [Joseph John] Thomson's theory of scattering

[21] We refer, for example, to an experiment which he suggested in late 1921 (Einstein, 1921b) and which we have discussed in Volume 1, Section V.1, and also to the experiment which Schrödinger devised in 1919 for the same purpose. (See Chapter I, Section I.5.)

[of electromagnetic waves of short wavelength], for this demands that the scattering of electrons, radiating as they do because of their forced vibrations when traversed by a primary X-ray, shall give rise to radiation of exactly the same frequency as that of the radiation falling upon them. Nor does any modification of the theory such as the hypothesis of the large electron suggest a way out of the difficulty. This failure makes it appear improbable that a satisfactory explanation of the scattering of X-rays can be reached on the basis of classical electrodynamics' (Compton, 1923b, p. 485).

The importance of Compton's experimental result was immediately recognized by Arnold Sommerfeld, one of the earlier strongest supporters of the electromagnetic wave theory of X-rays. He had the good fortune of just spending the winter semester of 1922/1923 as Carl Schurz Professor of Physics at the University of Wisconsin in Madison, and he quickly spread the news in America and, after his return to Munich, in Europe as well.[22] Thus, for example, he discussed the confirmation of the hypothesis of the light-quantum with Albert Einstein in August 1923, who was happy to hear details about the Compton effect, and he (Sommerfeld) included it in a prominent place in the next, fourth edition of his book *Atombau und Spektrallinien*, which appeared later in 1924 (Sommerfeld, 1924d). There he concluded Section 7, entitled '*Wellentheorie und Quantentheorie*' ('Wave Theory and Quantum Theory') and which was devoted to the quantum phenomena of light—such as the photoelectric effect and the Compton effect—with the words: 'We have revived Newton's corpuscles [of light]. Yet, we cannot dismiss the wave concept with these pseudo-corpuscles. The reason is that every spectroscopic measurement depends on interference phenomena, i.e., wave phenomena. While in quantum theory the wavelength constitutes a derived concept, it is the primary concept in experiments where frequency and energy are the derived quantities .... At the moment we have to admit the indispensability of wave theory [of light]; on the other hand, we are also sure of the indispensability of the quantum structure. In this question modern physics faces irreconcilable contradictions and has to admit a frank "*non liquet*" ["it is not clear"]' (Sommerfeld, 1924d, p. 59).

## II.2 The Side-Stages of the Development in Quantum Theory

### Early Centres of Quantum Theory: Berlin, Berne, Copenhagen, Munich

Quantum theory had originated in 1900 in Berlin with the theoretical work of Max Planck, based on the detailed data obtained by Heinrich Rubens and Ferdinand Kurlbaum at the *Physikalisch-Technische Reichsanstalt*, which was also

[22] The first mention of the Compton effect was contained in a letter which Sommerfeld wrote to Niels Bohr in Copenhagen; in it he talked about 'most interesting scientific news' which might imply that 'the wave theory of X-rays would finally have to be given up' (Sommerfeld to Bohr, 21 January 1923).

situated in Berlin.[23] It had received important support and extension from 1905 onwards by the theoretical investigations of Albert Einstein in Berne, especially through his proposal of the light-quantum hypothesis and his explanation of the low-temperature behaviour exhibited by the specific heats of solids. Walther Nernst (1864–1941)—from 1905 director of the physico-chemical institute at the University of Berlin—and his collaborators, notably Arnold Eucken (1884–1950), Frederick Alexander Lindemann (1886–1957) and Alfred Magnus (1880–1960), had (from 1909) obtained data at low temperatures which allowed them to check Einstein's predictions (Einstein, 1906g). Nernst and Lindemann found certain deviations (Nernst and Lindemann, 1911a, b) which ultimately resulted in an extension of the quantum theory of the specific heats of solids (Debye, 1912c; Born and von Kármán, 1913a). Thus in the first dozen years of the twentieth century two centres of quantum theory had emerged: Berlin, its birthplace, where Planck and Nernst resided, and the various places where Einstein worked, i.e., Berne (until the fall of 1909), Zurich (from October 1909 to April 1911), Prague (from April 1911 to August 1912), and again Zurich (from October 1911 to April 1914), until he finally came to Berlin in 1914. Besides Nernst and his school, Johannes Stark (1874–1957) had been the most active supporter of the quantum theory, starting in the fall of 1907—while at the University of Greifswald—with an analysis of the minimal wavelength of X-rays (Stark, 1907b), and later developing a quantum picture of the discrete radiation and ionization phenomena observed with canal rays, atoms and molecules (Stark, 1907c; 1908a, c, d). After moving (in 1909) to Aachen to assume a full professorship of experimental physics, Stark had continued his interest in the quantum theory, treating with it the problem of the creation of X-rays (Stark, 1909b, d).

In the period following the first Solvay Conference (30 October–3 November 1911) the interest in quantum theory and the phenomena associated with it grew rapidly beyond the formerly limited circle of Planck, Einstein, Stark and Nernst. It reached many places in Germany, in Western Europe, and in Scandinavia. Still, Berlin remained the centre of research in quantum physics, with Planck and Nernst continuing their respective work, and Emil Warburg (1846–1931)—since 1905 president of the *Physikalisch-Technische Reichsanstalt*—becoming involved in quantitative experiments on photochemical reactions (Warburg, 1911, 1912), which ultimately fully confirmed the quantum-theoretical photochemical equivalence law of Stark and Einstein (e.g., Warburg, 1918). The investigations of his student James Franck (1882–1964) and of Gustav Hertz (1878–1975), beginning in 1911 at the University of Berlin and aimed at determining the ionization potential of gases, had also brought about a substantiation of quantum-theoretical ideas (Franck and Hertz, 1911; 1913a, b, c; 1914a, b).

The situation, however, changed dramatically after the first paper 'On the Constitution of Atoms and Molecules,' written by Niels Bohr of Copenhagen, appeared in the July 1913 issue of the British journal *Philosophical Magazine*

[23] We prefer to stick to the conventional date for the beginning of quantum theory, in spite of Thomas S. Kuhn's plea for a later date (Kuhn, 1978). For details, see Volume 1, Chapter I.

(Bohr, 1913b). In it Bohr had proposed a theory of the discrete spectral lines emitted by atoms and molecules, based upon their composition of atomic nuclei and electrons and using quantum-theoretical hypotheses. He had immediately succeeded in explaining, in the same paper, the so-called Balmer spectrum of hydrogen. The subsequent systematic measurements of the characteristic X-rays emitted by different substances, first carried out in Manchester and continued in Oxford, provided an immediate proof of Bohr's atomic theory (Moseley, 1913; 1914b). The extension of this theory—reached by Arnold Sommerfeld in Munich, who took into account elliptic and even relativistic orbits for the electrons moving around the atomic nucleus (Sommerfeld, 1915b, c) had fully opened a wide, new field of application to quantum theory: the spectroscopy of atoms and molecules. Ever since Gustav Kirchhoff and Robert Bunsen had introduced spectral analysis in 1860, the experimentalists had collected data on the discrete radiation emitted by chemical substances; all this rich material was now suddenly declared to constitute an essential part of quantum physics. Spectroscopic results, old and new, would therefore dominate and even determine the progress of quantum theory during the next decade.

## Post-World War Centres and the Sommerfeld School

The crucial steps that rendered the study of atomic structure a part of quantum-theoretical research occurred during World War I. During this four-year period the pioneers, especially Bohr and Sommerfeld on the theoretical side and experimentalists like Friedrich Paschen in Tübingen, worked mainly alone, as most of their students and collaborators had gone away to do military service.[24] After the war was over regular university life started again in Germany, France and Great Britain. Soon an increasing number of lectures on various topics of quantum theory was offered, especially in Germany, ranging from heat radiation and the theory of solids to atomic structure. While formerly the only textbook available on the field was Max Planck's pioneering *Vorlesungen über die Theorie der Wärmestrahlung* (Planck, 1906; second edition, 1913), new books now appeared on the market, e.g., Artur March's *Theorie der Strahlung und der Quanten* (March, 1919) or Fritz Reiche's historical account *Die Quantentheorie: Ihr Ursprung und ihre Entwicklung* (Reiche, 1921). Both books not only treated radiation theory but covered the full range of quantum phenomena that had been discovered and discussed so far. The most successful book, however, was Arnold Sommerfeld's *Atombau und Spektrallinien* (Sommerfeld, 1919), in which the author summarized and explained the results of Niels Bohr's and his own researches on the quantum theory of atomic structure. This book, which quickly went through three editions (second edition: 1920, third edition: 1922), turned out to be a real bestseller, which was read and consulted not only by students of physics, but also of chemistry and the related fields. One would not be

[24] In the case of Bohr in neutral Denmark this is not strictly true. From 1916 he had acquired the assistance of Hendrik Kramers (1894–1952), who came from the likewise neutral Netherlands.

wrong if one claimed that *Atombau und Spektrallinien* was responsible for the fact that not only the number of lectures on quantum theory increased dramatically after its appearance, but also that most of the lectures offered were on atomic structure.

Shortly before Sommerfeld's book came out Niels Bohr had published two memoirs, entitled 'On the Quantum Theory of Line Spectra', in which he summarized the results of several years of hard work of his own on the foundations of this theory (Bohr, 1918a, b). Bohr's papers were difficult to understand; nevertheless, they became, together with Sommerfeld's popular book, the bible for a new generation of physicists who pursued the study of quantum theory. Bohr and Sommerfeld, the pioneers in the research on atomic structure, also became the champions of further development in this field and established influential schools in Copenhagen and Munich. Sommerfeld, a most effective and dedicated teacher, soon began to spread the influence of his school all over Germany. He secured for his disciples quite a few of the newly founded chairs in theoretical physics, where they then started to teach and promote quantum theory themselves. Thus Wilhelm Lenz (1888–1957) first went to Rostock in 1920 and then to Hamburg a year after, and Adolf Kratzer (born 1893) held the professorship of physics in Münster from 1922. A third centre of atomic theory, closely related to Copenhagen and Munich, came into existence in Göttingen after Max Born and James Franck accepted chairs in theoretical and experimental physics, respectively, at the university there in 1921.

The main centres of quantum-theoretical research in the early 1920s, Copenhagen, Munich and Göttingen, in each of which numerous doctoral students and research associates worked intensely on the latest and most difficult problems of atomic structure, also extended their influence to places where research on quantum-theoretical topics was carried out only occasionally, with few people involved. So the University of Münster, where Adolf Kratzer investigated the details of molecular spectra, could simply be regarded as a subsidiary of the Munich centre of atomic theory of his teacher Sommerfeld. A similar relationship existed between the theoretical institutes of the *Technische Hochschule* in Stuttgart and Munich. Peter Paul Ewald (1888–1985), who had also obtained his doctorate with Sommerfeld (in 1912) and had spent most of his early professional years in Munich (*Habilitation* in 1917), occupied the chair of theoretical physics in Stuttgart from 1921, becoming *Ordinarius* in 1922. Although he did not work specifically on problems of atomic structure—being involved rather with the atomic composition of crystal lattices—he employed Erwin Fues (born 1893), another former student of Sommerfeld's, as his assistant, and Fues did indeed devote his research to the explanation of spectra within the Bohr–Sommerfeld theory. Finally in Hamburg, Wilhelm Lenz held the chair of theoretical physics from 1921, attracting Wolfgang Pauli to his institute in 1922. Pauli kept up a sincere relationship with his revered teacher Sommerfeld, but also became close to Niels Bohr through several extended stays in Copenhagen; he thus represented a tie between the institutes in Munich and Copenhagen, even at times when the goals of Sommerfeld and of Bohr seemed to go in rather different directions. The University of Hamburg also gained a remarkable reputation in experimental atomic physics after Otto Stern (1888–1969) accepted an experimental professorship, endowed with a special laboratary for the molecular beam method, in early 1923. After the great success, which he had earlier achieved with this method in Frankfurt-am-Main through the discovery of the so-called Stern–Gerlach effect (in 1922), he would (in Hamburg) contribute many further important results. He created a flourishing experimental school of his own, attracting able collaborators from Germany and abroad, such as Immanuel Estermann (1900–1983) and the American Isidor Isaac Rabi (born 1898).

The great interest in the new quantum and atomic theory supported the establishment of separate chairs for theoretical physics at German universities. Thus the chairs of Ewald in Stuttgart and of Lenz in Hamburg were created.[25] In the latter case, of course, it also helped that the University of Hamburg had only been established shortly before; clearly they were trying to have modern science and atomic physics, which had to be considered among its most prominent representatives, gain a favoured place. However, at the old University of Tübingen, whose existence dates back to 1477, a position for a theoretical physicist was also installed when Friedrich Paschen invited Alfred Landé (1888–1975) to an extraordinary professorship in 1922. Landé, another recipient of a doctorate under Sommerfeld (in 1914), continued in Tübingen, in close association with the experimental physicist Ernst Back (1881–1959), with his successful work on the analysis of complex spectra and their Zeeman effects, resulting, e.g., in the famous interval rules (Landé, 1923a). Landé and Back, with the experienced Paschen backing them, made sure that Tübingen kept and retained its reputation as a Mecca of refined spectroscopy and of the quantum-theoretical analysis of complex spectroscopic data throughout the 1920s.

## Side-Stages: Breslau and Vienna

In contrast to the physics department at the University of Hamburg, that in Breslau represented neither a new place for work in quantum theory, nor did it possess any direct relations with the three leading institutes in Copenhagen, Göttingen and Munich. It was back in 1904 that Otto Lummer (1860–1925), who had worked since his student days at the University of Berlin and then at the *Physikalisch-Technische Reichsanstalt*, and who had participated (beginning in the 1890s) in the measurements of heat and blackbody radiation, and had finally demonstrated with Ernst Pringsheim (1859–1917) the first deviations from Wien's radiation law (Lummer and Pringsheim, 1899b), came to Breslau to accept the chair of physics and the directorship of the physical institute at the University of Breslau. The following year Pringsheim, a native of Breslau, joined him as

[25] The history of the theoretical chair in Stuttgart begins, in fact, a little before Ewald, as Erwin Schrödinger had held an extraordinary professorship earlier in 1921, before he went as *Ordinarius* for theoretical physics to the University of Breslau.

professor of theoretical physics. Both then continued their measurements on heat radiation, and their institute soon became a flourishing place for young talented physicists interested in quantum theory, such as Rudolf Ladenburg (1882–1952) (who took an assistantship with Lummer in 1907), Max Born (who returned in the summer of 1907 to his hometown) and Fritz Reiche (1883–1969) (who came from Max Planck in Berlin). While Born left after a year and a half to pursue his scientific career in Göttingen and then later in Berlin and Reiche also returned to Berlin in 1912, Ladenburg remained in Breslau until the outbreak of World War I. He then performed military service, first at the front and, after being seriously wounded, as head of the *Artillerie-Prüfungskommision* (Artillery Testing Commision) in Berlin, where several other quantum physicists (e.g., Max Born, James Franck and Alfred Landé) were also employed.

Immediately after the war Ladenburg returned to Breslau and resumed his earlier experimental investigations of the anomalous dispersion of light. He had originally interpreted the results obtained on the basis of classical radiation theory, but now he proposed to apply quantum theory; especially, he used Einstein's idea of the emission and absorption coefficients of atoms (Einstein, 1916d) in order to derive from the data the number of dispersion electrons (Ladenburg, 1921a). He thus established the quantum theory of dispersion of light by atoms, which was to play an important role in the development of quantum mechanics a couple of years later. Before that, however, he improved and extended his measurements. In the theoretical interpretation of the data he received the help of his friend Fritz Reiche, who in late 1921 accepted the call to the theoretical chair formerly held by Pringsheim (until his death in 1917) and then later for a short time by Erwin Schrödinger (from February to the fall of 1921). Ladenburg and Reiche formed a successful team of independent and original researchers in experimental and theoretical atomic physics; their work contributed in providing a proof of Bohr's theory of atomic constitution and paving a way beyond it. They also trained good students, such as Rudolph Minkowski (1895–1976), who went into quantum theory. In 1924 Ladenburg accepted an invitation to direct the physics department at Fritz Haber's *Kaiser-Wilhelm-Institut für physikalische und Elekrochemie*, where he continued to investigate dispersion phenomena. Reiche, on the other hand, remained in Breslau, where he further worked on the quantum theory of dispersion of light by atoms. His student Willy Thomas arrived at a quantitative relation for dispersion electrons (in June 1925 and independently of Werner Kuhn in Copenhagen: Thomas, 1925a); this relation, sometimes called the Thomas–Reiche–Kuhn sum rule, kept its validity even after quantum mechanics replaced the old quantum theory. Reiche stayed on in Breslau until the Nazi government forced him to give up his chair in 1934.

The physics at Breslau, developed in the institute of Ladenburg and Reiche, was soon incorporated into the main stream of quantum and atomic theory, which Niels Bohr dominated. This was not necessarily the case with research on quantum theory carried out after World War I at other places. They include especially Paris in France and distant Dacca in India. However, before we summarize the events at those places, we wish to look first at the development of physics in Vienna after World War I.

In the years after 1910 the University of Vienna had held out good prospects of becoming, if not an important centre, at least a productive side-stage in quantum-theoretical research. We have discussed earlier (in Chapter I, Section I.2) how Arthur Erich Haas, a doctoral candidate in the philosophical faculty, suggested in 1910 a quantum-theoretical treatment of Joseph John Thomson's model of atomic structure (Haas, 1910a). Haas' proposal was very sharp, representing the first push in that direction; with the help of little known assumptions he managed to explain what nobody else could at that time: the dimensions of the hydrogen atom and the limiting frequency of the Balmer series. Ultimately his success even stimulated Friedrich Hasenöhrl, the skeptical professor of theoretical physics. Hasenöhrl preferred, however, a less audacious application of quantum-theoretical rules to explain the discrete spectra of atoms and molecules (Hasenöhrl, 1911). Karl Herzfeld, a young student at the University of Vienna, also worked on the same problem as his professor, Hasenöhrl (Herzfeld, 1912a). In July 1913 Hans Thirring, Hasenöhrl's assistant, published a work which represented the next Viennese contribution to quantum theory: in it Thirring improved upon a calculation of Max Born and Theodore von Kármán concerning the specific heat of crystals (Thirring, 1913). Later in the same year Thirring submitted two further papers on the same subject (Thirring, 1914a, b), and on the basis of this work he obtained his *Habilitation* early in 1915. Erwin Schrödinger, who had suggested the problem to Thirring, was also interested in the quantum theory of the specific heats of crystals, but did not publish anything at that time on it. Then World War I broke out, and Hasenöhrl, Herzfeld—who had returned in 1914, after two years of study at Munich and Göttingen, to the University of Vienna and received his doctorate—, Schrödinger and Thirring were called to military service.

Besides keeping the young, promising Viennese quantum physicists away from their institutes, the war decisively changed the physics situation in Vienna in just one stroke: the death of Hasenöhrl deprived theoretical physics of its leader. The search for a worthy successor to Hasenöhrl took years. Among the strongest candidates proposed were Marian von Smoluchowski and Arnold Sommerfeld, but the former died in September 1917 and the latter declined after extended negotiations in Vienna. Finally Gustav Jäger, a senior contemporary from the days of Stefan, Loschmidt and Boltzmann, then professor of physics at the *Technische Hochschule* of Vienna, was appointed.[26] He accepted on condition that Hans Thirring would assist him in teaching modern physical theories. When

[26] Gustav Jäger was born on 6 April 1865 in Schönbach near Asch in Bohemia. After studying physics at the universities of Vienna (with Josef Stefan) and Berlin (with Hermann von Helmholtz and August Kundt), he became assistant to Stefan and later to Ludwig Boltzmann. In 1905 he was called to the *Technische Hochschule* in Vienna as professor of '*allgemeine und technische Physik*' ('general and technical physics'). As a true disciple of Loschmidt and Boltzmann he showed a lifelong interest in the kinetic theory of matter, contributing especially to the theory of 'ideal' liquids.

Franz Exner retired two years later, Jäger was happy to exhange this theoretical chair for an experimental chair at the University of Vienna, and he proposed that Thirring become his successor in physical theory.[27] Thirring was ultimately appointed, first in 1921 as *Extraordinarius* and six years later as *Ordinarius.*

When negotiating for the succession of Hasenöhrl, Sommerfeld had expressed the opinion that 'Vienna did not have to look elsewhere for a replacement to Hasenöhrl, since there already existed an able new generation in Vienna' (Flamm, 1958, p. 4). By the new generation he was referring to the many former students of Hasenöhrl. Of these, Thirring certainly had to be considered as the first candidate. He had studied with Hasenöhrl since 1907, obtaining his doctorate under Hasenöhrl's supervision in 1911. From late 1910 he had also been the official assistant at the theoretical institute. Having been called to military service in January 1915, he had not been sent to the front, and soon (in March 1915) returned to Vienna with an order to develop some war-related technical ideas of his own. He had therefore been able to remain at home during the entire war and was able to give lecture courses on theoretical subjects, at least from the winter semester of 1916/1917.[28]

As the expert on modern theories during the time of Jäger's theoretical professorship, Thirring lectured on such topics as X-ray interferences and crystal structure (announced for the winter semester of 1918/1919; see *Phys. Zs.* **19**, 511, 1918) or on general relativity theory (winter semester of 1920/1921; see *Phys. Zs.* **21**, 647, 1920). He still remained actively interested in the quantum theory of crystal lattices, on which he wrote several papers (e.g., Thirring, 1920), but from 1918 he shifted his main research interest to relativity theory.[29] He became an expert in this field which was, for instance, reflected in the organization of the two volumes of the *Handbuch der Physik* with which the chief editors, Hans Geiger and Karl Scheel, entrusted him, namely Volume III (*Mathematische Hilfsmittel der Physik*, 1928) and Volume IV (*Allgemeine Grundlagen der Physik*, 1929). He also wrote two articles for the *Handbuch*, one on '*Elektrodynamik bewegter Körper und spezielle Relativitätstheorie*' (Thirring, 1927), and the other on '*Begriffssystem und Grundgesetze der Feldphysik*' (Thirring, 1929). In the latter review paper he treated, in a systematic way, both the field theories in prerelativistic and in relativistic physics, displaying his pedagogical talents as well as a capacity for the consistent organization of the rich material.

Hans Thirring was a worthy occupant of the theoretical chair at the University of Vienna. He delivered fine lectures on all fields of classical and modern theory.

[27] Jäger retired from his chair in 1934. He died on 8 January 1938.

[28] According to the *Vorlesungsverzeichnis*, published in the *Physikalische Zeitschrift*, Thirring announced in the winter semester of 1916/1917 a course on partial differential equations of mathematical physics (*Phys. Zs.* **17**, 519, 1916), in the following summer semester of 1917 a course on hydrodynamics (*Phys. Zs.* **18**, 182, 1917) and in the winter semester of 1917/1918 one on physics of crystals (*Phys. Zs.* **18**, 530, 1917).

[29] We have mentioned Thirring's first important contributions to General Relativity Theory in Footnote 453, Section I.5.

However, with respect to original research on topics of atomic and quantum physics he stood far behind Sommerfeld, the earlier candidate for Hasenöhrl's succession. He was perfectly aware of this fact and soon tried to send his outstanding students—for example, Victor Weisskopf and Guido Beck—to places where they could learn about the most recent developments first hand and obtain stimulation and adequate guidance in their own research.

The failure of the Vienna theoretical physics institute to become a centre of atomic and quantum theory may be partly ascribed to unfavourable conditions. After World War I the Austro-Hungarian Empire had been completely disintegrated. Vienna became the capital of a small country, German-speaking Austria, with limited economic resources, a kind of German province that was forbidden by the victorious Allies to join the German *Reich*. Academic positions were not richly endowed, and Thirring was lucky enough to supplement his salary by earnings from his technical inventions, e.g., from the production of selenium cells for cinematographic use.[30]

After World War I several people, who would later become famous representatives of quantum theory, left Vienna. For example, Karl Herzfeld first went to Munich, assuming an assistant's position at the physico-chemical laboratory of Kasimir Fajans. He became *Privatdozent* for theoretical physics and physical chemistry at the University of Munich, and as such was associated with the *Institut für Theoretische Physik*. There he found himself among a group of congenial people, especially the young collaborators and students of Sommerfeld, who was simultaneously a teacher and a participant in the discussions on the most recent research at that centre.[31] There he also met his fellow-countryman Wolfgang Pauli, who had come in the fall of 1918 from Vienna to begin his study of physics under Sommerfeld's guidance. Several months later, in the spring of 1920, Erwin Schrödinger set out on his way to scientific fame, which led him abroad first to Jena, Stuttgart and Breslau, then to Zurich and finally to Berlin.

The losses mentioned could of course not be compensated. In contrast to Berlin and some other places in Germany, where several extremely bright young scientists from Hungary arrived during the 1920s—e.g., Cornelius Lanczos, Leo Szilard, John von Neumann and Eugene Wigner—Vienna did not attract talent from the neighbouring, former Habsburg, countries.[32] Still, one physicist, Adolf

[30]The patents, obtained for his technical ideas, and his reputation as an inventor later helped Thirring to overcome a difficult time. When in 1938 he was dismissed from his chair by the Nazi Government—Thirring had become an active pacifist in the 1920s and 1930s—he joined industrial firms as technical consultant (first the Elin AG, later the valve laboratories of Siemens and Halske, both in Vienna). After World War II he would resume his chair. He retired in 1959, being succeeded by his son Walter. Hans Thirring died on 22 March 1976 in Vienna.

[31]Herzfeld was promoted in 1923 to an extraordinary professorship. In 1926 he went to The Johns Hopkins University, Baltimore, first as Speyer Guest Professor, then as professor of physics. In 1936 he moved to the Catholic University of America in Washington, D.C., as head of the physics department. He retired in 1968 and died in June 1978.

[32]The reason for this fact must be sought in the unfavourable conditions of post-war Austria. It was a small country, which offered little hope for a future career in the academic and industrial professions. Germany, on the other hand, possessed many more universities; it had remained a large country with an industry that developed despite inflation and war reparations.

Gustav Smekal, whose research concentrated on the problems of quantum physics, returned to Vienna in those days.

## Adolf Smekal and Quantum Theory in Vienna (1919–1923)

Smekal was born in Vienna on 12 September 1895, the son of an army officer. After graduating from the *Realschule* in Olmütz, Moravia, he began his studies at the *Technische Hochschule* in Vienna in the winter semester of 1912/1913 and then proceeded after a year to the University of Graz.[33] He obtained his doctorate in 1917 from Graz with a theoretical dissertation on a problem of statistical mechanics. In it he arrived at the conclusion that quantum-theoretical hypotheses were not really necessary if one wanted to avoid the equipartition of energy. However, he soon changed his opinion and turned into a serious supporter of quantum theory, after he had gone to Berlin for postdoctoral studies.[34] There he attended the lectures of Albert Einstein, Max Planck, Heinrich Rubens and Emil Warburg, and made the personal acquaintance of these famous scientists. On learning about the most recent research on quantum theory—he also attended (in May 1918) the Wolfskehl lectures of Max Planck in Göttingen—Smekal started a period of industrious publishing, dealing in his papers with many of the burning problems of that time, e.g., with the adiabatic principle (Smekal, 1918a) or with an application of Planck's first quantum hypothesis (of 1900) to paramagnetism (Smekal, 1918c). The lively discussions going on then in Berlin, on Bohr's theory of atomic structure, stimulated Smekal to enter the main field of his research in the following years, i.e., the interpretation of the discrete X-ray spectra emitted by atoms, which had been discovered by Charles Glover Barkla and Charles A. Sadler back in 1908 (Barkla and Sadler, 1908b).

The explanation of discrete X-ray spectra on the basis of Bohr's theory of atomic structure was undertaken, immediately after the pioneering paper of Bohr (Bohr, 1913b) had been published, by Henry Gwyn Jeffreys Moseley in England; the results had brilliantly confirmed Bohr's theory for $K_\alpha$ and $K_\beta$, the strongest lines of the so-called $K$-spectra (Moseley, 1913; 1914b).[35] In the following years X-ray spectroscopy was greatly developed through the work of Maurice de Broglie (1875–1960) in Paris, Manne Siegbahn (1886–1977) and his students in Lund and Ernst Wagner (1876–1928) in Munich. On the other hand, Walter Kossel (1888–1956) in Munich had proposed a detailed theoretical picture of the origin of X-ray spectra in the spirit of Bohr's theory: the $K$-lines were supposed to arise through jumps of electrons to the lowest or $K$-orbit from higher-lying

[33] Graduation from a *Realschule* did not entitle the student at that time to study at an Austrian university. He could do so only after spending one year at the *Allgemeine Abteilung* (general department) of the *Technische Hochschule* in Vienna. (For details of Smekal's biography, see: Flamm, 1960, and Forman, 1975.)

[34] Extreme nearsightedness freed Smekal of any military service even during the war. It did not prevent him, however, from becoming an enthusiastic alpinist.

[35] For the details of Moseley's work we refer to Volume 1, Section II.3.

orbits, the $L$-lines from jumps of electrons to the first excited or $L$-orbit in an atom, etc. (Kossel, 1914b).[36] Arnold Sommerfeld's theory of relativistic fine structure (Sommerfeld, 1915c) had removed some of the remaining difficulties, but a violation of the spectroscopic combination principle seemed to remain (Kossel, 1916b; Sommerfeld, 1916c). In the middle of 1917 Peter Debye, in Göttingen, had published a paper in which he suggested that three electrons instead of two (as assumed by Bohr and his followers) might occupy the $K$-orbit; he had thus succeeded in explaining the observed $K_\alpha$-data if he assumed, in addition, that the lines emerged through an electron jumping from an excited $L'$-orbit (not identical with the $L$-orbit associated with the $L$-spectra) to the $K$-orbit (Debye, 1917). Lars Vegard (1880–1950) of the University of Oslo subsequently had extended Debye's idea and claimed that in the $L$-orbit there were maximally seven electrons, in the next higher ($M$-) orbit eight electrons, etc. (Vegard, 1917a, b). Such occupation numbers obviously contradicted not only the observed periodicity of properties of chemical elements, but also other facts and principles. At this stage Smekal entered the field.

He wrote the first publication on X-ray spectra that bore his name, together with Fritz Reiche's, in Berlin. The authors completed it in March 1918 and submitted it to the *Annalen der Physik* (Reiche and Smekal, 1918). They tried to improve on Debye's calculation of the $K_\alpha$-lines of atoms, and they obtained the result that no fit to the available data could be made with co-planar $K$- and $L$-orbits. They therefore proposed a spatial model of the atoms. Smekal soon composed a second paper on X-ray spectra, which he forwarded to the Vienna Academy of Sciences, where it was presented at the session of 6 June 1918 (Smekal, 1918b). Again he attempted to calculate the $K_\alpha$-lines including perturbation of the orbits inferred from neighbouring orbits; he arrived at a satisfactory agreement with the observed data, if he assumed that the lines were emitted by a transition of an electron from a supercharged $L$-orbit (i.e., a normal $L$-orbit occupied by three electrons) to the $K$-orbit. In Smekal's new approach the combination principle and the Bohr–Einstein frequency condition remained perfectly valid, in contrast to other theoretical treatments (Sommerfeld, 1916c; 1918b). In two following papers, submitted in February and March 1919 to the *Verhandlungen der Deutschen Physikalischen Gesellschaft* and the Vienna Academy of Sciences, respectively, Smekal pursued his previous analysis further: in the first paper he demonstrated how, by detailed fixing of super-charged orbits, the frequency condition was never violated (Smekal, 1919a); and in the second he argued again that atoms with co-planar electron orbits were not able to emit the observed X-ray spectra (Smekal, 1919b). The development of a new theory for that purpose would become the task for the next couple of years in Vienna.

[36] For details of the development of X-ray spectra and their interpretation, see Volume 1, Section III.3.

In the spring of 1919 Smekal accepted the position of an assistant to Professors Heinrich Mache and Ludwig Flamm at the *Technische Hochschule* in Vienna.[37] He quickly adjusted to the new environment and also became an active member of the circle around Hans Thirring. In the fall of 1920 he became assistant at Gustav Jäger's *II. Physikalisches Institut* and simultaneously *Privatdozent* at the University of Vienna. However, he retained his association with the *Technische Hochschule*, where he also lectured as *Privatdozent* from 1921.

Smekal quickly distinguished himself in Vienna as the most active representative and partisan of atomic and quantum theory. In the early 1920s he spoke frequently at the meetings of the *Gauverein Wien*, the Vienna Section of the German Physical Society; the latter was founded on 20 February 1920 and in the following meeting, four days later, Smekal was elected secretary of the section.[38] He gave his first talk on 29 April 1920, discussing the 'Deviations from Coulomb's Law in Close Vicinity of the Elementary Charges' ('*Abweichungen von Coulombschen Gesetze in großer Nähe der elementare Ladungen*,' Smekal, 1920b).[39] About a week later, on 6 May 1920, he discussed the 'Fine Structure of X-ray Spectra' ('*Feinstruktur der Röntgenstrahlen*,' Smekal, 1920c): in particular, he spoke about some details of the recent work by Arnold Sommerfeld, Walter Kossel and himself. In the following two presentations on 8 November 1920 and 1 March 1921 he again addressed topics of nuclear physics: in the first he discussed the theoretical consequences of Ernest Rutherford's recent discovery of a new helium isotope with mass 3 (Smekal, 1920e), and in the second

[37] After Gustav Jäger left the *Technische Hochschule* in 1918 to take over the theoretical chair at the University of Vienna, Heinrich Mache, a former occupant of the second chair at the *Technische Hochschule*, changed to the first chair. Ludwig Flamm, formerly assistant to Gustav Jäger, succeeded Mache in the second chair in the spring of 1919. Flamm, who had got to know Smekal during the Wolfskehl week in Göttingen in May 1918, had arranged the invitation to Smekal.

[38] Early in 1920 the German Physical Society adopted a new statute. This statute encouraged the constitution of local sections, besides the Berlin one (which had kept the old name '*Physikalische Gesellschaft zu Berlin*' and which had been the precursor of the *Deutsche Physikalische Gesellschaft* (German Physical Society), by stating: 'Members of the German Physical Society, who live in places or regions outside the suburban Berlin traffic, may unite into a *Gauverein* (local section) in order to pursue physico-scientific goals. Unions already existing for the same purpose may turn into *Gauvereine*, provided all their members have become members of the German Physical Society' (*Verh. d. Deutsch. Phys. Ges.* (3) **1**, p. 3, 1920). The first non-Berlin section, *Gauverein München*, was established on 14 January 1920, consisting of fourteen members of the German Physical society with Arnold Sommerfeld as chairman; next followed the *Gauverein Wien* (founded 20 February 1920), then the *Gauverein Frankfurt-Giessen-Marburg* or *Gauverein Hessen* (founded 24 January 1921), the *Gauverein Niedersachsen* (founded 9 July 1921), the *Gauverein Prag* (founded 2 February 1922), the *Gauverein Rheinland-Westfalen* (founded 21 June 1924), *Gauverein Thüringen-Sachsen-Schlesien* (founded 19 November 1924), the *Gauverein Baden-Pfalz* (founded 20 June 1925) and finally the *Gauverein Württemberg* (founded 28 November 1925).

The *Gauverein Wien* soon developed into one of the most active sections of the German Physical Society, holding many meetings (the titles or abstracts of the papers were published in the *Verhandlungen der Deutschen Physikalischen Gesellschaft*, third series) and gaining a large number of members.

[39] Smekal referred in his talk especially to a model of the $\alpha$-particle that had been suggested earlier by Wilhelm Lenz (1918); it yielded a repulsive force between two hydrogen nuclei proportional to $r^{-2.117}$, with $r$ denoting their distance.

he discussed the interpretation of the observed mass differences between radioactive substances and their decay products (Smekal, 1921d).[40] Also at the next meeting of the *Gauverein Wien* Smekal delivered a talk, this time dealing critically with Johannes Stark's recent attack on Bohr's quantum theory of spectral lines (Stark, 1920); he found several errors in Stark's arguments, while in other cases the experimental situation had not yet been clarified (Smekal, 1921f).[41] In later meetings Smekal turned to the discussion of the importance of adiabatic invariants (Smekal, 1921g) and to what he called 'An Attempt Towards a Unified Application of Quantum Theory' (meetings of 3 February and 8 March 1922; see *Verhandlungen* **3**, pp. 14 and 35: '*Ein Versuch einer einheitlichen Anwendung der Quantentheorie I, II*') or 'An Attempt Towards a General Application of Quantum Theory' (meeting of 6 March 1922, *Verhandlungen* **3**, p. 34: '*Ein Versuch einer allgemeinen Anwendung der Quantentheorie*'). With 'general application' he had in mind a method that would enable one to treat systems in quantum theory that are not multiply periodic.[42] From about the middle of 1922 Smekal spoke less frequently at the Vienna *Gauverein*; according to the *Verhandlungen* he presented talks on 9 May 1922 on 'Solidity and Molecular Forces' ('*Festigkeit und Molekularkräfte*'; see *Verhandlungen* **3**, p. 37) and on 11 June 1923 on 'Directional Quantization in a Magnetic Field' ('*Richtungsquantelung im Magnetfeld,*' Smekal, 1923b), discussing especially the recent mechanical considerations of Albert Einstein and Paul Ehrenfest connected with the Stern–Gerlach effect (Einstein and Ehrenfest, 1922).[43] After more than two years, on 16 and 23 November 1925, he surprised the Vienna physicists with an entirely new topic, speaking on the structure of real crystals (Smekal, 1925d, e).

The titles of the talks presented at the meetings of the *Gauverein* indicate the direction of Smekal's scientific involvement; he indeed worked on some of the burning problems of quantum theory and its application to atomic structure in the early 1920s. As for his most original contributions in this field, two must be mentioned particularly: an extended and detailed investigation of the X-ray spectra of elements, which he presented in many papers, and the result of a short note on the quantum theory of dispersion. He first continued his previous Berlin study of spatial atomic models by consideration of the stability of electrons in $L$-shells (Smekal, 1920a). In a series of three papers, entitled 'On the Fine Structure of X-ray Spectra' ('*Zur Feinstruktur der Röntgenspektren*') and submitted to

[40] In a paper entitled '*Über Rutherfords Entdeckung eines neuen leichten Atomkernes*' ('On Rutherford's Discovery of a New Light Atomic Nucleus') and published in *Naturwissenschaften* (Smekal, 1921b), Smekal had analyzed in full detail the energy balance in the nuclear reactions, which Rutherford had investigated during World War I and which had culminated in the discovery of the transmutation of a nitrogen nucleus into an oxygen nucleus (Rutherford, 1919).

[41] As we have mentioned in Volume 1, Section III.4, Sommerfeld immediately defended Bohr's theory against unjustified criticism (Sommerfeld, 1921c).

[42] Smekal submitted, in September 1922, a paper to *Zeitschrift für Physik*, in which he displayed his ideas concerning this problem (Smekal, 1922b; see also the addendum of March 1923: Smekal, 1923a).

[43] We have discussed these topics in Volume 1, Section IV.3.

*Zeitschrift für Physik* in August 1920 (Smekal, 1921a), February 1921 (Smekal, 1921c) and March 1921 (Smekal, 1921e), he tried to show that the data could be described by an extension of the Sommerfeld–Kossel theory, implying: '1. Quantization of the shell as a whole with *three* quantum numbers. 2. Admission of shell states, where *not all electrons behave as equal ones*' (Smekal, 1921a, p. 45). Although Smekal succeeded in describing the details of the observed spectra—e.g., the *L*-spectra of tungsten (Smekal, 1921c)—his theory did not deeply influence the later course of that subject. The competing proposals worked out by Gregor Wentzel, and especially by Dirk Coster (which added the concept of 'screening doublets' to the original 'relativistic doublets' of Sommerfeld), helped to prepare the spectacular success of Bohr's new theory (which was developed by Bohr from 1921), namely, the discovery of the element hafnium (Coster and Hevesy, 1923). However, Smekal's involvement in X-ray spectroscopy at least supported the development by stressing right from the beginning two points which were not obvious to his competitors: the rigorous validity of the combination principle and the necessity of introducing a third quantum number to describe the stationary states of electrons in inner shells of the atom.[44]

The second remarkable contribution of Smekal to quantum theory left a more lasting mark on the history of physics. In an extended '*Zuschrift*' to the *Naturwissenschaften*, dated 15 September 1923, he dealt with the quantum theory of dispersion (Smekal, 1923c). After analyzing the situation, he concluded that 'without taking into account the translational motion [of atoms *and* radiation, as Einstein had assumed in his 1916 derivation of Planck's radiation law (Einstein, 1916e)] no equilibrium of the dispersion radiation could be achieved' (Smekal, 1923c, p. 874). He noted: 'This conclusion puts us in a position to deduce, under the influence of monochromatic radiation, the existence of a *new type of quantum transition*, which might be called (for simplicity) translational quantum transitions in the following' (Smekal, 1923c, p. 874). The derivation of the new type of radiation was straightforward and implied only momentum and energy conservation in elementary scattering processes between atoms (with mass $M$ and initial velocity $v$) and a light-quantum with initial frequency $\nu$. Energy conservation then demanded the relation

$$\frac{Mv^2}{2} + E_m + h\nu = \frac{Mv'^2}{2} + E_n + h\nu', \tag{10}$$

where $v'$ denotes the final velocity of the atoms (which passes from the quantum state $E_m$ into the state $E_n$), and $\nu'$ is the frequency of the light-quantum after scattering. In the case of a negligible contribution from the atomic momenta the difference $(\nu - \nu')$ would be given, up to a sign ($\pm$), by the transition frequency $\nu_{mn}$ ($= E_m - E_n/h$).

The effect predicted by Smekal in 1923 would be discovered about five years later by Chandrasekhara Venkata Raman in India (Raman, 1928). Smekal would

[44] Smekal would be honoured for his contribution in clarifying the fine structure of X-ray spectra by the *Haitinger-Preis* (more accurately, half of it), in 1923.

then again submit a note to *Naturwissenschaften*, whose title '*Zur Quantentheorie der Streuung und Dispersion*' ('On the Quantum Theory of Scattering and Dispersion') would be similar to that of his earlier paper, and point out the relation of Raman's observation to his old theoretical result (Smekal, 1928). The Raman-Smekal effect, as it has sometimes been called, would play a great role in the analysis of molecular structure.

## The *Physikalische Berichte* and the *Encyklopädie* Article; Start of Otto Halpern

As the years went by, Smekal's former scientific productivity seemed to decline.[45] At the same time, however, he increased his activity in another field of work, namely the publication of reviews of current scientific literature for the German review journal *Physikalische Berichte.* The *Berichte* were published, beginning in January 1920, by the *Deutsche Physikalische Gesellschaft* and the *Deutsche Gesellschaft für technische Physik* with the goal of 'reviewing completely and rapidly the entire physical and physico-technical literature, in issues that would be distributed twice a month' (*Berichte* **1**, p. 1).[46] From the beginning Smekal's name was included in the list of permanent collaborators, together with the names of his Viennese colleagues V. F. Hess, K. W. F. Kohlrausch, K. Przibram and H. Thirring.[47]

Smekal began the new activity rather slowly. In the first volume one finds only three of his reviews: one on a paper by Gustav Jaumann from Brünn, dealing with the physics of continuous media (*Berichte* **1**, 1920, pp. 1371–1372); the second on an attempt by the Viennese physicist Friedrich Kottler to support a dualistic radiation theory, in which one assumes continuous wave fronts (as in electromagnetic radiation theory), on the one hand, with occasional local concentration of energy and momentum (as in light-quantum theory), on the other (*Berichte* **1**, pp. 1461–1462)[48]; and the third on Ludwig Hopf's dimensional

[45] We register only four papers in the years 1924 and 1925: two papers were directly related to the frequency-shifted dispersed radiation (Smekal, 1924, 1925a); a third paper treated Bose–Einstein statistics (Smekal, 1925b), and the fourth one the metastable states of atoms and molecules (Smekal, 1925c).

[46] The list of reviewed journals, reproduced in the front of the issues of Volume 1 of the *Berichte*, contained over 200 numbers, from *Abhandlungen der Berliner Akademie* to *Zeitschrift für wissenschaftliche Photographie.* Although the greater part consisted of journals in the German language, the major foreign scientific journals were covered as well. Still the *Ankündigung* stated: 'It is intended to further increase the already large number of journals reviewed, and especially to close, as soon as possible, the gap which exists at the moment, especially with respect to foreign literature; the goal is to cover without exception the total [scientific] world literature' (*Berichte* **1**, p. 1).

[47] To the first volumes of the *Berichte*, the Viennese Wolfgang Pauli, then in Munich, served as one of the most active contributors. Apart from his own he reviewed papers dealing especially with quantum-theoretical topics, e.g., those of A. Landé, M. Planck, R. Gans, C. G. Darwin, A. Smekal, G. Hettner, J. J. Thomson, J. M. Burgers, H. A. Kramers, A. Rubinowicz, G. Krutkow and I. Langmuir. After 1923 he discontinued his reviewing activity.

[48] The dualistic theory, which Kottler advocated, had been suggested in 1913 by Erich Marx from Leipzig; with it the latter wished to describe the results obtained by experiments on the accumulation of radiation energy and to provide simultaneously a foundation for Planck's radiation law (Marx, 1913).

considerations in radiation theory (*Berichte* **1**, 1920, p. 1621). In Volume 2 of 1921 he published over 30 reviews primarily on papers dealing with fundamental concepts and physical theories and with topics of atomic structure and spectral lines. Of course, he also reviewed, as was the custom then, some of his own papers, such as the ones on the fine structure of X-ray spectra; and, in rare cases, he reviewed the papers of his competitors like Lars Vegard (*Berichte* **2**, 1921, p. 953) and later Gregor Wentzel (*Berichte* **3**, 1922, p. 340).

The work for the *Berichte* acquainted Smekal thoroughly with most of the latest ideas in quantum theory: such as those put forward by recognized experts like Niels Bohr—Smekal reviewed, among others, Bohr's two fundamental memoirs of 1918 on the quantum theory of line spectra (Bohr, 1918a, b; see *Berichte* **2**, 1921, pp. 1191–1193), the Nobel lecture (Bohr, 1923b; see *Berichte* **5**, 1924, p. 651) and the 1923 article on the fundamental postulates of quantum theory (Bohr, 1923a; see *Berichte* **6**, 1925, pp. 491–493)—and those proposed by outsiders, like the Munich astrophysicist Robert Emden on light-quanta (*Berichte* **3**, 1922, pp. 744–745) or the Edinburgh mathematician Edmund T. Whittaker on the quantum mechanism in atoms (*Berichte* **4**, 1923, pp. 1097–1098). Reading the literature occasionally provided the stimulus for his own scientific investigations. Smekal also reviewed, in the summer of 1923, the paper of Charles Galton Darwin on the quantum theory of optical dispersion (Darwin, 1922; see *Berichte* **4**, 1923, pp. 788–789); a few months later he wrote his own paper on the same subject (Smekal, 1923c). But the continuous attention given to the ideas of other people might also have prevented him from developing his own original ideas.

Whatever really happened in these years—whether Smekal plunged into the reviewing activity because his own creative period in quantum-theoretical research had ended, or whether reviewing took up all his time—he performed, by his reviews of the most difficult papers for the *Berichte*, a valuable service to the physics community in the 1920s.[49] He rendered an even greater service by a monumental review article, entitled '*Allgemeine Grundlagen der Quantenstatistik und Quantentheorie*' ('General Foundations of Quantum Statistics and Quantum Theory') and written for the *Encyklopädie der mathematischen Wissenschaften* (Smekal, 1926b). This article ran over 350 printed pages and contained 797 footnotes referring to an even greater number of original papers. Smekal organized this vast material into 27 sections, which he grouped together in three large chapters: '*I. Die Entwicklung der klassischen Statistik zur Quantenstatistik. II. Allgemeine Grundlagen der Quantentheorie. III. Spezielle Anwendungen der Quantenstatistik.*' ('I. The Development of Classical Statistics into Quantum Statistics. II. The General Foundations of Quantum Theory. III. Special Applications of Quantum Statistics.') Smekal finished the major part of Chapter I in July 1923,

[49] We might mention that Smekal reviewed such decisive papers as that of Bohr, Kramers and Slater on the quantum theory of radiation (Bohr, Kramers and Slater, 1924; see *Berichte* **6**, 493–494, 1925), Schrödinger's third and fourth papers on wave mechanics (Schrödinger, 1926f, g; see *Berichte* **8**, 1330–1332, 1927) and several papers by Paul Dirac, e.g., the one on the quantum theory of emission and absorption of radiation (Dirac, 1927b; see *Berichte* **8**, 1333–1334, 1927).

proceeded with half of Chapter II in August 1924, and concluded the rest in June 1925. The article appeared in the last part (Part 3) of Volume V of the *Encyklopädie.* Arnold Sommerfeld, the editor of this volume, which was devoted to physical topics, praised Smekal's achievement in his *Nachwort*, claiming that 'with this last comprehensive article on quantum theory the real centre of present-day research in physics has been grasped and presented with a thoroughness that has not been achieved thus far' (Sommerfeld, in *Encyklopädie* **V/3**, p. 1215).

Smekal's article, forming issue No. 6 of Volume **V/3**, appeared on 15 July 1926. At about the same time another review article on quantum theory became available: it was Wolfgang Pauli's contribution '*Quantentheorie*' in Volume XXIII of Geiger–Scheel's *Handbuch der Physik* (Pauli, 1926b). Pauli had also completed his review article in the summer of 1925 (in September). Therefore, like Smekal, he could not include the decisive progress and breakthrough to a consistent quantum mechanics, achieved from the summer of 1925 to the spring of 1926 by Max Born, Paul Dirac, Werner Heisenberg, Pascual Jordan and Erwin Schrödinger. The two review articles on quantum theory by Smekal and Pauli had been written just too early to include the latest developments. However, Pauli's account was used by atomic physicists in the following years, while Smekal's was hardly referred to. The reason for this will be seen in the different aspects which were emphasized in the respective treatments. Pauli, in his *Handbuch* article, was able to restrict himself to the connections between quantum theory and atomic structure because Smekal had already covered the statistical aspects in an article entitled '*Statistische und molekulare Theorie der Wärme*' ('Statistical and Molecular Theory of Heat'), which appeared in an earlier volume (Volume IX) of *Handbuch der Physik.*[50] The statistical aspects certainly played a dominant role in the early phases of quantum theory, but now in the mid-1920s they had moved away from the centre of interest. Thus Pauli's conception represented the modern view of quantum theory in contrast to Smekal's old-fashioned view. Pauli further possessed a decisive advantage over Smekal due to the fact that he was a member of the then leading schools of quantum theory in Copenhagen, Göttingen and Munich; he knew about the trends first hand; he was active in determining the future of quantum theory and did not have to find out about it from the literature.[51]

[50] Volume IX of Geiger–Scheel's *Handbuch* was also published in 1926. Smekal's article of about 100 pages treated the kinetic theory of heat and its application to gases and solids, both in classical and in quantum theory (Smekel, 1926a).

[51] Characteristic of the author's difficulty to judge in Vienna future development is perhaps Section 23, '*Rückblick auf die Quantentheorie. Prinzipielle Schwierigkeiten und Axiomatisierungsversuche*' ('Retrospective View on the Quantum Theory. Principal Difficulties and Attempts at Axiomatization'). Smekal discussed there the trends in quantum theory in the early 1920s in a rather knowledgeable manner, expressing especially the opinion that one should 'in the present situation of the theory preferably refrain from a too narrow formulation of the quantum *postulates*' (Smekal, 1926b, p. 1130). Later development indeed took this direction, and one of the crucial steps in Werner Heisenberg's pioneering paper (Heisenberg, 1925c) consisted of a reformulation of the Bohr–Sommerfeld quantum condition in the spirit of the so-called dispersion-theoretic approach. But instead of quoting the papers of Hendrik Kramers, Max Born, Werner Heisenberg and others, who had worked with the

Smekal, on the other hand, would abandon the field of quantum theory and atomic structure and would turn to problems of crystal physics which did not excite many physicists.[52]

While Smekal slipped out of the research in quantum theory, his place in Vienna was taken over by his younger colleague Otto Halpern.[53] Halpern was born in Vienna on 25 April 1899. He studied at the University of Vienna and obtained his doctorate, under Hans Thirring, with a thesis on phosphoresis in 1922; he then served as assistant in the theory institute of his professor. Halpern addressed the Vienna *Gauverein* for the first time at the meeting of 20 June 1921, when he discussed the light-quantum hypothesis (*Verhandlungen* **2**, 45, 1921). About a year later he spoke twice, on the needle radiation (at the meeting of 26 June 1922; see *Verhandlungen* **3**, 50, 1922) and on the relation between classical and quantum statistics (at the meeting of 3 July 1922; see *Verhandlungen* **3**, 56, 1922). In 1923 and early 1924 he published a set of four papers in *Zeitschrift für Physik*, each of which dealt with an important problem of quantum theory. In the first he studied the influence of crossed electric and magnetic fields on the hydrogen spectrum and confirmed Paul Epstein's result that a sharp quantization, according to the rules of multiple periodic systems, was possible (Halpern 1923a); in the second contribution he discussed the application of the adiabatic hypothesis to a model of ortho-helium (Halpern, 1923b); in the third, which he submitted like the other two in August 1923, he calculated the quadratic Zeeman effect (Halpern, 1923c); and in the fourth, submitted in December 1923, he dealt with the thermal equilibrium of blackbody radiation and atoms (Halpern, 1924a).

---

dispersion-theoretic approach, Smekal referred to papers of H. A. Senftleben from Breslau, who had attempted to establish a completely new foundation of physics based on the assumption of discrete 'substance elements'—such as charge and mass of the electron and Planck's constant—and on giving up the continuous space-time description (Senftleben, 1924a, b; 1925). Senftleben's approach, though it contained interesting ideas and original insights, did not lead anywhere.

[52] Smekal became involved in the physics of crystals in 1922, when he wrote an article '*Technische Festigkeit und molekulare Festigkeit*' ('Technical Solidity and Molecular Solidity,' Smekal, 1922a). From 1925 he made the physics of solids his main field of scientific interest. He started a completely new career, which turned him into an authority on such technical questions as brittleness and the technology of pulverization.

In 1928 he accepted a call to the chair of theoretical physics at the University of Halle, where he also had available experimental facilities for research in solid state physics. After 1945 he came to West Germany, serving as professor at the *Technische Hochschule* in Darmstadt (1946–1949). In November 1949 he was appointed to the chair of experimental physics at the University of Graz. Smekal died in Graz on 7 March 1959.

[53] Smekal's reputation as a connoisseur of the literature was still high enough to make Hans Geiger and Karl Scheel entrust him with the organization of the contributions to the second edition of Volume XXIV of their *Handbuch der Physik*. This volume was split into two parts: Part I (entitled '*Quantentheorie*') contained, e.g., the article of Wolfgang Pauli on wave mechanics (Pauli, 1933c); Part II (entitled '*Aufbau der zusammenhängenden Materie*') contained, among others, the article of Max Born and Maria Goeppert-Mayer on crystal dynamics (Born and Goeppert-Mayer, 1933), the article of Arnold Sommerfeld and Hans Bethe on metal electrons (Sommerfeld and Bethe, 1933), and Smekal's review entitled '*Strukturempfindliche Eigenschaften der Kristalle*' ('Structure Dependent Properties of Crystals,' Smekal, 1933).

Halpern also performed research on quantum-theoretical problems in the following years. He published, later in 1924, an investigation on X-ray scattering by atoms, in which he found that 'the consistent application of the classical theory [of radiation] allows essentially a correspondence-like understanding of the effects postulated by the quantum theory [i.e., the light-quantum hypothesis]' (Halpern, 1924b, p. 171). It should be noted that Halpern did not publish any paper in 1925, at a time when quantum theory was converted into quantum mechanics.[54] However, once matrix mechanics had been formulated, he soon contributed the treatment of the rotator in the new theory (Halpern, 1926a, b). Three years later he joined his Vienna-born colleagues Pauli and Smekal in becoming the author of a review article on quantum theory; he wrote, together with his former teacher Hans Thirring, an account of the development which the 'new quantum theory' had pursued from 1926 for the *Ergebnisse der Exakten Naturwissenschaften* (Halpern and Thirring, 1929).[55]

In 1928 Halpern—who had unsuccessfully applied to the University of Vienna for the next step in his academic career, the *Habilitation*—left his hometown and went to Leipzig on a Rockefeller fellowship. From there, after two years, Werner Heisenberg recommended him for an associate professorship at New York University.[56]

Smekal and Halpern demonstrated perfectly the important qualities of the Vienna school of physics. They had received a thorough education in classical physics; they possessed a very high and quick intelligence in picking up topics lying at the forefront of the development in physics; they were industrious and they knew practically the entire literature on the topics on which they worked. What prevented them from becoming leading pioneers in the field of fundamental research was perhaps a lack of stubborn originality, which had distinguished their countrymen Wolfgang Pauli and Erwin Schrödinger. Being confined to Vienna, they also missed the constant stimulus and excitement that prevailed at the great centres of quantum theory at that time, as well as the critical discussions

[54] He did not submit Part II of his 1924 paper '*Zur Theorie der Röntgenstrahlstreuung*' ('On the theory of X-ray Scattering,' Halpern, 1924b) until June 1926 (Halpern, 1926c), that is, nearly two years after the first part. Neither did he speak in the meantime at a meeting of the *Gauverein*.

[55] Thirring alone composed the first part of the article, entitled '*Die Grundgedanken der neueren Quantentheorie*' ('The Basic Ideas of the More Recent Quantum Theory,' Thirring, 1928). The writing of review articles dealing with quantum theory had meanwhile become a speciality, if not a monopoly, of Viennese authors. Eugene Guth, another student of Thirring's, contributed at about the same time the article on '*Entwicklung und Grundlagen der Quantenphysik*' ('Development and Fundamentals of Quantum Physics') to Volume IV of the *Handbuch der Physik* (Guth, 1929).

[56] In New York Halpern investigated a variety of physical topics, including those of thermodynamics, optics, physical chemistry, atomic, electron, neutron and meson physics, becoming a lively member of the large group of theorists that gathered in New York and held weekly colloquia at Columbia University. During World War II he worked as a staff member at the Radiation Laboratory of the Massachusetts Institute of Technology, Cambridge (1941–1945), dealing primarily with problems of radar techniques. After the war he moved to the Lawrence Radiation Laboratory at Livermore, California (in 1947), which he left in 1961 because of invalidity due to an automobile accident. In 1965 he returned to Vienna, from where he later went to London. He died in London on 29 October 1982. (For details of his biography, see Harvey Hall, 1983.)

of all topics of atomic structure that were carried out there. Thus they were destined to be only side-figures in the development of quantum theory in the 1920s, the importance of which corresponded to the role of Vienna as a side-stage in this field of physics.

## Paris and Modern Physics from 1900

Like Vienna, Paris—that great west-European capital of a nation which had provided brilliant contributions to science over the centuries—also did not house a flourishing school of quantum theory in the first quarter of the twentieth century, in spite of the fact that the prevailing conditions had been much more favourable.[57] At the end of the nineteenth and the beginning of the twentieth centuries, Paris had been the centre of pioneering activities in many of the latest developments of physics. The most conspicuous contributions had been made in the field of radioactivity, where Henri Becquerel's initial discovery had been brilliantly extended by the work of Pierre and Marie Curie and André Louis Debièrne. The investigations of the Parisian scientists Aimé Cotton (1896–1951), Alfred Cornu (1841–1902), Henri Becquerel and Henri Alexandre Deslandres (1853–1948) on the recently (1896) detected Zeeman effect deserve high praise; their efforts constituted really pioneering work in spectroscopy in the days before 1900.[58] In the first decade of the twentieth century the physicists in Paris achieved two major successes in the kinetic description of matter. In 1905 Paul Langevin (1872–1946) extended the electron theory of matter to account for the magnetic behaviour of paramagnetic substances (Langevin, 1905a, b). Starting in 1908, Jean Perrin provided in a series of investigations the full experimental confirmation of Albert Einstein's theory of Brownian motion (Einstein, 1905c), obtaining a most reliable value for Avogadro's number, i.e., the number of molecules in a gram-atom of a substance (see, e.g., the review report of Perrin: Perrin, 1912).[59]

[57] While Paris physicists had excelled during the eighteenth century in the field of mechanics and had contributed to crucial experiments in electricity, they turned (after 1800) essentially to the topics of thermodynamics, electrodynamics and optics. Names like André Marie Ampère, Augustin Fresnel and Sadi Carnot may indicate the standard of Paris physics in the first half of the nineteenth century.

[58] We have discussed the French contributions to the Zeeman effect in Volume 1, Section IV.4. The earlier contributions of French scientists to the spectroscopy of atoms and molecules, notably by Éleuthère-Élie Nicholas Mascart (1837–1908), Paul Émile Lecoq de Boisbaudran (1832–1912), Cornu and Deslandres have been cited in Section II.1 of the same volume.

[59] Jean Baptiste Perrin was born in Lille on 30 September 1877 and educated at the *École Normale Supérieure*, becoming there an assistant in physics from 1894 to 1897. After receiving the degree of *Docteur-ès-sciences* in 1897 (with a thesis on cathode and X-rays)—Perrin had shown (in 1895 as the first experimentalist) that cathode rays bear a negative electric charge—he was appointed reader in physical chemistry at the University of Paris; in 1910 he was promoted to a full professorship, which he held until 1940. Then he went to the United States; he died in New York City on 17 April 1942.

Perrin's researches extended over many fields of experimental physics: he studied cathode rays and the action of X-rays on the conductivity of gases; he investigated fluorescence, the disintegration of radium and the emission and transmission of sound. For his experiments with colloids, especially on their Brownian motion, he received the Nobel prize in physics for 1926. He also acted most

In pure theory the great mathematician Henri Poincaré (1854–1912) had given—in several fundamental investigations, starting in 1900 and culminating with his long 1905 memoir to the *Circulo Matematico* of Palermo (Poincaré, 1906)—a consistent description of the motion of a fast electron (independently of Albert Einstein's simultaneous theory).

While the kinetic theory of matter and relativity theory had thus been represented by pioneering achievements in Paris, the same cannot be said about quantum theory, the other modern development of theoretical physics in the twentieth century. The first occasion on which a French scientist referred to it was at the meeting held in Brussels in the fall of 1911 on 'The Theory of Radiation and Quanta,' the so-called first Solvay Conference. French scientists—unlike their colleagues in Germany, Austria and Great Britain—had not shown any particular interest in the theory of heat radiation. However, they had worked on other phenomena relating to what one called 'quanta' in those days; these implied electrons, atoms and molecules and the corpuscular objects emitted from radioactive substances. This general definition of 'quanta' justified, of course, the invitation to Marie Curie, Paul Langevin, Jean Perrin and Henri Poincaré to attend the Solvay Conference, although none of them had contributed thus far to Planck's quantum theory. Langevin and Perrin even delivered reports on their work.[60] Perrin reviewed essentially the results of his experiments proving the reality of atoms (Perrin, 1912). Langevin, on the other hand, discussed in his report 'The Kinetic Theory of Magnetism and the Magneton' (Langevin, 1912). That is, he presented his own theory of paramagnetism which was based on the assumption of the existence of molecules endowed with a permanent magnetic moment (Langevin, 1905a, b); he also discussed the extension, which Pierre Weiss (1865–1940) in Zurich had given to Langevin's theory, in order to account for the phenomena of ferromagnetism (Weiss, 1907). Weiss later observed that the saturation magnetization of the various ferromagnetic substances under investigation were nearly always integral multiples (the integers assuming values like 3, 8, 9, 10, etc.) of a certain constant, whose value he had determined as being 1123.5 c.g.s. units; he then coined, for the ratio of this quantity and Avogadro's number, the name 'magneton' (Weiss, 1911a). When Weiss presented these results at the Karlsruhe *Naturforscherversammlung* of September 1911, Max Abraham mentioned the idea that 'the magnitude of the [elementary] magnetic moment [or magneton] was equal to the product of the quantum of action and the value

---

successfully as a science politician: e.g., he was the driving force behind the creation of the *Centre National de la Recherche Scientifique*, he founded the *Palais de la Découverte*, and he helped to establish many institutions of research, e.g., the *Institut d'Astrophysique* and the *Institut de Biologie Physico-Chimique* in Paris. He received many national and international honours, e.g., memberships in the *Académie des Sciences* (1923; 1938 president), and the Royal Society of London (Perrin had already won the Joule prize of this society in 1896).

[60] Besides the Paris scientists mentioned above, Marcel Brillouin and Maurice de Broglie were in the French delegation. (We have discussed the first Solvay Conference in some detail in Volume 1, Section I.6.)

of the specific [electric] charge' (Abraham, in Weiss, 1911c, p. 952). In the same discussion, Richard Gans proposed the assumption that a rotating electron (mass $m$, charge $e$) created the elementary magnetic moment or Weiss' magneton $\mu$, deriving the equation

$$\mu = n\frac{h(e/m)}{2\pi}, \tag{11}$$

which yielded Weiss' result approximately for $n = 0.1$ (Gans, in Weiss, 1911a, p. 952).[61] There was a difficulty in Gans' result as, on the basis of quantum theory, $n$ had to be an integral number rather than a fraction. Now at the Solvay Conference, starting from Arnold Sommerfeld's quantum hypothesis, Langevin obtained the same expression as Gans except that $n$ assumed the value $\frac{1}{8}$ (see Langevin, 1912, Section IV).[62]

The reports presented at the Solvay Conference on quantum theory, and their intensive discussion, left a deep impression on the participants from Paris. Henri Poincaré, who had asked penetrating questions in Brussels, immediately sat down to investigate a point of fundamental importance; in a paper presented to the *Académie des Sciences* on 4 December 1911 he gave an answer to the question whether or not quantum theory, which yielded Planck's law of blackbody radiation, could be formulated in terms of differential equations (Poincaré, 1911). The answer, which he explained in some detail in an extended paper that appeared in *Journal de Physique Théorique et Appliquée* (Poincaré, 1912a), was negative. Unfortunately, however, Poincaré died in July 1912, and with his death the French voice in the international concert of quantum physicists became silent for a while, and remained so even after the French edition of the proceedings of the Solvay Conference of 1911, i.e., the reports and discussions that were carefully prepared by Paul Langevin and Maurice de Broglie, was published (Langevin and de Broglie, 1912).

It was characteristic of the French situation that even the theoretician Langevin, who had acquainted Perrin—as the latter acknowledged in his Solvay report—with Einstein's theory of Brownian motion and had himself invoked the quantum hypothesis to explain Weiss' magneton, did not actively engage any further in quantum theory. Neither he nor Marcel Brillouin (1854–1948)—professor of physics at the *Collège de France*, who had raised deep and critical questions in the discussion of the Solvay reports in Brussels—picked up the important quantum-theoretical problems in the following years, such as the theory of the molecular spectra—whose experimental investigation had been a French

[61] Upon Gans' presentation, Weiss had remarked that Albert Einstein had already suggested the same explanation for the magneton.

[62] We have discussed Sommerfeld's specific quantum hypothesis (Sommerfeld, 1911b) in Volume 1, Section I.6.

speciality for a long time—or the theory of the specific heats of solids, or even the constitution of atoms. This fact appears all the more surprising as a procedure along lines similar to that of Langevin's treatment of the magneton problem might have offered an opportunity to attack successfully the problem of atomic structure and spectra. Jean Perrin, in a public address given in February 1901 to the students and friends of the University of Paris, had sketched a model of the atom, consisting 'on the one hand, of one or several masses bearing very strong positive electric charges—like positive suns, whose charge is very superior to that of a corpuscle—and, on the other hand, of a multitude of corpuscles—like small negative planets—and the entirety of these masses gravitate under the action of electric forces, such that the total negative charge exactly compensates the total positive charge to yield a neutral atom' (Perrin, 1901, p. 460). This exactly planetary model of the atom had been confirmed recently by Ernest Rutherford in Manchester (Rutherford, 1911b); and to this model a quantum hypothesis could be easily applied. But the French scientists left the pioneering deed to the Dane Niels Bohr, who developed from it a successful model of atomic structure and molecular constitution (Bohr, 1913b, c).

## Quantum Theory in Post-War Paris: Dauvillier, L. Brillouin and L. de Broglie

Perhaps it is not fair to accuse the French physicists of having failed to make fundamental contributions to the problems of atomic and quantum theory after 1911. After all, they participated eagerly in the second Solvay Conference at Brussels in October 1913, which was devoted to 'The Structure of Matter.' For this meeting Marcel Brillouin prepared a report on 'Crystal Structure and the Anisotropy of Molecules.'[63] The main issues debated at the second Solvay Conference were not quantum-theoretical problems of atomic structure, as one might have expected at that time, but the interference phenomena with X-rays, discovered the previous year by Walter Friedrich and Paul Knipping on Max von Laue's suggestion, and the subsequent development of the X-ray reflection technique by William Henry Bragg and his son William Lawrence Bragg. Bohr's work on the quantum theory of atomic structure, which had just appeared in print, was not even mentioned. Maurice de Broglie (1875–1960), who acted, as in 1911, also as a scientific secretary to the second Solvay Conference, became interested in Bragg's method for the purpose of obtaining the X-ray spectra of atoms in his laboratory; he introduced a trick which allowed him to get rid of spurious spectral lines arising from crystal imperfections (M. de Broglie, 1914). Unfortunately, World War I broke out, and de Broglie's studies of X-ray spectra, which constituted the next important contribution of the Paris scientists to quantum physics, were often interrupted.

[63] The French participants at the meeting were, besides Marcel Brillouin, Madame Curie, Georges Gouy and Paul Langevin.

During the war French scientists, like their colleagues in Austria, Germany and England, performed military service.[64] Hence their research activities were interrupted or greatly reduced. Meanwhile rapid progress took place in quantum theory, especially through the extension of Niels Bohr's theory of atomic constitution in the hands of Arnold Sommerfeld in Munich and the analysis of the theoretical foundations of quantum laws by Albert Einstein and Max Planck in Berlin, Paul Ehrenfest in Leyden and Niels Bohr in Copenhagen. News about these developments, in particular the papers published in Germany, reached the French scientists only after some delay (if at all). Thus they found a fully established theory of atomic constitution ready, when in late 1918 Paris returned to peace.

The outcome of the war considerably changed the conditions for the pursuit of science in France, and in general for the worse. France, although belonging to the official winners of the war, had suffered immensely from human and economic losses. The nation had become poorer, industry was near collapse, and as a consequence support for scientific research was reduced. In addition, international scientific relations and co-operation, which had largely been responsible for the tempestuous advance of physics in the previous twenty-five years, had been interrupted and were not fully resumed after the armistice in November 1918. The French scientists, for example, largely followed the official politics of the *Entente* powers in excluding their German and Austrian colleagues from international societies and conferences. This was meant as a punishment for the other side, but now turned into a disadvantage for French scientists. In the post-war years, the quantum and atomic theories were decisively promoted by German scientists in Germany. In Paris, scientists had to learn about this progress from published papers; they were cut off from the lively personal discussions and meetings that occurred in Germany, as for instance the so-called *Bohr Festival* of the summer of 1922 in Göttingen.[65] With limited financial resources available for scientific research, the French system did not adopt the 'German solution' in those days, namely the concentration on 'inexpensive' theoretical work instead of 'expensive' experiments. Students in Paris (and elsewhere in France) were simply not willing to devote their energy to theory, because there existed no opportunity to obtain appropriate positions in theoretical research in the French

[64] It seems, however, that French scientists had greater opportunities to escape the fighting at the front. There is no report of important casualties among French scientists, comparable to the losses of Moseley in England or Hasenöhrl in Austria. Maurice and Louis de Broglie, for example, worked—partly with Léon Brillouin—on wireless telegraphy.

[65] German scientists could not participate in the next two Solvay Conferences, the third in 1921 and the fourth in 1924. Only Albert Einstein from Berlin was invited but he did not attend. This fact might have contributed to the limited success of these meetings, as far as their role in the progress of quantum theory was concerned.

Personal relationships between French and German scientists did not flourish in these early post-war years. A rare exception was the relationship between Madame Curie, Paul Langevin and Albert Einstein, but in France the latter was not considered as a German.

university system.[66] As a consequence there were no large circles of students in Paris, eagerly competing and discussing the newest problems of quantum physics. Rather, a limited number of advanced students and young researchers, and also older professors like Marcel Brillouin and Paul Langevin, concerned themselves with the problems of quantum theory.

Still the French post-war literature on physics is not characterized by a lack of papers dealing with the problems of quantum theory. Indeed, the French journals, especially the *Comptes Rendus* of the *Académie des Sciences* and the *Journal de Physique et le Radium*, devoted considerable space to such papers, comparable to the situation in the leading German and English journals of those days.[67] It appears, however, that the authors of the papers worked out their results for themselves and in general, did not discuss them with colleagues before publication.[68]

The number of Parisian physicists involved in quantum theory during the early 1920s was small. Most of their names were familiar from pre-war times, especially those of Marcel Brillouin and Maurice de Broglie. The younger scientists were mostly related to them; thus Léon Brillouin (1889–1969) was the son of Marcel, and Louis de Broglie was the brother of Maurice. Léon Bloch (1876–1947) and his brother Eugène Bloch (1878–1944), who had made a reputation in ultraviolet spectroscopy, occasionally wrote on quantum theory.[69] Langevin, on the other hand, published very rarely, if at all; he concentrated on giving excellent lectures on all subjects of physical theory, including relativity and quantum theory, in his classes at the *Collège de France*. The largest group of people working on quantum-physical problems gathered around Maurice de Broglie in his laboratory. Maurice had been able to carry out some work on X-ray spectroscopy even during the war (M. de Broglie, 1916a, b), and by the end of the war he had become an authority in X-ray physics. Thus he was invited to present a report on the quantum relation, $h\nu = \varepsilon$, and photoelectric phenomena at the third Solvay Conference

[66] We have discussed this particular point in Volume 1, Section V.4.

[67] According to the rules of the Paris Academy of Sciences, notes were accepted which should not exceed two and a half pages in print. No author was allowed to present—via a member of the *Académie des Sciences*—more than five papers per year. The *Journal de Physique et le Radium*, whose sixth series started in 1920, contained longer articles, and it may be considered as representing the French counterpart of the German journal *Zeitschrift für Physik*; papers on atomic and quantum theory covered a large portion of the issues (one issue per month usually appeared).

[68] The situation in the early 1920s may have been similar to that of Walter Elsasser when he told about his experience of publishing notes in the *Comptes Rendus* during the 1930s. He recalled: 'After I had written my first note, I went to Louis de Broglie, who was closest to my own interests and who, as a member of the Academy, could submit it in my behalf. He told me that he had not the slightest wish to read my manuscript; if he ever had any desire to read what I had written, he could always do this later at his leisure in print. I should hand my manuscript to him before the start of the Academy's session, which took place at the same hour of the afternoon at the same day of each week. Then, a week later I went to the official printer of the Academy where an envelope with the proofs of my article was waiting for me. Having corrected these, I had only to await the next issue of the *Comptes Rendus* a week later' (Elsasser, 1978, p. 187).

[69] For example, Léon Bloch reviewed Niels Bohr's recent (1921) work on atomic structure in *Journal de Physique at le Radium* (Bloch, 1922).

in 1921 (M. de Broglie, 1923). He found able and devoted colaborators such as Alexandre Dauvillier (1892–1979), who joined de Broglie's laboratory after obtaining his doctorate in 1920. Louis de Broglie also performed experimental work, both with his brother and with Dauvillier.[70]

In the interpretation of their results the Paris X-ray spectroscopists occasionally showed a remarkable independence from the dogmatic views held in Copenhagen. Dauvillier's attribution, especially, (in May 1922) of two weak X-ray lines to the element with the atomic number 72 (Dauvillier, 1922a), which seemed to confirm an earlier claim of the Paris chemist Georges Urbain that the element should belong to the group of rare earths, aroused the anger of the Copenhagen establishment. Niels Bohr, in his recent theory of the periodic system, had postulated that the element number 72 should have properties similar to the metal zirconium. A few months later when Dirk Coster and George de Hevesy decided the question in favour of Bohr's theory (Coster and Hevesy, 1923), the Paris physicist appeared to be discredited. In the following years, in Copenhagen at least, one tended to neglect to some extent the news on quantum physics coming from Paris.[71]

The absence of French quantum physicists in meetings with their German colleagues, and some disagreements on the interpretation of atomic specta, should not be interpreted to mean that they followed an anti-Bohr course in quantum theory. On the contrary, the theory of the atomic structure of Bohr and Sommerfeld represented in Paris the unchallenged basis of any interpretation of atomic spectra; people did not worry, like Bohr himself, about the fundamental assumptions, but they occasionally adopted different opinions about the interpretation.[72]

The conformity, in principle, with Bohr was fully expressed in many papers, as well as in larger reviews and books published by the Paris quantum physicists as, e.g., in the book *La Théorie des Quanta et l'Atome de Bohr* of Léon Brillouin (1922b). Brillouin, who had studied in Paris and later in Munich with Arnold Sommerfeld, received his doctorate in 1920 with a thesis on the quantum theory of solids supervised by Paul Langevin (L. Brillouin, 1920). He continued to work on quantum-theoretical problems afterwards. In his 1922 book he tried to summarize the status of quantum theory: he devoted the first two chapters to blackbody radiation, then explained the quantum theory of atomic structure (Chapters III and IV), and finally the fundamental principles, such as the correspondence principle, the statistical nature of quantum laws and the theory of adiabatic

[70] We have described the work on X-ray spectroscopy that went on in M. de Broglie's laboratory in Volume 1, Section V.4.

[71] We have presented the debate on the element number 72 in Volume 1, Section III.5.

[72] For example, in 1924 Louis de Broglie and Dauvillier introduced certain subgroups of electrons existing in completed *L*-, *M*- and *N*-shells of atoms, thus deviating from Bohr's theory of the periodic system of elements (L. de Broglie and Dauvillier, 1924). Their doubts on the validity of Bohr's views, together with other evidence, led Edmund C. Stoner in England to propose a new scheme of electron arrangement in atoms (Stoner, 1924), which ultimately opened the path to the exclusion principle and electron spin. (See our discussion in Volume 1, Section V.4.)

invariants (Chapters V to VII). Brillouin displayed a perfect knowledge of the current literature, referring, e.g., to the latest results of Sommerfeld and of his student Werner Heisenberg on the anomalous Zeeman effect.[72a]

Two specific points from Brillouin's book might be mentioned here. First, he treated in detail the propagation of electromagnetic radiation in a dispersive medium, emphasizing the difference between phase and group velocity.[73] Second, he stressed the necessity of providing an understanding of Niels Bohr's postulates in the theory of atomic structure, i.e., to search for a 'general formulation of the quantized electrodynamics' ('*forme générale de l'électromagnetisme quantifié*'), calling this 'one of the most important theoretical problems which presently exists' (Brillouin, 1922, p. 63). Both points played a crucial role in the investigations of his colleague Louis de Broglie.

Louis de Broglie (born 1892) became interested in quantum theory while he was a student of history and law at the University of Paris. He had studied eagerly the reports and discussions of the first Solvay Conference prepared for publication by his brother Maurice and he embarked upon efforts to understand the 'mysterious' quanta of Planck and Einstein.[74] After the war he studied with Paul Langevin and worked on X-ray spectroscopy at his brother's laboratory. From the latter work and from conversations with Maurice, Louis de Broglie learned especially the necessity of always considering simultaneously the points of view of waves and corpuscles when dealing with X-rays.[75]

The wave-corpuscle connection appeared to be so natural to Louis de Broglie that he made it the basis of his first note on blackbody radiation. There he treated radiation as consisting of light-quanta, or 'light-atoms', having a very small mass (L. de Broglie, 1922b). In a subsequent note he showed how the interference effects of radiation follow if one assumed an 'agglomeration of light-atoms' (L. de Broglie, 1922c). About a year later he published three notes in quick succession on the quantum problem: in the first he associated relativistic massive particles with a wave phenomenon (L. de Broglie, 1923a); in the second he showed that the wave phenomenon could be interpreted as a phase wave guiding the propagation of the particle (L. de Broglie, 1923b); and in the third he was able to explain, by applying Fermat's principle of optics to the phase wave associated with an electron, how the quantization of electron orbits—postulated in the Bohr-Sommerfeld theory of atomic structure—arose (L. de Broglie, 1923c). He elaborated on these ideas in his doctoral thesis, which he defended before a committee of the University of Paris on 25 November 1924 and then published in early 1925 in the *Annales de Physique* (L. de Broglie, 1925).

[72a] Brillouin also quoted the new, third edition of *Atombau und Spektrallinien* (Sommerfeld, 1922d), which would soon be available in a French translation.

[73] L. Brillouin had investigated this topic in connection with classical theory (relativity theory) in a paper, which he composed in Munich under Sommerfeld's influence (L. Brillouin, 1914).

[74] For details of the de Broglie story we refer the reader to Volume 1, Section V.4.

[75] The corpuscular aspects of X-rays were often emphasized by Maurice de Broglie in publications, e.g., in his Solvay report of 1921 (M. de Broglie, 1923).

Louis de Broglie's hypothesis of matter waves, proposed in 1923 and eloborated on in detail in his doctoral thesis of 1924, constituted a genuinely fundamental contribution to quantum theory, whose importance was only gradually recognized by the community of quantum physicists. Thus the side-stage that was Paris eventually came into the limelight of quantum-theoretical development, at least for a short period.

## II.3 A Newcomer to Zurich: Erwin Schrödinger

On 9 December 1922 Erwin Schrödinger delivered his inaugural address at the University of Zurich, dealing with the fundamental question as to what is the structure of natural laws.[76] Schrödinger, who had already been appointed professor of theoretical physics more than a year earlier—on 15 October 1921—would spend nearly six years in the busy and flourishing city of a peaceful country, which was so different from the overcrowded post-war Vienna. In spite of the fact that Zurich lacked the greatness of the imperial tradition, it provided the newcomer with a stable background, giving him the time necessary to promote his own scientific work, far from any hectic activity, which he disliked. As a consequence, Schrödinger settled down and was able to make a discovery that established his reputation among the great scientists in physical theory.

### The Zurich Background: History of the City, University and E.T.H.

Zurich, the largest town in Switzerland—by the time of Schrödinger's arrival there were about 200,000 inhabitants—and capital of the Swiss canton bearing the same name, is situated about 400 metres above sea level at Lake Zurich, on both banks of the River Limmat emerging from the lake and the River Sihl flowing into the lake. Its origin dates back to the prehistoric times of the later Stone Age. In 58 B.C. a settlement of the Celtic Helvitii fell to the Romans, who established the custom station of Turicum, which they later fortified. Then the Alemanni took over the Roman place, introducing the name Zurich. King *Ludwig der Deutsche*, grandson of Charles the Great, built the *Fraumünster Abbey* for his daughter in A.D. 853. Zurich became an important centre for the church, politics and economics, due to its prime position at the end of a lake and on a river, a position where vital trade routes leading from France to Eastern Europe and from the North to Italy crossed. During the tenth century it acquired the rights of market; it became a free *Reichsstadt* (imperial city) in 1218 and joined the Swiss Confederation in 1351. From 1436 to 1450 Zurich, supported by Austria, became involved in a war with the Swiss League, an *entente* of other Swiss cities; it lost the war but rose again a few decades later by leading the Swiss in the successful

[76] Six years later Schrödinger would publish the text of the address of an article in the *Naturwissenschaften*, bearing the title '*Was ist ein Naturgesetz?*' ('What Is a Law of Nature?', Schrödinger, 1929a).

Burgundian War (Battle of Morat, 1476). On New Year's Day 1519 Huldreich Zwingli inaugurated the Swiss Reformation in Zurich, and the city grew into a sanctuary and a secure refuge for Protestants from all countries, be it from the England of Mary Tudor or from France. From 1555 the economic foundation of Zurich was broadened by the addition of the silk and textile industries. Besides trade and industry Zurich developed a rich cultural life, unique in Switzerland. To the old literary tradition acquired around 1300 (characterized by the name Hadlaub and the so-called Manesse manuscript), the names of the writers Johann Jakob Bodmer and Salomon Gessner and of the world-famous scholar Johann Kaspar Lavater were now added. They turned Zurich into one of the foremost intellectual centres of the German-speaking Europe during the eighteenth century. The literary fame continued throughout the nineteenth century, represented by such personalities as Heinrich Pestalozzi, Gottfried Keller and Conrad Ferdinand Meyer. At the same time, Zurich nurtured its role as a home for political refugees, especially from Germany after the 1848 revolution.

The intellectual reputation of Zurich further increased, when (in 1833) it obtained its own university, the second after Basle (founded in 1460) in Switzerland.[77] The university housed four faculties: theology, political science, medicine and philosophy.[78] Within the philosophical faculty science received a prominent place, as three of the first six professors elected were scientists, namely Lorenz Oken (called from Munich) in natural history, natural philosophy and the physiology of man, C. J. Loewig (from Heidelberg) in experimental and medical chemistry and R. Schinz (from Zurich) in zoology. The most dramatic development took place in the medical faculty, which soon attracted students from all over Europe, including many beautiful female Russian students.[79]

Physics made a comparatively modest start at the University of Zurich. The first lecture course in this field—one on gravitation, molecular forces, electricity, magnetism and electromagnetism—was announced already for the first semester by Gottfried Escher, a former theologian and now professor at the *Gymnasium.*

[77] The University of Berne followed in 1834; and in 1873 the ancient Calvinist Academy in Geneva (founded in 1559) became a university; in 1889 the University of Fribourg was established, and in the following year the University of Lausanne.

[78] The University of Zurich emerged from several schools, including a semi-private one for the training of medical doctors (from 1782), one for political science (from 1807), and one for veterinarians (from 1820). An *Industrieschule* and a *Gymnasium* also played a role in the early years of the university, as several professors and docents held joint positions at these institutions and at the university. (For details of the history of the foundation of the University of Zurich we refer to the article of Günther Rasche and Hans Staub, 1979a, and to the Einstein biography of Carl Seelig, 1960.)

[79] Carl Ludwig Schleich, one of the later famous medical students from Germany (he arrived during the 1870s) later gave a lively report of his Zurich years in his recollections (Schleich, 1920). The great Zurich medical professors in the nineteenth century included the surgeon Theodor Billroth, whom we have mentioned earlier (in Section I.1), and the physiologist Adolf Fick (1829–1901). The latter is known in physics through his exact formulation of the law of diffusion (in 1855). In the early twentieth century the medical faculty of the University of Zurich counted among its members, e.g., the psychiatrist Eugen Bleuler (1857–1939) and the psychoanalyst Carl Gustav Jung (1875–1961).

Since Escher lacked the ability necessary to present demonstration experiments as required, the course was cancelled. Then Albert Mousson, formerly teacher at the *Industrieschule*, took over Escher's position at the *Gymnasium* in 1834 and became *Privatdozent* at the university; from the winter semester of 1834/1835 he delivered lectures on theoretical and experimental physics.[80] While retaining his position at the *Gymnasium*, he became fully engaged in work at the university, establishing, e.g., the *Physikalische Kabinett* (founded in 1835). In 1836 he was promoted to an extraordinary professorship, and from 1855 onwards he occupied full professorships both at the university and at the newly created Swiss *Polytechnikum*. Though an excellent teacher and organizer—he also took care of the collection of instruments at the *Polytechnikum* and became involved in technical tasks, such as obtaining a copy of the Paris metre standard for Switzerland and arranging gas illumination for his hometown of Zurich—Mousson did very little scientific research in physics.[81] The university, therefore, called (in 1857) Rudolf Clausius, from 1855 professor of physics at the *Polytechnikum*, to occupy a second chair. Clausius extended, in his Zurich years, the previous work on thermodynamics, making it a consistent theory; he dealt in particular with the formulation and applications of the second law of thermodynamics, and he made fundamental contributions to the kinetic theory of heat. Thus the double professorships of Mousson and Clausius at the university and the *Polytechnikum* signified what may be called the actual beginning of physics in Zurich; in the future it would often happen that the close co-operation of physicists from both institutions would lead to results which were also respected abroad.

The Zurich *Polytechnikum*, the second institution of higher education in the city, had the organization of a *Technische Hochschule*, as it would be called in Germany from the second half of the nineteenth century. But in Zurich the *Polytechnikum* represented much more due to its prehistory. In 1798 Philipp A. Stapfer, Minister of Education of the Helvetian Republic, planned the establishment of a central Swiss *Hochschule*, similar in character to the *École Polytechnique* in Paris. Half a century later Henri Dufour, one of the creators of the new Swiss Federal Constitution, picked up this idea again, and after a few years in early 1854 both chambers of the Swiss Parliament (the National Council and the Council of States) decided to found an *Eidgenössische Technische Hochschule* (Swiss Federal Institute of Technology) in Zurich. The *Polytechnikum* was opened on 16 October 1855, comprising originally of five schools: the *Bauschule* (School of Architecture), the *Inginieurschule* (School of Engineering), the *Mechanisch-*

[80] Albert Mousson, born in 1805, had studied in Berne and Geneva (at the academies of these cities), in Göttingen and finally in Paris (astronomy with Louis Arago, physics with Pierre Louis Dulong and Henri Regnault, theoretical mechanics with Siméon Denis Poisson, and calculus with Augustin Louis Cauchy), without obtaining a degree. In 1832 he was appointed a teacher at the *Industrieschule* in Zurich.

[81] Among the best-known works of Mousson, one may cite the measurement of the pressure dependence of the melting point of ice. He was interested in a wide range of scientific problems in geology, meteorology and glaciology. He became famous among zoologists for his huge collection of snails.

*technische Schule* (Mechanical-Technology School), the *Chemisch-technische Schule* (Chemical-Technology School), and the *Forstschule* (School of Forestry).[82] Besides these schools, which prescribed a rather rigid curriculum for the students, a sixth department was added, in which courses in mathematics and science, literature, modern languages, history and history of art, economics and law were offered, as at other universities of those times. Future *Gymnasium* teachers would obtain their training and education in this department.

The Swiss *Bundesrat* entrusted one of the leading architects of the day with the planning and construction of a proper home for the *Polytechnikum.* On the recommendation of the exiled Saxonian composer Richard Wagner, another refugee from Germany, Gottfried Semper (1803–1879), who also became head of the School of Architecture, erected the buildings on the slopes of the *Zürichberg.*[83] The *Polytechnikum* gave a chair to another participant in the German Revolution of 1848, the famous writer and literary historian Friedrich Theodor Vischer (1807–1887) from Stuttgart.[84] The first presidents of the education authority (*Schulrat*), Johann Konrad Kern and Carl Kappeler, made full use of the opportunities in order to gain, for the Swiss *Polytechnikum*, European celebrities as well as promising young scholars and teachers.[85]

## Mathematics and Physics at the E.T.H. (1855–1900)

Because of the composition of the Swiss Confederation, whose major parts of population were German- and French-speaking, the principal courses at the *Polytechnikum* had to be offered simultaneously in German and in French. While as a rule many good German-speaking candidates were available and accepted positions as lecturers, it was different with the French-speaking candidates.[86] Take mathematics, for example. It had been easy to find a German-speaking professor to give courses on calculus; this was Joseph Raabe (1801–1859), then

[82] The Karlsruhe *Polytechnikum* served as a model for the new Swiss *Polytechnikum*; the Karlsruhe *Polytechnikum* had been founded thirty years earlier. For details of the early history of the Zurich *Polytechnikum* we refer to the article of Knus (1981) and to the relevent chapter in Seelig's biography of Einstein (1960).

[83] Semper, the creator of the famous opera house in Dresden, had actively participated, like his friend Wagner, in the 1848 revolution and lost, as a consequence, his position in Dresden. He then lived from 1849 in Paris and London.

[84] Vischer had been a liberal writer and, in 1848, became a member of the Frankfurt Parliament, belonging to its left wing.

[85] The highest authority behind the *Polytechnikum* was the Swiss *Bundesrat*, the Government of Switzerland. The *Bundesrat* established the *Schulrat*, which had to supervise the institution. The presidents of the *Schulrat* travelled around Switzerland, Germany and other European countries to meet candidates. On their recommendation the selected people were elected by the *Bundesrat.* The parole followed in the selection was, 'The right man at the right place,' irrespective of nationality and political or religious belief.

[86] The reason for this situation is not difficult to guess. On the one hand, the French-speaking Swiss represented only a fraction of the German-speaking Swiss, hence the natural resources for 'French' scholars were not big. On the other hand, good French professors hardly ever left their own country to find employment abroad.

occupying a mathematics chair at the University of Zurich. For the French courses, first Jean-René Servient (1823–1856) was elected, but he died before he was able to assume his duties; the teaching of the next appointed professor, Ami de Beaumont (1820–1866), did not satisfy the authorities, and he was released in 1857, upon which Edouard Méquet (1821–1897) took over the chair.[87]

The occupants of the German-speaking mathematics chair, on the other hand, certainly gave no reason for any complaint, but rather increased the fame of the young institution. When Raabe retired in 1857 because of illness, Julius Richard Dedekind (1831–1916), then *Privatdozent* at Göttingen, became his successor. Dedekind was a brilliant and well-organized lecturer, for students of engineering as well as for those of Department Six.[88] He also continued active mathematical research in Zurich, devising there his later (1873) published work on continuity and irrational numbers ('*Stetigkeit und Irrationalzahlen*'). When he returned in the spring of 1862 to his hometown of Braunschweig, to accept a chair at the *Technische Hochschule*, Elwin Bruno Christoffel (1829–1900) arrived in the fall of that year to replace him. Like Dedekind he exhibited great talent as a teacher, lecturing on all fields of mathematics, including the most advanced ones, such as the geometry of curved space, one of the particular fields of his own research.

The growth of mathematics at the *Polytechnikum* was immediately rewarded when a chair in higher mathematics was created, to which Friedrich Emil Prym (1841–1915), a young and promising pupil of Bernhard Riemann and recommended by Christoffel, was elected. Christoffel then became the essential organizer of the new, sixth school of the *Polytechnikum*, the *Schule für Fachlehrer* (School for High School Teachers), which was established in 1866 from a part of the former Department Six.[89] He devised the obligatory curriculum and created the *Mathematical Seminar* for advanced students, following the model of several German universities like Göttingen and Berlin. After the winter semester of 1868/1869 he left Zurich and accepted a professorship in Berlin at the *Gewerbe-Akademie*, which was later turned into the *Technische Hochschule*. Since at the same time Prym went to Würzburg, the two chairs became vacant; but again two worthy successors were found in Hermann Amandus Schwarz (1843–1921) and Heinrich Weber (1842–1913). In spite of further changes in the following decades, the *Polytechnikum* kept its reputation as an excellent school for mathematics. For example, at the turn of the century the mathematics stars in Zurich were Adolf Hurwitz (1859–1919) and Hermann Minkowski (1864–1909). Wilhelm Fiedler (1832–1912), from 1867 professor of descriptive geometry, and Carl Geiser (1843–1934), a student and nephew of the famous Swiss mathematician Jakob

[87] The details of the early mathematics at the *Polytechnikum* has been told by Max-Albert Knus in an article for the volume celebrating the 150th anniversary of Bruno Christoffel's birth (Knus, 1981).

[88] Together with Dedekind, Bernhard Riemann had been considered for the *Polytechnikum* chair. The president of the *Schulrat*, Kappeler, however found, when inspecting Riemann in Göttingen, that his teaching might not suit the engineers.

[89] The other parts of Department Six, like literature, history and political science, then formed a new Department Seven, entitled '*Freie Fächer*' ('Free Subjects').

Steiner and from 1873 the occupant of the chair of geometry, were also worthy representatives of their fields.

In October 1896, perhaps the most ingenious student of the *Eidgenössische Technische Hochschule* enrolled in *Abteilung VI a* (the School for High School Teachers in Science): it was Albert Einstein. From the winter semester 1896/1897 he attended the lecture courses given by the mathematicians Hurwitz (on calculus and differential equations), Fiedler (on descriptive geometry and projective geometry), Geiser (on analytic geometry, determinants, infinitesimal geometry and geometrical theory of invariants), Minkowski (on geometry of numbers, function theory, potential theory, elliptic functions, analytical mechanics, variational calculus, algebra, partial differential equations, applications of analytical mechanics), and Arthur Hirsch (on definite integrals and linear differential equations).[90] In spite of the broad programme in mathematics Einstein did not show himself to be an eager student of that field; his fellow-student and friend Marcel Grossmann, who dutifully sat through all courses, often had to supply the notes he had taken so that Einstein could learn from them. But Einstein pursued another course, as he later wrote, 'fascinated by Professor Geiser's lectures on infinitesimal geometry, which were genuine masterpieces of pedagogical art and which later helped me greatly in my struggle with general relativity' (Einstein, 1956, pp. 10–11). He admitted: 'I had excellent teachers (for example, Hurwitz, Minkowski), so that I could have gotten a sound mathematical education. However, I worked most of the time in the physical laboratory, fascinated by the direct contact with experience' (Einstein, 1949, p. 15).[91]

However, at that time the professors representing physics at the *Polytechnikum* by no means matched the quality and high standard of their colleagues in mathematics. Adolf Fisch, another fellow-student of Einstein's recalled:

> Experimental physics was not particularly favoured [at the E.T.H.]. In theoretical physics, the docent for electrotechnics, Professor Heinrich Weber, presented excellent lectures. However, being a typical representative of classical physics, he ignored everything that came after [the work of] Helmholtz. At the end of the lectures one knew the history of physics, but not its present and future. We, therefore, had to study privately the more recent literature on our own. (Seelig, 1960, p. 47)

Einstein indeed spent much time on reading the more recent physics literature, especially the works of James Clerk Maxwell, Heinrich Hertz, Hendrik Antoon Lorentz and Ludwig Boltzmann.

Heinrich Friedrich Weber (1843–1912) had been a pioneer of electrotechnics in Germany and Switzerland. He also acquired some merit through his original experimental research on several topics in physics: he studied the temperature dependence of the specific heats of several substances (H. F. Weber, 1872), and

[90] Einstein's lecture courses at the E.T.H. have been listed by Carl Seelig (1960, pp. 38–40).

[91] Einstein's preference for physics, especially experimental physics, may be understood, if we take into account his original intention to study engineering at the E.T.H. (See the recollections of Louis Kollros, one of Einstein's fellow students at that time: Kollros, 1956.)

he investigated (in 1878) the validity of Adolf Fick's law of hydrodiffusion. But his lectures on theoretical subjects failed to arouse Einstein's interest, who missed in them such items as Maxwell's electromagnetic theory of light. As a consequence he stopped attending the lectures, and Professor Weber, an authoritative man, obtained the impression that Einstein, though being 'a rather intelligent boy' ('*ein ganz gescheiter Junge*,' Seelig, 1960, p. 48), was simply lazy.[92] While he rejected the theoretical lectures of Weber, Einstein still went regularly to his laboratory courses, as well as to those of the French-speaking professor of the E.T.H., the Vaudois Jean Pernet.[93] He also prepared his diploma thesis on a problem of physics, heat conduction, under Weber's supervision. Weber, however, was not very impressed by his student; and, after Einstein had passed (in the early summer of 1900) the examination for the teacher's (*Fachlehrer*) certificate, he did not offer him the position of assistant, which would have opened a career in science. Weber rather accepted two engineering graduates. Einstein, who had to earn his living in the next couple of years from irregular work as a substitute teacher, after which he joined the Swiss Patent Office in Berne as a technical expert, commented shortly after Weber had died: 'Weber's death is good for E.T.H.' (Einstein to Zangger, summer, 1912; quoted in Pais, 1982, p. 45).

## Einstein, Debye and von Laue at the University of Zurich (1909–1914)

Einstein, who had not been offered (upon graduation) a research position at the E.T.H., was destined to play a decisive role in physics in Zurich; essentially he helped to establish a lasting tradition, in which the short interlude (from 1855 to 1867) with Rudolf Clausius was not successful. Einstein, upon obtaining the citizenship of Zurich in 1901, had to wait until June 1902 before he found stable employment at the Swiss Patent Office (*Eidgenössisches Amt für geistiges Eigentum*). The regular salary which he earned as a technical expert in this office—he was promoted in April 1906—enabled him to marry and have a family, while work at the office left him enough spare time for his own scientific studies in the fields of statistical mechanics, electrodynamics of moving bodies and quantum theory. When he quit the Patent Office position in October 1909, released with an appraisal for his good services, Einstein had become a physicist renowned for important contributions to the latest physical theories, as the creator of special relativity theory and the principal promoter of quantum theory; because of this research he also received earlier in 1909 his first honorary degree from the University of Geneva. Being acknowledged and recommended by such international celebrities as Max Planck and Henri Poincaré, he then started his

[92] A similar opinion on Einstein was expressed by Hermann Minkowski, who said, after he had learned about special relativity theory: 'It [the theory] came as an enormous surprise to me, because earlier [i.e., in Zurich] Einstein was really lazy. He did not care for mathematics at all.' (Minkowski in a conversation with Born. '*Sie war für mich eine gewaltige Überraschung. Denn früher war Einstein ein richtiger Faulpelz. Um Mathematik hat er sich überhaupt nicht gekümmert.*' See Seelig, 1960, p. 45.)

[93] It is reported that Einstein did not get along well with Pernet, because Einstein did not follow the conventional rules in solving problems but applied his own ideas (see Seelig, 1960, pp. 48–49).

scientific career as an extraordinary professor of theoretical physics at the University of Zurich, where in 1905 he had received his doctorate with a thesis on molecular dimensions (accepted by Professors Alfred Kleiner and Heinrich Burkhardt).

Alfred Kleiner (1849–1916) succeeded Albert Mousson upon the latter's retirement in 1878, first as an *Extraordinarius.*[94] In 1885 he was promoted full professor and director of the new physics institute in *Rämistraße.* He established his own experimental research programme in physics, systematically investigating the following topics: temperature dependence of specific heats and heats of solutions; dielectric behaviour of non-ideal insulators; frequency dependence of the resistance and inductance of coils; and the influence of intermediate matter on gravity. Kleiner was an experienced experimentalist, who constructed, for example, a quadrant electrometer of very high sensitivity; he also trained a number of talented students, such as Friedrich Adler (1879–1960), son of the Austrian Socialist Leader Victor Adler.[95] In 1907 Adler became *Privatdozent* at the University of Zurich; being very interested in the analysis of fundamental physical concepts on the basis of Ernst Mach's ideas, he presented these topics in lectures that were well organized and well received. Therefore, when in 1908 Kleiner proposed the establishment of an extraordinary professorship for theoretical physics at his institute, he first thought of Adler as the most worthy candidate. But Adler refused, claiming that Einstein had to be given priority because of his superior scientific work. Einstein was then elected to the new position.[96]

Although Einstein had not been Kleiner's first choice, both got along rather well. 'Professor Kleiner, the chief of our institute,' Einstein wrote in a letter to Jakob Laub—who had been his first collaborator in Berne—'is perhaps not a magnificent physicist, but he is a wonderful man, whom I enjoy [working with]. It seems to me that scientific distinction and the worth of a personality do not always go together' (Einstein to Laub, quoted in Seelig, 1960, p. 168). During the three semesters which he spent at the University of Zurich, he completely fulfilled his teaching tasks—which had been described as 'lecture courses in topics of theoretical physics, accompanied possibly by laboratory experience' (see Rasche and Staub, 1979a, p. 218)—by presenting in the winter semester of 1909/1910 two courses in mechanics and thermodynamics, respectively, in the

[94] Kleiner had originally studied medicine, graduating in 1872 and obtaining his *Dr. med.* in 1874 at the University of Zurich. Having been attracted to science and physics, during a period of study which he spent in Berlin, he became (from 1873 to 1879) assistant at the physical institute of the *Polytechnikum.* In 1875 he obtained his *Habilitation* at the *Polytechnikum* and the University of Zurich. He kept his position as lecturer at the *Polytechnikum* until 1885.

[95] Adler came to Zurich to study chemistry with Alfred Werner, but soon changed to physics. During his studies he got to know Albert Einstein. Adler obtained his doctorate in 1902 with Kleiner.

[96] Einstein had meanwhile become *Privatdozent* at the University of Berne, on the initiative of Paul Gruner (who from 1894 had been *Privatdozent* for mathematical physics and from 1906 professor of theoretical physics in Berne). His lectures, however, were not successful: in the winter semester of 1908/1909 he gave a course on radiation theory, and in the following semester the course had to be cancelled because only one student, Max Stern, enrolled. For further details of Einstein's first position on the academic ladder we refer to Seelig's biography (Seelig, 1960, especially pp. 152–156).

summer semester of 1910 two courses in mechanics and the kinetic theory of heat, and in the winter semester of 1910/1911 one course on electrodynamics and another on selected topics of theoretical physics. In addition, he regularly conducted a physical seminar and participated in the laboratory courses of Professor Kleiner (in the summer semester of 1910 and the winter semester of 1910/1911).

Einstein's teaching did not instantly turn the University of Zurich into a centre of theoretical physics, though his main courses attracted about two dozen students.[97] In the meantime, he continued his own scientific work on the foundation and application of quantum theory. He studied with Ludwig Hopf (1884–1939)—who, after getting his doctorate with Arnold Sommerfeld in Munich, came to Zurich as Einstein's assistant—details of the statistical description of blackbody radiation (Einstein and Hopf, 1910a, b); and he found a relation between the elastic constants of a solid and the frequency entering into the quantum-theoretical formula for specific heats (Einstein, 1910). He received famous visitors from abroad, like Walther Nernst, who brought (early in 1910) the first results of his specific heat experiments in Berlin; and Arnold Sommerfeld, who came to discuss a problem of relativity theory.[98] These visits further contributed to the opinion formed by the German theoretical physicists, that the Zurich *Extraordinarius* had to be regarded as one of the leading pioneers in their field. As such Einstein would be invited to prepare a report on the problem of specific heats for the international conference on 'Radiation Theory and the Quanta,' held in the fall of 1911 in Brussels. At the time of the conference, however, he had already left Zurich, having accepted a full professorship of theoretical physics at the German University of Prague.

With Einstein's departure from Zurich theoretical physics at the University of Zurich was not abandoned—as had happened earlier in the case of Rudolf Clausius. In the summer of 1911 the university immediately secured a worthy successor to Einstein in Peter Debye, who had been Sommerfeld's student and (from 1906 in Munich) assistant, and who the previous year had become *Privatdozent* at the University of Munich. Debye dutifully fulfilled his teaching tasks in theoretical physics, assisted Kleiner in the laboratory courses, and contributed original and weighty research on problems of quantum and atomic theory. He thus developed in Zurich a kinetic theory of insulators (Debye, 1912a); and it was there that he developed his extension of Einstein's theory of the specific heats of solids by including a full spectrum of quantized vibrations (Debye, 1912b, c).[99] Debye, upon refusing a call to the University of Tübingen, was

[97] Hans Tanner, later professor of physics and mathematics at the *Technikum* in Winterthur, has given a lively account of Einstein as a teacher in Zurich in Seelig's book (Seelig, 1960, pp. 170–179). Tanner also received the topic for his doctoral dissertation (on the kinetic theory of gases) from Einstein.

[98] For details of the specific heats story, see Volume 1, Section I.6.

[99] We have discussed Debye's early contributions to quantum theory in Volume 1, Section I.7. (Also see Debye's biography, given on p. 138 of Volume 1.)

appointed *Ordinarius* (in April 1912) in Zurich; but shortly afterwards he accepted a professorship at the University of Utrecht and left the city at the end of the summer semester of 1912.

Another promising young scientist, this time a student of Max Planck, Max von Laue, assumed the position in Zurich. Von Laue, after completing his doctorate in 1903, continued his studies in Göttingen (with Karl Schwarzschild in optics and Max Abraham in electron theory), and then returned to Berlin in 1905 as Planck's assistant where he obtained his *Habilitation* in 1906; in 1909 he went to Munich as *Privatdozent* and assistant to Sommerfeld. While he had originally worked primarily on optical problems and, after 1906, on special relativity, he suggested in the spring of 1912 the idea of obtaining X-ray interference patterns by crystals; on 21 April of that year Walter Friedrich and Paul Knipping, following Laue's suggestion, observed the first interference patterns proving simultaneously the wave nature of X-rays and the atomic constitution of crystals (Friedrich, Knipping and Laue, 1912). The new *Extraordinarius* Laue—or von Laue, as he became in 1913, when nobility was bestowed upon his father—made this discovery the topic of his inaugural address at Zurich, entitled 'On the Wave Nature of X-rays' ('*Über die Wellennatur der Röntgenstrahlen*,' see Rasche and Staub, 1979a, p. 213). He also planned to continue experimental research on X-ray interference phenomena and did so. Although this experimental work suffered because of insufficient financial support and personal assistance, von Laue succeeded in obtaining some valuable results in Zurich; he thus arrived, with his student Franz Tank, at an explanation of the shape of the observed interference spots.[100] In the summer of 1914 von Laue accepted the call to the theoretical chair at the University of Frankfurt-am-Main, starting his new appointment in the winter semester of 1914/1915. All attempts by the University of Zurich to win back either Peter Debye (who had meanwhile changed from Utrecht to Göttingen), or von Laue (who received the 1914 Nobel Prize in physics for his discovery of X-ray interference effects), unfortunately failed.

The loss of von Laue to Frankfurt must be considered as very serious for theoretical physics in Zurich, as only a few months before Albert Einstein had also left the city. Einstein had been awarded, early in 1912, the professorship of theoretical physics at the E.T.H., due mainly to the endeavours of his friend Marcel Grossmann (1878–1936)—who, in 1907, had succeeded his teacher Wilhelm Fiedler in the chair of geometry.[101] The new professorship offered an important advantage to Einstein, because it was not connected with any 'duties

[100] We have mentioned von Laue and Tank's result (Laue and Tank, 1913) in Section I.4. for further details of von Laue's scientific work in Zurich, see the article of Rasche and Staub (1979b).

[101] The chair for theoretical physics at the E.T.H. emerged from the chair for higher mathematics, founded back in 1865, which had fallen vacant in October 1902 after Hermann Minkowski had left. Initially, the professors of neighbouring fields, notably the physicist Pierre Weiss and the mathematician Adolf Hurwitz, tried to fill the gap, but the *Schulrat* felt that theoretical physics had to be represented officially at the E.T.H. After getting letters of recommendation from Marie Curie and Henri Poincaré in Paris, who wrote enthusiastically about Einstein and his achievements, the Prague professor was appointed to the chair. (The chair was established in the eighth department of the E.T.H., the *Schule für Fachlehrer in Mathematik und Physik*; thus Einstein became a professor essentially in the same department in which he had obtained his teacher's diploma.)

in delivering general lecture courses in physics to a large auditorium and conducting laboratory courses'; it rather corresponded, as the president of the *Schulrat* pointed out in his proposal to the Swiss government, 'in a very particular way to the characteristics of Einstein,' namely, 'The activity of this chair must essentially extend to the students of the higher classes of Department Eight [of the E.T.H.] and embrace the advanced topics of mathematical physics' (from the proposal, dated 23 January 1912, quoted in Seelig, 1960, pp. 231-232).

Einstein joined the E.T.H. and his old department, now under the direction of his friend and former fellow-student Marcel Grossmann, in the fall of 1912; during the following semesters he lectured on analytical mechanics and on thermodynamics (in the winter semester of 1912/1913), on mechanics of continua and on the molecular theory of heat (in the summer semester of 1913) and on electricity and magnetism, and on ray optics and diffraction (in the winter semester of 1913/1914). In addition, he held seminars for advanced students on topics of physical theory. He also promoted quantum theory with several investigations, e.g., with a paper on the specific heat of molecular hydrogen, written with his assistant Otto Stern (Einstein and Stern, 1913).[102] However, Einstein put the main emphasis in the second Zurich period on work towards the establishment of a fundamental theory of gravitation. He obtained essential help from Marcel Grossmann, who acquainted him with the advanced geometrical methods of Bernhard Riemann, Bruno Elwin Christoffel, Gregorio Curbastro Ricci and Tullio Levi-Cività; the two of them completed an 'Outline of a Generalized Theory of Relativity and a Theory of Gravitation' ('*Entwurf zu einer verallgemeinerten Relativitätstheorie und einer Theorie der Gravitation*,' Einstein and Grossmann, 1913). In the summer of 1913 Max Planck and Walther Nernst travelled to Zurich to offer Einstein a very special position in Berlin—membership of the Prussian Academy, together with a professorship at the University of Berlin that was free of any teaching obligations. The conditions proposed, including a handsome salary, were so favourable that Einstein accepted and left Zurich again in April 1914.

Einstein's second move from Zurich—he would return only once again for a longer period, namely in 1919, to deliver lectures on relativity theory—did not harm the development of physical research in Zurich in the same way as it had done in the case of Rudolf Clausius about half a century earlier. Still he left a gap in theoretical physics, which was not easy to fill. Experimental physics was much better off. On the one hand, the E.T.H. chair, previously occupied by Heinrich Friedrich Weber, had been taken in 1903 by the Alsatian Pierre Weiss. In Zurich he carried out, from 1905, systematic experimental studies of the magnetization of substances in the presence of strong magnetic fields. In 1907 he extended Paul Langevin's kinetic theory of diamagnetism and paramagnetism to describe the phenomena in ferromagnetic substances (Weiss, 1907). Further investigations led to the result that atoms and ions possessed magnetic moments, which are integral

[102] Otto Stern, after obtaining his doctorate, joined Albert Einstein in Prague. He also came to Zurich, where he got his *Habilitation* in physical chemistry in 1913 (at the E.T.H.). For his biography we refer to Volume 1, Section IV.3 (Footnote 699 on p. 433).

multiples of a fundamental constant—the (Weiss) magneton (Weiss, 1911a, b, c). Weiss' pioneering researches on the magnetic properties of matter not only provided a new proof of kinetic theory, but could be related—as we have already pointed out—to quantum theory.

## Physics in Zurich until 1920: Greinacher, Meyer, Bär, Ratnowsky, Wolfke

Weiss, who stayed in Zurich until 1919—he then accepted the directorship of the physics institute at the University of Strasbourg—remained completely absorbed in research on magnetism and would not embark on experimental research in other areas of modern physics. Such research existed, however, at the University of Zurich, where Heinrich Greinacher worked as assistant and *Privatdozent* from 1907.[103] Having already dealt, in Geneva, with the properties of polonium (Greinacher, 1906; Greinacher and Kernbaum, 1907), he now investigated in Zurich the special effects shown by radioactive substances and became one of the experts in this field—as his many contributions to the *Physikalische Zeitschrift*, then the foremost German scientific journal for radiation and radioactivity, indicate. In order to obtain an accurate and reliable method of measuring the intensities of X-rays and radioactive radiation, Greinacher designed what he called the *Ionometer* in 1913 (Greinacher, 1914). The instrument consisted of a special ionization chamber and an electrometer; since the latter needed a d.c. voltage of 200 to 300 volts, and the Zurich power stations supplied only 110 volts a.c., he devised a voltage multiplying circuit. In the following years he improved on the method, which was later called the Greinacher circuit (*Greinacher-Schaltung*); thus he reached by 1920 a voltage multiplication factor of 10.[104]

Greinacher's research and teaching represented modern experimental physics at the University of Zurich for years, if one neglects the X-ray investigations of the theoretician von Laue. After Alfred Kleiner's forced retirement in 1915, due to illness, Greinacher received the title of professor, but he did not succeed Kleiner as director of the physics institute.[105] Instead, another former student of Emil Warburg, Edgar Meyer, was called to this position.[106] Meyer had earlier

[103] Heinrich Hermann Greinacher was born on 31 May 1880 in St. Gallen, Switzerland. He studied physics in Zurich and Geneva—where he also attended the *Conservatoire*, obtaining in 1901 first prize as an instrumentalist—and Berlin (with Emil Warburg and Max Planck). He obtained his doctorate with Warburg in 1904, upon which he served as an assistant to Warburg, to Charles-Eugène Guye in Geneva (in 1905) and to Georg Quincke in Heidelberg (in 1906).

[104] Greinacher's method would be used in 1932 by John Douglas Cockcroft and Ernest Thomas Sinton Walton in Cambridge to accelerate protons producing the first artificial nuclear reactions (Cockcroft and Walton, 1932a, b).

[105] In 1924 Greinacher became full professor and director of an institute at the University of Berne. He continued to do research in radioactivity, nuclear and elementary particle physics; in 1926 he created the proportional counter and in 1934 the spark counter, instruments which registered charged particles. He retired in 1952 and died on 17 April 1974 in Berne. (For details of his biography we refer to the articles of Heinz Balmer: Balmer, 1970, 1974.)

[106] Edgar Meyer was born on 5 March 1879 in Bonn and studied at the universities of Munich and Berlin, receiving his doctorate in 1903, upon which he became an assistant to Emil Warburg.

been in Zurich, namely from 1907 to 1909, as assistant in Kleiner's institute. He had also received his *Habilitation* at the University of Zurich (in 1908), before joining Johannes Stark in Aachen (from 1909 to 1911). In 1911 he had been called as *Extraordinarius* to the University of Tübingen. Meyer had made a reputation for himself through fundamental experimental investigations on the nature of $\gamma$-rays (Meyer, 1910) and through tricky studies of the photoelectric effect with ultra-microscopic metallic particles (Gerlach and Meyer, 1913). From the experience gathered in the latter topic, Richard Bär, a later member of Meyer's institute (from 1917), profited greatly.

In Zurich the former mathematician Bär became a most able physicist.[107] In particular, he started experiments on the atomistic structure of electricity (Bär, 1918b), which soon brought him into a scientific debate with Felix Ehrenhaft in Vienna. Irene Parankiewicz, a student of Ehrenhaft's, claimed at that time that her measurements of the falling velocity of smallest charged spheres (made of sulphur, selenium or mercury), proved the existence of subatomic charges being as small as 1/1,800 of the value of the elementary quantum of electricity (Parankiewicz, 1917). Bär showed, however, by re-evaluating the same data—also taking into account Schrödinger's study of the Brownian motion and the error estimate contained in it (Schrödinger, 1915b)—that 'the charges of these smallest microscopic particles observed so far still do have an order of magnitude of Millikan's value of the elementary charge' (Bär, 1918a, p. 373). He then systematically refined the methods of measuring the shape and density of submicroscopic particles (Bär, 1919; Bär and Luchsinger, 1921), and finally summarized all research in his *Habilitation* thesis of 1922. He found that in evaluating the data obtained in the Ehrenhaft-Millikan experiments one had to take three effects into account: first, the Brownian motion of the particles used; second, the deviations from Stokes' law for very small particles; and third, the non-spherical shape of the particles. 'Then not the slightest indication follows for the existence of electric charges falling below the value of the elementary charge,' he concluded. 'The apparent subelectrons found by Ehrenhaft and his pupils are generated mainly by assuming too high a density [for the observed submicroscopic particles]' (Bär, 1922, p. 200).

Bär, who would later shift his research to studies of the electric conduction in gases, and finally, after 1928, contribute important results by applying the Raman Effect, has been called 'perhaps the most distinguished experimental physicist who ever worked at the University of Zurich, and never at the E.T.H.' (Rasche and Staub, 1979a, p. 215). Although in 1924 he received the call to the experimental chair at the University of Berne, he remained in Zurich.[108]

[107] R. Bär was born on 11 September 1892 in Basle. He studied mathematics in Zurich and Würzburg. After getting his doctorate in 1915 with Emil Hilb (in Würzburg), he became an assistant to David Hilbert in Göttingen.

[108] Bär was a modest man, who contented himself with receiving, in 1928, the personal title of a professor. (This corresponded neither to an *Extraordinarius* nor to an *Ordinarius*, but to what was called in Germany an '*außerplanmässiger Professor*'.) He died on 13 December 1940 in Zurich.

The excellent experimentalist Bär was not the only assistant at the physical institute of the University of Zurich. Meyer took over, for example, von Laue's former collaborator Franz Tank, who had obtained his doctorate in 1916. Tank became *Privatdozent* at the university in the summer of 1918, at the same time as Karl Wilhelm Meissner, whom Meyer had brought with him from Tübingen. In Zurich, Meissner continued spectroscopic research, especially on the fine structure of atomic lines.[109]

In comparison with the research of his effective assistants, Meyer's own scientific work, during the first Zurich years, suffered somewhat from his extensive teaching activities. He presented to his students, especially in the beginning, brilliant lectures supplemented by perfect experimental demonstrations.[110] In 1919 he entered into a series of experimental investigations on gaseous discharges, extending over several years, and dealing in particular with the problem of the spark potential. Later he returned to the topic of his doctoral dissertation, i.e., the study of the ozone content of the atmosphere. Altogether he kept the institute in *Rämistraße*—which was quickly restored after being partially burnt down in 1923—in excellent condition and trained many talented students, who made their careers in physics in Switzerland and abroad.[111]

While experimental physics in Zurich certainly received a fresh impetus in the second decade of the twentieth century, one may wonder what happened to theoretical physics after Einstein and von Laue had left. If one examines the announcements of lecture courses presented at the E.T.H. and the university, such as the ones published regularly in *Physikalische Zeitschrift*, one discovers that the teaching programme in theory went on uninterrupted at both institutions. At the university the main courses, as well as those on special topics, were delivered by Simon Ratnowsky, who came from Russia and had become *Privatdozent* in the spring of 1913.[112] At the E.T.H., on the other hand, the Pole Mieczyslaw Wolfke presented the theoretical lectures, beginning with the summer semester of 1914, for which he announced a course first on electric light sources and a second on heat radiation and quantum theory.

[109] K. W. Meissner was born on 15 December 1891 in Reutlingen. He studied at the universities of Tübingen and Munich, learning spectroscopy under the supervision of Friedrich Paschen.

From Zurich Meissner was called to an extraordinary professorship at the University of Frankfurt in 1925, becoming an *Ordinarius* in 1928 and director of the physical institute four years later. He left Frankfurt in 1937 and emigrated to the United States. There he assumed chairs at the Worcester Polytechnic Institute, Worcester, Massachussetts (1938–1941) and later at Purdue University, Lafayette, Indiana (1941–1959). He died on 13 April 1959 while travelling on a ship to Europe. (For his biography, see Czerny, 1959.)

[110] Later, after a complaint from the non-physics students, he lowered the level of his teaching. (See Rasche and Staub, 1979a, pp. 213–214.)

[111] Meyer retired in 1949 from his position. He died on 29 February 1960 in Zurich.

[112] Ratnowsky's lectures in the presence of von Laue were devoted to what was then considered special topics for advanced students: he treated, in the winter semester of 1913/1914, the electron theory of metals, and in the summer semester of 1914 statistical mechanics and quantum theory on the one hand, and magneto- and electro-optics, on the other.

Simon Ratnowsky was born on 21 September 1884 in Stawropol, Caucasus.

The two industrious docents managed, over a period of five years or more, to cover the main topics, then taught in theoretical physics, at regular intervals. Besides classical theory they devoted particular attention to modern subjects; thus they lectured in the summer semester of 1915 on heat radiation (Ratnowsky), on the mathematical theory of radioactivity and the inner structure of atoms (Wolfke), and on the symmetry principle and its applications to physical phenomena (Wolfke). Wolfke, who also became, in the fall of 1914 *Privatdozent* at the University of Zurich, presented the last two courses to the students of the E.T.H. *and* the university, and he continued to do so in the following years.[113] In the winter semester of 1915/1916 Ratnowsky, besides giving the main course on electrodynamics, announced two other special courses, one on electron theory and the other on the molecular theory of solids; Wolfke treated wave theory as the main topic, and modern theory of gravitation as a special course. Soon the most recent advances on the inner structure of atoms also received their place in the Zurich lectures, presented by Wolfke in the winter semesters of 1916/1917 and 1917/1918, and by Ratnowsky and Wolfke in the winter semester of 1919/1920.

The Zurich teaching program in theoretical physics no doubt competed favourably with that of most places in Germany, or in any other country. But how did Ratnowsky and Wolfke compare with their colleagues with respect to research? In the physics journals of those years one finds a number of contributions from both authors on problems of quantum theory and atomic structure. Thus Ratnowsky submitted, as early as March 1912, a paper to *Annalen der Physik*, entitled '*Die Zustandsgleichung einatomiger fester Körper und die Quantentheorie*' ('The Equation of State of Monatomic Solids and the Quantum Theory,' Ratnowsky, 1912). The purpose of the paper was 'to use Planck's theory of the [quantized] resonators . . . for the derivation of an equation of state of real monatomic solids,' namely, 'by modifying Planck's entropy equation which can only be considered as the equation of state of an "ideal" solid' (Ratnowsky, 1912, p. 639). By employing the earlier ideas of Gustav Mie and Eduard Grünneisen on the intermolecular forces in solids, he arrived at a quantum-theoretical equation of state for solids, which satisfied the known empirical data. In a further publication, submitted in January 1913 to *Verhandlungen der Deutschen Physikalischen Gesellschaft*, Ratnowsky extended his theory on the basis of Peter Debye's recent extension of Einstein's theory of the specific heats of solids (Debye, 1912c); he thus generalized the equation of state for a real solid by taking into account the full oscillation spectrum of a solid (Ratnowsky, 1913). In three papers, published between 1915 and 1918, he tried to establish several quantum-theoretical formulae in gas and solid state theory without explicitly invoking the quantum hypothesis (Ratnowsky, 1915, 1916, 1918). Ratnowsky did not argue in this work against the existence of the quantum of action in the description, but wished to have its origin in the inner constitution of atoms and molecules and in the

[113] We have given biographical data on Wolfke in Volume 1, p. 559 (Footnote 852).

particular changes of the electron-nucleus systems. 'We stick to the fact,' he ultimately concluded, 'that classical statistical mechanics, applied to molecular processes, leads to correct results, provided one does not neglect any detail in such processes' (Ratnowsky, 1918, p. 530).

Ratnowsky—who also supervised the thesis of Jan von Weyssenhoff dealing with the quantum theory of rotating molecular systems and the theory of paramagnetism (Weyssenhoff, 1916)—cannot be considered a highly successful or original and creative author in the field of quantum theory, although his papers display a solid knowledge of the theoretical methods, especially those of statistical mechanics. A similar statement may apply to the publications of his colleague Wolfke, who concentrated his research on two major problems of quantum theory. In a series of papers, published between 1913 and 1921, he tried to answer the question as to whether one could derive Planck's blackbody radiation formula by assuming localized light-atoms or quanta; he finally showed that the answer was positive, if one admitted the existence of spatially independent light-molecules of one, two, three, etc. light-atoms $h\nu$ for each frequency $\nu$.[114]

In a second series of papers Wolfke discussed the origin of the hydrogen spectra. He first showed, early in 1916, that the Balmer spectrum might possibly be emitted by the hydrogen molecule, because Debye's model for this system seemed to give rise to the same kind of radiation as Bohr's model for the hydrogen atom and even fitted the data better (Wolfke, 1916a). A few weeks later he withdrew this result, claiming that closer inspection of the most recent data on atomic constants preferred Bohr's hydrogen atom (Wolfke, 1916b). In the fall of 1919 he returned to the problem of hydrogen lines, dealing now with the many-line spectra. Again he took Debye's hydrogen molecule model as the basis for his considerations; however, he argued that the lines were not emitted from neutral hydrogen but rather from the hydrogen molecule ion; he then calculated the spectra by applying Bohr's quantum conditions to a system, in which two hydrogen nuclei rotate on a circular orbit around a fixed electron, finding fair agreement between experiment and theory (Wolfke, 1920).

### Environment after World War I: Weyl, Debye, Scherrer, Schrödinger

Towards the end of the second decade Ratnowsky and Wolfke came up against strong competition at their respective institutions, first from Paul Epstein, and then from Hermann Weyl. Weyl (1885–1955) had been professor of geometry at the E.T.H. from 1913, succeeding Carl Geiser.[115] In lecture courses he occasionally treated special topics of theoretical physics, for example, in the winter semester of 1914/1915 (potential theory and differential equations of mathematical physics) and in 1916/1917 (mathematical theory of the electromagnetic field). After Einstein had created General Relativity Theory, Weyl became strongly interested—like his former Göttingen teachers David Hilbert and Felix Klein—in this topic,

[114] We have discussed Wolfke's work on this problem in Volume 1, Section V.3.

[115] Weyl came from Göttingen. For his biography see Volume 1, p. 291, Footnote 471.

in which physical ideas united with elegant mathematical conceptions. In the summer semester of 1917 he gave a course entitled '*Raum-Zeit-Materie*' ('Space-Time-Matter'), which he published in the spring of 1918 as a book bearing the same title (Weyl, 1918c).[116] From now on he continued to lecture regularly on topics of gravitation and electrodynamics. But he became even more deeply involved in the problem of generalizing General Relativity into a unified theory of matter by his attempt to merge the theories of gravitation and electricity (Weyl, 1918a, b).[117]

While Weyl emerged as a principal and original contributor of fundamental ideas in General Relativity, Paul Sophus Epstein (1883-1966), a newcomer to Zurich recommended by Arnold Sommerfeld, and who obtained his *Habilitation* at the University of Zurich in the spring of 1919, had already acquired in Munich an excellent reputation in quantum theory: in 1916 he solved the problem of deriving the Stark effect of the hydrogen lines in the Bohr-Sommerfeld theory of atomic structure (Epstein, 1916a, c), and he composed further papers on the theory which confirmed the expectation that he would remain a highly original and productive author. Epstein began his Zurich career by presenting lectures on special theoretical topics, such as electron theory (in the winter semester of 1919/1920), but soon also took over the main courses, e.g., one on mechanics in the winter semester of 1920/1921. Thus he indeed challenged the monopoly held thus far by Ratnowsky.

Still more changes occurred in physics in Zurich. After Pierre Weiss left the E.T.H. in 1919 and returned to his native country—French Alsace—Peter Debye replaced him early in 1920 as director of the physics institute. Debye did not come alone from Göttingen, he brought with him his former Swiss student Paul Scherrer.[118] The latter joined the E.T.H. first as *Privatdozent* (spring of 1920), and soon advanced on the academic ladder to become *Extraordinarius* (still in 1920) and full professor (in 1924).

Debye had occupied Woldemar Voigt's chair from 1914, and had been extremely successful in Göttingen. He had shifted his interest during this period from kinetic theory to the structure of molecules and atoms, contributing well-recognized work on the dispersion of light by molecules (Debye, 1915a), on the explanation of the Zeeman effect (Debye, 1916a, b), on the fine structure of hydrogen-like atoms (Debye, 1916c), and on the X-ray spectra of atoms (Debye, 1917). While in Göttingen, he had also made an even more significant contribution

[116] It might be of interest to mention that in the preface to this book Weyl thanked Richard Bär, the former assistant of Hilbert and present associate at the physical institute of the University of Zurich, for help in correcting the proofs (Weyl, 1918c, p. vi).

[117] Weyl announced a course on the theory of the electromagnetic field in the winter semester of 1918/1919, and he lectured on electricity and gravitation during the summer semester of 1919.

[118] P. Scherrer was born on 3 February 1890 in St. Gallen. He first studied botany at the E.T.H. (1908-1909), and then changed to mathematics and physics. After spending one semester at the University of Königsberg (summer 1912), he completed his studies in Göttingen with Woldemar Voigt and Peter Debye. He obtained his doctorate in 1916 with a thesis on the Faraday effect of the hydrogen molecule. Two years later he became *Privatdozent* at the University of Göttingen.

to experimental physics: he had developed in 1915, together with Paul Scherrer, a new method of X-ray analysis of crystals; since it worked with small crystals—the so-called 'powder' method—and yet produced sharp interference patterns, it soon became the commonest method of structure analysis (Debye and Scherrer, 1916a, b).

Scherrer had not only worked experimentally, but also occasionally concerned himself with theoretical questions; thus he had treated, in the summer of 1916, the ideal gas as a multiply periodic quantum system, and used this model to derive the chemical constant of monatomic gases (Scherrer, 1916). Now in Zurich he shared with Debye research in and the teaching of experimental as well as theoretical physics.[119] They did not really push out Wolfke, the former docent for theoretical physics, because he was about to leave Zurich and return to his native country Poland. In 1920 he received a call to a chair at the University of Warsaw. Ultimately he accepted, a year later, the position of professor of experimental physics and director of the physical institute at the Technical University in Warsaw, where he developed into a successful experimentalist.

Compared with the total number of changes at the E.T.H., in connection with the coming of Debye and Scherrer, the situation at the University of Zurich remained essentially stable. Edgar Meyer and his associates still represented experimental physics in the 1920s—only Heinrich Greinacher would move in 1924 to Berne. It seemed rather certain that one of the two docents for theoretical physics, either Ratnowsky or Epstein, would be appointed to a professorship. The qualities of both Russians were well known to the physics community in Zurich. If Epstein possessed a lead by the strength and originality of his research, Ratnowsky balanced it by his teaching experience. Thus each of them could count on strong support in the faculty. But when the time for decision came, the chair was offered to neither of them, but rather to a third person, again a foreigner. The Austrian Erwin Schrödinger accepted and came to Zurich in October 1921.[120]

The decision in favour of Schrödinger must have created some surprise in Zurich, where his name was probably hardly known to most physicists. Schrödinger had, in early 1920, left his hometown of Vienna, where the prospects of obtaining a suitable academic position did not appear to be at all bright. To be more accurate, prospects had never been very good for professional physicists, either in Vienna or in the rest of Austria. Hans Thirring later recalled a conversation with Schrödinger that had taken place about 1911 on this subject; on Thirring's remark 'that there existed in any case the possibility of being called

[119] Debye and Scherrer worked very successfully together in many fields while they were at the E.T.H. They especially refined their technique of X-ray analysis and applied it to disentangle many chemical structures.

Scherrer stayed on in Zurich after Debye left for Leipzig in 1927. He gradually changed his interest to solid state physics and in the 1930s to nuclear physics. He retired in 1960 from the chair at the E.T.H., but continued to teach at the University of Basle. He died on 25 September 1969 in Zurich. For details of Scherrer's biography we refer to the obituary by P. Huber (1970).

[120] Ratnowsky, who in 1921 obtained the personal title of a professor, continued to teach at the University of Zurich until 1926, when he went to the *Kantonsschule* in Winterthur. He died in Winterthur on 6 February 1945.

to Germany, if all the academic positions in Austria were occupied,' Schrödinger had replied by shaking his head, 'Yes, but in that case one must certainly have done something very special' (Thirring, 1947, p. 107). After the end of World War I, with the Austrian Empire having shrunk to a small country, the situation became much worse. Schrödinger—with his previous prospects for the Czernowitz chair having disappeared—had therefore been perfectly happy when Max Wien, needing 'an assistant for his laboratory courses, who knew enough about modern theories to lecture on them' (Schrödinger, 1935, p. 87), invited him to his institute at the University of Jena. In April 1920, being newly married to Annemarie Bertel—his Salzburg acquaintance from 1913—he had moved to Germany and acquired a *Habilitation* in Jena. Although he had been promoted in September of the same year to *außerodentlicher Professor ohne Lehrstelle*—meaning a non-staff professor—he had already decided to change to the *Technische Hochschule* in Stuttgart as *Extraordinarius* for theoretical physics. Even Stuttgart had not been able to keep Schrödinger for longer than the winter semester of 1920/1921, and he moved on in the spring of 1921 to the University of Breslau as an *Ordinarius*. Then the offer from Zurich reached him.

Jena, Stuttgart and Breslau were not places where the physicists had particularly developed relations in physics with Zurich, and one may wonder how Schrödinger entered the competition for the chair at the university there. It is difficult to assume that this happened because of his record in teaching standard topics of theoretical physics.[121] One of the duties of the new Zurich professor was that 'he had to deliver in each winter semester a course on the mechanics of solids, four hours per week' (Rasche and Staub, 1979a, p. 218).[122] Schrödinger had apparently never given lectures on this topic; although he was certainly acquainted with it, his pre-war research on lattice dynamics was aimed at disproving the continuum theory of solids, which was an essential part of the standard mechanics of solids.

However, it may have counted in Schrödinger's favour that Edgar Meyer rated highly his contributions in describing and analyzing the fluctuation phenomena, because they constituted in a way the theoretical counterpart to his own careful and delicate experiments in that field. As we have mentioned earlier, Meyer's associate Richard Bär had, in his critical examination of subelectronic charges in 1918, made use of one relevant paper of Schrödinger. As Bär was still working on the problem in 1921, he might have counted on some theoretical advice from the new professor. The spectroscopist, K. W. Meissner, the other assistant of

[121] From the, not quite complete, lists of lecture courses held at the universities of Vienna, Jena and Breslau and at the *Technische Hochschule* in Stuttgart in the years in question, we can derive the following teaching activities of Schrödinger: he taught only special topics in Vienna (from the summer semester of 1914, interrupted by the war years), announcing, for example, in the summer semester of 1919 the theory of selected experiments (*Phys. Zs.* **20**, p. 286: '*Theorie ausgewählter Versuche des Meßpraktikums*'); in Breslau he presented during the summer semester of 1921 one course on general thermodynamics (a standard course), and a special course for advanced students, entitled 'Dynamics of Atomic Structure' ('*Dynamik der Atomstruktur*': see *Phys. Zs.* **22**, p. 255).

[122] The first course of this kind was announced by Schrödinger for the winter semester of 1922/1923, but he may have already presented it as the first Zurich course in the winter semester of 1921/1922.

Meyer, must have been interested in a more recent paper which Schrödinger published in the spring of 1921 in *Zeitschrift für Physik*, and in which he suggested an interpretation of the sharp subordinate series in the alkali spectra (Schrödinger, 1921a). Yet these two items, fluctuation phenomena and atomic structure, did not play a dominant role in Schrödinger's physical research at that time. He was rather deeply engaged in another field, which was not really considered fundamental physics even in those days, namely physiological optics and especially colour theory.

## Development of Colour Theory from Newton to Schrödinger

Colour theory is one of the central issues of physiological optics, i.e., the field of optics dealing with the perception or sensation of (mainly) the human eye. Aristotle had treated this question in Antiquity, but later interest discontinued in Europe until the seventeenth century. Isaac Newton, with his experimental analysis of light, then provided a physical definition of colours; he showed, in particular, the decomposition of 'white' sunlight into the spectral colours, and *vice versa* the restitution of white light from an additive mixture of spectral colours. Newton's identification of colour with the physical properties of light stimulated, in the early nineteenth century, harsh criticism of the poet and naturalist Johann Wolfgang von Goethe, who developed a totally different colour theory. While Goethe's ideas concerning the origin and decomposition of colours did not attract physicists in the following decades, his contributions served to establish what was called physiological optics—we remind ourselves here of his explanation of after-images, coloured shades and optical deceptions. The physiologists of the nineteenth century then classified many aspects of colour vision; for example, Johannes Evangelista Purkinje investigated the phenomena of colour observation in the dark, discovering the so-called Purkinje effect, i.e., the fact that blue remains perceptible longer as a distinct colour than does red when brightness is diminished.

The modern development in colour theory started with the English physicist and physician Thomas Young (1773–1829), who not only revived the wave theory of light, but also worked on the specific properties of the eye and on colour vision. His hypothesis, assuming that colour sensation depended on the presence in the retina of three kinds of nerve fibres responding to red, green and violet light, respectively, was extended half a century later by Hermann von Helmholtz (1821–1894). At about the same time A. Grassmann formulated the fundamental rules of the Newton–Young three-colour theory: each observed colour can be obtained from a mixture of three components; every mixed colour can be thought of as consisting of a saturated colour to which a certain amount of 'white' has been added; if one component of a mixture is continuously changed, so will the resulting mixed colour. James Clerk Maxwell and Helmholtz provided the full empirical foundation for these rules. On the other hand, they were attacked in the 1880s by the physiologist Ewald Hering (1834–1918), who wished to replace the three-colour theory by a system of the antagonistic colour pairs black–white, blue-yellow and red-green. Physicists tended to follow Helmholtz and Maxwell rather than Hering, partly because the latter's ideas resisted quantitative formulation.

One of the problems in colour theory constituted the proper choice of fundamental colours, which represent the corners of the so-called colour-triangle—a figure containing any observable colour in its interior as composed of a mixture of the fundamental colours. Franz Exner of Vienna, for example, devoted (shortly after 1900) a series of papers to solving this problem, which was intimately connected with the physiological problem of colour sensitivity of the human eye.[123] During the second decade of the twentieth century, A. H. Munsell in Boston and Wilhelm Ostwald in Leipzig developed an organization of the body colours or pigments based on a psychological point of view. Ostwald, for example, proposed a colour atlas containing 2500 colour points, each determined by a set of three co-ordinates—the black-content, the white-content and the colour-content (Ostwald, 1916, 1921). His proposal openly favoured Hering's theory over Helmholtz' and stimulated criticism by the physicists in Vienna.

K. W. Fritz Kohlrausch, who had in 1916 returned from military service to Vienna—because of an ear defect—and (besides fulfulling his university obligations as assistant and docent) taught colour theory at the *Kunstgewerbeschule*, formulated the official answer in a series of three papers that were published in 1920 in the *Physikalische Zeitschrift*, the same journal which carried Ostwald's articles (Kohlrausch, 1920a, b, c). In particular, he raised objections to several fundamental assumptions in Ostwald's system. He claimed that replacement of the standard parameters, saturation and brightness (which determined a colour in the Young-Helmholtz scheme) by the new parameters—i.e., black-content and white-content—could not really be justified on experimental grounds; on the other hand, the quantity brightness, which was avoided in Ostwald's scheme, could certainly be determined by a quantitative measurement of the re-emission function—i.e., the function expressing the attenuation of the energy distribution of the incident light when reflected from a coloured surface compared with the reflection from a normal white surface (Kohlrausch, 1920a). In the colour triangle of Newton and his followers the contributions, $r$, $b$ and $g$, from the three fundamental colours red, blue and green, fixed the colour value and saturation of any given colour, and Kohlrausch demonstrated—using Franz Exner's principle that the brightness of colours follows the additivity rule—that the quantity of brightness indeed could be found as a linear combination of the triple $r$, $b$, $g$ (Kohlrausch, 1920b).

Finally, after some detailed analysis of appropriate examples, Kohlrausch arrived at the conclusion that Ostwald's basic equation—namely, that the colour value, black-content and white-content always added up to unity—although

[123] Olga Steindler, a student of Exner's, published in 1906 a detailed study of colour sensitivity, of both the normal and the colour-blind eye, which provided a further test in favour of the Young-Helmholtz theory and against Hering's theory (Steindler, 1906).

providing a first rough approximation to reality, 'could not be considered to be an epistemological progress' (Kohlrausch, 1920c, p. 477).

Kohlrausch's complaints against Ostwald's colour theory were not shared by all physicists. Thus, for example, the geophysicist and meteorologist Otto Meißner (1881–1946) of Potsdam, who published between 1919 and 1923 several colorimetric investigations based on Ostwald's scheme in the *Physikalische Zeitschrift* claimed that 'some of his criticism obviously rests partly on misunderstandings' (Meißner, 1921, p. 268). Kohlrausch dismissed such opinions as being founded on inconsistent assumptions (Kohlrausch, 1921). The debate for and against Ostwald's colour theory continued in the following years. For example, Tadeusz Oryng of Warsaw concluded, after carefully examining the foundation of the opposing systems: for Ostwald the definition of 'metamer' colours—i.e., colours which appear the same but are composed differently—created great difficulties; in the system adopted by Kohlrausch the same applied to the brightness definition of very different colours (Oryng, 1925). However, another Viennese physicist and student of Exner made an important contribution to this problem and solved the difficulty.[124]

Simultaneously with Kohlrausch's first paper on colour theory in *Physikalische Zeitschrift* in early August 1920 an article by Erwin Schrödinger, entitled '*Theorie der Pigmente von größter Leuchtkraft*' ('Theory of Pigments with the Highest Luminosity') appeared in *Annalen der Physik* (Schrödinger, 1920c).[125] By analyzing the shape of the re-emission function—which according to the Vienna school allowed one to define quantitatively the property of brightness of a given colour—he demonstrated that a pigment (or body colour) under investigation could possess finite, maximum brightness under one condition: the re-emission function assumes for the entire visible spectrum the values 0 and 1 and shows only two discontinuities (i.e., jumps from 0 to 1, and *vice versa*) over the entire range. While these results practically agreed with the conclusions derived in Ostwald's scheme, Schrödinger still expressed hesitation in admitting Ostwald's concepts of 'purity,' 'black-content,' and 'grey-value' as experimentally well-defined quantities. He noted:

> Notwithstanding my high regard for Ostwald's valuable and laboriously achieved successes, I still hold, for example, his absolute determination of "purity" and "greyness" from the values of the re-emission at only *two*, though distinguished points of the spectrum (the maximum and the minimum) for nothing else but a good rule of thumb, which is not at all suited to an accurate determination of these concepts. (Schrödinger, 1920c, p. 621)

[124] With respect to Ostwald's theory Arthur Eduard August Koenig concluded, in his handbook article of 1929: 'His [Ostwald's] presentation fits nicely into the understanding of artists and painters. The great work, which is valuable for technology, deserves recognition, even if some reservations exist about the foundations of his theory, which has been sharply attacked by different physicists, while at the same time physiologists responded more positively and helped to improve the theory' (Koenig, 1929, pp. 153–154).

[125] Schrödinger submitted this paper on colour theory in December 1919, three months before Kohlrausch finished his first paper on colour theory raising strong criticism against Wilhelm Ostwald's colour theory (Kohlrausch, 1920a).

After this preliminary exercise in colour theory, Schrödinger turned to a more ambitious programme, the exact definition of colours including the property of brightness. He established in three extended papers, submitted in March 1920 again to the *Annalen der Physik*, what he called 'colour metric' ('*Farbenmetrik*,' Schrödinger, 1920d, e, f). A shorter note summarizing the main ideas had appeared earlier in *Zeitschrift für Physik* (Schrödinger, 1920b).

In his autobiographical note, presented in 1933 to the Nobel Foundation, Schrödinger stated that '[his] papers on colour theory originated from discussions with Kohlrausch and Exner and from reading Helmholtz' publications' (Schrödinger, 1935, p. 88). Exner, shortly before his retirement, had indeed devoted new efforts and experimental research to colour theory, dealing especially with the problem of a quantitative determination of brightness for different colours (Exner, 1920). His results encouraged Kohlrausch, an old friend and experimental adviser to Schrödinger, to confront the criticized Ostwald theory with the new Vienna conceptions. We do not underestimate the powerful influence of Exner and Kohlrausch in getting Schrödinger involved in colour theory; however, the main stimulus in attacking the difficult problem of defining the property of brightness in a consistent theory came from Helmholtz.

In a series of papers published in 1891 Hermann von Helmholtz had attempted to solve the problem of comparing the brightness of two different colours by 'a generalized application of Fechner's law to colour theory' using 'the shortest lines in the colour system,' defined in a suitable Riemannian space (Helmholtz, 1891a, b, c).[126] Helmholtz had prepared these studies by previously analyzing the validity of the so-called Weber–Fechner law in physiology, stating that the sensation of intensity is proportional to the logarithm of the stimulus intensity; he had especially shown that some apparent violations of the law might arise from the characteristic light ('*Eigenlicht*') of the retina (Helmholtz, 1890).[127] With the new confirmation of the Weber–Fechner relation in colour perception Helmholtz then turned to his main task, namely the interpretation of the colour system—in which each colour was given by three independent variables—as a three-dimensional Riemannian manifold. Helmholtz argued: 'We thus transfer the situation in the colour system (which is less accessible to intuition) to the situation (with which we are much more accustomed) of combining geometrical distances, defined in a centre-of-gravity construction' (Helmholtz, 1891a, p. 407 of reprint). This treatment demanded an appropriate extension of the Weber–Fechner law, because the sensation intensities now had to be defined in a three-dimensional space;[128] from careful experiments, partly carried out by his

[126] Helmholtz had been stimulated to do the new research on colour theory as a consequence of the preparation of a new, improved (second) edition of his monumental *Handbuch der physiologischen Optik*, originally published between 1856 and 1867.

[127] The logarithmic dependence had first been proposed by the physiologist Ernst Heinrich Weber (1795–1878), elder brother of the physicist Wilhelm Weber, as early as 1835, and had later been extended by Gustav Theodor Fechner (1801–1887), who founded a physiological scheme on it.

[128] Until then the Weber–Fechner law had only been applied to changes of sensation intensity in one dimension, for example, to acoustic sensations.

student Eugen Brodhun (1860–1938), Helmholtz derived an expression for the difference of sensation intensity, $dE$, in terms of an admixture of colour components, $dx$, $dy$, $dz$, to a given colour—characterized by the triple numbers $(x, y, z)$—namely,

$$dE = \text{const.} \sqrt{\left(\frac{dx}{x}\right)^2 + \left(\frac{dy}{y}\right)^2 + \left(\frac{dz}{z}\right)^2}. \tag{12}$$

Helmholtz checked Eq. (12) in a second paper, in which he considered all available data both for normal and colour-blind eyes (Helmholtz, 1891b). In the third contribution he finally discussed the crucial problem of determining most similar colours, or what he called the 'shortest colour sequences' ('*kürzeste Farbenreihen*'); for that purpose he referred to the concept of 'shortest distances' in a three-dimensional Riemannian space (Helmholtz, 1891c). In order to describe the observations he did not use Eq. (12), but generalized it slightly to

$$dE = \text{const.} \sqrt{\left(\frac{dx}{a + x}\right)^2 + \left(\frac{dy}{b + y}\right)^2 + \left(\frac{dz}{c + z}\right)^2}, \tag{12$'$}$$

where $a$, $b$, and $c$ denote constants. In this way he succeeded, e.g., in explaining the earlier findings of Ernst von Brücke; namely, if the intensity of light is gradually diminished for an extended spectrum, from a certain brightness onwards, only the colours red, green, and blue-violet remain visible.

Helmholtz had originally hoped that the line-element—i.e., the expression describing infinitesimal distances in the three-dimensional Riemannian colour space—exhibited a particularly simple structure when expressed in terms of the fundamental colours; however, he had not been able to achieve this goal (see Helmholtz, 1891b). 'This fact, and a computational error in Helmholtz' paper, whose discovery further weakens the connection [of the theory] to experiment, may be cited as the reason why his [Helmholtz'] intellectually interesting idea of a Riemannian geometry of colours has not been appreciated and—so far as I know—has not been followed further,' remarked Schrödinger nearly thirty years later (Schrödinger, 1920d, p. 401). Schrödinger, who had previously become acquainted with Riemannian geometry while working on problems of General Relativity, now tried to improve on Helmholtz' ideas by developing, in a set of three memoirs, the 'basic outlines of a theory of colour metric for daylight vision.' He submitted these papers in March 1920 to *Annalen der Physik*, where they were published in November of that year (Schrödinger 1920d, e, f). At the same time, he presented the main ideas on what he called 'colour metric' ('*Farbenmetrik*') in a short note sent to the *Zeitschrift für Physik*; this article appeared in June 1920 as Schrödinger's first publication on colour metric (Schrödinger, 1920b).

Careful study of Helmholtz' works gave Schrödinger the necessary push. In particular, he recognized 'what Helmholtz does not seem to have noticed—that his line element yields a brightness function, which runs contrary to experience:

namely, the cube root of the product of factors associated with the fundamental sensations [i.e., the sensation of the three-dimensional colours]' (Schrödinger, 1920d, p. 401). However, after some research he found a solution to this difficulty by observing 'that another form of the line element is possible—and one may certainly be able to improve upon it—which already in the simple form presented here allows one essentially to establish order and to remove disturbing contradictions [in Helmholtz' theory]' (Schrödinger, 1920, pp. 401-402).

## Colour Metric

The physicist Schrödinger approached the problem of 'colour geometry' ('*Farbenmeßkunst,*' Schrödinger, l.c., p. 402) from the strict point of view of an experimentalist, rather than from a physiological or phychological point of view which, for instance, had guided Wilhelm Ostwald. He did not consider, however, the experiments as *l'art pour l'art,* but stressed their 'totally singular, so to speak distinguished, position from the epistemological point of view' (Schrödinger, 1920d, p. 402). Such an opinion agreed fully with that taken by Schrödinger's teacher Franz Exner, who had assigned colour vision and colour theory a special role in the understanding of physical phenomena. Exner had treated, e.g., these topics in his *Vorlesungen über die physikalischen Grundlagen der Naturwissenschaften* (Exner, 1919).[129] In his view, the special role arose from the difficulty in sharply separating the phenomena of the external reality from the human sensation or reception of the phenomena, or in establishing the detailed and unique correlation between the external phenomena and the specific sensations. Exner had claimed:

> Perhaps one succeeds best in this problem [of correlating physical phenomena and human sensations] with respect to light sensations; hence we shall try to trace and indicate in this field the paths along which the stimulus from objective, physical processes is translated into physiological and psychological sensations. One will then recognize the great difficulties involved in dealing with those problems; but one will also gain the advantage of keeping a sharp eye on the borderline between the external and the internal world. (Exner, 1919, p. 615)

Schrödinger, because of his interest in the epistemological questions of physics, decided to engage himself in this interesting subject and to contribute, if possible, his own insights and results.

The problem posed in colour metric was to find a method of measuring quantitatively all three parameters characterizing a pigment colour. Evidently, the solution implied a distinct, objective decision about the equality or inequality of any two given colours. As Helmholtz had stated clearly in 1891, the main difficulty rested in the fact that the colour space was three-dimensional. Schrödinger systematically divided the entire task of colour metric into two parts. In Part

[129] Exner devoted four of the ninety-five chapters to colour vision and colour theory (Chapters 82 to 85).

I, which he called 'lower colour metric' ('*niedere Farbenmetrik*'), one restricted oneself to using only methods that allow one to decide whether two or more given colours are *equal*; in Part II, entitled 'higher colour metric' ('*höhere Farbenmetrik*') or 'genuine metric of colours' ('*eigenliche Metrik der Farben*'), one applied the method of greatest similarity—which had earlier been suggested by Hermann von Helmholtz. Mathematically different methods had to be used in both cases: for the lower colour metric an affine, linear space sufficed, while for the higher colour metric a genuine curved Riemannian space proved to be necessary.

Schrödinger devoted two of the three memoirs to displaying the lower colour metric. In the first memoir he laid down the definitions and assumptions of standard three-colour theory: he defined, for example, the concept 'colour' as denoting a group of identical looking 'lights' ('*Lichter*') having a characteristic spectral distribution function $f(\lambda)$—the wavelength $\lambda$ extending over the visible spectrum between 400 and 800 $\mu$m; he further assumed that the distribution function satisfied the addition and multiplication laws—the so-called Grassmann laws—and reproduced the known results by applying the mathematics of affine spaces (Schrödinger, 1920d). In the second memoir he dealt with the specific manifold describing the observed colours, i.e., a kind of cone (having the shape of a paper bag) whose apex coincides with the origin—denoting all colours having zero brightness—and on the base the pure spectral colours red, green and violet are ordered in their correct sequence in a closed curve, the extrema red and violet being connected by a straight line containing the purple colours, which fill the gap in the physical spectrum as opposed to the physiological spectrum (Schrödinger, 1920e). Schrödinger also showed in the same paper how any colour could be broken down into a given system of three non-coplanar vectors representing gauge colours, and how changing the gauge colours could be achieved by a co-ordinate transformation in the three-dimensional affine space; and, finally, he discussed the different types of colour blindness within the concepts of this theory.

The need to extend the colour metric of the affine space, or the lower colour metric to the higher colour metric, arose from a very simple practical question: How could one decide in the case of two given colours causing different sensations which was the brighter one? 'The relation of the lengths of the associated vectors does not primarily decide the case,' Schrödinger remarked in the introduction to his third memoir, explaining: 'We may indeed assign to the three colours, the gauge colours, three completely arbitrary vectors, e.g., associate quite a short vector with a very bright colour and at the same time a very long vector with a dark colour' (Schrödinger, 1920e, p. 482). As a consequence it was not possible to deduce any information about the magnitude of separation between two colours, from the distance between the associated points on or in the colour cone. Still the separation of two colours, including the separation of their respective brightnesses, made sense in human observation. To find an appropriate theoretical description, Schrödinger picked up the earlier idea of Hermann von Helmholtz stating that the use of the Weber-Fechner law—i.e., the law accounting for the intensity of sensation—in the three-dimensional colour space demanded the introduction of Riemannian geometry; the shortest connection joining two points representing given colours in this space, obviously invariants in this higher colour space, provided a physical measure for the separation of colours.

To determine the properties of the required Riemannian space, Schrödinger stated three necessary assumptions. First, he introduced the quadratic line element $ds^2$, i.e.,

$$ds^2 = \sum_{i=1}^{3} \sum_{k=1}^{3} a_{ik}\, dx_i\, dx_k, \tag{13}$$

with the coefficients $a_{ik}$ representing functions of the co-ordinates $x_i$ and $x_k$ ($i$, $k = 1, 2, 3$) which describe two colours. Second, he assumed that for two barely separable colours the $dx_i$, $dx_k$, and $ds^2$ constitute differentials, with $ds$ having the same value for each such colour-pair. 'The *Ansatz* [(13)] completely determines, as is known, a metric for the manifold of the number of triples $(x_1 x_2 x_3)$ in the general sense of Bernhard Riemann, if one interprets $ds$ as the *line element* of this manifold. Hence we are now and only now allowed to speak about a proper metric of colour space,' Schrödinger commented (Schrödinger, 1920e, p. 484). He further claimed, as the third assumption: 'The difference between arbitrary colours is judged according to the magnitude of $\int ds$, taken for the shortest connecting line (the geodetic line) drawn between the two colour points of the manifold, whose measure is given by Eq. [(13)]' (Schrödinger, 1920e, p. 485). Schrödinger finally suggested a suitable expression for the brightness $h$ of a colour: a linear function of the coefficients $a_{ik}$ and the co-ordinates of a colour.

The main task of colour metric then consisted of deriving the coefficients $a_{ik}$ of the line element, Eq. (13). Helmholtz had originally proposed the form

$$ds^2 = \frac{1}{3}\left(\frac{dx_1^2}{x_1^2} + \frac{dx_2^2}{x_2^2} + \frac{dx_3^2}{x_3^2}\right), \tag{14}$$

where the $x_i$ $(i = 1, 2, 3)$ denote the three fundamental co-ordinates; however, he had also noticed that Eq. (14) was contrary to observation (Helmholtz, 1891b). Schrödinger now found an additional argument against Eq. (14). It yields, in particular, the following expression for the brightness $h$ of a colour given by the co-ordinate triple $(x_1 x_2 x_3)$:

$$h = \text{const.} \sqrt[3]{x_1 x_2 x_3}\,. \tag{15}$$

If applied to the spectral distribution of sunlight, a rather unlikely brightness distribution resulted; and, moreover, Eq. (15) violated the empirical additivity rule for brightness. Schrödinger succeeded, however, in proposing a different, more suitable line element, namely,

$$ds^2 = \frac{1}{\alpha_1 x_1 + \alpha_2 x_2 + \alpha_3 x_3}\left(\frac{\alpha_1\, dx_1^2}{x_1} + \frac{\alpha_2\, dx_2^2}{x_2} + \frac{\alpha_3\, dx_3^2}{x_3}\right), \tag{16}$$

where $\alpha_1$, $\alpha_2$, $\alpha_3$ denote constants that had to be determined experimentally. With the line element, Eq. (16), the brightness of a colour given by the triple $(x_1x_2x_3)$ could be written as

$$h = \alpha_1 x_1 + \alpha_2 x_2 + \alpha_3 x_3 . \tag{17}$$

It obviously satisfied the additivity rule; and it also fulfilled the Weber–Fechner law, since for any two colours, given by the triples $(x_1x_2x_3)$ and $(x_1'x_2'x_3')$, respectively, the equation

$$s = \int_{x_1'x_2'x_3'}^{x_1x_2x_3} ds = \ln \frac{h}{h'} \tag{18}$$

relates the associated brightnesses $h$ and $h'$.

With his specific colour metric, Eq. (16), Schrödinger not only provided a reasonable fit to the available data, the fundamental gauge colours having been determined by Arthur Koenig by investigating colour-blind people, finding the three constants to assume the values $\alpha_1 = 43.33$, $\alpha_2 = 32.76$ and $\alpha_3 = 1$; he also showed how the theory worked in practical situations, such as heterochromous photography. Finally, he calculated geodetic lines and the surfaces of equal brightness in the colour spectrum. In a colour 'triangle' containing all colours with equal brightness the geodetic lines connecting the white point with the spectral colour points on the periphery are in general curved lines, with the exception of the shortest lines to the fundamental colour points which are straight. The colour on such a geodetic line does not change, but the addition of white moves a colour point from the periphery of the triangle to the central white point. Schrödinger thus easily explained the well-known phenomena stating that for a high admixture of white only a few colours can be separated by the human eye, with large gaps of colourless points lying in between. On the other hand, Schrödinger was fully aware of the shortcomings of the proposed theory. It provided, like many regularities noted earlier in colour vision, just an approximation to the real situation; especially, it failed where the Weber–Fechner relation between sensation stimulus and sensation intensity broke down. Schrödinger wrote at the close of the final, third memoir:

> The value of our theory seems to lie in the fact that it discloses the internal connection of all these approximate laws; and simultaneously, the intimate connection of the *deviation of the real colour manifold from our idealized "purely Fechnerian" colour manifold* which in most cases still await exact experimental investigation. (Schrödinger, 1920e, p. 520)

The work on colour metric, which he completed in Vienna, certainly interested his colleagues in Jena, a place of great tradition in the optical industry, but less so in Stuttgart. However, Schrödinger then moved to Breslau; and again he found himself in an institute, where research on optical phenomena played an important role, at least from the time Otto Lummer (1860–1925) and Ernst Pringsheim (1859–1917) arrived there earlier in the century. Lummer, a former assistant and collaborator of Hermann von Helmholtz at the *Physikalische-Technische Reichsanstalt*, had become a specialist in photometry and research on electromagnetic radiation from hot bodies (also blackbody radiation) in Berlin; and Pringsheim had been associated with him. In Breslau the two of them—Lummer from 1904 in the experimental chair, and Pringsheim from 1905 in the theoretical chair—continued their interest in the physics of radiation. Thus Lummer, when he took over, together with his colleagues Arnold Eucken and Erich Waetzmann—professors of physical chemistry and physics, respectively, at the University of Breslau—the planning of the eleventh edition of the renowned Müller-Pouillet's *Lehrbuch der Physik*, reserved mainly for himself the writing of the optical part, entitled '*Lehre von der strahlenden Energie*' ('Physics of Radiant Energy'). He enlisted Erwin Schrödinger as an assistant, to whom the sections on optical sensations were properly assigned. Schrödinger could finish his contribution, a handbook-like article of over one hundred printed pages—covering standard colour theory, colour sensations at daylight and at dawn, as well as questions of spatial and temporal fluctuations and the problem of comparing colours (Schrödinger's higher colour metric)—in 1925 (Schrödinger, 1926j).[130]

## Causality and Chance: The Nature of Physical Laws

Besides acquiring an expert on optical sensations and colour theory, in Schrödinger, the University of Zurich also acquired a theoretician, who displayed a deep interest in the philosophical problems of science, as well as in philosophy itself.[131] As we have mentioned earlier, in 1918 Schrödinger had hoped to obtain a physics chair at a quiet university—in Czernowitz in the eastern Polish part of the Habsburg Empire—allowing him to devote a good deal of time to philosophical studies. To some extent, Schrödinger's interest might be considered as a legitimate heritage from previous Vienna physicists, especially Ernst Mach and Ludwig Boltzmann. His immediate teacher and institute director, Franz Exner, had not restricted his activities to purely scientific questions either. He had, when surrounded by students and collaborators, freely discussed topics going beyond the frontiers of physical research, and he had used philosophical considerations in his lectures. This fact may be easily recognized from the written form of Exner's *Vorlesungen über die physikalischen Grundlagen der Naturwissenschaften*, the book

[130] Due to Lummer's death in 1925, the first half of Volume II on optics—to which Schrödinger contributed—would remain the only book for which he ultimately carried full responsibility. The other parts and volumes were edited by Erich Waetzmann (Volume I on mechanics in 1929), Karl Wilhelm Meissner (Volume II, second half, 1929), Arnold Eucken and Karl Herzfeld (Volume III on heat theory, 1925 and 1926, respectively), Siegfried Valentiner and Arnold Eucken (Volume IV on electricity and magnetism, 1932–1934), and Alfred Wegener and August Kopff (Volume V on physics of the earth and the universe, 1928).

[131] We have referred to the philosophical attitude of the physicist Schrödinger on various occasions in Chapter I.

which we have already mentioned earlier. Exner devoted the last part of this book, entitled '*Über Naturgesetze*' ('On the Laws of Nature') to analyzing the structure of the physical laws and he arrived at several remarkable conclusions.[132]

Starting from a discussion of the rules and customs of thought ('*Denkgewohnheiten*'), Exner pointed to the fact that physical laws originate basically as statements on averaged quantities derived from our experience; as a consequence they do not necessarily apply either to spaces larger or at times longer than our experience extends. He noted:

> Much more important than the question of the exact validity of physical laws in times and spaces transcending our means of observation—and which we can neither confirm or disprove—is the alternative question: Are the laws completely and unlimitedly valid within the range of times, spaces and masses accessible to us? In other words: Are the laws satisfied in the same manner, as they seem to be for factual observations, also applicable for arbitrarily short times? Or, in mechanical processes, are they valid if applied to arbitrarily small masses and to arbitrarily short distances? (Exner, 1919, p. 655)

Referring to the phenomena of Brownian motion, Exner suggested that the answer to the last question was perhaps 'No', and he further claimed that not only the intensity of very diluted light, but also the gravitational force emanating from heavy masses might show granular structure. Hence he argued: 'It might be conceivable . . . that the gravitational constant appears to us to be a constant only because of the fact that we are unable to measure it at arbitrarily short times' (Exner, 1919, p. 657).

The statistical nature of all natural laws, as assumed by Exner, demanded a modification of the adopted causality concept in physics. The latter had to be replaced by the statement: 'There must exist causes, which determine what happens *on the average*, but only this and not single events; they thus establish laws' (Exner, 1919, p. 664). Exner, therefore, proposed to separate two ranges of physical description: the macroscopic world, ruled by *law-like* relations, and the microscopic world—notably the world of atoms and electrons—ruled by *chance* and *accidents*.

Exner's thorough analysis of physical laws left a deep impression on Schrödinger, and he made it the topic of his inaugural address at Zurich, entitled '*Was ist ein Naturgesetz*?' ('What is a Law of Nature?').[133] 'Now what we call a "law of nature" is nothing more than any one of the regularities observed in natural occurrences,' he declared, in perfect agreement with his former teacher

[132] Exner had already treated this question in his inaugural address as *Rektor* of the University of Vienna back in 1908 (Exner, 1908). The address was entitled '*Über Gesetze in Naturwissenschaften und Humanistik*' ('On Laws in the Sciences and the Humanities') and was delivered on 15 October 1908. For a discussion of the address we refer to Paul A. Hanle's thesis (Hanle, 1975, pp. 36-41).

[133] The address was published more than six years later when Schrödinger thought it might become useful in the interpretation of quantum mechanics.

(Schrödinger, 1929a, p. 9; English translation, pp. 135–136). However, he did not stop at this point, he even sharpened Exner's statements by saying:

> Within the past four or five decades physical research has clearly and definitely shown that *chance* is the common root of all rigid conformity to Law that has been observed, at least in the overwhelming majority of natural processes, the regularity and invariability of which have led to the establishment of the postulate of universal causality. (Schrödinger, 1929a, p. 9; English translation, p. 136)

By 'overwhelming majority' he meant the processes between single atoms, which in huge concentrations co-operate to establish the known, causal physical laws, for example, the laws describing the behaviour of gases. While most physicists shared the belief that the processes between single atoms obeyed deterministic laws, Schrödinger joined Exner in denying the necessity for such a conclusion. So he suggested, for example: 'Naturally, we *can* explain the energy principle on the large scale by its already holding good in the single events. But I do not see that we are *bound* to do so' (Schrödinger, 1929a, p. 11; English translation, p. 143).

Schrödinger not only suggested an acausal nature of atomic processes; he also pleaded for a unified description of all physical phenomena in terms of an acausal, or not necessarily causal, interpretation. Thus he claimed, for example, that Einstein's law of gravitation could be perfectly explained as expressing the average behaviour of an enormously large number of atoms. He also considered the 'astonishing success' achieved at that time in the application of the laws of electrodynamics in analyzing the structure of atoms as constituting 'the most serious objection that can be advanced against the a-causal view' (Schrödinger, 1929a, p. 11; English translation, p. 146). Still he emphasized 'the serious intrinsic incoherences' plaguing the electrodynamical description of atomic structure, and concluded his address with the following words: 'I prefer to believe that, once we have disregarded our rooted predilections for absolute Causality, we shall succeed in overcoming these difficulties, rather than expect atomic theory to substantiate the dogma of Causality' (Schrödinger, 1929a, p. 11; English translation, p. 147).

Schrödinger's views on the nature of physical laws, especially his credo for an acausal interpretation of the phenomena, showed his Zurich colleagues that their choice of a new professor of theoretical physics had fallen on a rather independent thinker.[134] Having accepted the call in 1921 Schrödinger ceased to

[134] Exner's and Schrödinger's radical views were not shared by the majority of their colleagues at that time. We refer, at this point, to the discussion of the causality problem by Moritz Schlick from Rostock—a former student of Max Planck's—or by Walter Schottky from Würzburg (Schlick, 1920; Schottky, 1921). While the former stood firmly on the principle of causality, as extended in relativity theory, the latter admitted the possibility that causal laws might only determine the average behaviour of atoms; still, Schottky expressed the hope that the apparent acausality, inherent in some aspects of the quantum-theoretical description, might be dissolved in the future consistent theory.

move quickly from place to place—as he had done in the previous year—and settled in the friendly city of Zurich at the Limmat. He did not even consider the positions offered in his native country at various times. The main reason was later recalled by Mrs. Schrödinger, who reported: 'From Stuttgart he [Schrödinger] had three calls: to Vienna, Breslau and Kiel. We went to Breslau . . . because in Vienna there was a terrible financial situation. He asked for a salary which they could not pay. He couldn't go for this little salary. Thirring could take the job, because he had some connection with industry. That made it possible for him to live on this small salary' (A. Schrödinger, AHQP Interview, 5 April 1963, p. 3). Leaving inflation-ridden Germany in 1921 certainly improved Schrödinger's financial situation and enabled him to stay on in Zurich until a most prestigious position was offered to him in 1927, the chair of Max Planck in Berlin. Then Schrödinger did not resist and left Switzerland.

## II.4 Atomic Structure and Physiological Optics: General Research Programme of Schrödinger in Zurich (1922–1925)

Settling in Zurich in the fall of 1921 brought Schrödinger back to a steady and quiet life, similar to that which he might have experienced many years previously in pre-war Vienna. Forgotten were the quick moves from one place to another to obtain a better position on the academic ladder; he also left behind the increasing financial difficulties plaguing people in post-war Germany, notably inflation. Thus he wrote relieved, even a year after his arrival in Zurich, to Wolfgang Pauli, who was about to join Niels Bohr in Copenhagen: 'You will live better there. I hope you do not return for the time being from there; for, if I imagine having to go back right now to Germany, I'd be horrified; and you will feel the same way in half a year from now. One really needs some "*otium*" (leisure), i.e., free from care; and one lacks this if the price of butter or a regular meal at an inn depends on the exchange rate' (Schrödinger to Pauli, 8 November 1922). In Switzerland inflation did not exist; everything could be bought at a fixed price, though the salary of a university professor did not make Schrödinger a rich man in Zurich.[135]

Being provided with the necessary security and leisure, Schrödinger fulfilled his teaching duties at the university, presenting several courses each semester in addition to holding seminars and tutorials (*Proseminars*) in theoretical physics

[135] In a later letter to Sommerfeld in Munich Schrödinger wrote about a '"distinguished" Zurich ball, where the local plutocracy shows its snobbery in loges of 300 Swiss francs and where one belongs, as a simple professor, to the miserable lower class'; he added that he had 'sent his wife alone, under the protection of [Edgar] Meyer, the Bärs and the Scherrers,' to that ball, since 'the entrance ticket of 25 Swiss francs multiplied by *two* appeared to me to be too much for the questionable pleasure' (Schrödinger to Sommerfeld, 7 March 1925).

for less advanced students.[136] Between 1922 and early 1926 he contributed eighteen papers and elaborated on two extended review articles. As in the Vienna period, Schrödinger did not devote himself to one or even a few topics, but treated, simultaneously, a variety of problems in theoretical physics; thus he wrote four papers on atomic structure, three papers plus a handbook article on the quantum theory of specific heats, six papers on gas statistics, four papers plus a review article on colour vision and colour theory, and one paper on relativity theory.

From the diverse titles of these papers one can hardly derive a main guiding line to Schrödinger's researches during the early Zurich years; in this respect, this period seems to resemble the earlier period in Vienna. However, if one examines more closely the content of these publications, one will discover that the author indeed concentrated on one major field, namely kinetic theory and the modifications introduced in it by the requirements of quantum theory. For that purpose one must just put the papers on gas statistics and on the specific heats of solids and gases in one category, which will then represent half of all publications in that period. The emphasis on statistical problems will also be deepened if one adds to the published papers the many and extended notes and notebooks, which Schrödinger worked out during the same period and which deal in some detail with the problems of classical molecular statistics, degeneracy and quantum statistics. Hence it is no exaggeration if one claims that Schrödinger developed, in the four years between 1922 and the end of 1925, a systematic research programme on quantum statistical mechanics; a research programme, which would then pass over smoothly into another programme, that of establishing the theory of wave mechanics. The transition from one programme to the other would not only be smooth in time; but also the important limits leading to the new atomic theory would emerge from the specific quantum statistical treatment of systems of many degrees of freedom, especially gas models, as devised by Schrödinger. We shall, therefore, postpone the discussion of the work on quantum statistics to the following, last section, and deal here mainly with the more scattered research on other topics.

## Getting into the Theory of Atomic Structure

While the concern with problems of kinetic theory and quantum statistics must be considered to be the natural continuation of his earlier research, the interest in problems of atomic structure meant a novel feature in Schrödinger's scientific

[136] According to the *Vorlesungsverzeichnis*, published in advance in *Physikalische Zeitschrift*, Schrödinger presented the following courses: in the summer semester of 1922 one on electron theory (4 hours per week) and another on kinetic theory (4 hours per week); in the winter semester of 1922/1923 one on analytical mechanics (4 hours) and one on the special problems of electron theory (4 hours); in the summer semester of 1923 one on the theory of light (4 hours) and one on atomic structure and the periodic system of elements (2 hours, plus $1\frac{1}{2}$ hours on mathematical tools). In later semesters he especially taught, besides the standard theoretical topics, courses on classical and quantum statistical mechanics, about which detailed notebooks have been kept in the Schrödinger *Nachlaß*.

work. This interest was certainly enhanced by his leaving Vienna in 1920, since questions of the inner constitution of atoms and molecules had been neglected so far in the official physics community over there. The situation was totally different in those places in Germany where Schrödinger moved to in 1920 and 1921; in Jena, as well as in Stuttgart and Breslau there were colleagues, who were enthusiastic about the latest results obtained from quantum theory, and in particular its application to atomic and molecular structure, yielding an explanation of the observed spectra. Certainly Arnold Sommerfeld's popular and famous book *Atombau und Spektrallinien* of 1919 had contributed greatly to this situation. Schrödinger became personally acquainted with Sommerfeld in the fall of 1919, when he visited Munich and delivered a talk on needle radiation in the *Kolloquium.* There he also met Wolfgang Pauli, his younger fellow countryman, who studied with Sommerfeld at that time. Pauli and Schrödinger established a friendship and exchange of scientific ideas orally and by written communications, which would last to the end of their lives.[137]

*Atombau und Spektrallinien* and the relationship with Sommerfeld and his circle—a relationship which was continued in Stuttgart, where Sommerfeld's former student Paul Ewald would succeed him as *Extraordinarius* at the *Technische Hochschule*—was not the only source from which Schrödinger derived his knowledge on the quantum theory of atomic structure. In Breslau he met Rudolf Ladenburg, who had connections with Berlin and Copenhagen, and who contributed actively to important problems in that field. Schrödinger soon took the opportunity to approach Niels Bohr and Hendrik Kramers, sending them copies of his first contribution to the theory of atomic spectra (Schrödinger, 1921a). In a letter to Niels Bohr he added a request, writing: 'Your important communications, in the Danish Academy from 1918, are very difficult to obtain in Germany. In Breslau we have but *one* copy that I know of, that is the one of Mr. Ladenburg. Would it be possible for you to let me have a copy?' He explained the reason for this request in the following words:

> Having been made Professor Clemens Schaefer's successor in the chair of theoretical physics at Breslau, I keep a college [give a course] on the atom model just now, and so you may imagine how thankful I should be to possess your important papers, especially because Sommerfeld's beautiful work (which is of course our standard-work, at present in Germany on all questions of the atom) differs, as you well know, in many respects widely from your own, most fascinating mode of viewing the problems. (Schrödinger to Bohr, 7 February 1921; original letter published in N. Bohr: *Collected Works 4*, p. 737).

[137] The first letter kept in the Pauli *Nachlaß* is the one written by Schrödinger from Jena, dated 20 July 1920. In it Schrödinger acknowledged a letter from Pauli, who had told him that he had started to work on the quantum theory of paramagnetism; Schrödinger also expressed interest in hearing more about Pauli's results in future; he finally made some comments on the problem of diamagnetism, on which Pauli had worked a little earlier and had just completed a paper (Pauli, 1920a).

The next letters, which have also been kept in the *Nachlaß*, Schrödinger wrote on 13 February 1921 from Stuttgart, and on 7 November 1921 from Zurich.

Bohr sent Schrödinger the requested copies of his papers (i.e., Bohr, 1918a, b) and also that of Kramers' doctoral thesis (Kramers, 1919), and wrote a letter to Breslau:

> Your paper in the *Zeitschrift für Physik* has of course, interested me very much. By the way, some time ago I made exactly the same consideration and carried out the necessary calculations; for example, I reported, in a lecture presented to the Physical Society here in December of last year... (as an illustration the value of) the radius which such an idealized spherical charge distribution must assume in order to account exactly for the [observed] terms of the sharp subordinate series of lithium. (Bohr to Schrödinger, 15 June 1921).[138]

Schrödinger's paper, to which Bohr referred, had been submitted in January 1921 to *Zeitschrift für Physik*, where it was published in the third issue of Volume 4 appearing several weeks later (Schrödinger, 1921a). In it Schrödinger proposed an 'attempt at a model-like interpretation of the terms of the sharp subordinate series,' especially in the case of the alkali spectra. In a paper comparing the data on spark-spectra of alkaline earths with the arc-spectra of alkali elements, which appeared a few months earlier in *Annalen der Physik*, Erwin Fues, a student of Sommerfeld, had drawn attention to the following fact: 'Inclusion of the $s$-term with the azimuthal quantum number 1 seems to demand that it [i.e., its main quantum number] is given by an integral number; the results of the following discussion, however, lend support to the old view that the running numbers of the $s$-term are odd multiples of $\frac{1}{2}$' (Fues, 1920, p. 1).[139] The way out of this inconsistency seemed somehow obvious: the mutual interactions of the electrons surely caused perturbations of the series electrons in all atoms except the hydrogen atom. However, as Schrödinger said in the introduction to his paper: 'The nature of the perturbation is not known; it must be considerable, at any rate; the orbit [of the series electron] cannot, even in an approximation, be *a Keplerian ellipse*' (Schrödinger, 1921a, p. 347).

Schrödinger now proposed that the perturbation arose from the fact that the orbit of the series electron closely approached the inner electron shell. He therefore estimated first the perihelion distance of an elliptic orbit, denoted by the couple of azimuthal and radial quantum numbers (1, 1); then he found the

[138] Bohr mentioned that he had not been able so far to publish his considerations due to overwork; but he elaborated on it in later papers (e.g., Bohr, 1921e) and especially in his 1922 lectures in Göttingen.

[139] Erwin Fues was born in Stuttgart on 17 January 1893. He studied physics from 1912 to 1914 at the universities of Berlin and Munich, and, after military service, from 1916 at the universities of Tübingen (1916–1918) and Munich (1918–1920). He received his doctorate with Arnold Sommerfeld in Munich; in 1922 he became an assistant of Paul Ewald at the *Technische Hochschule* in Stuttgart, and two years later he obtained his *Habilitation* there. From 1925 to 1927 he held a fellowship of the International Education Board, visiting Zurich and Copenhagen. In 1929 he became full professor of physics at the *Technische Hochschule* in Hanover, from where he moved in 1934 to Breslau (at both the university and the *Technische Hochschule*), in 1943 to the University of Vienna and in 1947 to the *Technische Hochschule* in Stuttgart.

distance in question to be smaller than the radius of the inner shell in the case of alkali atoms, say for sodium, if he applied Alfred Landé's recent cubic theory of atoms (Landé, 1920e). 'This is rather remarkable,' he concluded, 'since it means that *hydrogen-like* orbits, even in the most general sense of the word, *cannot at all be built up* here, for the azimuthal quantum number 1 and non-vanishing radial quantum numbers, just because of their collision with the screening sphere consisting of eight electrons [in the interior of the atoms]. In this, it seems to me, must necessarily be contained the key explaining the special case of the $s$-term. Since the spectroscopic systematics, and in particular the selection principle, hardly leave any doubt about the fact that the $s$-term has the azimuthal quantum number 1, the conclusion seems inevitable that the *s-orbits penetrate into the neighbouring inner shell, and even lie partly in its interior*' (Schrödinger, 1921a, p. 348).

In order to evaluate quantitatively the perturbation of the orbit of the series electron, Schrödinger neglected the deformation of the inner shell caused by the impinging outer electron, and replaced the action of the inner shell on the electron by that of a sphere, on whose surface the electric charge is uniformly distributed. Then the orbit of the series electron could be separated into two pieces: an outer one—i.e., lying outside the inner shell—created by the attraction of a unit charge, and an inner one, created by the attraction of an effective charge that was—in the case of sodium—nine times as big. Evidently, the curvature of the inner elliptic orbit was much larger than that of the outer. By applying standard Sommerfeld theory to both pieces in a straightforward way, Schrödinger obtained the following expression for the energy term of the series electron

$$W = -\frac{Z^2 e^2}{2a_0}\frac{1}{(n^*)^2}. \tag{19}$$

In Eq. (19), $Ze$ denotes the nuclear charge in the interior of the inner shell—Schrödinger took, in the case of sodium, the total nuclear charge; $a_0$ is the radius of the first Bohr hydrogen orbit, and $n^*$ represented an effective total quantum number, which assumed in Schrödinger's calculation the values,

$$n^* = 1.26;\ 2.26;\ 3.26;\ \ldots \tag{19a}$$

He concluded: 'The suspicion that this type of orbit is associated with the $s$-term, therefore, is very suggestive' (Schrödinger, 1921, p. 354).

Schrödinger's work on the sharp subordinate term of alkali atoms may be considered his 'journeyman's piece' ('*Gesellenstück*') in the theory of atomic structure. It emerged from a simple physical idea, and helped to remove a principal difficulty in the existing Bohr–Sommerfeld theory. By applying simple (and a little hand-woven) mathematical methods—very much resembling the methods Schrödinger had used nine years earlier in his first theoretical paper on the contribution of metallic conduction electrons to magnetism (Schrödinger,

1912a)—he obtained a valuable result for the interpretation of atomic spectra. True, this result was not free of arbitrary assumptions, and Wolfgang Pauli, to whom Schrödinger had sent a copy of the finished paper, immediately emphasized one of these assumptions in a letter, namely: Why did the series electron of an alkali atom in the ground state not make use of the circular orbit, characterized by the pair of azimuthal and radial quantum numbers $n = 1$ and $n' = 0$? This orbit would not penetrate into the interior of the atom, and hence would not give rise to a non-integral effective quantum number $n^*$ in the energy term.

In his reply Schrödinger could not give a really good answer. Instead he referred to the investigations of Landé, who had claimed that even the noble-gas-like alkali atoms tended, due to energetic considerations, to develop elliptic instead of spherical shells (Landé, 1920e). Schrödinger wrote:

> Perhaps it is a general feature that the outermost external system [of electrons in an atom] tends, due to elbow room, to assume the elliptic situation even in the exterior. At any rate, *if* my view is at all correct, then the normal orbit [ground state] *must*, of course, . . . be identical with my lowest orbit $n = 1$, $n' = 1$, or perhaps even be identical with $n = 1$, $n' = 2$, and nothing else; at least for alkali atoms. For the alkaline earth atoms the normal state is $\frac{3}{2}S$ (singlet lines). . . . For the alkaline earth triplets I would like to put *one* electron on the circular orbit ($n = 1$, $n' = 0$)—the total elbow room does not exist anymore; the *other* should run on an orbit of the new [penetrating] type. Altogether, triplet and singlet terms would be different because of the different behaviour of the second electron, and thus the occurrence of *two* series systems, instead of a single one, might be easily explained. (Schrödinger to Pauli, 13 February 1921)

Schrödinger confessed, however, that he had not yet been able to carry out the necessary, quantitative calculations for the spectra of alkaline earth atoms.

The content of the letter confirmed what could already be discerned from the published paper: Schrödinger had fully acquainted himself with the most advanced literature on atomic structure existing at the beginning of the 1920s. This knowledge allowed him, during the following years, to publish from time to time a calculation or an idea, which promoted progress in atomic and quantum theory. As for his initial paper on the $s$-terms of alkali atoms, it was properly cited and advertized by Niels Bohr, the pioneer and guiding scientist in this field, in subsequent lectures and publications; first, in an address on 'The Structure of the Atom and the Physical and Chemical Properties of the Elements,' which he delivered before the Physical and Chemical Societies in Copenhagen on 18 October 1921 (Bohr, 1921e), and then, before a larger audience in the famous lectures presented at Göttingen in the summer of the following year. Especially in the fifth lecture, presented on 20 June 1922, Bohr remarked:

> Schrödinger attempted to explain the large deviations of the $s$-terms [of the sodium atom] from the hydrogen terms by the assumption that all $s$-orbits reach into the region of the innermost electron group. Also he arrived at the conclusion that the normal state is characterized by the quantum number $2_1$. However, as with the

> binding of the third and seventh electron, we must assume that, because of the peculiar stability of the electron configuration already present, the eleventh electron will now be bound in a new type of orbit, namely, in a $3_1$-orbit. (Bohr, 1977, p. 494)

Still Bohr's new, slightly different assignment of the principal quantum number—3 instead of 2—did not change the result obtained by Schrödinger for the *s*-term energy, because the latter depended crucially on the inner parts of the orbit, and 'There the orbit will differ only little from the $2_1$-orbit' (Bohr, 1977, p. 494).[140]

## Schrödinger's First 'Student': Erwin Fues

Bohr's Göttingen lectures were also attended by Erwin Fues, who had meanwhile moved to Stuttgart. Less than four months later he submitted a memoir in two parts, entitled '*Die Berechnung wasserstoffunähnlicher Spektren aus Zentralbewegungen der Elektronen*' ('The Computation of Non-Hydrogen-Like Spectra from Central Motions of the Electrons') to *Zeitschrift für Physik* (Fues, 1922a, b). He picked up the idea of penetrating orbits, but improved on the calculation of the energy states by removing Schrödinger's approximation. Fues, instead of ignoring the action of the inner electrons on a sphere, took into account a cubic distribution of eight electrons in the inner *L*-shell. He thus found an improved expression for the 'quantum defect' (*Quantendefekt*) $q$,

$$q = n - n_{\text{eff}}, \tag{20}$$

describing the difference between the real principal quantum number and the effective number entering the Rydberg–Ritz-like formula, Eq. (19). 'This yields,' he concluded, 'an association of optical terms with totally different quantum numbers than was hitherto assumed; one must, for example, denote the term $1.5S$ of sodium by $3_1$, and the term $2P$ by $3_2$' (Fues, 1922a, p. 372). In a footnote Fues added that he had arrived at this conclusion 'before obtaining knowledge of Bohr's [new] interpretation of these terms' and stated: 'Bohr has given in Göttingen also a very simple method of how to estimate the quantum numbers of non-hydrogen-like atoms on the basis of his model conceptions. By this fact, however, the considerations presented here have not been rendered superfluous, because they attempt, within well-determined precision limits, a *quantitative* description of terms' (Fues, 1922a, footnote 2 on pp. 272–273).[141]

By carrying out completely Schrödinger's original suggestion, Fues succeeded in establishing mathematically an important detail of Bohr's visionary conception of the theory of the periodic system of elements. In Part II of his memoir he

[140] In Bohr's notation the main number denotes the principal quantum number, and the lower index the azimuthal quantum number.

[141] Independently of Fues, A. Th. van Urk in Leyden also worked on an improvement of Schrödinger's theory of penetrating orbits. He especially improved the evaluation of the expression for the quantum defect; in this way he not only obtained a better fit of the alkali spectra, but also succeeded in deriving the new principal quantum numbers of Bohr, namely, 3 for sodium and 4 for potassium (Urk, 1923).

demonstrated that 'all terms of the visible sodium spectrum can be explained, at least up to an accuracy of 3 percent, by calculating the energies of orbits of electrons attracted by a central potential' (Fues, 1922b, p. 9); hence, 'the proof has been provided that the conception, formed recently about the origin of non-hydrogen-like series spectra, is correct' (Fues, *loc. cit.*). But the idea of penetrating orbits also worked in the cases of the homologous spectra, namely the spark spectra of the singly-ionized magnesium and the doubly-ionized aluminium (Fues, 1923). Finally, a year later, Fues extended this successful description to the higher alkali element cesium; he again demonstrated that Bohr's views on the quantum orbits of its series electrons could be justified, provided he took into account relativistic corrections to the electron's motion inside the inner shell (Fues, 1924).

With the new work Fues obtained his *Habilitation* at the *Technische Hochschule* in Stuttgart. Two years later he went to Zurich on a fellowship of the International Education Board and, after having followed Schrödinger's ideas for years, was thus able to participate in the latter's work on wave mechanics. Fues would in particular calculate the spectra of diatomic molecules with the undulatory mechanical methods and submit, as early as April 1926, a paper containing the results to *Annalen der Physik*, indeed the first contribution to the new theory that did not come from Schrödinger himself (Fues, 1926a). Thus Fues became a real disciple of Schrödinger.

Before Schrödinger settled in Zurich, he had had hardly any chance to train students. In Vienna, where he became *Privatdozent* in early 1914, the outbreak of World War I soon interrupted his teaching career. In addition, Schrödinger had retained a position at an institute, whose director, Franz Exner, was about to retire. However, despite this unfavourable situation he provided guidance in at least one case, i.e., in the experimental investigation on the Schweidler fluctuations of the radioactive decay conducted by Miss Bormann.[142] The post-war period in Vienna—as well as the subsequent periods which Schrödinger spent in Jena, Stuttgart and Breslau, respectively—did not last long enough to allow him to have advanced students. In Zurich, however, for the first time, conditions in that respect seemed to be ideal. Yet, even there we do not know of any doctoral candidate of Schrödinger. This is quite surprising, in spite of the fact that the University of Zurich did not enjoy a high reputation among students studying theoretical physics.[143] After all, some advanced students must have participated in Schrödinger's *Seminar*, which was regularly announced every semester. He also did his share in examining candidates for the philosophical doctorate at his institution. Evidently, none of them got a dissertation topic from him to work on. Schrödinger explained this fact later by admitting that he 'was very bad in collaborating, unfortunately, even with students' (Schrödinger, 1935, p. 87). In

[142] We have discussed Schrödinger's theoretical work on this topic, and Miss Bormann's experimental tests of it, in some detail in Section I.5.

[143] Going a little farther back in history, even Einstein, Debye and von Laue during their occupation of the theory chair, had not, promoted any of their students to doctorate.

agreement with this statement, he had no doctoral students in Zurich, and none later in Berlin or Graz.

Moreover, in Zurich Schrödinger wrote all his papers alone, except for one on a statistical problem, in which he was joined by his old Viennese friend and collaborator Fritz Kohlrausch (Kohlrausch and Schrödinger, 1926). When Fues arrived in Zurich in the spring of 1926, he would not be drawn into collaboration with Schrödinger, e.g., on the topic of the latter's next investigation in wave mechanics; fortunately, he was far enough advanced to settle on a problem of his own, which he solved and published alone, although he did acknowledge Schrödinger's interest and help in it (Fues, 1926a, p. 396). The next young physicists to come to Schrödinger were Walter Heitler (1904–1981) and Fritz London (1900–1954), both arriving in the spring of 1927.[144] And they would not collaborate with Schrödinger either, but worked out a successful application of wave mechanics to describe covalent chemical binding (Heitler and London, 1927). In the case of Heitler, whose earlier interests had been in physical chemistry, one might understand why they did not work together; but the fact that Schrödinger did not work with London is rather remarkable, as the latter had recently become interested in an idea which Schrödinger had earlier proposed in a paper published in 1922.

## A Letter from Fritz London Concerning an Extension of Weyl's Theory

Fritz London mentioned the idea in detail in a letter written to Schrödinger in December of 1926, a few months before he came to Zurich. He wrote:

> Very Respected Professor: Today I must talk to you seriously. Do you know a certain Mr. Schrödinger who described, in the year 1922, a "noteworthy property of the quantum orbits"? Do you know this man? What, you assert, you know him rather well, you even then were, when he wrote the paper, with him and were implicated in this work? This is really shocking. Hence you did know, already four years ago, that one does not possess any rods and clocks for the definition of an Einstein–Riemannian measure in the continuum description, which results from analyzing atomic processes; hence, one must see whether perhaps the general principles of measurement arising from Weyl's theory of distance transfer might help. And you did realize already four years ago that they even help very well. Because, while usually nonsense results if one applies Weyl's distance transfer . . . , you demonstrated that for discrete real orbits [of the electrons in atoms] the gauge factor reproduces itself (for [the particular value of the constant] $\gamma = h/2\pi i$) on a spatially closed path; and you did realize especially then, that on the $n$th [quantum] orbit the unit of measure swells and shrinks $n$ times, exactly as in the case of a standing wave describing the position of charge. You therefore demonstrated that Weyl's theory becomes reasonable—i.e., *will lead to a unique measure determination*— only if combined with quantum theory; and one has no other choice, if the whole world of atoms represents a continuum process without any identifiable fixed point.

[144] Heitler obtained his doctorate with Karl Herzfeld at the University of Munich in 1926. London was senior to him and had already received his doctorate with a dissertation on a philosophical topic in 1921; he then continued to study physics in Göttingen and Munich, before accepting an assistant's position with Paul Ewald in Stuttgart in 1926.

[And, continuing this tone of pleasantry, he concluded:] This you did know and did not state or express a word about it. This has never happened before.... And yet, you removed in this work not only the incurable confusion of the Weyl theory. You even did have the resonance nature of the quantum condition in your hands, long before de Broglie; and you pondered whether to take $\gamma = h/2\pi i$ or $\sim e^2/c$. Will you now quickly confess that you held, like a priest, truth in your hands, that you kept it secret; and will you now tell everything you know to your contemporaries? (London to Schrödinger, December 1926)[145]

With these words London referred to an investigation, which Schrödinger had submitted early in October 1922 to *Zeitschrift für Physik*, bearing the title "*Über eine bemerkenswerte Eigenschaft der Quantenbahnen eines einzelnen Elektrons*' ('On a Notable Property of Quantum Orbits of a Single Electron,' Schrödinger, 1922c). London claimed that this paper contained essential, if not decisive, ideas of undulatory mechanics, which Schrödinger would introduce more than three years later. London even jokingly implied that Schrödinger had in 1922 concealed results from his contemporaries. In order to see whether such an 'accusation' was justified, one must analyze more closely the origin and content of Schrödinger's paper on the notable property of quantum orbits.

The first hint on the subject of the paper may be found in a letter which Schrödinger wrote, early in November 1921, shortly after he had arrived in Zurich. Pauli had previously presented to him a copy of his review article on relativity theory for the *Encyklopädie der mathematischen Wissenschaften* (Pauli, 1921b); and Schrödinger wanted to thank his colleague for the valuable gift. After doing so in his letter, he continued: 'By the way, may I bother you, if you can spare the time, with the following, probably very stupid question; this question occurs to me from time to time...: Are electron orbits geodetic lines or not?' (Schrödinger to Pauli, 7 November 1921). Schrödinger went on to argue that the answer should be 'No.' 'Because,' he wrote, 'if I put into the field of a charged, massive [atomic] nucleus, an electron and a neutral mass point at the same point in the same direction and with the same initial velocity, then they exhibit totally different motions' (Schrödinger to Pauli, *loc. cit.*). Hence, he concluded, the law for a geodetic line did not in reality play a fundamental role, as in nature non-charged smallest particles did not exist.

The question asked by Schrödinger seemed to be very simple, and his answer certainly satisfied Pauli as much as the additional remark: 'The law [of the geodetic line]... just applies approximately to the centre-of-mass motion of the larger, on the whole, non-charged systems in fields and on orbits, such that the dimensions of the system can be neglected as being too small' (Schrödinger to Pauli, 7 November 1921). However, what law then replaced the law for the geodetic line of General Relativity Theory in the case of the inner motions of

[145] The letter is not dated, but can be put in relation to the other letters contained in the scientific correspondence; there is a letter from London to Schrödinger, dated 1 December 1926, which Schrödinger answered on 7 December. London referred to both letters in the letter quoted above. (See also Raman and Forman, 1969, pp. 305–306, especially footnote 43.)

electrons in an atom? Schrödinger tried to answer this question in the paper, on the notable property of the single electron orbits, of October 1922. For a theoretical scheme to deal with this problem he referred to Hermann Weyl's unified theory of gravitation and electricity (Weyl, 1918a, b; 1919b).

Erwin Schrödinger had been following Weyl's work on the generalization of Einstein's theory of gravitation for years. He had studied Weyl's book on that topic, *Raum-Zeit-Materie* (Weyl, 1918c), soon after its publication.[146] The elegance of Weyl's mathematical methods attracted Schrödinger, who also liked the emphasis on variational principles and geodetic lines in Weyl's presentation. In the third edition of the book in the fall of 1919, Weyl included his new attempt at a unified theory of gravitation and electricity. This attempt consisted of an extension of the foundations of geometry beyond those given by Bernhard Riemann decades earlier; and in a derivation from the new, generalized world geometry both gravitational phenomena and electromagnetic phenomena followed. Weyl discussed the outlines of his theory in the later sections of *Raum-Zeit-Materie*, beginning with Section 35 (in the fourth edition), entitled 'The Metrical Structure of the World as the Origin of Electromagnetic Phenomena' (Weyl, 1921).

He opened this section with the remarks: 'To be able to characterize the physical state of the world at a certain point of it by means of numbers we must not only refer the neighbourhood of this point to a co-ordinate system but we must also fix on certain units of measure .... This idea, when applied to geometry and the conception of distance after the step from Euclidean to Riemannian geometry has been taken, effected the final entrance into the realm of infinitesimal geometry. Removing every vestige of ideas of "action at a distance," let us assume that world-geometry is of this kind; we then find that the metrical structure of the world, besides being dependent on the quadratic form [i.e., the metric tensor $g_{ik}$], is also dependent on a linear differential form $[\sum_{i=0}^{4}]\phi_i\,dx_i$' (Weyl, 1921; English reprint, p. 282).

The new geometrical structure of the physical space immediately suggested, 'not only to identify the coefficients of the quadratic groundform $[\sum_{i,k=0}^{4}]g_{ik}\,dx_i\,dx_k$ with the potentials of the gravitational field, but also to identify the coefficients of the linear ground form $[\sum_{i=1}^{4}]\phi_i\,dx_i$ *with the electromagnetic potentials*' (Weyl, 1921, English reprint p. 283). That is, the linear form $\phi_i\,dx_i$ $(=\phi_0\,dx_0 + \phi_1\,dx_1 + \phi_2\,dx_2 + \phi_3\,dx)$, which determined in Weyl's world geometry the change of the measure $l$ of a distance (*Strecke*) in what he called 'congruent transference' (Weyl, *loc. cit.*, p. 308) through the equation

$$dl = -l\,\phi_i\,dx_i, \tag{21}$$

also described the electromagnetic field. The components of the latter's tensor, therefore, could be obtained from the relation

$$F_{ik} = \frac{\partial\phi_i}{\partial x_k} - \frac{\partial\phi_k}{\partial x_i}, \tag{22}$$

where the $\phi$'s simply represented the electromagnetic potentials.

[146] We have mentioned this fact in Section I.5.

In Section 65 of his relativity article (Pauli, 1921b) Wolfgang Pauli discussed Weyl's theory. He sent a copy of it to Schrödinger. The latter wrote back, in the letter of 7 November 1921 to Pauli: 'Because of the lack of time I have unfortunately not yet come to the deeper and more difficult parts [of Pauli's article].' After he had studied these, especially Section 65, Schrödinger felt that he would try to answer his own question, posed in the same letter to Pauli, namely: What kind of world line might determine the path of charged particles? In the course of this investigation he arrived at an interesting connection between the world-line of an electron in Weyl's unified theory and the Bohr–Sommerfeld orbits of electrons in atoms (on the basis of the quantum theory of atomic structure).

Weyl had previously attempted to derive from his theory particular solutions describing a spherical electron and its motion (see, e.g., Weyl, 1918a), and he summarized both the successes and difficulties of this approach in the later editions of *Raum-Zeit-Materie.* Especially, he discussed the hypothesis of Bohr's quantum theory of atomic constitution, according to which electrons circulating on individual stationary states should not emit radiation. 'Our field-equations make assertions only about *the possible states of the field,* and *not about the conditioning of the states of the field by the matter,*' he concluded, adding: 'This gap is filled by the *quantum theory* in a manner of which the underlying principle is not yet fully grasped' (Weyl, 1921; English reprint, p. 303). More than a year later Schrödinger claimed that he might have found a key to close this gap.

## A Notable Property of Quantum Orbits

Schrödinger was fairly well acquainted with the geometrical methods of Hermann Weyl, since he himself had used essentially the same methods in dealing with the problems of the colour metric. He started out from an equation in Pauli's relativity article, i.e., the integrated form of Eq. (21) for the transference or the displacement of the measure $l$,

$$l_{P'} = l_P \exp\left\{-\int_P^{P'} \phi_i \, dx_i\right\}, \tag{23}$$

expressing $l$ at point $P'$ in terms of its value at point $P$, and an exponential factor containing a path integral (extending from $P$ to $P'$) over the linear form $\phi_i \, dx_i$.[147] By introducing explicitly the usual electromagnetic potentials, i.e. the Coulomb

[147] We have corrected a mathematical sign in Pauli's equation (Pauli, 1921b, p. 761).

Equation (23) is not contained in Weyl's book; hence we may indeed conclude that Schrödinger's renewed interest in Weyl's work on generalized gravitation theory arose from his reading of Pauli's handbook article.

potential $V$ and the axial vector potential $\mathbf{A}$ ( $= A_x, A_y, A_z$), as

$$\phi_0 = \gamma^{-1} eV, \qquad \phi_1 = -\gamma^{-1}\frac{e}{c}A_x,$$
$$\phi_2 = -\gamma^{-1}\frac{e}{c}A_y, \qquad \phi_3 = -\gamma^{-1}\frac{e}{c}A_z, \tag{24}$$

where $\gamma$ denotes a universal proportionality constant yet to be determined, the exponential factor on the right-hand side of Eq. (23) could be written in the form

$$\exp\left\{-\int_P^{P'} \phi_i\, dx_i\right\} = \exp\left\{-\frac{e}{\gamma}\int_P^{P'} (V\,dt - A_x\,dx - A_y\,dy - A_z\,dz)\right\}. \tag{25}$$

In the introduction to his paper on 'a notable property of the quantum orbits' of October 1922, Schrödinger then claimed: 'The property of the electron orbits which appears notable to me, as announced in the title [of the paper], is the following: the "true" quantum conditions, i.e., those which suffice to fix the energy and therefore also the spectrum [of the atom], are also just sufficient *to turn the exponent of the distance factor* [(25)] *into an integral multiple of* $\gamma^{-1}$ $h$ *for all approximate periods of the system*' (Schrödinger, 1922c, p. 14).

Schrödinger proved this assertion by considering in detail five examples of electron orbits in atoms, as described by the Bohr–Sommerfeld theory: (i) the unperturbed Kepler orbit; (ii) the action of a constant magnetic field on the orbit (Zeeman effect); (iii) the action of a constant electric field on the orbit (Stark effect); (iv) the combined action of parallel electric and magnetic fields on the orbit; and (v) the relativistic Kepler orbit. In all cases he found the result by fairly elementary calculations. For example, in the case of a non-perturbed Kepler motion of the electron he derived from the single quantum condition—i.e., the relation $2\tau\bar{T} = nh$, where $\tau$ denoted the period, $\bar{T}$ the average kinetic energy of the electron, $n$ the integral quantum number and $h$ Planck's constant—and the relation between the average kinetic energy and average potential energy—i.e., $\bar{T} = \frac{1}{2}e\bar{V}$—the equation

$$e\tau\bar{V} = e\int_0^{\tau} V\,dt = nh\,. \tag{26}$$

As a consequence the distance factor (25) became in this case $-nh\gamma^{-1}$ for an entire period $\tau$.

While the result, Eq. (26), had already been derived earlier by the Dutch physicist Adriaan D. Fokker—as Schrödinger had learned in a letter from Hermann Weyl (see footnote 1 on p. 14, *loc. cit.*)—Schrödinger succeeded in verifying the same to be true in all cases of multiply periodic electron orbits.[148] He thus

[148] For example, in the case of the (normal) Zeeman effect—with the Larmor precession frequency $\nu_L = \tau_L^{-1} = eH/(4\pi mc)$—the quantum condition was $Hf(e/c) = n'h$, with $H$ the strength of the magnetic field, $f$ the projected orbit area on a plane perpendicular to the magnetic field vector, and $n'$ an integral number. Hence the exponential factor (25) became $\exp\{-\gamma^{-1}Hf(e/c)\}$ or $\exp\{-\gamma^{-1}n'h\}$. In the other cases straightforward calculations also yielded the desired result.

summarized the results in the following words: 'If the electron in its orbit carried along with it a "distance," which is transferred unchanged during the motion, then the measure of this distance would—if one started from an arbitrary point on the orbit—always be multiplied by an integral multiple of $\exp\{h/\gamma\}$, whenever the electron returns approximately to its initial position and simultaneously to its initial state of motion' (Schrödinger, 1922c, p. 22).[149]

'It is hard to believe,' Schrödinger concluded, 'that this result is but an accidental mathematical consequence of the quantum conditions and does not possess any deeper physical meaning' (Schrödinger, 1922c, p. 22). While he did not want to imply that the electron actually carried a 'distance' with it, he suggested that perhaps the electron showed some kind of 'adjustment' ('*Einstellung*') such as Weyl had proposed in order to interpret the meaning of 'distance transferences', namely 'that the electron cannot adjust with an arbitrary velocity, but that the adjustment must rather display a certain dependence on the quasi-periodic orbit cycles' (Schrödinger, 1922c, p. 23). At the end of his paper, Schrödinger finally discussed the values which the constant $\gamma$ might assume, suggesting as possible choices $h$ and $e^2/c$. Taking $e^2/c$ would yield the very high value of exp(1,000) for the exponential factor (25). If he chose, however, for $\gamma$ the purely imaginary value,

$$\gamma = \frac{h}{2\pi\sqrt{-1}}, \tag{27}$$

the exponential factor (25) became equal to unity, hence 'the measure number of a distance carried along [by the electron] would be reproduced after each quasi-period' (*loc. cit.*, p. 23). He cautiously added the remark: 'I do not dare decide whether such an interpretation might make sense in the framework of Weyl's world geometry' (Schrödinger, 1922c, p. 23).

The sentences just quoted represented how Schrödinger interpreted his own results in October 1922. At that time he did not think of contradicting Weyl's hypothesis about electron's 'adjusting' to the measuring rod; and he did not suggest any physical meaning connected with the imaginary factor in the exponent of Eq. (27), such as the phase of a wave or anything vaguely reminiscent of it. Nor did he talk of the electron as being in resonance when moving on a Bohr orbit. Instead, he appeared to be rather concerned with the question of the particular value of the factor $\gamma$. Thus, after proposing the values $h$, $e^2/c$ and $-ih/\sqrt{2\pi}$, he added: 'By the way, one must bear in mind that $e$, $h$ and $c$ do not represent the only universal constants known to us. Further, if one takes into consideration the (usual) constant of gravitation $\kappa$ and any universal mass, e.g., the electron mass, then $e^2/\kappa m^2$ = pure number $\simeq 10^{40}$. Hence $he^2/\kappa m^2$ is a "universal quantum of action" of the order of magnitude $10^{13}$ erg $\cdot$ sec. It must

[149] Schrödinger had to add the word 'approximate' initial position, since in the case of relativistic motion, the position after a full period did not coincide with the initial position.

be borne in mind with this that, from dimensional arguments alone, one *cannot* derive *anything* with any certainty' (Schrödinger, 1922c, p. 23).

This is as far as Schrödinger's considerations extended in the fall of 1922. Had he known more, had he even in mind speculative suggestions on how to give the 'notable property' of quantum orbits a suitable interpretation, Schrödinger certainly would not have hesitated to phrase them properly. As it was, Schrödinger's paper did not stimulate a wider audience; his Viennese colleague Adolf Smekal wrote an extended review of it for the *Physikalische Berichte* (published in Volume **4**, No. 18, pp. 1082–1083), and that was about all that happened. In this review Smekal drew attention to an article by William Wilson (1875–1965) of London that was published a few months later in the *Proceedings of the Royal Society of London*; in it Wilson proposed an extension of the quantum conditions in a general relativistic scheme, i.e.,

$$\oint\left(p_k + \frac{e}{c}A_k\right)dq_k = n_k h\,, \tag{28}$$

where $p_k$, $q_k$ and $A_k$ denoted the components of momentum, position and of the electromagnetic potential, respectively (Wilson, 1923, Eq. (2a) on p. 481). It was also possible to derive, Smekal pointed out, the specific results of Schrödinger by applying Eq. (28).

Smekal's hint concerning Wilson's treatment illustrates a rather general situation in those years. That is, during the first half of the 1920s a considerable number of papers were published in physics journals dealing with possible connections between the two great theoretical schemes, relativity theory and quantum theory. The above-mentioned papers of Schrödinger and Wilson, therefore, constituted only a small fraction of the contemporary work in this field. For example, early in 1925 three papers, two by Cornelius Lanczos (1893–1974) of Frankfurt and another by Ernst Reichenbächer of Wilhelmshaven, were submitted to *Zeitschrift für Physik* dealing with the same problem as Schrödinger (Lanczos, 1925b, c; Reichenbächer, 1925).[150] In spite of the fact that both authors followed a different approach to that of Schrödinger, it must be noted with surprise that they did not cite the earlier paper.[151]

[150] We have already discussed Lanczos' work in Volume 3, Section V.2.

Reichenbächer (1881–1944) was a *Gymnasium* teacher, and later (since 1929) also *Privatdozent* at the University of Königsberg, having been promoted in 1944 to an extraordinary professor. From 1916 he contributed to the discussion of a unified field theory of gravitation and electromagnetism (Reichenbächer, 1917); he then discussed in 1920 aspects of Weyl's theory (Reichenbächer, 1920).

[151] Lanczos attempted to derive the conditions for the structure of radiation in a finite, general-relativistic universe, with the aim of proving that electrons in quantum orbits do not radiate. Reichenbächer, on the other hand, studied the four-dimensional world lines corresponding to Bohr orbits; he arrived at the conclusion that the hypothesis of a proper time, $h/(m_e c^2)$, for electrons (mass $m_e$) might be responsible for the existence of quantum conditions.

## Light-Quantum, Doppler Effect and Compton Effect

Lanczos' and Reichenbächer's articles mark the continuing interest of physicists in the problem of connecting quantum and relativity theory; an interest that had begun soon after the establishment of general relativity theory. An early example constituted a paper by Karl Försterling of Königsberg, entitled '*Bohrsches Atommodell und Relativitätstheorie*' (Bohr's Atomic Model and Relativity Theory,' Försterling, 1920), in which Försterling examined the question of the compatibility of the two theories.[152] 'Quantum theory and relativity seem to be occasionally considered as opposing fields, because quantum theory is at variance with Maxwell's equations, the starting point of relativity theory,' he stated introducing the problem and then added: 'Since, however, the principle of relativity claims to be universally valid, quantum theory must nevertheless yield to it' (Försterling, 1920, p. 404). He then demonstrated how Bohr's frequency postulate, which indeed contradicted classical electrodynamics, fitted smoothly with relativistic invariance, in such a way that the specific line shifts followed, as predicted by special and general relativity theories. Wolfgang Pauli, in his review of Försterling's paper for *Physikalische Berichte*, then criticized the derivation given for the transverse Doppler shift of the spectral lines by the remark: 'It should, however, be noted that the transformation formula, used by the author for the emitted energy is valid only if in [the rest system of the atom] $K'$ no momentum is emitted at all' (see *Phys. Ber.* **2**, issue No. 9 of 1 May 1921, p. 489).[153] About a year later Schrödinger filled the gap in Försterling's proof; he published the complete solution of the problem of the Doppler effect within the quantum theory of atomic spectra in a larger note, '*Dopplerprinzip und Bohrsche Frequenzbedingung*' ('Doppler's Principle and Bohr's Frequency Condition'), submitted on 7 June 1922 to the *Physikalische Zeitschrift* (Schrödinger, 1922a).

The fact that Försterling had just succeeded in obtaining the transverse Doppler effect appeared only a 'little encouraging' ('*wenig ermutigend*,' Schrödinger, 1922a, p. 301) to Schrödinger, since this effect represented only a small correction to the classical result arising from relativity theory. The neglect of momentum transfer in the emission process, as pointed out by Pauli, even made this small achieved result useless. 'This restriction will, however, not be valid in any frame of reference; rather, the emitted quantum $h\nu$ should always, according to Einstein's foundation of radiation theory ([Einstein, 1917a])—i.e., in any frame of reference—carry along the momentum $h\nu/c$, the *biggest* amount which can be connected with this energy-quantum,' argued Schrödinger (Schrödinger, 1922a,

[152] K. Försterling was born on 23 April 1885 in Wernigerode, Harz. He studied physics, obtaining his doctorate at the University of Göttingen with Woldemar Voigt in 1909. He became *Privatdozent* for theoretical physics at the *Technische Hochschule* in Danzig in 1914; seven years later he moved to Jena, where he was appointed extraordinary professor in the fall of 1921, that is to say, replacing Schrödinger. In 1924 he accepted the call to a chair at the University of Cologne. He retired in 1951 and died on 20 June 1960 in Cologne.

[153] Besides the transverse Doppler effect, Försterling derived the gravitational shift of the spectral lines by applying a general relativistic transformation to the energies of Bohr's frequency condition.

p. 301). He was also happy to be able to demonstrate in his note, 'that the "velocity jump" thus arising yields, with Bohr's frequency condition, exactly the Doppler shift, that is, in all details, as it is required in relativity theory' (*loc. cit.*, p. 301).

It may perhaps be of some interest to observe that Schrödinger did not primarily intend, with his note, to establish the Doppler effect in Bohr's theory of atomic structure; Schrödinger came across the problem when systematically studying the theory of molecular spectra. The then available theory took into account three terms contributing to any energy state of a given molecule: first, the electronic term; second, the term arising from the oscillations of the nuclei; and third, the term due to the rotation of the molecule. In principle, Schrödinger argued in 1922, there must exist a fourth translational term, smaller than the others, which does not appear quantized but leads to a finite broadening of all spectral lines.[154]

Schrödinger proceeded in a straightforward way in solving his problem. In the special case of an atom moving in the same direction as the momentum transferred by it to the emitted quantum of radiation, he applied the laws of energy and momentum conservation, i.e.,

$$\Delta\nu = \frac{1}{h}\left(\frac{m}{2}v_1^2 - \frac{m}{2}v_2^2\right) \tag{29}$$

and

$$mv_1 = \frac{h\nu}{c} + mv_2\,, \tag{30}$$

where $m$ denotes the mass of the atom or molecule, $v_1$ and $v_2$ its velocity before and after the emission of a light-quantum having energy $h\nu$ and momentum $h\nu/c$. On inserting Eq. (30) into Eq. (29), he found the relation

$$\frac{\Delta\nu}{\nu} = \frac{v_1 + v_2}{2c} \tag{31}$$

yielding the elementary, classical Doppler effect.

In attacking the more general situation Schrödinger first defined the energies of the atomic system in its proper frames of reference before and after the emission of radiation, $E_1\cdot(1 - v_1^2/c^2)^{-1/2}$ and $E_2\cdot(1 - v_2^2/c^2)^{-1/2}$, respectively. Consequently he wrote Bohr's frequency condition as

$$h\nu = \frac{E_1}{\sqrt{1 - v_1^2/c^2}} - \frac{E_2}{\sqrt{1 - v_2^2/c^2}}\,. \tag{32}$$

[154] We may assume that Schrödinger's interest in the problem of molecular spectra emerged from his interest in the problem of specific heats on the one hand, and from the preparation of his lecture courses on atomic theory on the other; for example, he announced, for the summer semester of 1921, a course on 'dynamics of atomic structure' at the University of Breslau. (See *Phys. Zs.* **22**, p. 255, 1921: '*Dynamik der Atomstruktur*'.)

For the most general case, in which the initial and final velocity vectors made angles $\theta_1$ and $\theta_2$ with the momentum vector of the emitted radiation, momentum conservation then provided the two equations

$$\frac{E_1 v_1 \cos\theta_1}{c^2\sqrt{1 - v_1^2/c^2}} = \frac{E_2 v_2 \sin\theta_2}{c^2\sqrt{1 - v_2^2/c^2}} + \frac{h\nu}{c} \tag{33a}$$

and

$$\frac{E_1 v_1 \sin\theta_1}{c^2\sqrt{1 - v_1^2/c^2}} = \frac{E_2 v_2 \sin\theta_2}{c^2\sqrt{1 - v^2/c^2}}. \tag{33b}$$

By combining Eqs. (32) and (33a), (33b) he finally obtained the relation

$$\nu^* = \nu\sqrt{\frac{c - v_1\cos\theta_1}{\sqrt{c^2 - v_1^2}} \cdot \frac{c - v_2\cos\theta_2}{\sqrt{c^2 - v_2^2}}}, \tag{34}$$

connecting the frequency $\nu$ (shifted by the Doppler effect)—the classical plus the relativistic corrections—to the 'rest-frame' frequency $\nu^*$, defined by

$$\nu^* = \frac{E_1^2 - E_2^2}{2h\sqrt{E_1 \cdot E_2}}. \tag{34a}$$

Schrödinger's derivation of the full, relativistic Doppler effect for spectral lines emitted by atoms and molecules was immediately recognized as an important contribution to atomic theory. Niels Bohr mentioned it in his next review article, '*Über die Anwendung der Quantentheorie auf der Atombau. I. Die Grundpostulate der Quantentheorie*' ('On the Application of Quantum Theory to Atomic Structure. Part I. The Fundamental Postulates'), submitted in November 1922 to *Zeitschrift für Physik* (Bohr, 1923a).[155] He stated especially: 'The question [i.e., the Doppler shift in spectral lines] is discussed in an interesting way by Schrödinger (*Phys. Zs.* **23**, 301, 1922) in connection with the idea put forward by Einstein, that the emitted radiation is entirely directed' (Bohr, 1923a; English translation, p. 28, footnote ‡). However, he immediately criticized the derivation also, stating: 'Apart from the fact that this idea is far removed from the presentation here given of the actual applications of the quantum theory, it may be recalled that, simply because of the smallness of the mass-ratio of the negative and positive particles of the atom, an eventual change of momentum can have no observable effect on the spectra of isolated systems' (Bohr, *loc. cit.*) Bohr did not accept Einstein's light-quantum hypothesis as being consistent with his scheme of atomic theory at that time, in which light had to be described by classical electrodynamics; hence, he demanded a different procedure in order to derive possible effects on the spectral lines, arising from the relative motions of atoms and molecules.

[155] Schrödinger's note in *Physikalische Zeitschrift* had appeared in the issue of 1 August 1922.

Schrödinger's Zurich colleague Peter Debye, on the other hand, showed no such scruples; in his paper, entitled '*Zerstreuung von Röntgenstrahlen und Quantentheorie*' ('X-ray scattering and Quantum Theory') and submitted in March 1923 to *Physikalische Zeitschrift*, he made use of exactly the momentum equations (33) and arrived at the theoretical prediction of the Compton effect (Debye, 1923; especially, Eqs. (II) and (III) on p. 162).[156] Again a few months later, on 15 September 1923, Adolf Smekal in Vienna signed his note '*Zur Quantentheorie der Dispersion*' ('On the Quantum Theory of Dispersion'), which he sent to *Naturwissenschaften* and in which he generalized Schrödinger's procedure in predicting another effect—which would be observed several years later by Chandrasekhara Venkata Raman—namely, the existence of incoherent, secondary radiation arising from the scattering of light by atoms (Smekal, 1923c).[157]

In writing the paper of 1923 on the scattering of X-rays, Debye did not present himself as an unshakable partisan of the light-quantum hypothesis; he rather considered his theoretical investigation 'as an attempt to derive, for the time being, from the two assumptions, "quanta" and "needle radiation", as detailed conclusions as possible, with the sole use of as general laws as possible, i.e., the laws of "energy conservation" and "momentum conservation" '; he thereby followed the goal, 'to obtain a deeper insight into the quantum laws, especially with respect to their connection with wave optics' (Debye, 1923, p. 166). The publication of Arthur Holly Compton's experimental results in the May issue of *Physical Review* (Compton, 1923b)—Debye's theoretical study appeared in the issue of 15 April of *Physikalische Zeitschrift*—then seemed to justify fully the light-quantum-like derivation of the Compton effect. When the latter was confirmed beyond doubt, towards the end of 1923, Niels Bohr's negative attitude towards the light-quantum was in serious trouble. However, early in 1924 the Copenhagen master of atomic theory found a way out, which he presented in a paper with his collaborator Hendrik Kramers and the American visitor John Clarke Slater, submitted in January and February 1924 to *Philosophical Magazine* and the *Zeitschrift für Physik*, respectively (Bohr, Kramers and Slater, 1924). The essential content of the new radiation theory expounded in the paper presented the conclusion that one might be able to reconcile Compton's observation with the wave nature of radiation; that is, one could avoid the use of the light-quantum hypothesis if one assumed that energy and momentum were not strictly valid, on the statistical average only, in processes involving the interaction of light and matter.[158] What appeared as a novel idea in Copenhagen, however, had been known, and even publicly advocated for years in Zurich.

[156] We have discussed the discovery of the Compton effect and its independent prediction by Debye in Volume 1, Section V.1.

[157] We have discussed Smekal's paper in Volume 1, Section V.1. It might be interesting to note that Smekal reviewed earlier Schrödinger's note on the Doppler effect (Schrödinger, 1922a) for the *Physikalische Berichte* (see Volume **3**, No. 23 of 1 December 1922, p. 1207).

[158] We have treated the origin and the content of the radiation theory of Bohr, Kramers and Slater in some detail in Volume 1, Section V.2.

## Statistical Energy and Momentum Conservation

In the fourth edition of his book *Raum-Zeit-Materie*, whose preface was signed in November 1920, Hermann Weyl had stated, in discussing the structure of matter: 'It seems that the *theory of statistics* plays a part . . . which is fundamentally necessary. We must here state in an unmistakable language that physics at its present stage can in no wise be regarded as lending support to the belief that there is a causality of physical nature which is founded on rigorously exact laws . . . . *Statistical physics*, through the quantum theory, has already reached a deeper stratum of reality than is accessible to field physics [i.e., general relativity and electrodynamics]; but the problem of matter is still wrapped in deepest gloom' (Weyl, 1921, English reprint, p. 311). Such suggestions, vague as they were, appealed greatly to Weyl's colleague Erwin Schrödinger, the former student of Franz Exner. He had not only worked on statistical methods for a decade, but he had also developed certain ideas on how to proceed in atomic theory.

In a letter to Pauli, dated 8 November 1922, Schrödinger first gave hints of these ideas, starting with a discussion of his own treatment of the Doppler effect. He admitted that he had based the successful derivation completely on the light-quantum hypothesis, and added:

> On the other hand, there must also exist a spherical wave [in the emission process], otherwise no interference and diffraction would arise. This is the fatal dilemma exactly. I believe, for my part, that in the emission process—*horrible dictu* [it is horrible to say it]—the conservation law of energy-momentum is violated. What then do all the so-called "$h\nu$-relations" tell us? (I mean, the experimental ones!) They tell us that there exists some *equivalence* between electrons and atoms having a definite energy and radiation of a definite *frequency*, an equivalence with respect to the creation of a certain effect in an atomic system. Nothing else. That is, in the situation in which the effect is caused by radiation, exactly the same "quantum" of radiation energy is used for that purpose, as must be possessed by the electron, if the *latter* should create this effect: This *we assume as an obvious prerequisite*—however, it *does not* appear obvious to me. (Schrödinger to Pauli, 8 November 1922)

One might, he argued in order to illustrate his point, open a door either by a projectile having definite energy or momentum, or by just shaking it statistically with a certain average frequency. In the latter case the same energy might not always be necessary to achieve this goal, except when the door-opening mechanism followed strictly causal and conservative laws. 'But the Devil knows whether our atomic systems also change according to *necessity* only, or are determined by *strictly causal* laws, not to mention whether or not they behave like conservative mechanical systems.' He, Schrödinger, believed that the ' "solution" to the terrible contradictions of quantum theory' had to be sought along the path suggested by Franz Exner, i.e., by assuming an essential statistical nature of the physical laws. He concluded: 'For example, I consider it to be probable that one has to assume, say, in the present case, that the spherical wave is emitted and yet a recoil exists.

Contradiction? Why? Can it not be imagined that the conservation of energy-momentum is a macroscopically valid average relation, which is unknown to atomic physics in the same way as the second law of thermodynamics? At least, it *may* be that way—and I can see nearly no other way out' (Schrödinger to Pauli, 8 November 1922).

Schrödinger presented his doubts on the validity of energy and momentum conservation, when applied to processes involving a single or a few atoms, in his Zurich inaugural address which he delivered a month later (Schrödinger, 1929a). But he did not apply them in his own research on quantum or atomic theory. However, when in the spring of 1924 Bohr, Kramers and Slater published their new radiation theory, he reacted quite enthusiastically. 'With the greatest interest I have just read the fascinating turn of your ideas in the May issue of *Philosophical Magazine*,' he wrote soon afterwards to Bohr, continuing: 'This turn touches me extremely sympathetically. Being a disciple of the old Franz Exner, I have long since become acquainted with the idea that our statistical laws are probably not based on "microscopic" lawfulness, but perhaps on "absolute chance," and that perhaps even energy and momentum conservation are only statistically valid' (Schrödinger to Bohr, 24 May 1924). Some time later he composed an extended note, entitled '*Bohrs neue Strahlungshypothese und der Energiesatz*' ('Bohr's New Radiation Hypothesis and the Energy Conservation Law,' Schrödinger, 1924d), in which he stated that the 'new conception . . . demands equally great interest from the point of view of the physicist and the philosopher,' because: 'By it the opinion, which has been put forward by several authors, i.e., that the single molecular process might perhaps not be determined uniquely and causally through "laws," is realized for the first time' (Schrödinger, 1924d, p. 720)

In his note, published in the *Naturwissenschaften* issue of 5 September 1924, Schrödinger first summarized the main features of the Bohr–Kramers–Slater theory: the hypothesis of the virtual radiation field of the atoms, containing the frequencies of all spectral lines emitted by them, with the associated transition probabilities; the validity of energy conservation on the statistical average over many atomic processes; the new approach to the phenomena of dispersion of light by atoms essentially on the basis of the concepts of wave theory; and the explanation of the Compton effect and of the natural line-widths of spectral lines.[159] Then he added a few considerations of his own, in which he attempted to illustrate both the consequences and the origin of statistical energy conservation.

Schrödinger especially calculated the energy fluctuation $\Delta$ of a gram-atom of hydrogen, obtaining the following value per second

$$\Delta\,(\mathrm{sec}) = \sqrt{\frac{\varepsilon RT}{\tau}}, \tag{35}$$

[159] Paul Hanle, in his Ph.D. thesis, has provided a detailed discussion of Schrödinger's response to the Bohr–Kramers–Slater theory (Hanle, 1975, Chapter 6).

where $\varepsilon$ denoted the energy of a characteristic line, say the red hydrogen line ($= 3 \times 10^{-13}$ erg), $\tau$ the corresponding duration time (about $2 \times 10^{-8}$ sec), $R$ the universal gas constant and $T$ the absolute temperature. For a temperature of 3000° absolute he found for $\Delta$ (sec) the value 1900 erg; if taken in relation to the total energy (taken as $\frac{1}{2} RT$, or $2.5 \times 10^{11}$ erg in the above case) a relative fluctuation of $1.5 \times 10^{-8}$ per sec or $0.87 \times 10^{-4}$ per year resulted. Thus the fluctuation turned out to be relatively small for a macroscopic sample of hydrogen atoms—it depended on the square root of the mass considered—while it became large (i.e., of the order of unity) for a single atom in the time $\tau$. Of course, $\tau$ depended strongly on the frequency of the atomic lines, hence for very hard X-rays considerable fluctuations might be expected. But since lines of such frequency could be excited only at very high temperatures (of millions of degrees), the author concluded: 'Without anticipating a more accurate computation of special cases or an ingenious experimental method (which might be devised by someone), we may therefore be in a position to state, as the comforting result of our first rough estimate, that the new conceptions of the radiation processes are not contradicted in this respect by present experience' (Schrödinger, 1924d, p. 724).

In the second half of the paper Schrödinger went on to derive some physical consequences from the above considerations. He recalled first the fact that energy conservation formed, together with the second law, the basis of thermodynamics, the most characteristic description of macroscopic systems. From the later nineteenth century, the concept of entropy had been found to break down in the realm of atoms and molecules; and in the early twentieth century Marian von Smoluchowski had shown how the reversibility governing the processes between single molecules did not contradict the occurrence of entropy increase in processes involving a large number of molecules, because the description of macroscopic phenomena was *not too tightly related* to that of microscopic phenomena. The new Exner–Bohr view of statistical energy conservation seemed to reveal an even deeper gap between the macroscopic and the microscopic laws of nature. If energy conservation was not valid in atomic processes, the following general situation would then arise in physics: 'A *closed* system then exhibits for relatively *short* times only, *approximately* a definite average behaviour. In the limit $t = \infty$ its behaviour will be *completely undetermined*, because its *energy content* deviates from the initial value in an open cone of dispersion according to the $\sqrt{t}$-law. We cannot decrease the dispersion unless we enlarge the *extent* of the system, or consider the latter as the partial system of a more comprehensive system ("heat bath")' (Schrödinger, 1924d, p. 724). Or, to put it another way—more in the style of Clausius' formulation of thermodynamical laws: 'A certain stability of the processes in the world *sub specie aeternitatis* [in the eyes of eternity] can only exist due to the *connection* of each single system with the rest of the entire world. The disconnected single system would represent, from the point of view of conformity, chaos. The connection is necessary to establish a lasting *regulation*; without it the single system would, as far as energy is concerned, wander around at random' (Schrödinger, 1924d, p. 724).

Hence, according to this interpretation, energy conservation in the natural processes represented an exception rather than the rule. It was satisfied in some rare limiting cases, surrounded by a sea of dissipative systems. The fact that coupling to larger systems created a certain stability in physics showed some resemblance, as Schrödinger pointed out at the end of his note, to social, ethnical and cultural phenomena.

In the fall of 1925 none of the atomic physicists, not even Niels Bohr—whose ideas stimulated Schrödinger's response—shared those thoughts.[160] His Zurich colleagues may have been even less involved. For example, Debye, although open to possible violations of energy and momentum conservation in atomic processes—as expressed in the conclusion of his 1923 paper on X-ray scattering—did not wish to pursue any speculations away from physics. The most likely person to sympathize with Schrödinger's proposal might have been Hermann Weyl; however, Weyl at that time was deeply absorbed in a variety of his own activities devoted to the foundations of mathematics, the analysis of the space problem, the development of a theory of continuous groups, and the philosophy of mathematics and science. Wolfgang Pauli in Hamburg, on the other hand, who had previously been the recipient of Schrödinger's speculations, did not react—as far as we can see from the published scientific correspondence which contains no letters from the years 1924 and 1925; at that time (of the Innsbruck conference) he had become a staunch opponent of the violation of energy and momentum conservation in atomic processes.[161]

## Scientific Exchange with Sommerfeld

While the exchange of scientific ideas with Pauli was interrupted, Schrödinger came into closer contact with the latter's teacher Arnold Sommerfeld. In November 1924 he received, as a present from Sommerfeld, a copy of the new, fourth edition of *Atombau und Spektrallinien* (Sommerfeld, 1924d), and he immediately returned his heartiest and warmest thanks in a letter.[162] 'You really are an unselfish

[160] In Copenhagen, Bohr and Kramers studied Schrödinger's note carefully. Among his notes Kramers kept an unpublished manuscript, entitled, '*Über eine mögliche Bedeutung des Versagens des Energieprinzips bei Atomprozessen*' ('On a Possible Significance of the Failure of Energy Conservation in Atomic Processes'); in it Schrödinger's ideas were discussed as well as the objections raised by Albert Einstein, at the *Naturforscherversammlung* of September 1924 in Innsbruck, against the Bohr-Kramers-Slater theory. Kramers was of the opinion that Schrödinger's considerations on energy fluctuations did not draw any conclusions on the description of macroscopic systems. The real situation was rather that the new radiation theory really leads to an energy increase in macroscopic systems, but this energy increase could be balanced by an equal energy decrease following from a suitable modification of the theory. (See Helge Kragh: *Die Entwicklung des Bohrschen Komplementaritätgedankens*, unpublished manuscript, especially pp. 26–27.)

[161] We have discussed Pauli's objection to the Bohr-Kramers-Slater theory in Volume 1, Section V.5. We might mention that Schrödinger also participated in the Innsbruck *Naturforscherversammlung*. (See Schrödinger to Sommerfeld, 21 July 1925.)

[162] Sometime earlier Sommerfeld—at Schrödinger's request—wrote an official letter to him, with the intention of helping him establish a systematic series of lecture courses in all the fields of theoretical physics at the University of Zurich. Schrödinger, as we have mentioned in the previous section, had

teacher to all of us, using your strong teaching abilities to enable as many as possible to elaborate further along the path you have prepared,' Schrödinger wrote enthusiastically (Schrödinger to Sommerfeld, 19 November 1924). Sommerfeld's book immediately provided a stimulus for Schrödinger's work, as he continued in his letter: 'On p. 737 I found the *Motto* to a little note, which I was just about to submit, namely on the rotational heat of $H_2$, with half-integral quantum numbers for molecular rotation' (Schrödinger, 1924f, p. 342).[163]

Contact with Sommerfeld continued during the following year and more. Less than four months later, on 7 March 1925, Schrödinger again expressed his thanks in a letter. 'Today I sat for the whole morning studying the beautiful new papers which you have been kind enough to forward to me, together with those of [Gregor] Wentzel, and studying the corresponding parts of your book,' he opened the letter (Schrödinger to Sommerfeld, 7 March 1925). Schrödinger then went on to discuss several still controversial issues in atomic spectroscopy, such as the relativistic doublet problem and the relation of X-ray spectra to optical spectra. Ultimately, he felt stimulated by an illustration in Sommerfeld's book to discuss a new problem. The illustration (see Sommerfeld, 1924d, p. 691) showed the irregular dependence of the frequency difference within a given multiplet of the so-called Bergmann series of the barium atom on the running-term number.[164]

Schrödinger connected this illustration with a criticism of another theoretical work. Thus he wrote to Sommerfeld: 'You know how Born and Heysenberg [sic] [(Born and Heisenberg, 1924b)] calculate the deformability of the alkali ions from the deviation of the terms [of the alkali atoms] from those of the hydrogen. I have now checked the situation and find that the result fits [the experiments] only very roughly if one wants to call this a fitting at all.' Schrödinger, the former occupant of an experimental assistant's position, showed himself to be really annoyed by the arguments of the pure theoreticians from Göttingen, for he continued: 'I cannot find any reason why one is allowed—on p. 395, line 6 from bottom [of Born and Heisenberg, 1924b]—to speak about "inaccuracy of data" in the case of spectroscopy! The apparently small discrepancies of the table [on the same page of Born and Heisenberg, *loc. cit.*] are slaps in the face of experimental accuracy' (Schrödinger to Sommerfeld, 7 March 1925).

---

been charged with presenting a course on mechanics *every* winter semester, and this duty collided with such a plan. 'I wish, in particular, that [Simon] Ratnowsky alternates with me in delivering the course on mechanics, thus enabling me to present my coherent course on theoretical physics,' Schrödinger explained to Sommerfeld, adding: 'A commission has been set up on this problem, which will soon meet, and for this purpose your letter (together with one of similar content by Planck), will be very helpful to me' (Schrödinger to Sommerfeld, 19 November 1924).

[163] Schrödinger submitted the paper referred to above on 24 November 1924 to *Zeitschrift für Physik* (Schrödinger, 1924f). We shall treat the content of this paper in the next section.

[164] According to the usual theory of atomic constitution the frequency difference should decrease continuously with the increasing running number.

Born and Heisenberg had, in their theoretical analysis of the deformability of noble-gas-like ions (published in May 1924: Born and Heisenberg, 1924b), calculated the polarizability $\alpha$ of the noble-gas-like cores of alkali ions from spectroscopic data by means of a Rydberg-Ritz formula, i.e.,

$$\nu + \Delta\nu = \frac{R}{(n + \delta_1 + \delta_2/n^2)^2}, \tag{36}$$

where $R$ denotes the Rydberg constant, $n$ the principal quantum number, and the constants $\delta_1$ and $\delta_2$ (denoting deviations from the hydrogen terms) were proportional to $\alpha$, i.e.,

$$\delta_1 = -\frac{3}{4a_1^3k^5}\alpha, \tag{36a}$$

and

$$\delta_2 = \frac{1}{4a_1^3k^3}\alpha, \tag{36b}$$

where $a_1$ is the radius of the hydrogen atom and $k$ the azimuthal quantum number. They had found, by inserting data from the series of alkali atoms or the corresponding alkaline-earth ions (in the latter case the expressions for $\delta_1$ and $\delta_2$ had to be multiplied by the factor 4!), 'that, as far as can be expected in the light of the inaccuracy of the data, the ratio $\delta_2/\delta_1$ in general comes out all right' (Born and Heisenberg, 1924b, p. 395). Schrödinger did not agree with this conclusion at all. He explained to Sommerfeld:

> Actually the situation is as follows. If one calculates, e.g., the polarizability of the Na-core from the deviation of the $3d$-term from the [corresponding] hydrogen term, and then computes (with the result obtained) the deviation which has to be expected for the $4f$-term from the $H$-term, the latter turns out four times too large compared with reality. Or, to express it another way (as I have actually calculated): if the deviation from the hydrogen term is caused by the polarization of the core, then nine times the $3d$-term should deviate from Rydberg's constant only seven times as much as sixteen times the $4f$-term. In reality, the deviation of the former is twenty-six times as large. (Schrödinger to Sommerfeld, 7 March 1925)

The problem attracted Schrödinger, because it was related to the only detailed investigation into atomic structure he had ever been involved in (see Schrödinger, 1921a, discussed earlier). In the spring of 1925 he felt that he should also try to improve on the hydrogen-like terms of the same atomic structures, namely of the alkali atoms. From the above-mentioned failure of the Born–Heisenberg approach he immediately derived one consequence: 'The core must be, therefore, much less polarizable in the cases of orbits with larger periods than for orbits of high frequency.... This looks like a (normal) dispersion effect' (Schrödinger to

Sommerfeld, 7 March 1925). On the other hand, he concluded, from irregularities in the *d*- and *f*-terms of the lithium atom, that the situation of anomalous dispersion also occurred, which he interpreted by assuming, 'that the orbit of the series electron also possesses higher harmonics.' He now proposed to explain the shape of the illustration in Sommerfeld's book using the concept of such higher harmonics. 'By the way, we are always dealing with cases,' he continued, 'where the core still contains one or several electrons, which are originally co-ordinated with the series electron. This may result in the fact that it [i.e., the core] possesses eigenfrequencies which are sufficiently low, so that they can be overtaken by one of the first, still stronger harmonics of the orbit of the series electron' (Schrödinger to Sommerfeld, 7 March 1925).

Since the quantitative calculation of these ideas promised to involve tricky calculations and some insecure assumptions, Schrödinger doubted whether he should go ahead. 'Besides, one has the definite feeling with all classical dispersion calculations, that they are only makeshift,' he concluded (Schrödinger to Sommerfeld, 7 March 1925). Sommerfeld, in a sympathetic letter dated 20 March 1925, encouraged Schrödinger to pursue this problem. As Schrödinger expressed it, some months later, he had received 'an injection of *Münchenin*' ('*eine Injektion Münchenin*'), otherwise he felt 'the child would never have been born' ('*das Kind wäre nie geboren worden*,' Schrödinger to Sommerfeld, 21 July 1925).

The paper, which Schrödinger submitted on 7 April 1925 to *Annalen der Physik*, had the title '*Die wasserstoffähnlichen Spektren vom Standpunkte der Polarisierbarkeit des Atomrumpfes*' ('The Hydrogen-like Spectra from the Point of View of the Polarizability of the Atomic Core') and appeared in the June issue (Schrödinger, 1925c). It complemented the 1921 paper with a full treatment of *all* alkali and alkali-like terms, as the former paper had dealt with non-hydrogen-like terms and the new one with hydrogen-like terms. Schrödinger first approached the problem in the same way as Born and Heisenberg had done a year previously, namely with the hypothesis that the field of the series electron created a dielectric polarization of the core, which in turn gave rise to a force term proportional to $1/r^5$ in the equation of motion (of the series electron). 'While the assumption of a constant polarizability of the core [as described by the above term] reproduces perfectly the behaviour of the term values within a single series, it is totally insufficient, if one wants to conclude from the behaviour of one series that of another.' Schrödinger summarized the result of comparing this theory with the data, adding: 'The fact is that . . . the "outer" orbit series approach the similarity to hydrogen very much *faster* than would be expected on the basis of that assumption' (Schrödinger, 1925, p. 48). He further noticed that the introduction of half-integral values for the quantum number $k$ improved *in no case* the constancy of the polarizability as derived from a fit of Eqs. (36) to the data; in the example of a well-examined series of the cesium atom he rather found: 'The constancy of the numbers gets essentially worse' (Schrödinger, 1925c, p. 51).

The most peculiar phenomena happened, however, in the cases of the *d*- and *f*-terms of the sodium and lithium atoms where the calculations yielded *negative* polarizability—i.e., the deviation of the terms from the corresponding hydrogen terms went in the opposite direction. 'It seems that, so far, the shortfall of the hydrogen term in these cases has been assumed, partly explicitly and partly tacitly, as not being real, but caused by errors in the wavelength determination or in the calculation of the series limits,' Schrödinger stated, and continued: 'This is perhaps still barely possible, but I do not consider it to be probable' (Schrödinger, 1925c, p. 52). Now, on the contrary, if further experiments confirmed the fact, then it might be explained by some kind of *anomalous* dispersion. This interpretation, Schrödinger continued, seemed to require the existence of very high frequencies in the orbits of the series electron; and he proposed to take for that purpose the higher harmonics of the orbit frequencies or their quantum-theoretical counterparts. In addition, the assumption of different polarization mechanisms of the core other than the one used by Born and Heisenberg might provide a more suitable situation in applying the anomalous dispersion scheme.

As a particular example in carrying through his ideas, Schrödinger discussed the irregularities in the alkaline-earth spectra and their homologues, where the atoms or ions possessed alkali-like cores. He especially analyzed, in detail and quantitatively, the case of the Bergmann series of ionized aluminium (Al II), obtaining an excellent fit to the data by taking a 'resonance frequency' with a wave number of 46,000 $cm^{-1}$. He summarized:

> A peculiar perturbation of the Bergmann series of Al II discovered by Paschen can be interpreted such that the polarizability is very large for exciting frequencies that are a little smaller than the resonance frequency of the core, while it turns negative above the resonance frequency. For the exciting frequency one has to take the most probable transition frequency from the Bergmann orbit in question; and for the resonance frequency one has to take the transition frequency of the sodium-like core, corresponding to the sodium doublet. (Schrödinger, 1925c, p. 69)

In his new paper on the alkali spectra and the like, Schrödinger tried to go as far as possible in rescuing the Bohr–Sommerfeld concepts of atomic structure, as he desired 'not to let the light extinguish, which had shone through Sommerfeld's calculation on the structure of the Ritz term formula' (Schrödinger, 1925c, p. 45). He was little bothered by the fact that parts of his analysis were contained in an earlier paper by Douglas Rayner Hartree of Cambridge, and which he had overlooked (Hartree, 1924). He just knew that he had successfully carried through the idea hinted at in Sommerfeld's book.[165]

[165] Sommerfeld said, in discussing the figure on page 691 of *Atombau und Spektrallinien*: 'The occurrence reminds us of the critical periods of waves or oscillations of springs. Indeed one has the impression, when looking at Figure 137 below, that a similar situation might occur in atomic mechanical systems: for certain values of the period of the series electron [there exists] a reaction in the atomic core, which leads to anomalous $\Delta\nu$, and which might possibly originate from a kind of resonance betwen the series electron and the atomic core' (Sommerfeld, 1924d, p. 690).

## Work on Colour Vision

Schrödinger occasionally referred to the fact that at times he turned his attention, during the first half of the 1920s, from problems of quantum physics to those of colour theory, out of sheer despair arising from the difficulties of the theory of atomic structure. (See, e.g., Schrödinger to Wien, 28 May 1925.) This statement implies that in those years he treated simultaneously questions in two such different fields of physics. But it cannot be understood—except perhaps for one situation ocurring in late 1925—in the way that Schrödinger concentrated during all these years his highest devotion to quantum-theoretical problems of atomic structure and, if he failed to make progress in these, he simply relaxed by looking at the less strenuous problems of colour theory. From the above discussion of his specific work on atomic spectra we derive the following conclusions: first, he only concerned himself with a few selected topics, actually with only a small fraction of the great number of topics that were investigated at that time by his many contemporaries; hence, he could be considered only as a minor, occasional contributor in this field, although admittedly *all* his contributions displayed considerable quality. Certainly, the specific results he achieved, e.g., in the theory of the Doppler effect for spectral lines, were assessed and more or less accepted as successful results. On the other hand, his continuous concern with problems of colour vision and colour theory must not be seen as an escape from despair, but may easily be interpreted as a natural consequence of Schrödinger's writing an extended article on '*Gesichtsempfindungen*' ('Visual Sensations') for the second volume of the new (eleventh) edition of *Müller-Pouillet's Lehrbuch der Physik* (Schrödinger, 1926j). In this article of ultimately 105 pages, organized in 44 sections, Schrödinger discussed, after a general introduction (Section 21), the conceptions of human light perception in day time (Part A, Sections 22–39), at dawn (Sections 40–47), the temporal and spatial psychological effects of light action (Sections 48–56), and finally threshold perceptions and specific problems of comparing different schemes of colour vision and colour theory (Sections 57–64). He finished the work for the article in the summer of 1925, although the Volume 2, Part 1 of *Müller-Pouillet* containing it, was not published until a year later.[166]

In preparing this article and studying the relevant literature, Schrödinger came across several questions which required further investigation. This led him to submit, between the fall of 1924 and the spring of 1925, three papers presenting clarifications of older controversies and several new aspects. The first paper of this series appeared in the *Naturwissenschaften* issue of 7 November 1924 and was entitled '*Über den Ursprung der Empfindlichkeitskurven des Auges*' ('On the

[166] The references in Schrödinger's review article include papers up to early 1925, e.g., his own article on subjective star colours published in the 1 May issue of *Naturwissenschaften* (Schrödinger, 1925a). The preface to Volume 2, Part 1 of *Müller-Pouillet's Lehrbuch der Physik*, eleventh edition, was signed by Arnold Eucken and Erich Waetzmann on 23 August 1926.

Origin of the Sensitivity Curves of the Eye,' Schrödinger, 1924e).[167] In this article Schrödinger attempted to explain the detailed structure of the normal sensitivity of the human eye for colours, on the one hand, and for low brightness, on the other. He took the duplicity theory of vision as a basis, i.e., he started from the assumption that the human eye contained two mechanisms to receive light and colour sensations. One mechanism, provided by the 'cones,' was responsible for colour perception and functioned only for higher intensity, i.e., in daylight. However, the second mechanism, provided by the 'rods,' served as a highly sensitive instrument also acting for very low light intensity, e.g., at dawn.[168] While the details of the sensitivity curves for the perception of the fundamental colours—Schrödinger supported, as we have discussed earlier, the three-colour theory of vision of Thomas Young and Hermann von Helmholtz—seemed to be easily explained by referring to the spectral distribution of the solar light, the sensitivity curve of the rod mechanism was shifted towards the blue end of the spectrum.[169] 'Where does this notable displacement of the rod curve towards shorter wavelengths come from,' asked Schrödinger and added: 'I do not remember having found anywhere an attempt at an explanation for this. Being fully aware of the fact that I possess only a part of the knowledge necessary to judge this case, I still want to discuss here several possible explanations, even if this only serves the purpose of directing other people's attention to it [that problem], who are more qualified than I to judge' (Schrödinger, 1924e, p. 926).

The first possibility accounting for the observed difference was, of course, to assume different mechanisms for the reception of light sensations by cones and rods, respectively. It seemed, however, that Franz Exner had already decided against this possibility by showing that the reception mechanisms in both cases were essentially electromagnetic resonators in the respective nerve cells (Exner, 1922). 'A real explanation, which does not refer to unknown organic structural conditions,' Schrödinger concluded, 'can be found, in my opinion, only in the fact that the rod apparatus had originated under the influence of *another type of illumination light* having a different energy distribution curve from that of the cone apparatus' (Schrödinger, 1924e, p. 927). The most obvious illumination light, to which the rods were adapted, was the light at night, i.e., the light of the moon and the stars; however, this light had more or less the same energy distribution as daylight, hence one had to search for another explanation. One could think of the following hypothesis: (i) in the history of nature the rod mechanism had been developed earlier than the cone mechanism, and (ii) the

[167] It can be seen that this article owes its origin to the preparatory work for the review article, because it refers to various literature which is quoted not only in the review but also in the previous, tenth edition of *Müller-Pouillet's Lehrbuch.* (See Schrödinger, 1924e, p. 926, footnote 1, left column.)

[168] Cones and rods refer to the microscopic structure of nerves existing in the human eye.

[169] The curve of the rods, which does not allow for the differentiation of colours, would be expected to agree with the average curve of the three cone mechanisms responsible for receiving the colour sensations red, green and blue. Actually its maximum was shifted by more than 30 $\mu$ towards the blue end of the spectrum.

sun was hotter at that time, hence the maximum of its energy distribution was then shifted to shorter wavelengths. While the first part of this hypothesis seemed to be very plausible, as the lowest animals in the biological sequence do not possess any colour sensitivity, a considerable shift of the solar energy distribution during the past billion years could be excluded on the basis of astrophysical considerations. 'The third and most probable explanation appears to me to lie in the *green-blue colour* exhibited by *water* in thicker layers,' Schrödinger finally argued. 'For an aquatic animal living at some depth below the surface the composition of solar light must indeed be approximately changed in the way needed for an explanation.... The rod apparatus would therefore, according to this hypothesis, be an *older* visual organ, which had originated at that time in acquatic life' (Schrödinger, 1924e, p. 926).

Schrödinger's explanation of an older phylogenetic origin of the rod mechanism was supported by the previous investigations of the Munich biologist Carl von Hess and his Zurich colleague, the opthalmologist Alfred Vogt.[170] This explanation, moreover, constituted 'strong support for the duplicity theory [of visual perception],' because it took into account the existence of two independent mechanisms of light perception, especially the formation of the younger cone mechanism for animals living on dry land, as well as the more primitive rod mechanism; the new mechanism was then responsible for visual perception in day time; 'the old rod mechanism, on the other hand, took over the role of an organ at dawn, for which purpose it was particularly suited from the beginning through its great adaption width' (Schrödinger, 1924e, p. 929).

A further paper on a related subject, entitled '*Über die subjektiven Sternfarben und die Qualität der Dämmerungsempfindung*' ('On the Subjective Colours of Stars and the Quality of Sensation at Dawn'), constituted a reply to a question that had been asked by Kurt Felix Bottlinger (1888-1934). The Babelsberg astronomer had, in a short note appearing in one of the February 1925 issues of *Naturwissenschaften*, emphasized the discrepancy between the colour vision of stars and of terrestrial light sources having the same temperature, and he closed with the remark: 'It would be interesting to hear once an opinion from the physiological side on this undoubtedly remarkable phenomenon, which has so far never been discussed according to my knowledge' (Bottlinger, 1925a, p. 180). Schrödinger's answer appeared in the 1 May issue of the same journal (Schrödinger, 1925a). He claimed that the noted discrepancy might be explained by a complex co-operation of effects from both mechanisms, the cone and the rod mechanisms, at comparatively low light intensities. He especially argued that the

[170] A. Vogt (1879-1943) had expressed similar views on an older origin of the rod mechanism in his inaugural address, delivered on 1 December 1923 at the University of Zurich. Schrödinger did not attend the lecture, however, and did not know of Vogt's considerations when publishing his own paper in October 1924. He, therefore, apologized in a later paper, stating: 'With pleasure I seize the opporunity of stating the *priority of Vogt* over me on the question of the phylogenesis of the rod apparatus, which had unfortunately escaped me when composing the note quoted' (Schrödinger, 1925a, footnote 1 on p. 376).

white colour point did not coincide in the two mechanisms; thus the white point of the rod mechanisms was seen as some blue point by the cone apparatus—a fact which Schrödinger confirmed by simple experiments carried out at Peter Debye's institute.[171] Consequently, white stars having the surface temperature of the sun appeared during the night as yellow, while stars much hotter than the sun seemed to be white. Still another reason caused the colour shift of yellow stars to red ones: any yellow colour point possesses in colour theory an admixture of fundamental red, and it is this component which remains perceptible, according to the Bezold-Brücke phenomenon.[172]

By publishing these ideas on colour vision and light perception Schrödinger wished to bring them into public discussion, hoping that people would respond and criticize; and he more or less succeeded. F. Hauser pointed out, in connection with the first paper, 'that the spectrum of the diffuse celestial light must exhibit a considerable brightness shift towards the blue end as compared to the spectrum of direct sunlight' (Hauser, 1925, p. 197). Hauser did not wish to contradict the hypothesis of the later phylogenetic origin of the cone mechanism, but rather to provide a further argument, besides the one given by Schrödinger, as to why the rod mechanism was not completely dropped in the later evolutionary development. Bottlinger, on the other hand, confirmed Schrödinger's explanation of subjective star colours (Bottlinger, 1925b, c). In any case, the desired discussion of the problems and answers took place, and Schrödinger confidently incorporated his results into the review article for *Müller-Pouillet's Lehrbuch.*

## Last Papers on Colour Theory

At about the same time Schrödinger became involved in a renewed discussion on the problem of colorimetry, especially after Tadeusz Oryng had partly contradicted, in the 1 February issue of *Physikalische Zeitschrift*, the criticism raised by Fritz Kohlrausch against Ostwald's colour theory.[173] Schrödinger put himself on the side of his friend in a note '*Über Farbenmessung*' (On Colour Measurement'), which appeared three months later in the same journal (Schrödinger, 1925b). He also used this to clarify the principles of colour determination, declaring right at the beginning: 'The method of measuring and defining a colour, which must be considered accordingly to the classical papers of Maxwell, Helmholtz, A. König, Dieterici and Kries as a secured possession, does *not* rest, as one might believe from the critical remarks of Mr. Oryng, on heterochromous photometry, but solely on the statement of the indiscernibility of two colour fields, on a so-called colour equation' (Schrödinger, 1925b, p. 349). 'This has

[171] Schrödinger thanked his colleague from the E.T.H. explicitly in a footnote to his paper (Schrödinger, 1925a, footnote 1 on p. 374).

[172] Similarly, the appearance of yellow-green, green-yellow and blue-green stars as saturated green sources could be explained by the Bezold-Brücke phenomenon. (We have described it briefly in Section II.3, p. 300.)

[173] We have discussed these points in Section II.3.

absolutely nothing to do with heterochromous photometry,' Schrödinger repeated, 'and just as little with a definite colour theory' (Schrödinger, 1925b, p. 350).[174] With respect to a check of Wilhelm Ostwald's claim that a given pigment colour will be uniquely determined by measuring the co-ordinates of the two points, he referred to Kohlrausch's investigations which had settled the case against Ostwald.[175] Schrödinger argued further that Ostwald's concept of the 'metamerism' of two colours, (i.e., the fact that two colours having very different composition might cause the same sensation) was not very helpful, as in general there existed for each colour sensation infinitely many possibilities of composition. Schrödinger noted:

> The confusion, however, which the new phrase has created in *reality, surpasses* every expectation: since the new notion has been given, everything which has formed (since Newton and Maxwell) the *experimentally secured* foundation of all colour determination—and, as we explicitly wish to note, of Ostwald's as well as of the classical determination—is seriously considered very doubtful and as depending only on the positive outcome of future experiments. (Schrödinger, 1925b, p. 351)

Finally, he replied to Oryng's claim, that all classical colour determination depended heavily on the colour perception of the individual observer, with the following strong remarks:

> Is it really impossible to come to an understanding with those who adhere totally or partially to Ostwald's absolute colour determination, at least insofar that one cannot determine the *appearance* of a light more absolutely than it *really is*, i.e., only in the approximation in which the colour vision of normal persons coincides? (Schrödinger, 1925b, p. 351)

All the differences between individual observers did not render the classical methods in colour theory useless, but represented 'the really existing variation range, even of the normal visual organs, which cannot be abolished and which removes the possibility of an absolute determination—as strictly as for a purely physical quantity—in the case of colour' (Schrödinger, 1925b, pp. 351-352). What Ostwald and his partisans demanded could therefore never be realized. Hence Schrödinger did not bother to describe Ostwald's theory in his review article but rather declared in his introduction that he was omitting it 'due to certain doubts which have been raised against it' (Schrödinger, 1926j, p. 457).

Later in 1925 Schrödinger again submitted a further, longer paper on colour theory; it was communicated to the Vienna Academy at the session of 17 December 1925 and had the title '*Über das Verhältnis der Vierfarben- zur Dreifarbentheorie*' ('On the Relation of the Four-Colour Theory to the Three-Colour Theory,'

[174] In a footnote (No. 2 on p. 350) Schrödinger added that the entire problem considered by Oryng belonged to what he himself had earlier called the 'lower' colour metric.

[175] Oryng had, in his paper, pleaded for new checks. (See Oryng, 1925, p. 186.)

Schrödinger, 1925f).[176] The so-called four-colour theory, especially that which Ewald Hering had advocated, rested on the basis of three pairs of contrasting sensations: red-green, yellow-blue and white-black. Schrödinger touched only briefly on this topic in his review article, stating essentially its principles and explaining its advantages and difficulties (Schrödinger, 1926j, pp. 494-497).[177] Since the four-colour theory and its derivatives—like Ostwald's colour theory—'was well liked in artistic circles' (Schrödinger, *l.c.*, p. 496), Schrödinger, in his new paper, did not wish to argue for or against it. Schrödinger wrote in the introduction:

> We are dealing with the mere statement that on a purely formal level the relation between the two theories—the three-colour theory and the four-colour theory—can be thought as an extremely simple one, namely as a mere *transformation of variables*. The situation is not particularly deep from the mathematical point of view; nevertheless, it has up to now never been stated, according to my knowledge, with full clarity; certainly many have not even recognized the point, otherwise it might have given another direction to the discussion [on the respective validity of the two theories]. (Schrödinger, 1925f, p. 472)

In order to find the desired simple relation explicitly, Schrödinger started from the standard colour triangle in the three-colour theory, an equilateral triangle with the fundamental colours red, green and blue as corners and the white point in the centre; the spectral colours lay on a curve in the interior of this colour triangle with only the wavelengths from 545 to 670 falling on the side connecting the red and green corners. Of course, Schrödinger continued, one can also describe the spectral colours and all the other colours (e.g., the pigment colours) by taking a different triangle, which is created from the standard one by a linear co-ordinate transformation. He then constructed what corresponded to the colour triangle in Hering's four-colour theory in the following way: two sides were formed by the lines joining the blue and white points and the red and white points, respectively; the third side then coincided with a part of the line connecting the colours of zero brightness ('*Alychne*', Schrödinger, 1925f, p. 477). The new triangle thus possessed the white point as a corner; the second corner lay on the line connecting red and white away from any real colour and was called 'original

[176] Clearly, Schrödinger wrote this paper *after* the completion of his article for the Müller-Pouillet *Lehrbuch*, because he did not refer to it in the review although he considered the results to be important. (See below at the end of this section.)

[177] Schrödinger devoted Section 39, '*Die psychologische Farbenordnung*: *Herings Theorie der Gegenfarben. v. Kries' Zonentheorie*' (The Psychological Colour Organization: Hering's Theory of Opposite Colours. Von Kries' Theory of Zones') to the four-colour theory and its modification by Johannes von Kries (1853-1928).

Among the difficulties of four-colour theory Schrödinger mentioned especially that 'the theory fails completely in explaining the defects and anomalies of colour perception' (Schrödinger, 1926j, p. 496). He also conceded however that 'it cannot be doubted at all that Hering's conceptions connect much better than those of Helmholtz with the immediately perceived psychological organization of colours; hence they are much better suited to provide a view on the manifold of sensation qualities, which every non-colour blind person finds in his consciousness' (Schrödinger, 1926j, p. 496).

green' ('*Urgrün*'), and the third corner *nearly* coincided with the fundamental blue point and was called 'original blue' ('*Urblau*') in Hering's theory.

To represent the content of the four-colour theory, one has to convert the new triangle to a quadrangle, which is easily obtained in a straightforward procedure: first, one transforms the triangle white–original green–original blue into a rectangular one with the two equal smaller sides white–original green and white–original blue; these sides then constitute a rectangular co-ordinate system in the plane, with the white point as origin and original blue and original green making the unit distances. Opposite the original blue and original green are the corresponding counter-colours 'original yellow' and 'original red' at negative unit distances, respectively. The two pairs original blue–original yellow and original green–original red, together with the third pair black–white, then allow one to describe the colours quantitatively in Hering's theory also and to establish a mathematical relation between the two theories.

In his paper Schrödinger presented several constructions, in which the relations between the co-ordinates of a given colour in the two schemes could be expressed particularly simply. He also derived an important conclusion on the nature of the blue. The 'fundamental blue' of the three-colour scheme and the 'original blue' of the four-colour scheme nearly coincided, with the latter corresponding to a point (or pigment colour) of *no brightness*. Schrödinger therefore tended to support the opinion that both were identical.[178] If it really coincided, an important physiological conclusion could be derived: 'Then the three "components" of the visual organ are still probably *not of the same type*; the idea of the *symmetric role* of the three components [i.e., red, blue and green] suggested by the three-colour theory fails, and one might have to keep this [fact] in mind when searching for the physiological foundation' (Schrödinger, 1925f, p. 488).[179]

Finally, Schrödinger did not avoid the temptation to give the above results a phylogenetic interpretation, sketching the following picture of colour vision. Early in biological evolution animals developed an organ to receive light of a certain (comparatively large) frequency interval; in a second step a refinement of that organ occurred, which enabled the animals to differentiate between lights of shorter and longer wavelengths, respectively. But, 'since for the development of an individual differentiation ability for single frequencies the prerequisites were lacking, the most probable was the formation of a gross differentiation ability; i.e., there was established a particular sensation element making it possible to denote the *dominance* of the short-wavelength or long-wavelength components, as compared to the "normal" constitution of light (sunlight)' (Schrödinger, 1925f,

[178] It may be of some interest to mention in this connection, the fact that blue had no brightness was one of the starting points of J. W. von Goethe's colour scheme, which actually initiated the development of physiological and psychological colour theory, leading ultimately to Hering's four-colour theory. Schrödinger never mentioned, even in his review article, Goethe's contribution; hence the fact had probably escaped him.

[179] Especially, one might be unable to find an elementary physiological mechanism for creating the blue sensation.

p. 489). This sensation element, Schrödinger claimed, led to a psychological registration of the blue–yellow sequence containing the neutral white point; hence dichromatic colour vision, which is typical of many animals and also of a part of colour-blind human beings, was established. The full tri-chromatic vision occurred in a third phylogenetic step, in which a further differentiation was introduced dissolving the basic sensation of yellow into a sequence from red to green. All these evolutionary steps, Schrödinger emphasized, can still be traced in the human eye; the most peripheral zones of the retina exhibit a gross sensitivity to light with no colour differentiation; the inner ones register the blue–yellow difference, and only the innermost regions near the *fovea* display the full sensitivity to human colour sensations.

With such general considerations Schrödinger concluded his last contribution to colour theory which he considered, in retrospect, as his most valuable contribution when judging his entire work in this field. 'The only important one, it appears to me, is the latest obtained re-recognition of the real meaning of the three- and four-colour schemes and their connection with the phylogenesis of colour vision' he wrote in his autobiographical sketch for the Nobel lectures (Schrödinger, 1935, p. 88). However, in the development of Schrödinger the scientist, his concern with problems of colour theory and colour vision should not be forgotten.

## II.5 Problems and Extensions of Statistical Mechanics

'In scientific work I have . . . never followed a strong line, that is, a programme guiding me for a long period of time,' Schrödinger once wrote in an autobiographical note (Schrödinger, 1935, p. 87). This statement might certainly be consistent with the variety of problems Schrödinger discussed in his papers during the first half of the 1920s. However, in spite of that, one finds a series of ten articles and notes, plus an extended handbook article, contributed between 1921 and early 1926, which really indicate a stronger line in Schrödinger's research; they are all concerned with problems of statistical mechanics. Furthermore, statistical mechanics and the kinetic theory of matter constituted the most favoured topics which the new professor announced for his lecture courses at that time.[180] Hence, he kept up a continuing, even deepening interest in such topics, and it may be worthwhile to present here a systematic overview of the particular problems on which Schrödinger concentrated his attention, and also to indicate the possible coherence of these problems, and see whether they formed a part of a greater guiding programme. This overview is all the more justified as

[180] According to the previews of the lecture courses given in the *Physikalische Zeitschrift*, Schrödinger announced, e.g., lectures on kinetic theory and closely related subjects for the winter semester of 1920/1921 (at the University of Jena), for the winter semester of 1921/1922 (at the University of Breslau), and for the summer semester of 1922 (at the University of Zurich). Although he did not deliver the first two lecture courses at the announced places, one can still conclude that statistical mechanics and its applications represented Schrödinger's most favourite topic.

Schrödinger's work on statistical mechanics ultimately provided him with the starting point for getting into and developing wave mechanics.[181]

## Isotopes and Gibbs' Paradox

Schrödinger had brought the problem for his first publication in this field from Vienna. It had emerged from a debate on the nature and the separability of different isotopes of the same chemical element, which Schrödinger's colleagues George de Hevesy and Friedrich Paneth of the *Institut für Radiumforschung* had carried out with Kasimir Fajans (1887–1975), then at the *Technische Hochschule* in Karlsruhe. Hevesy and Paneth had concluded, on the basis of electrolytical investigations, that the different isotopes—say, Ra E and bismuth—were chemically completely equivalent, i.e., 'isotopic elements may really replace each other completely, as far as their chemical mass action is concerned' (Hevesy and Paneth, 1914, p. 805). Fajans, however, had argued rather, 'that the isotopes must be considered, as far as their chemical mass action is concerned, as independent parts' (Fajans, 1914, p. 940). He had, in the course of his thermodynamical treatment of the problem, drawn attention to an important fact: namely, that one must abandon, in the theoretical discussion of the thermodynamical behaviour of isotopes 'the usual auxiliary tools, like semi-permeable walls or solvents which dissolve two substances to a different degree' (Fajans, 1914, p. 937). Still, this criticism did not convince the Viennese scientists, and the debate had dragged on for more than a year without producing a satisfactory clarification of the problem—although their opponents finally agreed that their respective views led to a consistent interpretation of the available experimental results.[182] After World War I interest in the properties of isotopes increased. Hevesy, then experimenting in Budapest (at the chemical institute of the Veterinarian School) and in Copenhagen (at Niels Bohr's invitation), respectively, had continued the earlier work on separating isotopes. Twice he had gone to Vienna to report to his old friends on the new results at the meetings of the *Gauverein* of the German Physical Society, held on 18 March and 22 November 1920.[183]

[181] Paul A. Hanle devoted a full chapter of his Ph.D. thesis, entitled 'Schrödinger's Quantum Statistics of the Ideal Gas,' to discussing important aspects of this work (Hanle, 1975; see also his article on the same topic published in *Archive for the History of Exact Sciences*: Hanle, 1977).

In our discussion we shall, as pointed out above, especially expand on the connections between the various problems treated by Schrödinger, on the internal continuity and systematics of his approach to statistical mechanics over a period of several years, and, as usual, on its embedding into the related work of his contemporaries.

[182] Hevesy and Paneth had first answered by stating that with respect to electrical properties their simplified description—in comparison with Fajans' treatment—was valid in any case: 'A mixture of isotopes is something different, in principle, from a mixture of two different elements, even if their chemical properties are as close as they can be' (Hevesy and Paneth, 1915, p. 51). See also the following papers: Fajans, 1916; Hevesy and Paneth, 1916.

[183] See the announcements of Hevesy's talks, entitled '*Die Trennung der Isotope durch Zentrifigugieren*' ('The Separation of Isotopes by Centrifugal Action') and '*Die Isotopentrennung bei Quecksilber*' ('The Isotope Separation in the Case of Mercury'), in *Verhandlungen der Deutschen Physikalischen Gesellschaft* (3) **1**, pp. 47 and 97 (1920).

Schrödinger had had occasion to attend the first of these meetings while he was still in Vienna.[184] About a year later he submitted a paper, '*Isotopie und Gibbssches Paradoxon*' ('Isotopy and Gibbs' Paradox'), to *Zeitschrift für Physik* (Schrödinger, 1921b). Schrödinger started explicitly from the previous discussion between de Hevesy and Paneth on the one hand, and Fajans on the other, summarizing the result as 'thermodynamical diversity of isotopes in principle, in spite of their complete or nearly complete chemical replaceability' (Schrödinger, 1921b, p. 163). Thus far he partially supported the view promoted by the Karlsruhe physico-chemist. However, he contradicted a result which Fajans had obtained, stating that the energy set free by the mixing or diffusion of the volumes, containing different isotopes of the same chemical element, was proportional to the difference of their respective atomic weights—hence as a consequence, it should become arbitrarily small as the mass difference disappeared.[185] Schrödinger rather claimed:

> With respect to this result it appears to me of interest to point to the fact that one can conduct—in an experiment, which is very simple to imagine as a thought experiment—a reversible, isothermal process of mixing two (chemically not reacting) gases; hence the full amount of [mixing] energy can be gained as long as a finite difference of the molecular weights exists at all; and this energy amount is not at all related to the magnitude of this difference, assuming the value for chemically different gases—which is well known from the foundations of thermodynamics. (Schrödinger, 1921b, p. 163)

This independence of the mixing energy from the weight difference of the isotopes obviously gave rise to the situation which Josiah Willard Gibbs (1839–1903) had discussed at the very end of his *Elementary Principles in Statistical Mechanics* (Gibbs, 1902, Chapter XV). Hevesy and Paneth had already referred, in their first paper on the nature of isotopes, to this so-called Gibbs' paradox claiming: 'It is evident that, viewed from this most general standpoint [i.e., from the point of view of statistical mechanics], isotopes of elements are not identical' (Hevesy and Paneth, 1914, p. 804, footnote 1).

In proving his assertion Schrödinger used ordinary thermodynamical argumentation.[186] He followed Fajans' warning not to use the concept of semipermeable walls, 'whose existence for isotopes might be rightly doubted' and rather applied—again as Fajans had done—'a sufficiently extended homogeneous gravitational field' (Schrödinger, 1921b, p. 163). Thus he started with two identical volumes of magnitude $V_0$, each of which was filled with one gram-atom of different isotopes, having molecular weights $M$ and $M'$, at the same temperature. He then brought the volume containing isotopes of mass $M'$ ($< M$) isothermally to a higher position (the volume $V_0$ was thereby increased to $V_1$) and established

[184] Schrödinger presented his work on the colour metric at the previous meeting of the *Gauverein*, in which de Hevesy was also present and accepted as a member. (See *Verhandlungen*, *loc. cit.*, p. 46.)

[185] Fajans had indicated the outlines of the proof for that statement—following a consideration of his student Michael Polanyi—in his first contribution to the isotope discussion. (See Fajans, 1914, p. 937.)

[186] It is very likely that Schrödinger arrived at the result given in his paper while preparing his lecture course on general thermodynamics for the summer semester of 1921 at the University of Breslau.

a connection with the vessel containing molecules of mass $M$ using a very thin tube (whose volume can be neglected); as a consequence a distribution of the isotopes in agreement with the barometric formula was reached. Finally, he reduced, again isothermally, the upper volume to zero, thus producing a homogeneous mixture of all molecules $M$ and $M'$ in the lower volume $V_0$. The energy consumed in the entire process could be calculated and found to be zero. In order to arrive at the situation of the usual mixture of isotopes, Schrödinger then simply let the volume $V_0$ expand to $2V_0$. 'We . . . thus can win *from the reversible mixture at constant volume altogether the energy amount* $RT \ln 2$, at the expense of the heat of the surroundings, just as is well known from treating any two chemically different gases (which do not react with each other) by the method of semipermeable walls,' Schrödinger stated in concluding his demonstration (Schrödinger, 1921b, p. 165).

## Extension of the Theory of the Specific Heats of Solids

More than a year later, in September 1922, Schrödinger sent a further paper, dealing with a problem of the kinetic theory, to the *Zeitschrift für Physik* (Schrödinger, 1922b). The author derived the stimulus for the new work from two papers on the theory of solids that had been published in the same journal more than a year previously. Max Born and E. Brody of Göttingen had then tried to explain the deviations of the specific heats of monatomic solids from the classical value (according to the rule of Pierre Louis Dulong and Alexis Therèse Petit) at extremely high temperatures, by taking into account finite amplitudes of the lattice vibrations (Born and Brody, 1921a, b). Schrödinger did not doubt their results but claimed: 'That the proof is carried out [in Born and Brody's paper] in quantum theory, is—at least for the present purpose—not necessary, since only such temperatures come into play for which the classical theory agrees with the quantum theory' (Schrödinger, 1922b, p. 170). Born and Brody has also seen the point and had written: 'One can obtain this formula, which is valid for higher temperatures, without the quantum theory from classical statistical mechanics. However, the calculation for this does not become simpler at all; hence, we have preferred the [quantum-theoretical] method which leads, for all temperatures, to a successful result' (Born and Brody, 1921a, p. 136, footnote 1). Whilst the proof of the Göttingen physicists demanded an extension of the usual quantization rules in applying to finite amplitude oscillations—which at least implied a new hypothesis—Schrödinger found that he could derive the same result 'almost at once by looking at a canonical ensemble' (Schrödinger, 1922b, p. 170).

He thus proceeded from a general formula

$$\exp\left(-\frac{\psi}{T}\right) = \int \cdots \int \exp\left(-\frac{E}{T}\right) dq_1 \cdots dp_n\,, \tag{37}$$

describing a canonical ensemble at temperature $T$, with $\psi$ denoting the free energy and $E$ the energy function, i.e.,

$$E = \tfrac{1}{2}\sum_{l}(p_l^2 + \omega_l^2 q_l^2) + f_3(q_i) + f_4(q_i), \tag{37a}$$

depending (in the case of a solid) on the momentum co-ordinates, $p_i$, and the position co-ordinates, $q_i$, of the lattice points. The functions $f_3$ and $f_4$ were cubic and biquadratic forms of the position co-ordinates and described the anharmonicity of the interatomic forces in the lattice. By expanding (carefully and in a straightforward way) the exponential in the integral on the right-hand side of Eq. (37), Schrödinger arrived, in a few lines, at the expression for the free energy of one gram-atom of the solid, i.e.,

$$\psi = 3RT \ln\frac{\nu}{kT} + CT^2, \tag{38}$$

where $R$ and $k$ denote the universal gas constant and Boltzmann's constant, respectively, $\nu$ an average oscillation frequency ($\nu^n = \nu_1 \cdot \nu_2 \cdots \nu_n$), and $C$ a constant depending on $f_3$ and $f_4$. With Eq. (38), he immediately obtained Born and Brody's main formulae for the energy content $U$ and the specific heat of the solid, namely,

$$U = \psi - T\frac{\partial\psi}{\partial T} = 3RT - CT^2 \tag{39}$$

and

$$C_v = \frac{dU}{dT} = 3R - 2CT. \tag{40}$$

These results had been found by Born and Brody only after pages of complicated calculations.

In his note Schrödinger followed this demonstration with a second part, in which he again simplified a calculation of Born and Brody, namely the evaluation of the quantum-theoretical perturbation energy terms, which must be added in the case of finite amplitude anharmonic oscillations to the standard (harmonic) energy of the solid. Born and Brody had applied a systematic quantum-theoretical perturbation scheme for multiply periodic systems—following Karl Schwarzschild's previous procedure (Schwarzschild, 1916c)—and Born had later extended that scheme with Wolfgang Pauli (Born and Pauli, 1922).[187] Schrödinger now proposed to avoid this lengthy procedure by referring to a trick that had been suggested by Niels Bohr and which made use of Paul Ehrenfest's adiabatic

[187] We have discussed the work of Born and Pauli in Volume 1, Chapter IV.2.

hypothesis.[188] For this purpose he assumed that the perturbed system (with anharmonic energy terms) is established from the unperturbed system (the solid exhibiting only harmonic oscillations) by an infinitely slow adiabatic transformation, such that the Hamiltonian function at time $t$ assumes the form

$$H_{\text{pert.}}(t) = H_0(p, t) + \frac{t}{\tau} f(p, q), \tag{41}$$

with $H_0(p, t)$ being the Hamiltonian describing the elastic vibrations and $f(p, q)$ the inelastic terms, which are fully turned on after the (large) time $\tau$ when $f(p, q)$ becomes equal to $f_3 + f_4$ (see Eq. (37)). The correction term, $\Delta E$, to the energy can then be obtained from the equation

$$\Delta E = \frac{1}{\tau} \int_0^\tau f(p, q)\, dt, \tag{42}$$

where $p$ and $q$ represent the variables of the actual system at the instant of time $t$. Schrödinger evaluated Eq. (42) for the situation considered by Born and Brody in their second paper (Born and Brody, 1921b), and found after three pages of easy calculation all the results of the previous authors, even correcting for an error in one of the main formulae.[189]

It was not Schrödinger's intention to replace (by his contribution) the methods developed by Born and Brody in Göttingen. 'The great significance of the method of Born and Brody, which has led to the highly remarkable, important and generalized investigations of Born and Pauli Jr., is not questioned, of course, by them [i.e., Schrödinger's short and clever evaluations],' Schrödinger admitted (Schrödinger, 1922b, p. 172). That is, unlike the Göttingen scientists, he did not expect to establish a new, fundamental scheme, but simply hoped to contribute a modest share to the specific questions of statistical and kinetic theory.

The paper of September 1922 showed explicitly Schrödinger's continuing interest in the problem of the specific heats of solids. It further showed that he followed rather carefully the literature in this field, which provided him once in a while with the opportunity to contribute a small piece of work improving an earlier result. 'Rarely is my word the first [answer to a question], but it is often the second, and it is stimulated by the desire to contradict or to correct, even if later the systematic continuation [of my answer] turns out to be far more essential than the contradiction which only served to start the whole thing,' as he later characterized his procedure (Schrödinger, 1935, p. 88). This characterization, to

[188] Bohr had sketched the procedure, in Section 3 of his first memoir of 1918 on the quantum theory of line spectra (Bohr, 1918a).

[189] Actually, it turned out that neither Born and Brody's original formula for the second-order correction (Born and Brody, 1921a, p. 149, Eq. (27)), nor Schrödinger's improved result were correct. (See Born and Brody, 1922a, p. 205 and Schrödinger's *Erratum* in *Z. Phys.* **11**, p. 396.)

some extent, also applied to another note on a statistical problem, which again arose from a study of the literature.

## Thermal Equilibrium between Light and Acoustic Rays

In an issue of the French journal *Annales de physique*, which appeared in early 1922, Léon Brillouin (1889–1969) had published an article with the long title '*Diffusion de la lumière et des rayons X par un corps transparent homogène—Influence de l'agitation thermique*' ('Propagation of Light and X-Rays through a Transparent Homogeneous Medium—Influence of Thermal Agitation,' Brillouin, 1922a). The Paris physicist had discussed there, on the basis of classical wave theory, the scattering of light and X-rays (of wavelength $\lambda$) at a grating established in a homogeneous isotropic medium by an acoustic wave (of wavelength $\Lambda$). He had especially shown that the electromagnetic waves were partly reflected at the fronts of the acoustic waves, provided the Bragg-like condition,

$$2\Lambda \sin\theta = \lambda\,, \tag{43}$$

was satisfied. ($\theta$ denoted the angle between the normal of the incident wave and the front of the acoustic wave.) He had further calculated the intensity of the reflected radiation, the intensity and polarization of the propagating light, and he had finally suggested testing the theoretical results with the observations from X-ray scattering, including even their temperature dependence. With the last point of his paper Brillouin had come across one of Schrödinger's old problems, as the latter had some years earlier discussed the influence of temperature on X-ray patterns obtained from regular crystals. As we have mentioned in Section II.4, Schrödinger had, early in 1914, developed a theory of the intensity of the so-called Laue spots, arriving at a somewhat different result from that of Peter Debye (Schrödinger, 1914b; Debye, 1914a). In 1922, his criticism of Debye's work had been supported by Brillouin, who claimed that his own results '*were not entirely in agreement with those that P. Debye had given* on the same problem' and that moreover, the previous author's formulae '*were not consistent with those of Rayleigh and Einstein* concerning the propagation of light' (Brillouin, 1922, pp. 120–121).[190] In other words, Debye's theory of the intensity of X-ray interference patterns was at variance with the systematic scattering theory of radiation developed by Lord Rayleigh, and especially with the description of opalescence of radiation by Albert Einstein (Einstein, 1910a).

All of the conclusions of the Paris physicist—his improved treatment of the intensity problem for X-ray interference patterns, and the connection of this problem to the problem of opalescence (a typical fluctuation phenomenon)—interested Schrödinger, who had evidently continued to follow the literature on

[190] Brillouin found that Debye had restricted himself merely to calculating the intensity of the centre of the X-ray spot and had not properly discussed the propagation of the X-rays through the crystal.

his old problems.[191] Still, his main concern shifted to a new aspect of the problem, when he responded to Brillouin's work in a note, '*Über das thermische Gleichgewicht zwischen Licht- und Schallstrahlen*' ('On the Thermal Equilibrium between Light and Acoustic Rays'), submitted in February 1924 to *Physikalische Zeitschrift* (Schrödinger, 1924b). 'For me the interest in the reflection process under consideration lies in the fact that in it two purely periodical, strictly "sinusoidal," phenomena of very different frequency interact energetically with each other,' he stated, and continued, 'From the point of view of quantum theory one must apply the quantum laws to such a process; and one may further expect that the conclusions of the quantum theory agree, at least as far as the frequencies and the directions of the rays are concerned, *exactly* with those of the classical theory—analogously with the fact that for a *harmonic* oscillator, and only for this, the frequency emitted agrees with the classical one' (Schrödinger, 1924b, p. 90). In other words, Schrödinger wished to reconsider Brillouin's problem from the point of view of quantum theory, especially by applying the light-quantum hypothesis which he had earlier used successfully (in 1922) in dealing with the Doppler effect of atomic radiation, and which Arthur Holly Compton and Peter Debye had applied (several months later) even more successfully in explaining the particular frequency shift of X-rays when scattered by electrons bound in atoms—the Compton effect. Like Brillouin, however, Schrödinger also wanted to deal with the statistical aspects of the phenomenon of interacting light and acoustic waves, aspects which had previously been discussed in the case of the Compton effect by Wolfgang Pauli, again on the basis of the light-quantum hypothesis (Pauli, 1923b).

In order to derive Brillouin's reflexion formula, Eq. (43), in quantum theory, Schrödinger introduced two quantum hypotheses: 'Both the light ray (frequency $\nu$, propagation velocity $v$ and wavelength $\lambda$) and the acoustic ray ($N$, $V$, $\Lambda$) can, being "harmonic oscillators," each absorb or emit an energy quantum, $h\nu$ or $hN$, respectively' (Schrödinger, 1924b, p. 90). The interaction between two rays then led to the result that the light ray nearly retained its energy ($h\nu' = h\nu \pm hN \approx h\nu$), but changed its momentum drastically, due to the vector addition rule. In the approximation, $\lambda' = \lambda$, Schrödinger immediately derived Brillouin's Eq. (43). He also easily evaluated the correction to this result for $\lambda' \neq \lambda$, obtaining the improved equation

$$2\Lambda \frac{\sin\theta \mp \beta}{1 - \beta^2} = \lambda\,, \tag{44}$$

where $\beta$ denotes the ratio of velocities, $V/v$.[192]

[191] We might mention at his point that Brillouin's new treatment represented a definite progress over Schrödinger's earlier work, which had been restricted—as far as rigorous results were concerned—to the one-dimensional situation.

[192] Besides Eq. (44), Schrödinger found two further equations, one for determining the angle $\theta'$ of the scattered X-ray ($\theta' \neq \theta$ for $\lambda \neq \lambda$), and the other for the frequency $\nu'$. (See the first two equations (3) in Schrödinger, 1924b, p. 91.)

In order to describe the equilibrium between optical and acoustical waves in an isotropic medium, Schrödinger investigated the number of elementary processes which occurred when the light-quanta of energy $h\nu$ collided with an $m$-quantum acoustical wave propagating towards it, and the number of inverse processes occurring when the light-quantum collided with a receding $(m-1)$-quantum acoustical wave. He assumed—similar to that which Pauli had done in the case of the interaction between electrons and light-quanta—that the number of processes in question was proportional to the numbers of light-quanta $J_\nu$ and $J_{\nu'}$ existing in the given frequency intervals and propagating in the given directions, and to the numbers of $m$-quantum or $(m-1)$-quantum acoustic waves, $N_m$ and $N_{m-1}$, respectively. He thus arrived at the equilibrium conditions

$$\frac{N_m J_\nu}{\nu^3} = \frac{N_{m-1} J_{\nu'}}{\nu'^3}; \tag{45}$$

or, by inserting the equilibrium distribution of acoustic waves at temperature $T$, i.e.,

$$\frac{N_m}{N_{m-1}} = \frac{\exp(-mhN/kT)}{\exp[-(m-1)hN/kT]} = \frac{\exp(h\nu/kT)}{\exp(h\nu'/kT)}, \tag{46}$$

at the equation

$$\frac{J_\nu \exp(h\nu/kT)}{\nu^3} = \frac{J_{\nu'} \exp(h\nu'/kT)}{\nu'^3} = \text{const.}, \tag{47}$$

where $k$ denotes the Boltzmann constant.[193]

Equation (47) was equivalent to the radiation law of Wilhelm Wien rather than that of Max Planck. Pauli had derived the same result from similar assumptions, but then he had modified the expressions for the numbers of the processes and inverse processes such that he obtained the experimentally satisfied radiation law. Schrödinger now followed the procedure of his younger colleague, which implied abandoning the complete statistical independence of light-quanta.[194] He especially added to the previous number of elementary processes, which he now called 'independent processes' ('*selbständige Prozesse*'), further 'dependent processes' ('*unselbständige Prozesse*,' Schrödinger, 1924b, p. 92)—whose ratio to the previous processes should by $J_{\nu'}/(\alpha\nu'^3)$—and similarly he augmented the independent inverse processes by appropriate dependent inverse processes, having the ratio $J_\nu/(\alpha\nu^3)$. As a consequence the new equilibrium equation replacing Eq. (47) now became

$$\frac{J_\nu \exp(h\nu/kT)}{\nu^3(1 + J_\nu/\alpha\nu^3)} = \frac{J_{\nu'} \exp(h\nu'/kT)}{\nu'^3(1 + J_{\nu'}/\alpha\nu'^3)} = \text{const.}, \tag{48}$$

[193] The independence of the constant of the temperature $T$ followed by applying Wien's displacement law. (See Schrödinger, 1924b, p. 93.)

[194] We have discussed Pauli's treatment in Volume 1, Section V.1.

which indeed agreed with Planck's law, if the constant was put equal to $\alpha$ and it assumed the value $2h/\nu$.

Having obtained the desired result, Schrödinger did not stop, however. 'The above description is rather unsatisfactory because of the many assumptions and the limited cogency,' he claimed. 'Hence we shall present still another one, which is perhaps more satisfactory and, by the way, also different in its content' (Schrödinger, 1924b, p. 93).[195] The new derivation rested on the 'actually already well-known derivation of the radiation law by "quantization of an ether block" due to Jeans, Debye and Rubinowicz, the importance of which has again recently been brought to light by W. Bothe' (Schrödinger, 1924b, p. 93).

The method of dealing with the 'ether block' had been first suggested by James Hopwood Jeans (1877-1946) in 1905 for the purpose of establishing the classical Rayleigh–Jeans radiation law (Jeans, 1905b), and had been extended by Peter Debye in 1910 to quantum theory (Debye, 1910).[196] After the discovery of the Compton effect Walther Bothe (1891-1957) became interested in the problem of light-quanta. Especially in a paper on '*Die räumliche Energieverteilung in der Hohlraumstrahlung*' ('The Spatial Energy Distribution in Cavity Radiation'), which was received on 23 October 1923 by the *Zeitschrift für Physik* and published a few weeks later, he revived the idea of multiple quanta and used it when dealing with cavity radiation (Bothe, 1923e).[197] He thus arrived, in a straightforward way, at Planck's law and the fluctuation expression obtained by Einstein (Einstein, 1909a), provided he took into account 'the tested general foundations of quantum theory' including 'the "quantization of the pure cavity radiation" according to Debye and Rubinowicz' (Bothe, 1923e, p. 151). Schrödinger extended the method of his predecessors slightly, when he wrote: 'On account of this conception one must not assume that in a bundle of light rays the individual light-quanta only propagate side by side, but that they are also gathered in a particular way to form $1, 2, 3, \ldots$-quantum "rays"; we imagine that it also makes sense to talk about a "zero-quantum ray," so to speak, an ether resonator, which *can* oscillate, but is momentarily in a *non-excited* state' (Schrödinger, 1924b, p. 93).

With the new assumption he again looked at the equilibrium problem of the elementary processes determining the interaction between light and acoustic waves. In particular, he now considered the passage of a quantum of radiation $h\nu$ from an $l$-quantum light ray in a certain direction to an $n$-quantum light ray of another given direction under the action of an $m$-quantum acoustic wave; and the inverse process, in which the quantum $h\nu$ was transferred from an $(n+1)$-quantum ray to an $(l-1)$-quantum ray under the action of an

[195] Schrödinger remarked, in the text and in a footnote (on p. 92), that neither his *Ansatz* leading to Eq. (48) was cogent, nor was Pauli's leading to the same result.

[196] Adalbert Rubinowicz (1889–1974) had repeated Debye's derivation of Planck's law in 1917, having based his approach on the quantization method for systems which had several degrees of freedom, developed earlier by Arnold Sommerfeld and Max Planck (Rubinowicz, 1917a).

[197] We have mentioned in Section II.3 that Mieczyslaw Wolfke had already earlier made use of the idea of multiple quanta.

$(m-1)$-quantum wave. By demanding thermal equilibrium (at temperature $T$), he obtained the number of $l$-quantum light rays in a given volume ($\Omega$), i.e.,

$$n_l = \frac{2\pi\Omega \exp(-lh\nu/kT)}{v^3}\left[1 - \exp\left(-\frac{h\nu}{kT}\right)\right], \tag{49}$$

where $v$ denotes the velocity of the light wave in the medium considered. From Eq. (49) he immediately derived Planck's law for the energy density of blackbody radiation, provided he summed over 0, 1, 2, . . .-quantum rays. In comparing the new derivation with the previous one, Schrödinger noted explicitly the following interpretation: 'To the processes which we distinguished ... as "independent" and "dependent" processes, respectively, there correspond, according to the conception of this section, transitions into a previously zero-quantum ray on the one hand, and into an arbitrary positive ($n > 0$)-quantum ray on the other hand' (Schrödinger, 1924b, p. 94).

## Outlines of a Book on 'Molecular Statistics' and Related Unpublished Notes

The three papers discussed so far arose from rather different stimulations. Still, they formed part of Schrödinger's strong involvement at that time in the general statistical description of matter. One can even say that they indicated, together with a few other publications—which we shall mention later—only the visible tip of the iceberg of Schrödinger's very deep-rooted interests in that field. From the unpublished notes we learn that he planned, e.g., to write a comprehensive text on the kinetic theory of matter, as there already exists a draft of the detailed organization for a book on '*Molekularstatistik*' ('Molecular Statistics').[198] Schrödinger divided the envisaged material into two parts, which he called, respectively: 'First Book: Theories, which assume only continuous changes of the models; Second Book: Theories, which also assume jump-like, non-mechanical changes of the models.' In the first book, dealing obviously with the classical statistics of molecules, he introduced six chapters devoted to: an introduction (Chapter I); the thermodynamical equilibrium between two arbitrary mechanical systems (Chapter II); applications of canonical distribution (Chapter III); general thermal properties of the mechanical models (Chapter IV); the theory of Brownian motion and of diffusion (Chapter V); and the theory of osmotic pressure (Chapter VI). The second book was intended to give a full account of the available quantum-theoretical statistical methods in eight chapters; after an introduction outlining briefly the existing principles of quantum theory of multiply periodic systems (Chapter VII), there followed a treatment of the equilibrium of linear

[198] The outline of the chapters and sections of the book run over 17 pages, which have been filed on AHQP Microfilm No. 40, Section 1. Like many, or most of the manuscripts, this draft is also undated. From the detailed titles of chapters and sections, however, one may conclude that the outline had been written between 1922 and early 1925, most probably not later than 1924.

oscillator systems (Chapter VIII) and of arbitrary systems (Chapter IX), then a discussion of the theory of solids (Chapter X) and of gases (Chapter XI), and finally chapters on gas dissociation and vaporization (Chapter XII), on gas degeneracy and the theory of the chemical constant (Chapter XIII), and on the theory of heat radiation (Chapter XIV).

For most chapters of the planned book Schrödinger also gave the titles of the individual sections. Thus he organized, e.g., the material of Chapter III (in the classical part of the book, entitled '*Anwendungen der kanonischen Verteilung*' ('Applications of the canonical distribution') into nine sections, including the following: 5. Langevin's theory of paramagnetism; 6. Weiss' theory of ferromagnetism; 7. Anomalies of the specific heats of ferromagnetic substances; 8. Debye's theory of dielectric substances; 9. Deviations from the ideal gas state—Van der Waals' equation of state. In Book 2 on quantum statistics he worked out the details of the organization of Chapters VII to X giving, e.g., the following details for Chapter VIII, entitled '*Gleichgewichte linearer Oszillatorensysteme*' ('Equilibrium of linear oscillator systems'): 1. $N$ linear oscillators in energy exchange. (Most probable distribution.) Comparison with the classical result. Prospect of an explanation of the thermal behaviour at low temperatures; 2. Most probable energy distribution between two oscillator systems of different excitation. Interpretation of a parameter as temperature; 3. Calculation of the temperature from a hypothesis about the entropy. Connection between entropy and probability; 4. Retrospective view of an analogous connection in the classical theory; 5. Investigation of the exchange. Translation into a "box problem." Average distribution numbers; 6. Proof of the smallness of fluctuation for certain conditions; 7. Insensitiveness of the entropy function; improvement of the definition of entropy. He gave a similar detailed structure to Chapter X on solid state, in which he introduced, for example, as Section 9 the topic 'High temperatures. Second approximation,' indicating in parentheses: 'belongs more to the classical theory' ('*gehört mehr in die klassische Theorie*'). He had shown exactly that point in his paper of September 1922, dealing with the specific heats of solids at high temperatures and the results of Born and Brody (Schrödinger, 1922b).

Besides the draft of the chapters and sections of the book on molecular statistics several notes and memoranda have remained and which stem from the same period, i.e., had been written roughly between 1923 and 1925.[199] Some represent incomplete parts of a larger scheduled manuscript, e.g., a set of two pages, entitled '*Versuch, Wechselwirkungen zu berücksichtigen*' (Attempt to take interactions into account), namely interactions between the molecules of a gas, whose statistics one is dealing with; or three pages of an '*Einführung der relativistischen Mechanik und magnetischer Kräfte in die Statistik eines Systems von Massenpunkten*' ('Introduction of relativistic mechanics and of magnetic forces into the statistics

[199] All of these manuscripts are undated; however, from the references given in some of them, and from some dates given on papers and sheets put into the manuscripts, one may get some idea of the time of their origin.

of a system of mass points').[200] In the latter manuscript Schrödinger examined the influence of magnetic fields on the statistical properties of relativistic particles.[201] He showed especially that for electrons in electric and magnetic fields—such as those that existed in atoms when exposed to external fields—the canonical equations of motion remained valid.[202] He also derived an expression for the velocity distribution law, namely

$$f(v)\, d\Omega = \beta \exp\left( -\frac{E_{\text{kin}} + U}{kT} \right) d\Omega\,, \tag{50}$$

where $\beta$ denoted the ratio of the velocity of the electrons $v$ to $c$, the velocity of light of *in vacuo*, $E_{\text{kin}}$ and $U$ their kinetic energy and *electrical* potential energy, respectively. The magnetic field only entered the phase-space volume element $d\Omega$ $(= dx\, dy\, dz\, dp_x\, dp_y\, dp_z)$ through the momentum components, i.e.,

$$p_x = m\dot{x} + \frac{e}{c} A_x\,, \tag{51}$$

where $A_x$ is the $x$-component of the magnetic vector potential, and $m$ and $e$ are the mass and charge of the electron.

More extensive manuscripts dealt, e.g., with the chemical constant and gas degeneracy,[203] with the quantum statistical method of Charles Galton Darwin and Ralph Fowler,[204] or they contained fairly detailed lecture notes on quantum statistics. Thus a notebook of 46 pages on a '*Vorlesung über Quantenstatistik*' ('Lecture on Quantum Statistics') was devoted to the following topics: definition of entropy; arbitrary entropy function and weights (*Beliebige Entropiedefinition und Gewichte*); knowledge of energy states (*Kenntnis der Energiestufen*); several degrees of freedom, periodic motion ('*Mehrere Freiheitsgrade, periodische Bewegung*'); conditionally periodic systems ('*Bedingt periodische Systeme*'); general theory of specific heats, taking into account the quadratic terms ('*Allgemeine*

[200] Both manuscripts have been filed on AHQP Microfilm No. 40, Section 1. The second manuscript must be dated after early April 1924, because it refers to a paper published in the 1 April 1924 issue of the *Physikalische Zeitschrift.*

[201] Erich Kretschmann, in an earlier paper, had doubted the general applicability of Liouville's theorem to ideal gases of relativistic atoms (Kretschmann, 1920); in 1924 he found, however, that it was nevertheless exactly valid for relativistic electrons if under the action of electric and magnetic fields (Kretschmann, 1924).

[202] Hence Schrödinger confirmed Kretschmann's result of 1924, quoted in Footnote 201.

[203] Later we shall return, in more detail, to the manuscript entitled '*Chemische Konstante und Gasentartung. II*'.

[204] Only five pages have remained of an English manuscript 'On Quantum Statistics,' filed on AHQP Microfilm No. 40, Section 3. Schrödinger tried 'to obtain the same purification of arguments [including especially the replacement of the unallowed use of Stirling's formula in the quantum statistical calculation by the method of steepest descent] by a more straightforward way of reasoning' (loc. cit., p. 1).

*Theorie der spezifischen Wärme unter Rücksicht auf die quadratischen Glieder*'); and gases ('*Gase*,' i.e., their rotational and vibrational heats).[205]

## Handbook Article on Specific Heats and a Scattered Publication

One may wonder why these efforts did not result in Schrödinger's writing one of the handbook articles on the kinetic theory of matter, classical statistics or quantum statistics. Indeed his plans and outlines agreed rather well, if not coincided, with some of the reviews given by his contemporaries in the physics handbooks and encyclopedias which were published in the 1920s. For example, what Schrödinger sketched in his notebook on quantum statistics showed a considerable similarity with what Adolf Smekal presented, under the title '*Allgemeine Grundlagen der Quantenstatistik und Quantentheorie*' ('General Foundations of Quantum Statistics and Quantum Theory'), for the *Encyklopädie der mathematischen Wissenschaften* (Smekal, 1926b).[206] On the other hand, many topics in Schrödinger's detailed outline of the chapters and sections of a proposed book on '*Molekularstatistik*'—which we have mentioned earlier—were discussed in a review article, which his other Viennese fellow-countryman Karl Herzfeld had composed for the third volume of *Müller-Pouillet's Lehrbuch der Physik* (Herzfeld, 1925).[207] It seems that all available requests for handbook articles had

[205] A further manuscript contains an outline of a lecture course or a review article on '*Quantenstatistik*,' mentioning the topics: '1. *Klassische Statistik* (*Allgemeine Mechanik, Liouvilles Theorem, Gastheorie, Festkörper, Strahlung, Schwankungen, Osmotischer Druck verdünnter und konzentrierter Lösungen, Ausdehnung der Mechanik auf den Fall eines Magnetfeldes*); 2. *Quantenstatistik*: *Grundlagen der Quantentheorie, Resonatorengleichgewicht, A priori-Gewicht, Strahlungstheorie, Spezifische Wärme der Gase* (*Rotationswärme, Schwingungswärme*), *Spezifische Wärme der Festkörper, Gasgleichgewichte* (*Dissoziation*), *Chemische Konstante, Dampfdruckkurve, Gasentartung, Intensität der Spektrallinien, besonders Bandenlinien, Schwankungen, Gleichgewicht zwischen Elektronen und Strahlung* (*Komptontheorie*), *Dipoltheorie der Dielektrika, Paramagnetismus* (*Curiesches Gesetz*), *Ferromagnetismus, Offene Fragen*.' There follow sketches of details, e.g., five pages on the foundations of exchange statistics (i.e., the statistical method dealing, e.g., with the exchange of radiation quanta between molecules); ten pages on proving the identification of the most probable and the average distribution numbers; four pages on fluctuation, etc. Altogether these notes extend over 100 pages (filed on AHQP Microfilm No. 40, Section 3, Subsections 2–8). Several sheets, letters and notes, which Schrödinger had kept with the manuscript, show dates between October 1923 and October 1924.

[206] See the previous footnote for the outline of Schrödinger's quantum statistics. Smekal's article is organized into three parts. Part I covers '*Die Entwicklung der klassischen statistischen Mechanik zur Quantenstatistik*' ('The Development of Classical Statistical Mechanics Towards Quantum Statistics'), Part II the '*Allgemeinen Grundlagen der Quantentheorie*' ('General Foundations of Quantum Theory'), and Part III '*Spezielle Anwendungen der Quantenstatistik*' ('Special Applications of Quantum Statistics'). In particular, the contents of Parts I and III agree fairly well with those discussed by Schrödinger in his outlines of quantum statistics and in his lecture notes on the same topic.

[207] Herzfeld's article is organized in seven chapters: Chapter I deals with '*Kinetische Theorie der Gase in elementarer Darstellung*' ('An Elementary Presentation of the Kinetic Theory of Gases'), Chapter II with '*Allgemeine Statistische Mechanik*' ('General Statistical Mechanics'), Chapter III with '*Gase*' ('Gases'), Chapter IV with '*Der feste Körper*' ('Solid Body'), Chapter V with '*Theorie der Flüssigkeiten*' ('Theory of Fluids'), Chapter VI with '*Theorie der Lösungen*' ('Theory of Solutions') and Chapter VII with '*Theorie der Schwankungen*' ('Theory of Fluctuations').

The similarities of the topics treated by Herzfeld and those which Schrödinger planned to cover, in the first book ('*Erstes Buch*') on classical molecular statistics, become even more evident by examining the details in Herzfeld's article for *Müller-Pouillet's Lehrbuch der Physik*.

already been sent out to other authors, and Schrödinger was indeed forgotten. Still, he possibly considered writing a separate book on kinetic theory and quantum statistics, but evidently did not find the time to do so. Thus the only larger publication, which resulted from his efforts with these topics, was a review article on specific heats, entitled '*Spezifische Wärme (theoretischer Teil)*' ('Specific Heats, Theoretical Part') for the *Handbuch der Physik* (Schrödinger, 1926k).[208]

Schrödinger had already published two articles on the problem of specific heats in his Vienna days (Schrödinger, 1917a, 1919b).[209] Especially, the second article in *Physikalische Zeitschrift* constituted a veritable review devoted to the specific heats of solids, and the new article which was contributed to the *Handbuch der Physik* could partly be considered as a modernized extension of this review. The additional parts, namely 25 pages (constituting about half of the article) treated the problem of the specific heats of gases; here Schrödinger discussed the interaction energy of gases (Section 2), the partition function (Section 3), the molecular energy of gases (Section 2), the partition function (Section 3), the molecular models of Bohr theory (Section 4), the separation of the terms (rotational, vibrational, electronic) in the partition function (Section 5), the classical theory of specific heats (Section 6), the quantum theory of the vibrational (Section 7), of the rotational (Section 8) and of the electronic heats (Section 9), the excited states of the molecules (Section 10), the refinement of the molecular models (Section 11), and finally the problem of quantizing translational motion and gas degeneracy (Section 12). Schrödinger took into account the literature published up to the summer of 1925, also referring to a recent paper of his own on the specific heat of the hydrogen molecule (Schrödinger, 1924f).[210]

Schrödinger's published papers on selected problems of quantum statistics appear as a rather restricted, even modest, excerpt from the large and ambitious programme indicated in the unpublished notes and manuscripts. He submitted in 1924 two papers to *Zeitschrift für Physik* dealing with the question of specific heats. The first one, a two-page note entitled '*Bemerkungen zu zwei Arbeiten des Herrn Elemér Császár über Strahlungstheorie und spezifische Wärme*' ('Remarks on two Papers of Mr. Elemér Császár on Radiation Theory and Specific Heats'), was received by the *Zeitschrift* on 20 May and published in the issue of 2 July 1924 (Schrödinger, 1924c); the second, '*Über die Rotationswärme des Wasserstoffs*' ('On the Rotational Heat of Hydrogen'), was received on 24 November and published in the issue of 31 December 1924 (Schrödinger, 1924f). Both owed their origin to the author's careful reading of the scientific literature.

[208] This article constituted a contribution to Volume X of the *Handbuch*, edited by Hans Geiger and Karl Scheel, which appeared in 1926. The experimental aspects of specific heats were treated by Karl Scheel; other articles in the same volume were: '*Zustand des festen Körpers*' ('State of the Solid Body') by E. Grüneisen, '*Zustand der gasförmigen und flüssigen Körper*' ('State of the Gaseous and Fluid Bodies') by J. D. van der Waals Jr., and '*Die Bestimmung der freien Energie*' ('The Determination of the Free Energy') by F. Simon.

[209] We have discussed these articles in Section I.5.

[210] From the quoted literature one concludes that Schrödinger finished the *Handbuch* article essentially in the summer of 1925.

Császár of the University of Budapest had in two papers, published in April and November 1923 in *Zeitschrift für Physik*, discussed a derivation of Planck's radiation law and its application—according to Peter Debye's theory—to the specific heats of solids (Császár, 1923a, b). The Hungarian author had proposed for that purpose to use a modification of Planck's quantum hypothesis, namely: 'The absorption occurs in a completely continuous way; the same applies to the emission, except however in the energy region between 0 and $(n+1)\varepsilon$—including the value $(n+1)\varepsilon$—in which region an oscillator can emit energy only at such a moment, when its energy has become an integral multiple of the energy-quantum $\varepsilon = h\nu$; in the case of emission the oscillator should emit its entire energy' (Császár, 1923a, p. 345).[211] He had thus arrived at a slight change in Planck's formula, replacing the expression of the energy density $E_\lambda$ for the wavelength $\lambda$ at temperature $T$ by

$$E_\lambda^{(n)} = \frac{c_1}{\lambda^5} \frac{1}{\exp(c_2/\lambda T) - 1} \cdot (1 + \alpha), \tag{52}$$

with

$$\alpha = \frac{\exp(c_2/\lambda T) - c_2/\lambda T - 1}{(c_2/\lambda T)\exp\{(n+1)c_2/\lambda T\}}. \tag{52a}$$

(The constants $c_1$ and $c_2$ are the original constants of Planck's law, and $n$ is an integer.)

Equation (52) evidently coincided with Planck's formula for the special value $n = \infty$, while it passed for $n = -1$ into the classical formula of Rayleigh and Jeans. It opened especially the opportunity to account for small deviations from Planck's law, if such would be detected in future.[212]

In the second paper Császár had derived, on the basis of the same quantum hypothesis, a modification of Peter Debye's formula for the specific heats of solids (Császár, 1923b). He had obtained, in particular, two formulae by taking $n = 0$ and 1, respectively (Eqs. (I) and (II) on pp. 216 and 217), and had shown, by comparing with the experimental data, that the new formula (II) allowed one to describe the observations 'in any case as well as Debye's formula' ('*in jeden Falle so gut wie die Debyesche*'), while in some cases 'formula (I) offers some advantage' ('*zeigt die Formel (I) einige Vorteile*,' Császár, 1923b, p. 220). From these results he concluded that 'only those atoms behave in a discontinuous manner, which are in such a state as assumed by all atoms when the entire [solid] body is cooled to nearly absolute zero temperature; the others follow the rules of classical physics' (Császár, 1923b, p. 220).

[211] Evidently, Császár went (with this quantum hypothesis) beyond Planck's second hypothesis, which assumed continuous absorption of radiation energy by the oscillating electrical system, but discontinuous emission for *all* energy regions (Planck, 1911a).

[212] For $n$ finite, but large, a small deviation can be properly described by Eq. (52).

Such a strong statement aroused Schrödinger's interest, who was concerned with the foundations of classical and quantum statistics. Császár's result appeared to him 'highly strange' ('*höchst befremdlich*'), hence he tried to test whether or not it agreed with the 'usually stated fundamental assumptions of statistical mechanics' (Schrödinger, 1924c, p. 173). He arrived at the following conclusions. First, he demonstrated that the energy distribution of the oscillator system given by Császár did not correspond to definite statistical weights for the individual states, but that his ratio of weights for the discontinuous and the continuous parts of the energy spectrum depended on the temperature, which obviously contradicted the principles of statistical mechanics. Second, he argued that the formula for the specific heats of solids following from Császár's quantum hypothesis was more complicated than the latter's expressions (I) or (II); hence, he held 'the quantitative agreement with Debye's formula, i.e., with experience, . . . to be improbable' (Schrödinger, 1924c, p. 174).[213]

## Specific Heat of the Hydrogen Molecule

The second paper on specific heats, published in 1924, dealt with the hydrogen molecule. It again arose from Schrödinger's general concern with kinetic theory and quantum statistics.[214] On 19 November 1924 Schrödinger, who had previously received a complimentary copy of Arnold Sommerfeld's new, fourth edition of *Atombau und Spektrallinien* (Sommerfeld, 1924d), wrote to the author:

> On page 737 I found the *Motto* for a small paper which I just planned to submit: rotational heat of $H_2$ with half-integral quanta for molecular rotation . . . . The result is that the agreement with Eucken's data is almost as good or as bad, as in the case of integral quanta, and that the moment of inertia is reduced by a factor *3:2* to about $1.4 \times 10^{-41}$ [c.g.s. units], which appears to me to be agreeable, because the moment calculated from the many-line spectrum [of molecular hydrogen], about $1.8 \times 10^{-41}$, should be *larger*, not *smaller* than the theoretical value because it corresponds to the excited state. (Schrödinger to Sommerfeld, 19 November 1924)

[213]Schrödinger explicitly calculated the energy content of a solid on the basis of Császár's hypothesis—or a slight modification of it—in his unpublished manuscript on '*Chemische Konstante und Gasentartung. II*'. The complicated expression which he obtained led him to remark: 'The calculations of Mr. Császár are wrong, he gets incorrect distribution numbers' (AHQP Microfilm No. 40, Section 1).

Császár replied to Schrödinger's criticism in a further paper, '*Die statistische Verteilungsfunktion in der Strahlungs- und Atomwärmetheorie*'; he especially claimed that Schrödinger had not evaluated the consequences from his hypothesis correctly, and he maintained the claim of good agreement between his formulae (for the blackbody radiation and the specific heats of solids) and the experimental data (Császár, 1925).

[214]Schrödinger treated the specific heats of gases, for example, in his '*Vorlesung über Quantenstatistik*' mentioned earlier. (See, especially p. 45 of the manuscript, filed on AHQP Microfilm No. 40, Section 2.) By comparing the text of the lectures with that of the article of November 1924, one may conclude that the lecture notes were written *before* Schrödinger obtained the results of the published paper; hence the latest date for the lecture course would be the winter semester of 1924.

Ever since the first experimental observation of the specific heat of [molecular] hydrogen at low temperatures by Arnold Eucken (Eucken, 1912), the theoretical explanation of the results had presented difficulties.[215] Especially, the dumb-bell model of Bohr, with two electrons rotating in a central plane perpendicular to the line joining the two hydrogen nuclei, had failed to solve the problem, whether one assumed total angular momentum to be 0 or 1 (in units of $h/2\pi$). Then Adolf Kratzer had suggested, in connection with his detailed analysis of molecular spectra, the possibility of half-integral quantum numbers (Kratzer, 1922). Sommerfeld had referred exactly to this possibility in the footnote of his book mentioned by Schrödinger; he had stated: 'Possibly the smallest value [of the rotational quantum number] is $m = \frac{1}{2}$. Hence one would also have to calculate the specific heat of rotation of the $H_2$-molecule in general by using the half-integral $m$' (Sommerfeld, 1924d, p. 737, footnote 1). Schrödinger picked up the idea and sent off his paper on the rotational heat of $H_2$ a few days later (Schrödinger, 1924f).

He started from a model of the hydrogen molecule in which the system of the two nuclei performs an ordinary rotation about the axis with half-integral quantum number $(2n - 1)/n$, $n$ being an integer. Hence he obtained the (quantized) energy values for the rotation $E_{\text{rot}}$,

$$E_{\text{rot}} = \frac{(2n-1)^2}{4}\frac{h^2}{8\pi^2 J}, \tag{53}$$

where $J$ denoted the moment of inertia of the molecule. Each particular value of $E_{\text{rot}}$, Eq. (53), corresponded to two dynamically different states of the molecule; for example, the lowest value $(\frac{1}{4}h^2/8\pi^2 J)$ to situations in which the nuclear angular momentum $(\frac{1}{2}h/2\pi)$ was parallel or antiparallel to the angular momentum of the electron (also $\frac{1}{2}h/2\pi$). Schrödinger then associated with each energy state $E_{\text{rot}}^{(n)}$ a particular statistical weight, $g_1, g_2, g_3, \ldots$, and found the quantum-theoretical '*Zustandssumme*' $\Sigma$ to be (with $x' = h^2/32\pi^2 JkT$)

$$\Sigma = g_1 \exp(-x') + g_2 \exp(-9x') + \cdots + g_n \exp\{-(2n-1)^2 x'\} + \cdots. \tag{54}$$

From $\Sigma$ he calculated the partition function ($\psi = R \ln \Sigma$) and with it the rotational term of the specific heat, $C_{\text{rot}}\,( = (\partial/\partial T)(T^2(\partial\psi/\partial T))$, i.e., up to the third term,

$$\frac{C_{\text{rot}}}{R} = \frac{g_2}{g_1}x^2\exp(-x)\frac{1 + 9\dfrac{g_3}{g_2}\exp(-2x) + \dfrac{4g_3}{g_1}\exp(-3x)}{\left[1 + \dfrac{g_2}{g_1}\exp(-x) + \dfrac{g_3}{g_1}\exp(-3x)\right]^2}, \tag{55}$$

with $x = 8x'\,( = h^2/4\pi^2 JkT)$. 'The form [of the dependence of $C_{\text{rot}}$ on the temperature] depends, of course, essentially on the weight numbers $g_1, g_2 \ldots$,' Schrödinger concluded and continued: 'The theoretical requirements are however at this point [i.e., with respect to a determination of the $g_i$], as is known, so

[215] We have discussed the early history of the problem in Volume, Section I.7.

uncertain that we rather tried to fit the weight ratios to the observed data as well as possible, as has already been proposed by Eucken' (Schrödinger, 1924f, p. 344).

In detail Schrödinger found, by using the low temperature data (between 35° and 110° absolute), the value 565 K for the characteristic temperature $\Theta(=h^2/4\pi^2 Jk)$, and, by also fitting the higher temperature data, the following ratios for the statistical weights: $g_1:g_2:g_3 = 1:2:4$. Another, similarly good description he obtained with $\Theta = 544$ K and the ratios $g_1:g_2:g_3 = 4:7:17$. The moments of inertia derived from these $\Theta$-values were $J = 1.43 \times 10^{-41}$ and $1.48 \times 10^{-41}$ c.g.s. units, respectively. They seemed to agree fairly well with the values derived from different experiments. Thus, from the data on the vapour-pressure of hydrogen, $J$ was determined to have the value $1.43 \times 10^{-41}$ or even lower (Eucken, 1920; Eucken, Karwat and Fried, 1924). Wilhelm Lenz, on the other hand, had deduced the higher value $J' = 1.85 \times 10^{-41}$ by analyzing the many-line spectrum of hydrogen (Lenz, 1919); also, this value did not contradict Schrödinger's because, as he noticed '$J'$ is now associated with the *excited* molecule, and one should expect $J' > J$, not $J' < J$, since in the excited state the distance of the [hydrogen] nuclei will increase rather than decrease' (Schrödinger, 1924f, p. 347). This result evidently favoured the model with half-integral quantum numbers for the rotation over the previous model treated with integral quantum numbers; the latter yielded $J$-values larger than Lenz' $J'$-value.[216]

Thus far the result of Schrödinger's new calculation appeared to improve agreement with data. However, an important problem of the theory remained to be solved. As Schrödinger concluded: 'I have not been able to provide a reasonable theoretical foundation for the weight sequences 1:2:4 or 4:7:17' (Schrödinger, 1924f, p. 347). On theoretical grounds one would expect a greater step from $g_1$ to $g_2$ and a smaller one from $g_2$ to $g_3$, but Schrödinger also recalled the fact that 'exactly the weight ratios of the *lowest* quantum states belong to those properties, about which today we do not have any knowledge founded on experiment, hence the theoretical conclusions in this field are certainly much less secure than for higher levels, where the correspondence argument settles the situation practically in a unique way' (Schrödinger, 1924f, p. 347).

The paper on the specific heat of molecular hydrogen exhibits many of the features characterizing the work of Schrödinger: he again took a well-known theoretical model and thoroughly discussed the consequences from a definite (new) assumption; he then compared the theoretical results in the most suitable way—i.e., in a way in which the assumed hypothesis could be submitted to the sharpest test—with the available data (a method usually followed by critical experimentalists when checking a theory); thus, he derived conclusions about the validity of a theory which were as reliable as possible in the situation considered. In addition to these specific features, Schrödinger's attempt to deal with his topics in an encyclopedia-like fashion also becomes obvious, including

[216] For example, Fritz Reiche had calculated in 1918 $J$-values between 2.095 and $2.214 \times 10^{-41}$ for the hydrogen model with integral quantum numbers for the rotation (Reiche, 1919a).

a discussion of all the relevant experimental and theoretical results available in the literature.[217]

## The Problem of Gas Degeneracy

The above paper belongs to the works which might also constitute part of a handbook article.[218] The same might be said about the article on '*Gasentartung und freie Weglänge*' ('Gas Degeneracy and the Mean Free Path'), which Schrödinger had submitted about a year before, i.e., in December 1923, to the *Physikalische Zeitschrift* and which was published in the issue of 15 January 1924 (Schrödinger, 1924a). The problem treated by Schrödinger again belongs among the topics that he mentioned explicitly in the outline of his planned book on '*Molekularstatistik*,' where he introduced, as Chapter XIII: '*Die Frage der Quantelung der Translationsbewegung* (*Gastentartung*). *Theorie der chemischen Konstante*' ('The Problem of the Quantization of the Translational Motion (Gas Degeneracy). Theory of the Chemical Constant').[219] Schrödinger gave some details of this chapter in his manuscript '*Chemische Konstante und Gasentartung. II*' ('Chemical Constant and Gas Degeneracy. II').[220] Especially, he devoted a few pages there to the problem of gas degeneracy while he treated, in other parts, mainly the problems involved in the statistical counting of identical particles, on which he would publish another paper in July 1925 (Schrödinger, 1925d).

The connection between the two problems, the chemical constant and gas degeneracy, might appear on first inspection a little surprising. Yet, historically, both topics originated from the same source, namely the so-called third law of thermodynamics or third heat theorem. Walther Nernst (1864–1941), the inventor of the theorem, had discussed many of its consequences in a lecture '*Über neuere Probleme der Wärmetheorie*' ('On More Recent Problems of Heat Theory'), presented on 26 January 1911 to the Prussian Academy of Sciences (Nernst, 1911a). He pointed out for the first time in that lecture the necessity of quantizing,

[217] Schrödinger then added a note, dated 19 December 1924, to the proofs of his paper, in which he also discussed the results of a report published by Herbert Stanley Allen of the University of St. Andrews, Scotland, in the July issue of the *Proceedings of the Royal Society of London* (Allen, 1924). Allen had studied the regularities of the many-lined spectrum of hydrogen and had derived, from the so-called Fulcher bands, the following values for the moments of inertia: $J'_a = 1.761 \times 10^{-41}$ for the initial state, and $J'_e = 1.827 \times 10^{-41}$ for the final state. While Schrödinger happily pointed out that both values were bigger than his $J$-value for the ground state of the hydrogen molecule, he also realized: 'Yet it remains strange that ... all ordered series [analyzed by Allen] can be represented as transitions from a *smaller* to a *larger* moment of inertia (in emission), which might possibly shake the plausibility argument: $J'$ (excited) $> J$ (normal).'

[218] We might mention at this point that Schrödinger's approach in 1924 did not finally solve the specific heat problem of molecular hydrogen; one still had to wait for the advent of the new quantum mechanics, *and* also include the spin of the hydrogen nuclei plus the relative distribution of ortho- and para-hydrogen at different temperatures (Dennison, 1926, 1927b).

[219] See the manuscript filed on AHQP Microfilm No. 40, Section 1.

[220] This manuscript has been filed on AHQP Microfilm 40, Section 1. The first part, '*Chemische Konstante und Gasentartung. I*' has not been filed (perhaps Schrödinger did not keep it).

besides the vibrations, the rotations of di- and poly-atomic gas molecules also. Finally, he addressed the problem of the translational motion of gas molecules. 'The usual kinetic theory of gases operates with velocities directed along straight lines, notably with values occurring in ballistics and known—even with a larger order of magnitude—from astronomy,' Nernst had said, and added: 'One cannot speak here of an extrapolation of the laws of mechanics; such an extrapolation would make sense only for enormously highly heated gases, and, if the straight line velocity of gas molecules then becomes comparable to the velocity of light [*in vacuo*], then also in this situation a deviation from the laws of mechanics should be expected with very high probability, and hence a breakdown of the present kinetic gas theory' (Nernst, 1911a, p. 87).

In Nernst's opinion the expected deviations could be described by applying the quantum hypothesis to the translational motion of gas molecules. Hence his Berlin talk of early 1911 immediately stimulated not only the development of a quantum theory of molecular rotations (and an explanation of the infrared spectra of diatomic molecules: Bjerrum, 1912), but also attempts at a quantization of the translational motion of atoms, molecules and electrons.[221] The main result of these attempts were the theory of the so-called chemical constant, developed between 1911 and 1919 by Otto Sackur, Hugo Tetrode and Otto Stern, and several theories of ideal gases at low temperatures which exhibited deviations from the classical formula.[222] Fritz Reiche, Schrödinger's successor in the chair of theoretical physics at Breslau, had given a fine summary of these results in his book *Die Quantentheorie*: *Ihr Ursprung and ihre Entwicklung* (Reiche, 1921; English translation: *The Quantum Theory*, 1922), notably in chapter V, entitled '*Das Eindringen der Quanten in die Gastheorie*' ('The Intrusion of Quanta into the Theory of Gases'). There he discussed the problem of gas degeneracy in Section 3 ('*Die Entartung der Gase*,' 'The Degeneracy of Gases'), after a section on the rotational heat of diatomic gases (Section 1) and another one on the infrared rotational spectra (Section 2), and before Section 4 on the chemical constant ('*Die chemische Konstante einatomiger Gase*,' 'The Chemical Constant Monatomic Gases').

Reiche had emphasized that 'the attempts to apply it [i.e., the quantum theory] to the translational energy of gases rest upon a much more insecure basis [than the attempts at a quantization of the rotational energy]'; and also that this step was necessary to remove 'the hitherto exceptional position occupied by

[221] We have discussed this extension of the quantum theory in Volume 1, Section I.7.

[222] The classical equation of state for the ideal gas contradicted Nernst's heat theorem because it yielded, for the expression $dA/dT$ (with $A$ denoting the work of expansion of a gas from volume $V_1$ to $V_2$ and $T$ the temperature)

$$d\int_{V_1}^{V_2} \frac{p\,dV}{dT} = R \ln \frac{V_2}{V_1}$$

($R$ being the universal gas constant). Evidently this result did not go to zero for $T \to 0$.

monatomic gases, whose molecules contain only translational energy . . . , for they, too, must succumb to the quantum law' (English translation, p. 79). He had then displayed the main aspects and results of the quantum theories of ideal monatomic gases proposed by Otto Sackur, Hugo Tetrode, Willem Hendrik Keesom, Wilhelm Lenz and Arnold Sommerfeld, Paul Scherrer, Max Planck and Walther Nernst.

In December 1923 Schrödinger joined in at essentially this stage of development. Citing Reiche's book as the source of reference to the earlier literature (see Schrödinger, 1924a, p. 41, footnote 1, left column), he first stressed the central role played by a characteristic temperature $\Theta$—similar to the one occurring in Debye's theory of the specific heats of solids—in all the previous discussions of gas degeneracy. This temperature was given by the relation

$$\Theta = \frac{h^2}{8ml^2k}, \tag{56}$$

where $h$ and $k$ denoted Planck's and Boltzmann's constant, respectively, $m$ is the mass of a gas molecule, and $l$ a characteristic length.

The above-mentioned theories now split into two different classes, in which $l$ and $\Theta$ assumed very different orders of magnitude. Thus the theories of Tetrode, Keesom, Sommerfeld and Lenz, and Nernst yielded $l \sim (V/N)^{1/3}$, i.e., a characteristic length having the same order of magnitude as the average distance between the molecules; on the other hand, the theory developed by Paul Scherrer in Göttingen—now Schrödinger's colleague in Zurich—rather suggested $l \sim V^{1/3}$, i.e., a macroscopic characteristic length (Scherrer, 1916). Schrödinger did not find both results acceptable. The first, in particular, led to comparatively high characteristic temperatures, i.e., $\Theta \sim 1\ \mathrm{K}/M$ (where $M$ denoted the molecular weight of the gas under consideration), which should give rise to degeneracy effects inconsistent with the available data. The second class of theories did not cause such difficulties, as it implied a much smaller $\Theta$; but they resulted in the unsatisfactory conclusion that 'twice the amount of gas in exactly the same state would not exhibit exactly the same physical properties' (Schrödinger, 1924a, p. 42); hence, quantum effects would show up only for a very diluted gas in very small volumes, i.e., as a kind of surface effect. To escape these difficulties, Schrödinger now proposed a different order of magnitude for $l$, lying between the previously suggested possibilities. He especially argued: 'For me, therefore, there has never existed any doubt about the fact that, e.g., in the gas model of Scherrer one should logically take for $l$ a quantity of the order of magnitude of the mean free path [of the molecules]' (Schrödinger, 1924a, p. 42).

The problem then arose of how to achieve the desired result. After demonstrating that a method of Sackur (1912) had failed—because it also implied a dependence of the chemical constant on the mean free path, which contradicted the very concept of the chemical constant—Schrödinger went on 'to present a rough sketch of a theory, which follows closely the work of Scherrer and Brody and which . . . demonstrates that the identification of the characteristic length with the mean free path is not at variance with experience' (Schrödinger, 1924a, p. 42).[223] That is, on the one hand he hoped to avoid the difficulty with Sackur's theory of 1912; and on the other, he hoped that the use of the mean free path in Eq. (56) would reduce the characteristic temperature to the order of $10^{-4}$ K for gas molecules, thus being consistent with the known observations.[224]

Schrödinger's main idea was rather simple. He considered a system (model) of $N$ gas atoms (mass $m$) in a volume $V$ at temperature $T$ and imposed on their velocity $v$ the quantum condition

$$2mv_n\lambda = nh \qquad (n = 1, 2, \dots), \tag{57}$$

where $\lambda$ denotes the mean free path of the atoms. To obtain the quantum-theoretical sum, $\sum \exp(-\mathscr{E}/kT)$—summed over all quantum-theoretical energies of the gas model—he made the following assumptions: for each single atom one knows (i) its position in one of the $Z$ space cells of equal size (into which the volume $V$ is subdivided in a quantum-theoretical treatment), (ii) its quantized velocity according to Eq. (57), and (iii) the space angle $k_n$ (into which its motion is directed). For the purpose of statistically counting the states he further defined *different* states in the model: two states could be distinguished, if at least one atom possessed either a different quantized velocity or moved in a different space angle—Schrödinger called two space angles different if the difference of the two velocity vectors associated with atoms under consideration had the absolute magnitude $v_1 = h/2m\lambda$—or were in one of the quantum-theoretically different volume cells.[225]

After calculating the quantum sum over all different states, Schrödinger obtained the following partition function $\psi$

$$\begin{aligned}\psi &= Nk \ln\left[\frac{\pi}{2}\frac{V}{\lambda^3}\sum_{n=1}^{\infty} n^2 \exp\left(-n^2\frac{\Theta}{T}\right)\right] \\ &\approx \tfrac{3}{2}Nk\ln(V^{2/3}T) + Nk\ln\frac{(2\pi mk)^{3/2}}{h^3}\end{aligned} \tag{58}$$

involving the characteristic temperature

$$\Theta = \frac{h^2}{8m\lambda^2 k}. \tag{59}$$

[223] Schrödinger was aware of the fact that other colleagues had already considered this possibility and had claimed that it led 'unfortunately to enormous contradictions with experience'; and also, that a previous attempt by Sommerfeld and Lenz in this direction had produced the wrong high energy limit of the ideal gas equation. (Sommerfeld, 1914b, Section I.8, found $pV = 3RT$ instead of $pV = \frac{3}{2}RT$.)

[224] Schrödinger sketched a preliminary version of this proposal in the manuscript '*Chemische Konstante und Gasentartung. II*', which we have already quoted several times. We therefore propose to date this manuscript earlier than the submission date of the paper on gas degeneracy, i.e., *before* the middle of December 1923.

[225] Max Planck had (in 1916) introduced the quantum-theoretical phase space cells—i.e., $\Delta p_x\,\Delta p_y\,\Delta p_z\,\Delta x\,\Delta y\,\Delta z = h^3$—into gas theory (Planck, 1916b).

Further, the mean free path had disappeared in the final expression for $\psi$,[226] and also for the entropy $S$, i.e.,

$$S = \tfrac{3}{2}Nk \ln(V^{2/3}T) + Nk \ln \frac{(2\pi mek)^{3/2}}{h^3}, \tag{60}$$

where $e$ denotes the base of the natural logarithm. The chemical constant would, therefore, not depend on $\lambda$.

The relation for the characteristic temperature, Eq. (59), was indeed the one which Schrödinger had expected to find; by inserting $\lambda \approx 10^{-5}$ cm it yielded $\Theta = 10^{-4}$ K. Schrödinger also confirmed a suggestion that Keesom had made in his talk at the 1913 Wolfskehl week in Göttingen, namely, that a degenerate electron gas would not contribute to the specific heats of metals (Keesom, 1914b) by taking a mean free path of $10^{-7}$ to $5 \times 10^{-8}$ cm for electrons in metals, $\Theta$ assumed values between 4,500 and 18,000 K.[227]

## Planck's Problematic Division by $N$!

Schrödinger devoted the most detailed discussion to another result of his calculation. The formula for the entropy, Eq. (60), deviated from the one which Max Planck had been propagating for years by an additional term, $Nk \ln(N!)$. Hence the statistical probability of Schrödinger's gas model differed from Planck's thermodynamic probability by a factor of $1/N!$. The difference, however, did not distrub Schrödinger at all, because he had doubted the correctness of this factor for some time.

As early as December 1900, Max Planck had introduced the factor $N!$, referred to above, into his derivation of the blackbody radiation law on the basis of a statistical distribution of $N$ equal energy packets of radiation among the resonators in the cavity (Planck, 1900f). Then Otto Sackur and Hugo Tetrode had followed a similar counting method in their papers on the chemical constant (Sackur, 1911; Tetrode, 1912). Finally, Planck had used the same division by $N$! for like ('*gleichartige*') particles in his quantum theory of monatomic gases (Planck, 1916b). Then, however, Paul Ehrenfest and Victor Trkal, in a paper on

[226] The approximation leading to the final expression for $\psi$ implied only the replacement of the sum $\sum_{n=1}^{\infty} n^2 \exp(-n^2\Theta/T)$ by the integral

$$\left(\frac{T}{\Theta}\right)^{3/2} \int_0^\infty x^2 \exp(-x^2)\, dx = \left(\frac{T}{\Theta}\right)^{3/2} \frac{\sqrt{\pi}}{4},$$

a step which could be justified for all temperatures.

[227] Electrons at the surface of the metals, however, had a different mean free path, $\lambda \sim 10^{-5}$ cm; hence, a characteristic temperature of about 1 K followed, and these surface electrons behaved like an ideal gas at normal temperatures, in agreement with the Richardson effect, i.e., the emission of electrons from hot surfaces.

'Deduction of the Dissociation Equilibrium from the Theory of Quanta and a Calculation of the Chemical Constant Based on This'—communicated in February 1920 to the Amsterdam Academy of Sciences (Ehrenfest and Trkal, 1920) and also published in a German version a year later in *Annalen der Physik* (Ehrenfest and Trkal, 1921)—had called the introduction of the factor $(N!)^{-1}$ 'obscure' (Ehrenfest and Trkal, 1920, p. 162). They had emphasized that the dependence of the thermodynamic entropy of a gas on $N$—the number of like molecules—'can only be satisfactorily settled by utilizing a process in which $N$ changes reversibly and then comparing the ratios of the probability with the corresponding differences of entropy' (Ehrenfest and Trkal, 1920, p. 163). By thus following the statistical rules of Boltzmann, Ehrenfest and Trkal had arrived at expressions for the probability which did not contain the dividing factor; as a consequence, their expression for the entropy of a monatomic gas was determined only up to a constant and especially did not involve the criticized factor. Planck had responded to this attack by stressing the fact that in the Ehrenfest–Trkal treatment identical states were counted more than once (Planck, 1922a). While Planck had again insisted on the physical indistinguishability of two states, in which identical particles exchange only their co-ordinates and momenta, Karl Herzfeld and David Enskog had tried to prove the necessity of the division by $(N!)$ by referring to different arguments (Herzfeld, 1922; Enskog, 1923).

Schrödinger treated the problem of the factor $(N!)^{-1}$, and the related problem of the definition of the absolute entropy, in his unpublished manuscript on '*Chemische Konstante und Gasentartung. II*'. He sketched there in detail Planck's procedure of defining the thermodynamic probability and noted several shortcomings. For example, he found that Planck had chosen his elementary phase-space cells to be so small that 'under normal conditions only about one-thousandth of a molecule is distributed in a "cell"'; hence, Stirling's formula could not be applied—as Planck had done throughout—and Planck's formulae were, in Schrödinger's opinion, 'adjusted with much skill and without any prejudice as regards the correct application of analysis, such that the right result follows.'[228]

Being convinced about the mistakes in Planck's treatment of statistical methods, Schrödinger tended to the opinion that Ehrenfest and Trkal's doubts against Planck's factor $(N!)^{-1}$ were justified. He was therefore not at all surprised when he found, in his paper of December 1923, that in his thermodynamic probability for the ideal quantum gas this factor was not included, and he defended the consequences of this result. Thus the entropy of $N$ condensated (degenerate) atoms in his model was not zero—as in Planck's—but rather $k \ln(N!)$. Schrödinger argued:

> This value for the entropy of condensation in the state free of heat (which appears strange on first inspection) follows—it seems to me—quite cogently, if one ponders

[228] Schrödinger had already been concerned with erroneous applications of Stirling's theorem earlier in Vienna. (See the discussion in Section I.5.)

on the criticism of Ehrenfest and Trkal. $N!$ *is* just the permutation number for $N$ (nearly) 'frozen' condensed atoms. I claim that Planck's theory cannot avoid the perhaps a little unfortunate necessity that a finite value of the entropy must nevertheless be ascribed to a system, from which one is unable to 'squeeze out' any amount of heat; this necessity occurs in the case of *mixed crystals* to which Planck's theory must ascribe—as we know—the 'absolute entropy' $k \ln[(N_1 + N_2)!/N_1!N_2!]$ (with $N_1$ and $N_2$ denoting molecular numbers). Moreover, it appears to me rather an advantage of our (or, if we have understood Ehrenfest correctly, of *his*) interpretation, that the mysterious special situation of the mixed crystals is thus removed. (Schrödinger, 1924a, p. 44)[229]

In spite of favouring the point of view taken by Ehrenfest and Trkal, Schrödinger thought about a '*Versuch der Korrektur des Planckschen Fehlers*' ('Attempt to Correct Planck's Error') in his unpublished notes. But he did not get very far in 1923 in rescuing Planck's method. About two years later, however, in a paper entitled '*Bemerkungen über die statistische Entropiedefinition beim idealen Gas*' ('Remarks on the Statistical Entropy Definition of an Ideal Gas'), which he sent to the Prussian Academy of Sciences (Schrödinger, 1925d) and which was presented by Max Planck to the meeting of 23 July 1925, he proposed several possibilities for the definitions of the entropy, including Planck's. What had caused Schrödinger to undertake these renewed efforts?

## Towards a New Statistics?

In August 1924 the *Zeitschrift für Physik* had published an article '*Plancks Gesetz und Lichtquantenhypothese*' by Satyendra Nath Bose (1894–1974), in which Bose suggested a new derivation of Planck's radiation law based on a new statistical method for light-quanta (Bose, 1924a). Albert Einstein had not only translated the paper by the Indian physicist, but also immediately applied his method in developing a new theory of ideal gases, in a communication presented to the Prussian Academy of Sciences on 10 July 1924 and entitled '*Quantentheorie des einatomigen idealen Gases*' ('Quantum Theory of the Monatomic Ideal Gas,' Einstein, 1924c). He had shown in this theory, which extended in two further contributions to the Prussian Academy of Sciences (Einstein, 1925a, b), that a monatomic gas indeed exhibited, at very low temperatures, deviations from the (classical) ideal gas law or degeneracy, and that the entropy of the gas was described essentially by Planck's method.[230] Since these new results in quantum

[229] Schrödinger also discussed the problem of mixed crystals in his unpublished manuscript '*Chemische Konstante und Gasentartung. II.*'

In order to remove the special situation for mixed crystals in his theory, Planck had previously suggested (in the fourth edition of his *Vorlesungen über die Theorie der Wärmestrahlung*, p. 213) that it be assumed that the absolute entropy of mixed crystals will also become zero in a temperature region where the laws of thermodynamics cease to be valid. Evidently, this seemed to Schrödinger an unsatisfactory way out of the difficulty.

[230] We have treated Bose's radiation theory and Einstein's quantum theory of the ideal gas in Volume 1, Chapter V.3.

statistical theory were obviously important and had to be taken seriously, Schrödinger renewed his previous attempts to correct 'Planck's errors' and arrive at a more consistent derivation of the latter's entropy expression. He summarized the results of this endeavour in his new paper in the summer of 1925.

Schrödinger first discussed three entropy definitions for a quantum-theoretical gas of $N$ atoms, which possess discrete energy states denoted by 1, 2, 3, . . .—with $N_i$ atoms being in the $i$th state—in a given volume having a given total entropy. The first entropy definition was given by the relation

$$S_{\mathrm{I}} = k \ln W_{\max}, \tag{61}$$

where $W_{\max}$ was the maximum of the permutation number expression

$$W = \frac{N!}{N_1!\, N_2! \cdots N_i! \cdots}. \tag{62}$$

He immediately criticized this definition; it could apply only at very low temperatures, because: 'As soon as the number of occupiable energy states is *larger* than the number of molecules—and this number is quite extraordinary larger, except for quite extremely low temperatures—$W_{\max}$ will become equal to $N!$; hence, $S_{\mathrm{I}}$ would be independent of the energy content $E$, which is meaningless' (Schrödinger, 1925d, pp. 434–435).[231] Hence he proposed a second entropy definition, namely,

$$S_{\mathrm{II}} = k \ln \sum W, \tag{63}$$

in which the sum is taken over *all* permutation numbers, Eq. (62). The definition, Eq. (63), coincided for very large molecule numbers with $S_{\mathrm{I}}$, Eq. (61), because in that case 'the sum of all other permutation numbers surpasses the maximum number only by a factor which, although very large, is yet so small compared with $W_{\max}$ that one can totally neglect its logarithm against that of the latter' (Schrödinger, 1925d, p. 435). Finally, he introduced a third definition of the entropy, $S_{\mathrm{III}}$, by taking a large number $n$ of gas systems, each consisting of $N$ atoms, etc., by

$$S_{\mathrm{III}} = \frac{S_{\mathrm{I}}^{(1)} + S_{\mathrm{I}}^{(2)} + \cdots + S_{\mathrm{I}}^{(n)}}{n}, \tag{64}$$

with $S_{\mathrm{I}}^{(i)}$ the entropy of the $i$th system according to the first definition. Thus Schrödinger established on safe grounds the fact that the average entropy per system, $S_{\mathrm{III}}$, coincided with $S_{\mathrm{II}}$.[232]

Schrödinger claimed that, since $S_{\mathrm{II}}$ could be considered as the correct continuation of $S_{\mathrm{I}}$ into the region of higher temperatures (it agreed with $S_{\mathrm{I}}$ for lower ones), the second entropy definition, Eq. (63), should be taken as the basis of

[231] The fact that, for normal temperatures, the occupation numbers of most states were very low (i.e., small fractions of unity) had already been noticed by Schrödinger in 1923. (See above.)

[232] In a large number $n$ of gas systems one could indeed apply the definition of $S_{\mathrm{I}}$ for the total ensemble, as now all the $nN_i$ could be made so large that Stirling's formula applied.

all further considerations. Still, it differed from Planck's; to arrive at the latter one had to take[233]

$$S = S_{II} - k \ln(N!). \tag{65}$$

Schrödinger noted:

> I do *not* intend *here*—as I have done elsewhere [Schrödinger, 1924a]—to fight this interpretation [of Planck], but *will consider it to be the correct one.* I wish to attempt a demonstration that, *if* one assumes it, one is almost *automatically* led to that definition of the entropy of the ideal gas, which has recently been introduced by A. Einstein [1924c, 1925a, b] and soon thereafter, apparently independently, by A. Schidlof [1924], on the basis of a type of statistics, which at first inspection seems to be very strange. (Schrödinger, 1925d, pp. 436–437)

Schrödinger was more ambitious than just taking Planck's subtraction procedure for granted, especially since he did not agree with its justification. He rather claimed: if one wants to count two states, which are related to each other by molecules exchanging their roles, as one state, then it would not suffice to divide the permutation number by $N!$. '*In order that two molecules are able to exchange their roles,*' he wrote, '*they must really have different roles*; otherwise we *have* just not counted such states as being different in their earlier enumeration; hence, we need not and also should not "correct away" a multiplicity, which had not previously existed for us at all!' (Schrödinger, 1925d, p. 437).

In following this concept one had to adopt a new procedure: one had to divide in the earlier expression, Eq. (63), each permutation number $W$ by itself; hence, each term in the sum should be replaced by unity, and the new, fourth entropy definition

$$S_{IV} = k \ln(\text{number of all distributions for a given total energy } E) \tag{66}$$

resulted. This entropy definition, Schrödinger claimed, would now be identical with the one proposed by Einstein and Schidlof; besides it agreed at low temperatures with Planck's definition, Eq. (65).[234] Schrödinger remarked: '*The transition from* [(*63*)] *to* [(*66*)] *is therefore really nothing more than the logical continuation of Planck's division—which is assumed for high temperatures to be statistically*

[233] Evidently, the entropy definition $S_{II}$ agreed with Schrödinger's earlier definition of 1923, Eq. (60).

[234] For low temperatures all molecules will be in different quantum states, hence the $N_i$ are either 1 or 0. Hence the correction of Planck's statistics, along the lines of Schrödinger's idea, becomes superfluous.

To show agreement with Einstein's new statistics, Schrodinger referred to the fact that in counting the distributions for $S_{IV}$, one might restrict oneself to those distributions which correpsonded to the most probable distribution to selected discrete energy groups ('*passend gewählte Stufengruppen*', Schrödinger, 1925d, p. 438) i.e., groups of states in which the energy is essentially the same; and that Einstein took essentially the number of these distributions to obtain his thermodynamic probability. (See Schrödinger, 1925d, pp. 437–438. We shall deal with the relationship of Schrödinger's ideas to Einstein's new gas satistics in more detail in Section III.1.)

*founded*—to the region of temperatures which are so low that the occasional *multiple occupation* of single quantum states cannot be neglected any longer' (Schrödinger, 1925d, p. 437). Planck's entropy for the ideal gas, he argued, did not really go to zero at absolute zero temperature but to a negative value, $-k\ln(N!)$, an impossible result since the permutation number $W$ had to be 'at any rate a *natural number* [an integer];' thus 'one is led nearly cogently to Einstein's form of statistics if one wants to give a statistical foundation for Planck's "division by $N!$"' (Schrödinger, 1925d, p. 438).

In closing the paper Schrödinger hinted at a 'totally different path' ('*ganz anderen Weg*,' Schrödinger, 1925d, p. 438), on which one might also obtain the desired result, namely the method of dealing not with the energy states of individual gas atoms but from the very beginning with only the energy states of the whole system of $N$ atoms. This method had already been applied earlier, especially by Max Planck in the 1916 paper on the quantum theory of monatomic gas (Planck, 1916b). When studying that method in the summer of 1925, Schrödinger discovered several difficulties, which he considered 'so large as to render impossible an execution, free of arbitrariness, of the beautiful idea: i.e., do not begin with the quantization of single molecules but with that of the whole gas' (Schrödinger, 1925d, p. 440). However, several months later, in December 1925, he had overcome these large difficulties; he would then present a solution to the problem of deriving Einstein's statistics on the basis of the 'beautiful idea' in two papers (Schrödinger, 1926a, b). We shall postpone the discussion of this important work, which is intimately connected with the origin of wave mechanics, to the next chapter.